Pest and Pathogen Control: Strategic, Tactical and Policy Models

Wiley IIASA International Series on Applied Systems Analysis

1 CONFLICTING OBJECTIVES IN DECISIONS
Edited by David E. Bell, *University of Cambridge,*
Ralph L. Keeney, *Woodward-Clyde Consultants, San Francisco,*
and Howard Raiffa, *Harvard University.*

2 MATERIAL ACCOUNTABILITY
Rudolf Avenhaus, *Nuclear Research Center Karlsruhe,*
and University of Mannheim.

3 ADAPTIVE ENVIRONMENTAL ASSESSMENT AND MANAGEMENT
Edited by C. S. Holling, *University of British Columbia.*

4 ORGANIZATION FOR FORECASTING AND PLANNING: EXPERIENCE IN THE SOVIET UNION AND THE UNITED STATES
Edited by W. R. Dill, *New York University,*
and G. Kh. Popov, *Moscow State University.*

5 MANAGEMENT OF ENERGY/ENVIRONMENT SYSTEMS: METHODS AND CASE STUDIES
Edited by Wesley K. Foell, *University of Wisconsin, Madison.*

6 SYSTEMS ANALYSIS BY MULTILEVEL METHODS
Yvo M. I. Dirickx, *Twente University of Technology,*
and L. Peter Jennergren, *Odense University.*

7 CONNECTIVITY, COMPLEXITY, AND CATASTROPHE IN LARGE-SCALE SYSTEMS
John Casti, *New York University.*

8 PITFALLS OF ANALYSIS
Edited by Giandomenico Majone
and E. S. Quade, *International Institute for Applied Systems Analysis.*

9 CONTROL AND COORDINATION IN HIERARCHICAL SYSTEMS
W. Findeisen, *Institute of Automatic Control, Technical University, Warsaw,*
F. N. Bailey, *University of Minnesota,*
M. Brdyś, K. Malinowski, P. Tatjewski, *and* A. Woźniak, *Institute of Automatic Control, Technical University, Warsaw.*

10 COMPUTER CONTROLLED URBAN TRANSPORTATION
Edited by Horst Strobel, *University for Transportation and Communication "Friedrich List," Dresden.*

11 MANAGEMENT OF ENERGY/ENVIRONMENT SYSTEMS: NATIONAL AND REGIONAL PERSPECTIVES
Edited by Wesley K. Foell, *University of Wisconsin, Madison, and* Loretta A. Hervey, *International Institute for Applied Systems Analysis.*

12 MATHEMATICAL MODELING OF WATER QUALITY: STREAMS, LAKES, AND RESERVOIRS
Edited by Gerald T. Orlob, *University of California, Davis.*

13 PEST AND PATHOGEN CONTROL: STRATEGIC, TACTICAL AND POLICY MODELS
Edited by Gordon R. Conway, *Imperial College of Science and Technology, London.*

Dedicated to

K. E. F. Watt
J. E. van der Plank
and the memory of
G. Macdonald

whose pioneering work in insect pest, plant-pathogen, and human-pathogen modeling, respectively, has inspired the contributors to this volume.

13 International Series on Applied Systems Analysis

Pest and Pathogen Control: Strategic, Tactical, and Policy Models

Edited by

Gordon R. Conway

Centre for Environmental Technology
Imperial College of Science and Technology
London

A Wiley–Interscience Publication
International Institute for Applied Systems Analysis

JOHN WILEY & SONS
Chichester–New York–Brisbane–Toronto–Singapore

Library of Congress Cataloguing in Publication Data:
Main entry under title:

Pest and pathogen control.

(International series on applied systems analysis; 13)
Based in part on papers presented at a conference sponsored by International Institute for Applied Systems Analysis and held Oct. 22–25, 1979.
"A Wiley–Interscience publication. International Institute for Applied Systems Analysis."
Includes biblographical references and index.
1. Pest control—Mathematical models—Addresses, essays, lectures.
2. Micro-organisms, Phytopathogenic—Control—Mathematical models—Addresses, essays, lectures.
3. Micro-organisms, Pathogenic—Mathematical models—Addresses, essays, lectures.
4. System analysis—Addresses, essays, lectures. I. Conway, Gordon. II. International Institute for Applied Systems Analysis. III. Series.
SB950.P425 1984 632′.9 83–16962
ISBN 0 471 90349 3 (U.S.)

British Library Cataloguing in Publication Data:
Pest and pathogen control.—(Wiley IIASA international series on applied systems analysis ; 13)
1. Pest control—Mathematical models—Congresses
2. System analysis—Congresses
I. Conway, Gordon R. II. International Institute for Applied Systems Analysis
632′.9′02851 SB950.A2

ISBN 0 471 90349 3

Printed in Great Britain at The Pitman Press, Bath

The Authors

R. M. ANDERSON

Department of Zoology
Centre for Environmental Technology
Imperial College of Science and Technology
London SW7 2AZ
UK

H. J. AUST

Phytopathologie und angewandte Entomologie Tropeninstitut
Justus-Liebig Universität
D-6300 Giessen
FRG

J. U. BAUMGAERTNER

Entomologisches Institut
Zürich
Switzerland

A. A. BERRYMAN

Department of Entomology
Washington State University
Pullman
WA 99164
USA

M. H. BIRLEY

Department of Medical Entomology
Liverpool School of Tropical Medicine
Pembroke Place
Liverpool L3 5QA
UK

G. CARLSON

North Carolina State University
Raleigh
North Carolina
USA

N. CARTER

Department of Theoretical Production Ecology
Agricultural University
Wageningen
The Netherlands

J. R. CATE

Entomology Department
Texas A & M University
College Station
TX 77843
USA

H. N. COMINS

Environmental Biology Department
Research School of Biological Sciences
Australian National University
Canberra ACT
Australia

G. R. CONWAY

Centre for Environmental Technology
Imperial College of Science and Technology
London SW7 2AZ
UK

B. A. CROFT

Department of Entomology
Michigan State University
East Lansing
MI 48824
USA

G. L. CURRY

Biosystems Division
Industrial Engineering Department
Texas A & M University
College Station
TX 77843
USA

R. DAXL

UNDP/FAO
Managua
Nicaragua

H. EL-SHISHINY

Academy of Scientific Research and Technology
Cairo
Egypt

A. P. GUTIERREZ

Division of Biological Control
University of California
Berkeley
CA 94720
USA

M. P. HASSELL

Department of Zoology
Imperial College at Silwood Park
Ascot,
Berks SL5 7PY
UK

B. HAU

Phytopathologie und angewandte Entomologie Tropeninstitut
Justus-Liebig Universität
D-6300 Giessen
FRG

C. S. HOLLING

International Institute for Applied Systems Analysis
A-2361 Laxenburg
Austria

A. S. ISAEV

Institute of Forest and Wood
Siberian Branch of the Academy of Sciences of the USSR
Krasnoyarsk
USSR

R. G. KHLEBOPROS

Institute of Forest and Wood
Siberian Branch of the Academy of Sciences of the USSR
Krasnoyarsk
USSR

J. KRANZ

Phytopathologie und angewandte Entomologie Tropeninstitut
Justus-Liebig Universität
D-6300 Giessen
FRG

E. KUNO

Entomological Laboratory
College of Agriculture
Kyoto University
Kyoto 606
Japan

K. MATSUMOTO

Fukushima Agricultural Experiment Station
Tomita-cho
Koriyama
Fukushima-ken 963
Japan

G. F. MAYWALD

CSIRO
Division of Entomology
Long Pocket Laboratories PMB3
Indooroopilly
Queensland 4068
Australia

L. V. NEDOREZOV

Institute of Forest and Wood
Siberian Branch of the Academy of Sciences of the USSR
Krasnoyarsk
USSR

G. A. NORTON

Environmental Management Unit
Imperial College at Silwood Park
Ascot
Berks SL5 7PY
UK

R. RABBINGE

Department of Theoretical Production
Ecology
Agricultural University
Wageningen
The Netherlands

J. E. RABINOVICH

Centro de Ecologia
Instituto Venezolano de Investigaciones Cientificas
Aparatado 1827
Caracas 101
Venezuela

U. REGEV

Economic Research Center
Desert Research Institute
Ben Gurion University
Beersheba
Israel

F. H. RIJSDIJK

Laboratory of Phytopathology
Agricultural University
Wageningen
The Netherlands

F. RODOLPHE

Laboratoire de Biometrie
INRA CNRZ
78350 Jouy en Josas
France

R. RODRIQUEZ

The Agricultural University
Santillo
Mexico

P. L. ROSENFIELD

UNDP/World Bank/WHO Special Program for Research and Training in Tropical Diseases
Geneva
Switzerland

C. A. SHOEMAKER

Department of Agricultural Engineering
Cornell University
Ithaca
NY 14850
USA

R. W. SUTHERST

CSIRO
Division of Entomology
Long Pocket Laboratories PMB3
Indooroopilly
Queensland 4068
Australia

B. R. TRENBATH

Centre for Environmental Technology
Imperial College of Science and Technology
London SW7 2AZ
UK

P. E. WAGGONER

Connecticut Agricultural Experiment Station
New Haven
CT 06504
USA

C. J. WALTERS

International Institute for Applied Systems Analysis
A-2361 Laxenburg
Austria

S. M. WELCH

Department of Entomology
Kansas State University
Manhattan
KA 66506
USA

J. C. ZADOKS

Laboratory of Phytopathology
Agricultural University
Wageningen
The Netherlands

Preface

The origins of this volume lie in the pioneering work of C. S. Holling and his colleagues in the application of systems analysis to the management of the spruce budworm, a major pest of the North American forests. His team, drawn largely from the Institute of Animal Resource Ecology at the University of British Columbia in Canada, worked at IIASA from 1973 to 1976 as part of its ecology/environment program, the overall aim of which was to develop mathematical tools and policy analysis addressed specifically to ecological problems. Holling (1978) describes some of the main achievements, and an account of some specific results of the budworm work is currently in preparation (Baskerville *et al.* 1982).

In 1976 it was appropriate to bring together the growing number of workers engaged in systems analysis in pest management, in order to exchange experiences and promote further development in this field. Accordingly, a pest management conference was held in October of that year at IIASA that brought a number of teams together for the first time — groups from the USA, the UK, the USSR, and Japan, as well as the IIASA team from Canada. The proceedings of this conference were edited by Norton and Holling (1979).

The success of this conference suggested a need to maintain communication among the participants so that they could remain aware of each other's subsequent progress. Thus, with IIASA's support, a Pest Management Network was created with the aim of maintaining contacts by distributing working papers rapidly and by holding workshops and conferences.

A second conference held at IIASA from 22 to 25 October 1979, attended by some 50 scientists, drew on a wider range of expertise than the first and widened the subject matter beyond insect pests (the topic of the first meeting) to include plant pathogens and human disease organisms. Despite the fact that these problems share a common set of dynamic ecological, socioeconomic, and mathematical properties, there has been little interchange of information and experience from one of these fields to another (see, however, Cherrett and Sagar 1977), and the 1979 conference provided a much needed opportunity to bring together workers who have specialized in the applications of systems analysis and mathematical modeling in these three fields.

The present volume arises from that conference; it consists partly of papers based on the presentations at the conference and partly on additional, solicited work. Participants and authors were chosen from as wide a range of countries as possible. The aim has been to produce an up-to-date volume representative of the state of the art in pest and disease modeling throughout the world.

IIASA was the host for the conference, and joined several of its National Member Organizations in supporting it: the French Association for the Development of Systems Analysis, the Max Planck Society for the Advancement of Sciences of the Federal Republic of Germany, the Foundation IIASA–Netherlands, the Japan Committee for IIASA, the Committee for IIASA of the Czechoslovak Socialist Republic, the Austrian Academy of Sciences, the Royal Society of the UK, and the National Academy of Sciences of the USA.

The editor and authors of the volume are particularly grateful to the staff of IIASA, especially Dr Roger Levien, Dr Andrei Bykov, Dr Matthew Dixon, Dr Oleg Vasiliev, Professor Hugh Miser, and Mrs Gabriele Orac for their support and encouragement of this venture and the hard work they put into making the conference and the eventual publication of this volume possible.

The editor also acknowledges gratefully an individual study award from the Ford Foundation, which provided time and partial support for editing this book.

Gordon R. Conway
December 1983

REFERENCES

Baskerville, G., W. Clark, C. S. Holling, D. D. Jones, and C. Miller (in preparation) *Ecological Policy Design: A Case Study of Forests, Insects and Managers*.

Cherrett, J. M., and G. R. Sagar (eds) (1977) *Origins of Pest, Parasite, Disease and Weed Problems*. Proc. 18th Symp. Br. Ecol. Soc. (Oxford: Blackwell).

Holling, C. S. (ed) (1978) *Adaptive Environmental Assessment and Management* (Chichester: Wiley).

Norton, G. A., and C. S. Holling (eds) (1979) *Pest Management*. IIASA Proceedings Series No. 4 (Oxford: Pergamon).

Contents

1 Introduction

Gordon R. Conway

1.1. INTRODUCTION

Long and bitter experience has taught man that to understand and control the pests and diseases that affect him and his crops and livestock is a complex, and seemingly never-ending, task. Indeed, for much of human history such afflictions were attributed, at best, to vaguely defined conditions of the environment and, at worst, to the malevolence of gods and evil spirits. It was only in the nineteenth century that the true causes and the contributing factors became known, through the efforts of a growing band of agriculturalists, medical practitioners, and naturalists who painstakingly pieced together the factual evidence from the field and laboratory. Out of their work there emerged a sound biological basis for the control of pests or diseases, and a set of decision rules, as we would now call them, based on the manipulation of the target of attack and its environment. A good paradigm of this progression is the history of malaria control, Harrison (1978).

In the immediate post-World War II years, however, the mass production and marketing of a range of synthetic insecticides, fungicides, herbicides, antibiotics and other medicines suddenly made the task look easy. Spectacular successes were achieved, a number of which persist to the present day. But, within only a few years, the development of resistance particularly to insecticides and antibiotics, the disturbing effects of pesticides on the natural regulating processes of pest populations, and the growing evidence of unwelcome side effects from the use of antibiotics and other drugs, reminded us that the problems of pest and disease control remain complex. Malaria, for example, once thought to be readily eradicable in large parts of the world and brought very close to this point in countries such as India and Sri Lanka, has resurged with a vengeance, bringing over 30 million cases a year to India alone since the mid 1970's (Sharma, 1982).

This rediscovery of the difficulties inherent in pest and disease control spurred research in a number of directions: first into the ecological basis of pest and disease relationships, building on prewar experience in the light of modern concepts and techniques; second, and arising out of the first, into methods of control that integrate chemical and biological approaches; third, into improving the resistance of crop plants and animals, including man, to pests and diseases; and fourth, and most recently, into the socioeconomic component of pest control. All four of these lines of enquiry are represented in this volume, but our principal concern is with a fifth—the application of the approaches and techniques of systems analysis and mathematical modeling to the comprehension of pest and disease systems and their control. As the subsequent chapters demonstrate, this approach is increasingly providing a natural framework for integrating the other lines of inquiry.

In this introductory chapter I first give a brief historical review of mathematical modeling in pest and disease control, and then discuss some aspects of the utility of systems analysis and mathematical modeling in this field. I argue that in terms of providing a better understanding of the nature of pest and disease systems this approach must be judged on how successful it is at raising and answering a set of key questions for each system. There is no ready guide to such questions, but a simple model provides a starting point for their identification.

1.2. MATHEMATICAL MODELING IN PEST AND DISEASE CONTROL

The 1940s saw not only the discovery of modern pesticides and drugs but also the birth of operations research and systems analysis, and it was therefore perhaps natural that, two decades later, a number of people should start looking to the latter to find remedies for the deficiencies of the former. Concurrently, there was also a revival of the 1930s interest in both the mathematical theory of animal population dynamics and in the mathematical theory of epidemics. Finally, the growth of modern computing facilities meant that complex pest and disease systems could be efficiently modeled and analyzed, and in some cases mimicked in considerable detail. A review of the outcome of these endeavors is one of the aims of this book.

1.2.1. Pest Models

The first major application of systems analysis to pest control was the use of operations research in the insecticidal control of locusts between 1940 and 1960 (Roy *et al.*, 1965; Lloyd, 1959; Rainey, 1960). However modern applications largely stem from the work of Watt (1961a,b). Both of Watt's first papers employed relatively simple population models to demonstrate the value of modeling in determining broad strategies of control. Subsequently, Watt applied a systems analysis technique (dynamic programming) to the problem of optimal pest control (Watt, 1963). Watt's work immediately stimulated further work in this field, but at the time of my first review in 1972 only a handful of papers were being published each year (Conway, 1973). It is only in the last few years that interest has grown dramatically, with major efforts devoted to the analysis and modeling of the pest complexes on cotton, rice, alfalfa, apples, and olives and of specific pests such as the mountain pine beetle, the cereal leaf beetle, the sugar cane froghopper, and the cattle tick. Complementing this work there have also been numerous theoretical studies on the optimization of insecticide applications, biological control dynamics, pesticide resistance, and the economics of control. The appropriate references can be found in this volume and in general reviews by Conway (1977a,b), Ruesink (1976), Shoemaker (1977), and Wickwire (1977).

1.2.2. Plant Disease Models

Plant disease modeling can be said to have begun with the publication of Van der Plank's major work in 1963, in which he used simple exponential and logistic models to describe and analyze disease epidemics. Much of the subsequent work has been devoted to improved models of epidemic progress and to modeling such components as sporulation, infection, incubation and latent periods, and airborne dispersal. In terms of control, attention has centered on forecasting the dynamics of host resistance using regression analysis, and on the production of crop loss models. In recent years, however, a number of complex simulation models have been developed for specific pathogens including potato late blight, southern corn leaf blight, and barley powdery mildew. General reviews include those by Kranz (1974) and Zadoks (1971).

1.2.3. Human Disease Models

Mathematical modeling in human disease control has a much longer history, going back to Bernoulli in the eighteenth century, who analyzed the dynamics of smallpox and demonstrated the advantages of inoculation (Bernoulli, 1760). In the mid-nineteenth century, Farr (1840) produced a model for a smallpox epidemic but the next major landmarks were at the turn of the century. Hamer (1906) built simple mathematical models based on the principle that the course of epidemics is a function of the contact rate between susceptibles and infectious individuals, and Ross (1911) produced a fairly elaborate model of the transmission of malaria. These deterministic models were further elaborated by Kermack and McKendrick in the 1920s and 1930s, who established the importance for epidemiology of the existence of a critical threshold value in the density of susceptible hosts (Kermack and McKendrick, 1927–33, 37–39). In the post-war years Macdonald (1957) improved on Ross's malaria model and also worked on helminth and other vector-borne diseases (Macdonald, 1961, 1965). Since the 1940s the work of Bartlett (1960) and Bailey (1975) has tended to focus attention on developing stochastic models. Most recently there have been extensions of this work but also interest in deterministic models (Anderson, 1982; Anderson and May, 1979; May and Anderson, 1979; Berger *et al.*, 1976; Ludwig and Cooke, 1975; Waltman, 1974).

1.2.4. Comparisons and Contrasts

This volume brings together practitioners in these three fields and one major aim is to present a comparison of the different approaches and techniques used, since both the traditions and the directions of current work are divergent. Of these fields, human disease research, has maintained the strongest tradition of simple analytical modeling and provides powerful insights (see Chapter 8), while the development of complex simulation models has progressed further for plant disease and insect pests. The use of models in evaluating and promoting integrated control and in incorporating socioeconomic components is most advanced for insect pests. However, plant disease models have perhaps been the most successful so far in terms of practical implementation.

1.3. THE UTILITY OF SYSTEMS ANALYSIS

The basis for comparison and evaluation of the work described in this volume must be its utility: on the one hand to research—gaining insight, comprehension, and truth—and on the other to management—the satisfactory solution of pest and disease problems. It is not necessary for me here to make the general case for the utility of systems approaches and mathematical modeling in understanding biological problems. It has been done on many occasions in the past 20 years. Instead, in this introductory chapter I confine myself to some questions of methodology, while in the later chapters introducing each subsequent part of the volume, I highlight those findings that provide specific examples of the practical success of systems analysis and modeling in pest and pathogen control.

It is a common claim that models, in particular large-scale computer simulation models, are pre-eminently useful as a research tool because they help to identify "gaps in knowledge". The discipline of building a model, considering elements and interconnections, forces attention on the absence of apparently critical information. Experiments are conducted, new information is incorporated in the model, new gaps are discovered, and so on. "One of the attractions of a simulation model is that it is never concluded." There are, however, dangers here; in particular that the building of a 'complete' model becomes an end in itself, attention being diverted to detail rather than focusing on the identification of the key processes that give the system its distinctive and important properties. Ecologists are increasingly recognizing that it is not the overall number of connections between elements that determines the essential behavior of ecological systems, but rather the quality and functional form of a few critical relationships (see May, 1979).

If, indeed, we assume that the essential dynamic behavior of complex ecological systems can be understood and predicted through a limited number of key processes and that, similarly, improved management of such systems can be achieved by altering a few key decisions, then the primary task of analysis becomes one of identifying and quantifying these key processes and decisions.

1.3.1. Identification of Key Questions

The problems inherent in this task have exercised me and my colleagues in recent years (Walker *et al.*, 1978; Gypmantasiri *et al.*, 1980; KKU-Ford Cropping Systems Project, 1982a,b). One problem is that the task is intrinsically multidisciplinary in nature; the analysis of pest and pathogen systems, for example, involves the basic disciplines of ecology, economics, and sociology, as well as the more specialized disciplines of entomology, plant pathology, agronomy, medicine, epidemiology, etc. How can such a diversity of disciplines be brought together for fruitful analysis? One solution is to engage the disciplines in modeling the ecological system using large-scale computer simulation. We are not convinced, however, that this is a particularly appropriate or efficient procedure for identifying the key relationships in an ecological system; although it is invaluable in comprehending the full effects and significance of such relationships once they have been identified. Our alternative approach has been to adopt a fairly flexible procedure of systems analysis, based on interdisciplinary seminars and workshops. The aim of this is to produce a set of mutually agreed key questions, the answers to which, hopefully, will identify and illuminate the key processes and decisions in the system. Eventually the questions are turned into hypotheses and the role of modeling, along with field and laboratory experimentation, field surveys, and trials, becomes one of refining and testing these hypotheses. The approach has some similarities with the workshop methods of Holling and his co-workers with their emphasis on such techniques as "bounding" and "looking outwards" (Holling, 1978).

1.3.2. Key Questions in Pest and Pathogen Systems

As yet it has not been possible to provide a ready guide to the form and nature of key questions in pest or pathogen systems. But as a starting point I offer a very simple conceptual model of such systems (Figure 1.1) that focuses on the

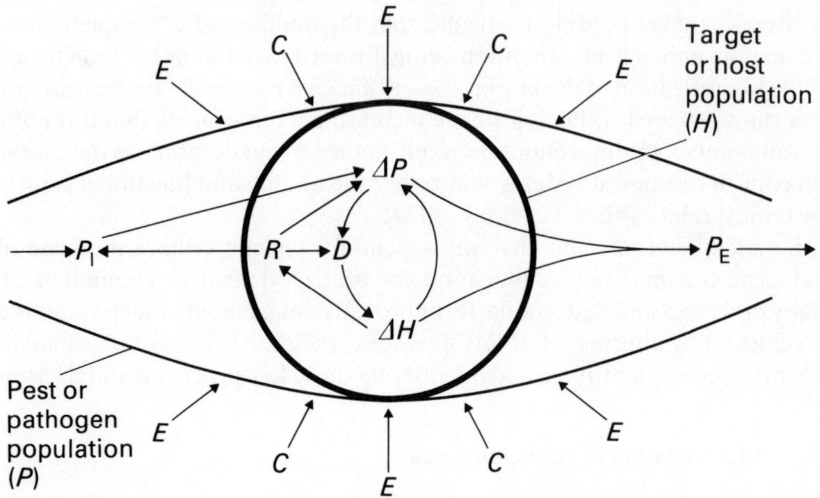

FIGURE 1.1 "Eye model" of pest and disease systems. P_I, P_E, immigrant and emigrant pest or pathogen populations; ΔP, ΔH, population change; R, resistance or immunity; D, damage; E, C, environmental and control factors.

key relationships present in the coupled system of a pest or pathogen population and a target or host population. This "eye model" highlights:

(1) *Immigration* (P_I): Why is a particular target or host population invaded rather than another and is the timing and size of immigration important for subsequent events?

(2) *Population change* (ΔP and ΔH): What are the key features in the relationship between changes in the pest or pathogen population and that of the target or host? Change is here explicitly intended to include changes in numbers and biomass, genotype, and phenotype both in space and time.

(3) *Damage* (D): What is the pattern of damage caused by the pest or pathogen to the target or host population in terms of development, growth, and survival, and to what extent can the target or host compensate for damage caused?

(4) *Resistance or immunity* (R): What part does this play in reducing immigration or damage and in affecting the rate of change of the pest or pathogen population?

(5) *Emigration* (P_E): Is emigration, and hence the probability of further spread, significantly affected by the interaction between the target or host and the pest or pathogen population

(6) *Environment (E) and Control (C)*: To what extent are all of these processes affected by environmental and man-made control factors, i.e., those factors outside the two-population system?

These questions are readily identifiable as the classical concerns of epidemiology, but I believe they are central to all pest and pathogen systems.

They, of course, represent broad issues; key questions, in our experience, are more tightly focused and their definition and formulation require deeper analysis of the system.

The broad questions listed above are all tackled in one form or another in this volume, although terminology and emphasis differ. For example, human and plant disease epidemiologists talk of R, the multiplication rate of the disease, while pest workers, tend to utilize r, the intrinsic rate of increase (See Chapter 2). The relation between immigrants and the outbreak phase has been particularly emphasized by plant disease workers, through the construction and utilization of forecasting models. By contrast, the dynamics of population change have been given more detailed attention in work on insect pests. Perhaps the most neglected issue in all fields is the damage relationship; surprisingly it is still common to find very detailed, quantified knowledge of pest and disease populations alongside the most rudimentary understanding of the damage function. This tends to lead to unproductive arguments on the relative virtues of different control methods, conducted without the economic data necessary to judge their real practical work.

1.4. STRATEGY, TACTICS, AND POLICY

Any judgment of the usefulness of systems analysis and mathematical modeling in helping us acquire a better understanding of the nature of pest and pathogen systems is inherently difficult and, at times, subjective. By contrast, the evaluation of the direct utility of this approach to practical management and control, in some ways, is easier, providing that we are very clear as to the nature of the help that we are expecting.

I have suggested elsewhere that models in applied ecology can be useful divided into strategic, tactical and policy models (Conway, 1977b). This is not a rigid categorization; there is clearly much overlap between the different kinds, but it does serve to focus on the level of help that we are expecting of models. Far too often models are criticized, quite unfairly, by practitioners in pest and disease control for not performing in ways which the model builders had, in any case, not intended them to perform. For example, they may be criticized for not providing quite specific advice on day to day control, whereas they may have been intended and constructed only with a view to establishing general guidelines. A clear exposition and shared understanding, at the outset, of the aims of a particular model will help to avoid this kind of problem and set a fairer basis for evaluation.

Strategy, tactics, and policy are distinguished primarily in terms of the geographic scale and the timespan over which they operate. Perhaps the best way of illustrating the differences is by military analogies. In fact, such analogies are not too far removed from our present concern; pest and pathogen control can be regarded as a form of military contest between two opposing sides, a theme well developed in Harrison's (1978) history of malaria control.

Military policy is determined by governments and is concerned with medium- and long-term courses of action. A government may adopt, for example, long-term policies of appeasement or of nuclear deterrence; it may form aggressive alliances, or develop a civil defence force for its protection. Medium-term policies will arise in a state of war. Then it may be policy to attack the industrial base or disrupt the shipping routes. Such decisions can only be taken at the national level and are based on considerations of the long-term costs and benefits to the society as a whole. By contrast, strategic decisions can be made by military commanders. Given a policy of taking a particular town a commander may chose, for example, between a strategy of "blitzkrieg" or of siege; he may decide to use tanks or an air attack, or employ a fifth column or some combination of these. The area of operation is local and the timescale of costs and benefits short. Tactical decisions are then made by the commander or his subordinates within the strategic guidelines. How many tanks or planes are to be used; what size and number of shells and bombs; at what time of the day should the attack be launched? Contemporary international relations and modern weaponry make decisions at all these levels exceedingly complex and it is thus not surprising that the use of systems analysis and mathematical modeling as an aid to decision making has developed much further than in the field we are considering.

The analogous scale for pest and pathogen control is shown in Table 1.1. As in the military field, *policy is determined primarily by decision makers at national and international levels of government, and policy models are used to weigh the costs and benefits of government intervention in pest and pathogen control and the form it should take*. The models attempt to answer such questions as: what are the priorities in human disease control; can this pest be controlled by government regulations or direct action or is it best left to the efforts of individual farmers; what fiscal or other methods are required to make the pesticide industry function in the national interest; what long-term regulation of pesticide use is required? *Strategy overlaps policy but is primarily the province of the farmer or medical practitioner and strategic models are intended to provide broad guidelines on how a particular pest or disease problems should be tackled*. For example, are pesticides or chemotherapy indicated; is vaccination or host resistance appropriate; is biological control feasible in both ecological and economic terms, or is a mix of strategies appropriate? *Tactical models, by contrast, aim to offer detailed advice on control for a specific location and more particularly at a specific time*. For example, they provide information on whether spraying is required in any particular year and on when and how much to spray. They also can determine the optimal mix of control strategies for a particular farm or for a particular patient.

TABLE 1.1 Hierarchical arrangement of decision-making models in pest and pathogen control.

<table>
<tr><td>Level</td><td>Field or individual</td><td>Farm or subpopulation</td><td>Region or settlement</td><td>National</td><td>International</td></tr>
<tr><td>Decision maker</td><td>Farmer, medical or veterinary practitioner (or patient)</td><td>Farmer, medical or veterinary practitioner</td><td>Regional agricultural service or health board</td><td>Department/ ministry/agency of agriculture or health</td><td>FAO, WHO</td></tr>
<tr><td rowspan="3">Appropriate models</td><td></td><td></td><td colspan="3">POLICY</td></tr>
<tr><td></td><td colspan="2">STRATEGY</td><td></td><td></td></tr>
<tr><td colspan="2">TACTICS</td><td></td><td></td><td></td></tr>
</table>

To date, most applications of systems analysis to pest and pathogen control have concentrated on strategic and tactical questions. There have only been a limited number of attempts to address policy issues, and this is reflected in the relative paucity of chapters on this topic included in this volume. In practice, as indicated in Table 1.1, there is an overlap of models at different levels and a considerable overlapping and interdependency of the decision makers involved. This too has received little attention or explicit analysis in the context of pest and pathogen control although several chapters span different levels and the issue of interdependency is at least raised implicitly.

The volume is divided into three parts, focusing successively on strategic, tactical, and policy models. A more logical sequence would be to place policy first but the sequence here reflects more closely the state of the art of systems analysis in this field. Strategic modeling has received most attention and is best developed while policy models are still the "Cinderella" of the field.

REFERENCES

Anderson, R. M. (ed) (1982) *The Population Dynamics of Infectious Diseases: Theory and Applications* (London: Chapman and Hall)

Anderson, R. M. and R. M. May (1979) Population biology of infectious diseases, Part 1. *Nature* 288:361–7.

Bailey, N. J. J. (1975) *The Mathematical Theory of Infectious Diseases* 2nd edn (London: Griffin).

Bartlett, M. S. (1960) *Stochastic Population Models in Ecology and Epidemiology* (London: Methuen).

Berger, J., W. Buhler, R. Repges, and P. Tautu, (Eds.) (1976) *Mathematical Models—Medicine*, Lecture Notes in Biomathematics 11. (New York: Springer).

Bernoulli, D. (1760) Essai d'une nouvelle analyse de la mortalité causée par la petite verole et des advantages de l'inoculation pour le prevenir. *Mém. Math. Phys. Acad. Roy. Sci. Paris* 1–45.

Conway, G. R. (1973) Experience in insect pest modeling: a review of models, uses and future directions, in P. W. Geier, L. R. Clark, D. J. Anderson, and H. A. Nix (eds) *Insect: Studies in Population Management*. (Ecol. Soc. Australia) Memoirs 1.

Conway, G. R. (1977a) The utility of systems analysis techniques in pest management and crop production, in J. S. Packer (ed) *Proc. XV Int. Congress of Entomology* (Maryland: Entomol. Soc. America) pp541–52.

Conway, G. R. (1977b) Mathematical models in applied ecology. *Nature* 269:291–7.

Farr, W. (1840) *Progress of Epidemics*, Second Report of the Registrar General of England and Wales (London: HMSO) pp91–8.

Gypmantasiri, P., *et al.* (1980) *An Interdisciplinary Perspective of Cropping Systems in the Chiang Mai Valley: Key Questions for Research*. (Thailand: Faculty of Agriculture, University of Chiang Mai).

Hamer, W. H. (1906) Epidemic disease in England. *Lancet* 1:733–9.

Harrison, G. (1978) *Mosquitoes, Malaria and Man: a History of Hostilities since 1880* (New York: Dutton).

Holling, C. S. (1978) *Adaptive Environmental Assessment and Management* (New York: Wiley).

Kermack, W. O. and A. G. McKendrick. (1927–33) Contributions to the mathematical theory of epidemics, Parts I–III *Proc. R. Soc. A* 115:700–21; 138:55–83; 141:94–122.

Kermack, W. O. and A. G. McKendrick. (1937–39) Contributions to the mathematical theory of epidemics, Parts IV and V. *J. Hyg., Camb.* 37:172–87; 39:271–88.

KKU-Ford Cropping Systems Project (1982a) *An Agroecosystem Analysis of Northeast Thailand* (Khon Kaen, Thailand: University of Khon Kaen)

KKU-Ford Cropping Systems Project (1982b) *Tambon and Village Agricultural Systems in Northeast Thailand*(Khon Kaen, Thailand: University of Khon Kaen)

Kranz, J. (Ed.) (1974) *Epidemics of Plant Diseases: Mathematical Analysis and Modelling*. Ecological Studies 13: 7–54 (Berlin: Springer-Verlag).

Lloyd, N. H. (1959) Operational research on preventative control of the new locust. *Anti-Locust Bull.* 35.

Ludwig, D. and K. L. Cooke (eds) (1975) *Epidemiology*. (Philadelphia: Society for Industrial and Applied Mathematics)

Macdonald, G. (1957) *The Epidemiology and Control of Malaria* (London: Oxford University Press).

Macdonald, G. (1961) Epidemiological models in studies of vector borne diseases. *Publ. Hlth. Rep.* 76:753–64.

Macdonald, G. (1965) The dynamics of helminth infections, with special reference to schistosomes. *Trans. R. Soc. Trop. Med. Hyg.* 59(5):489–506.

May, R. M. (1979) The structure and dynamics of ecological communities, in R. M. Anderson, B. D. Turner, and L. R. Taylor (eds) *Population Dynamics* 20th Symp. Br. Ecol. Soc. (Oxford: Blackwell).

May, R. M. and R. M. Anderson, (1979) Population biology of infectious diseases: Part II. *Nature*, 280:455–461.

Rainey, R. C. (1960) *Operational Research (on Locusts)*, Rep 7th Commonwealth Entomol. Conf. pp152–6.

Ross, R. (1911) *The Prevention of Malaria* 2nd edn (London: Murray).

Roy, J., *et al.* (1965) *Final report of Desert Locust Operational Research Team* (FAO: UNSFIDLIOPIS) Vol. 1, p285.

Ruesink, W. G. (1976) Status of the systems approach to pest management. *Ann. Rev. Entomol.* 27–44.

Sharma, U. P. (1982) Observations on the incidence of Malaria in India. *Indian J. Malariology* 19: 57–58.

Shoemaker, C. A. (1977) Pest management models of crop ecosystems, in C. A. Hall and J. W. Day (eds) *Ecosystem Modelling* (New York: Wiley)

Van der Plank, J. E. (1963) *Plant Diseases: Epidemics and Control* (New York: Academic Press).

Watt, K. E. F. (1961a) Use of a computer to evaluate alternative insecticidal programmes. *Science* 133:706–7.

Watt, K. E. F. (1961b) Mathematical models for use in insect pest control. *Can. Entomol.* 93:suppl 19.

Watt, K. E. F. (1963) Dynamic programming, "look ahead" programming, and the strategy of insect pest control. *Can. Entomol.* 95:525–36.

Walker, B. H., G. A., Norton, G. R., Conway, H. N. Comins, and M. Birley. (1978) A procedure for multidisciplinary ecosystem research, with reference to the South African savanna ecosystem project. *J. Appl. Ecol.* 15:481–502.

Waltman, P. (1974) *Deterministic Threshold Models in the Theory of Epidemics*, Lecture Notes in Biomathematics (New York: Springer).

Wickwire, K. (1977) Mathematical models for the control of pests and infectious diseases: A survey. *Theor. Pop. Biol.* 11:182–238.

Zadoks, J. C. (1971) Systems analysis and the dynamics of epidemics. *Phytopathology* 61:600–10.

PART ONE

STRATEGY

2 Strategic Models

Gordon R. Conway

2.1. INTRODUCTION

As I suggested in the introductory chapter, strategic models are intended to provide the broad guidelines for the best way to tackle a particular pest and disease problem. They flow most directly out of the kind of key question analysis I described earlier and focus on the basic decisions that have to be made by the farmer, farm advisor, or medical practitioner. The main purpose of such guidelines is to ensure that the strategies of pest and disease control that we adopt are appropriate to the life strategies of the pest or pathogen populations being controlled and to the dynamics of the economics of damage and loss. Their main usefulness is in determining the relative reliance that should be placed on different control techniques from the wide range potentially available. What, for example, are the merits of pesticidal control compared with host resistance or biological control, or of chemotherapy compared with vaccination or sanitation? And what general levels of control should be aimed for, consonant with the dynamics of the populations of the pest or pathogen and its target or host? In particular, at the present time, strategic models have great value in determining the general form of the mix of strategies in integrated control packages [e.g., Anderson's use of models to indicate integrated control of human diseases (Chapter 8) and a similar use by Curry and Cate for control of the cotton boll weevil (Chapter 10)].

To answer these questions strategic models focus on the core variables of the "eye model" (Figure 2.1). As such they are either exclusively biological, drawing on ecology, genetics, and physiology; or where the damage relationship is the focus, they are a blend of the biological and economics dynamics.

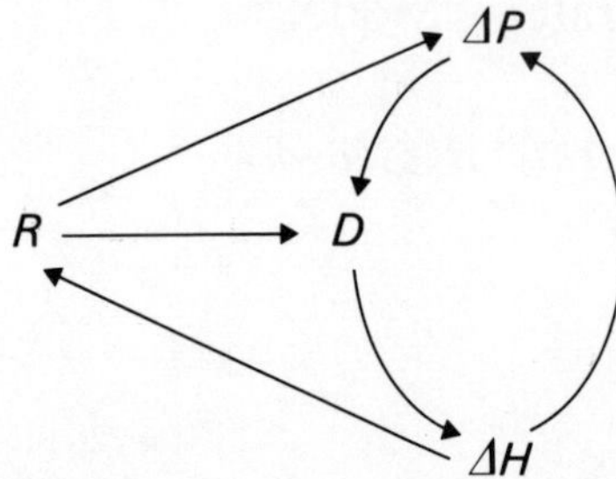

FIGURE 2.1 The core variables of pest and disease control. ΔP, ΔH = change in the pest or disease or host or target population; D = damage; and R = resistance or immunity.

2.2. PEST AND PATHOGEN POPULATION CHANGE, ΔP

The majority of strategic models so far developed have concentrated on the dynamics of the pest or pathogen populations, seeking to uncover the patterns of growth and regulation and to show how these are relevant for control. Population change for organisms with nonoverlapping generations, typifying, for example, most insect pests of temperate regions, can be described by the function:

$$N_{t+1} = F(N_t), \tag{2.1}$$

where N_{t+1} and N_t are numbers at times $t + 1$ and t, while for continuously overlapping populations such as are common in human and plant pathogens, the derivative:

$$\mathrm{d}N/\mathrm{d}t = f(N) \tag{2.2}$$

is more appropriate. N is variously defined depending on the focus of attention (Table 2.1).

TABLE 2.1 Definitions of N, the basic unit of population growth.

Insect pests	Number or density of individuals
Crop pathogens	Numbers of lesions or proportion of plants diseased
Human microparasites	Numbers of humans infected
Human macroparasites	Density of parasites per host

The function $F(N_t)$ or $f(N)$ may be a linear function of N, as for example, in the case of exponential growth:

$$\mathrm{d}N/\mathrm{d}t = rN. \tag{2.3}$$

This relation has been used to describe the early stages of plant disease epidemics (e.g., Trenbath's modeling of rust populations in Chapter 9); more realistically, it is likely to be a nonlinear or density-dependent function. The simple logistic is one such function,

$$dN/dt = rN(1 - N/K), \tag{2.4}$$

which is commonly used to describe the growth of insect populations, and plant and human disease epidemics. Another density-dependent form of relation used for insect populations is

$$N_{t+1} = \lambda N_t^{1-b}, \tag{2.5}$$

which was first applied to the spruce budworm (Morris, 1963; May *et al.*, 1974) and has been subsequently used in a number of pest models (Conway *et al.*, 1975; Sutherst *et al.*, 1979). Other forms of relationship are discussed by Bellows (1981) and May (1976).

2.2.1. Rates of Change

The parameters in these equations, together with a number of derived parameters, form the basic descriptors of population change and are much used in shorthand discussion of the dynamics of a particular pest or disease population. The parameters r, the intrinsic, and λ, the finite, rate of growth, for example, provide indicators of the potential for explosive growth in a population. Thus r is van der Plank's infection rate for pathogen populations, which measures the rate of growth of disease lesions on a plant or population of plants.

Alternatively, growth may be described by the basic reproduction rate $R_0 = N_{t+1}/N_t$, which is easily conceptualized for nonoverlapping generations as the ratio of numbers in succeeding generations, but may also be derived for overlapping generations on the basis of an estimated mean generation time. In human disease modeling a closely related parameter, the basic reproductive rate R, is employed (see Chapter 8), but here N_t equals either one infected host (microparasites) or one adult parasite (macroparasites), and N_{t+1} is either the number of new infected hosts produced during the infectious life of the initial host or the number of offspring produced during the reproductive lifespan of the original parasite.

In plant disease epidemiology R is the basic infection rate. This is different from, although akin to the microparasitic concept in human pathology. A fungal spore attacking a plant creates a lesion or area of damage. Within the lesion the fungus will multiply, eventually producing new spores which produce new lesions, on the same or on a different plant. However, there is a latent period between the inception of the lesion and its infectivity and thus R is defined as:

$$R = rN_t/(N_{t-p}),$$

where p is the latent period. The parameters r and R, however they are measured, provide an indication of the degree of effective control required and the likely success of different control strategies (see Chapter 8).

2.2.2. Life Strategies

A particularly powerful concept in population dynamics is Macarthur and Wilson's postulate of r and K selection (Macarthur and Wilson, 1967). As Conway (1976) and Southwood (1977) argue, it has relevance to strategic pest and disease modeling, providing useful insights into the life history strategies of pests and diseases and pointers to appropriate strategies of control.

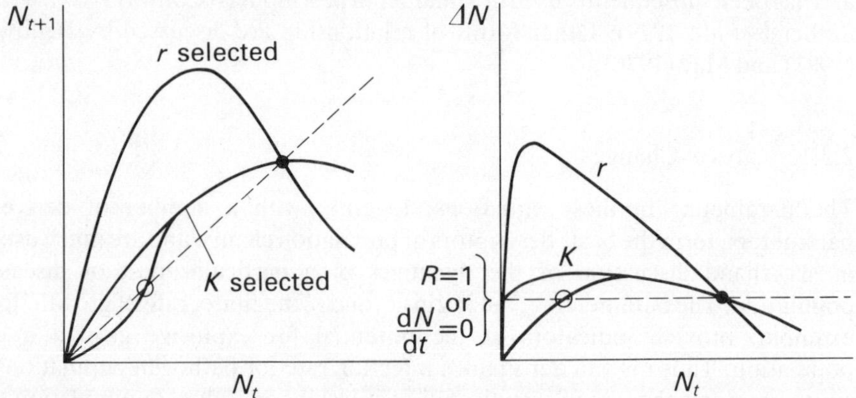

FIGURE 2.2 Reproduction curves for r and K selected pest and pathogen populations. • stable; ○ unstable equilibria.

Depending on the form of $F(N)$ (Figure 2.2), we can distinguish between r-pests such as the desert locust, or r-pathogens such as the cereal rusts or the measles virus, which are essentially opportunists, characterized by high potential rates of population increase and strong dispersal and host-finding ability; and, at the other extreme, K-pests such as the codling moth and K-pathogens such as the tapeworm, occupying more specialized niches, with lower potential rates of increase and showing greater competitive ability. The damage caused by r-pests is more a function of the numbers typifying r-pest outbreaks, while K-pests acquire pest status largely because of the character or quality of the damage they cause. In between the two extremes, Southwood (1977) hypothesizes an intermediate category of species (at least for insects) that are normally regulated at various levels by parasites and predators.

Each of these ecological strategies clearly invites an appropriate control strategy. In practical terms it becomes possible to draw up an r–K profile of the pests and diseases for the crop or host being attacked (Conway, 1981) and on this basis to deduce a preliminary framework for an integrated control program for the whole pest and disease complex, consisting of the relative priorities to be accorded different control strategies and their interactions (Table 2.2).

TABLE 2.2 Principal control techniques appropriate for different pest strategies.

	r-pests	Intermediate pests	*K*-pests
Pesticides	Early wide-scale applications based on forecasting	Selective pesticides	Precisely targeted applications based on monitoring
Biological control		Introduction or enhancement of natural enemies	
Cultural control	Timing, cultivation sanitation, and rotation →		← Changes in agronomic practice, destruction of alternative hosts
Resistance	General, polygenic resistance →		← Specific, monogenic resistance
Genetic control			Sterile mating technique

As I have argued, *r*-pests and pathogens are the most difficult to control Conway (1981). Their frequent invasions and the massive damaging outbreaks they are capable of producing require a fast and flexible response. Natural enemies cannot provide efficient control and thus pesticides remain the most appropriate means of attack. But because *r*-pests can rebound after even very heavy mortalities, pesticides have to be used more or less continuously. Resistant varieties of crops and animals may provide longer-term protection if the resistance is of a general, polygenic type providing, through broad physiological mechanisms, an overall reduction in the rate of increase of the attacking pest or pathogen. Many *r*-pests, such as the cereal rusts, regularly produce new races that can overcome narrowly based resistance (see Chapter 9). Cultural control may also be of help where it serves to reduce the likelihood and size of the pest or pathogen invasion. Early or late planting, for example, may prevent crop damage, while the production of weed-free seed and mechanical cultivation may prevent the buildup of *r*-selected weeds.

By contrast, *K*-pests and pathogens are theoretically more vulnerable. They may be forced to low population levels and, in some situations, eradicated, and for this reason they are the most suitable targets for the sterile mating technique. Pesticides may also be used where small populations cause high losses, such as when fruit is blemished, but the timing and application needs to be precise. However, because *K*-pests or pathogens are specialists, occupying relatively narrow niches, they are generally most vulnerable to strategies aimed directly at reducing or eliminating the effective pest niche. Resistant animal breeds or crop varieties are a powerful approach to *K*-pests and pathogens since permanent success can often be attained with simple monogenic resistance. Cultural control may also be very effective; small changes in agronomic practices such as planting density or pruning may render a crop unattractive or the *K*-pest or pathogen may be greatly reduced by eliminating its alternative wild host.

Biological control has its greatest payoff against intermediate pests and pathogens and this must be the preferred strategy against such species. If pesticides *have* to be used, they should be selective either in their mode of action or in the way they are applied, so that only the target pests are affected rather than their natural enemies and those of other pests. It also follows, of course, that when *r*, *K*, and intermediate pests or pathogens coexist the pesticide applications against the *r* or *K*-pests or pathogens should not interfere with the biological control of the intermediate species. Indeed, the successful control of the intermediate pests or pathogens on a crop or host should be seen as the initial target, with which the strategies for control of the remaining *r*- and *K*-pests or pathogens are then integrated.

Experience suggests that this model provides a powerful and reasonably realistic framework for practical pest control. Nevertheless the concept of *r*–*K* selection is only a metaphor for the considerable complexity inherent in real-life strategies of organisms and is open to criticsm on a number of grounds (Horn, 1978, Stearns, 1976, 1977). Translating the model into action thus requires careful observance of the dynamics of each pest and pathogen and its responses to different control actions (Perrin, 1980).

2.3. DYNAMICS OF CONTROL

The frequent failure of pesticides to provide satisfactory control has been due, in part, to poor application and targeting of pesticides (not discussed in this volume), and to the development of pesticide resistance (see Chapter 28). It has been due also to certain kinds of density-dependent response that are evoked in pest or pathogen populations by pesticide kills. Taking for example eqn. (2.5) as a model of density dependence, the form of response depends on b, the characteristic return rate. For values of b between 0 and 1 (typical of *K*-pests at their stable equilibrium; see Figure 2.2) the population recovers from mortality in an undercompensating manner, returning smoothly to its former equilibrium (Figure 2.3A). However, for $b > 1$ (typical of *r*-pests at their equilibrium) the response is to overcompensate, and the population rebounds to even higher levels than before, and oscillates before returning to its original equilibrium (Figure 2.3B).

The boomerang effect produced by an overcompensating response is illustrated further by Isaev and his colleagues in Chapter 3. They conceptualize regulation as the outcome of the interaction between two populations. These

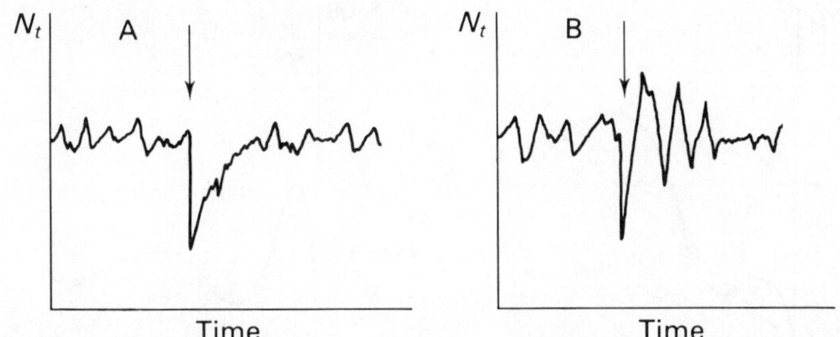

FIGURE 2.3 (A) Undercompensating and (B) overcompensating responses of pest or pathogen populations to perturbation, as for example caused by pesticide mortality.

may be age classes acting as competing subpopulations in the pest population or, the more common treatment, a pest population and the predator or parasite population that attacks it. Isaev and his colleagues show how relatively simple graphical techniques (phase portraits) can be used to lay out the complex dynamics of interacting population models so that the likely consequences of a perturbation, as for example caused by a pesticide kill, can be readily determined. Boomerang effects arising from pesticide control are relatively well known, but Isaev and his colleagues also describing the conditions under which biological control—the release of predators or parasites—may also produce an unexpected increase in the pest population.

2.3.1. Multiple Equilibria

Chapter 3 also provides examples of the frequently unexpected and even more dramatic response to control that arises when pest populations can occur at more than one equilibrium level (Figure 2.4). A pest population, may be normally kept at a low stable equilibrium by its natural enemies, but under certain circumstances can then escape to a new, higher equilibrium. The pattern is well illustrated by the spruce budworm (Peterman *et al.*, 1979), whose critical natural enemies are believed to be bird predators. When the dynamics of the budworm are plotted as a function of forest maturity (Figure 2.5), a threshold T_2 value of forest maturity (roughly leaf area per tree) is revealed. Above this threshold both equilibria are possible. If, for example, good weather favors a rapid increase in budworm numbers, or if there is a wave of immigrants, the population passes through the breakpoint and rapidly converges on the upper equilibrium. Above some upper threshold T_1 of forest maturity only the upper equilibrium exists. Peterman *et al.* (1979) argue that the use of pesticides holds the budworm population artificially just below the breakpoint, thus creating a perpetually unstable situation.

This phenomenon of thresholds and their implications for pest control is further discussed by Berryman in Chapter 4, drawing on his studies of the

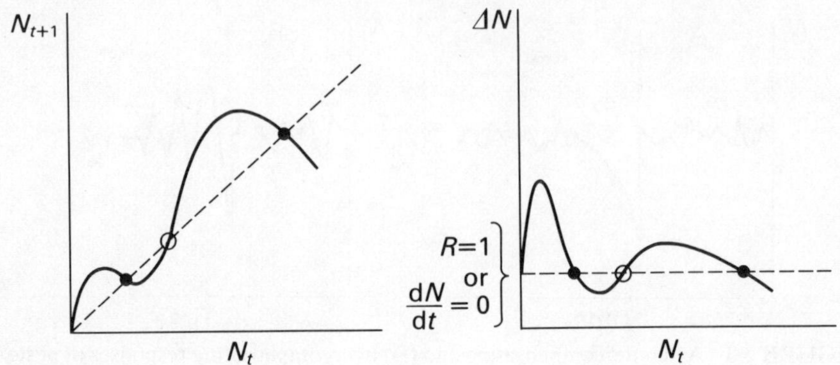

FIGURE 2.4 Multiple equilibria in a pest or pathogen population (after May 1977). • stable; ○ unstable equilibria.

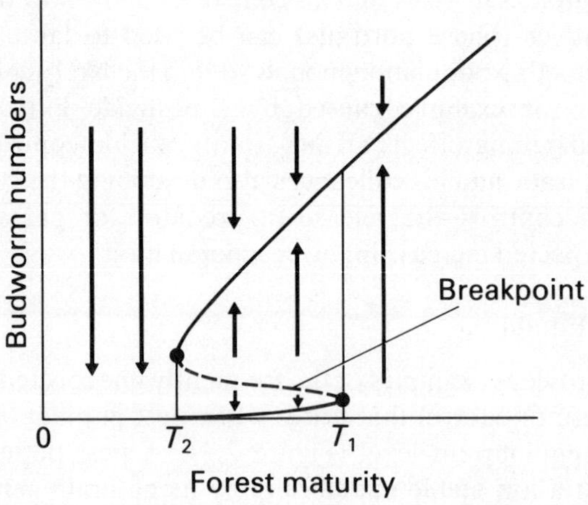

FIGURE 2.5 Manifold of spruce budworm population dynamics (after Peterman *et al.*, 1979).

mountain pine beetle. In this pest a population density threshold exists, above which the population is able to overcome the more resistant trees and hence has greater available food supply which, in turn, permits further population growth. The beetle is thus able to 'escape' from its normal endemic state in a manner analogous to that of the spruce budworm. The existence of thresholds such as these is clearly of critical importance in practical pest management (see Chapter 24) and Berryman argues that where modeling can help is in providing the pest manager with estimates, for each location, of the likelihood of a pest population breaking through the threshold. He goes on to show how the thresholds and the high-risk zones in the critical variables can be presented in simple graphical form and can hence be used as a basis for management.

2.4. BIOLOGICAL CONTROL

In addition to the spruce budworm and the mountain pine beetle it is known, or at least suspected, that multiple equilibria are to be found in the eucalyptus psyllid, *Cardiaspina albitextura* (Southwood and Comins, 1976), the European sawfly, *Diprion hercyniae* (Southwood, 1977), and in the reduviid vector of Chagas' disease, *Rhodnius prolixus*, discussed by Rabinovich in Chapter 5.

Clearly the existence of such low stable equilibria in pest populations provides a potentially powerful basis for control, provided that the damage produced by pest populations at these levels is acceptable. One cause of a low stable equilibrium may be the particular form of the functional response produced by the pest's natural enemies. This response measures the number of pests attacked per parasite or predator as a function of the pest population density. Three types of functional response exist (Figure 2.6) but only the sigmoid type III response is apparently capable of producing low stable equilibria.

In the case of the reduviid bug *Rhodnius* the laboratory experimental data are unfortunately not clear cut, but there are good reasons for expecting a type III functional response in its parasitoid enemy in the natural habitat of rural dwellings in South America. On this assumption Rabinovich's model in Chapter 5 suggests two widely separate equilibria of 20 and about 3000 bugs per dwelling, separated by an unstable equilibrium of just under 2000 bugs.

In Chapter 6, however, Hassell argues that in the case of synchronized, specific parasitoids a sigmoid response cannot stabilize because of the time delay between changes in parasitoid density and host mortality in the next generation. In the case of general predators such as the birds attacking the spruce budworm, it may be that they switch from one prey species to another and hence produce rapid changes in numbers. But Hassell maintains that for parasitoids, and perhaps also for predators, equilibria are produced by the spatial heterogeneities in the prey or host, by the natural enemy distributions, and, in particular, by the habit of nonrandom search by the enemy when the prey or host is patchily distributed. High searching efficiency, coupled with this pattern of search, will produce the low stable equilibria observed in nature and are, he maintains, a convincing explanation of biological control successes. On these assumptions, Hassell also demonstrates the further point that the simultaneous introduction of many species of parasitoids for biological control is a sound strategy to adopt. As Hassell admits, however, parasitoids exhibit rather simpler dynamic behavior than predators, and Gutierrez argues that much more detailed modeling of predator growth, development, and feeding behavior is required before their role in pest control can be properly evaluated. He describes how this can be done in Chapter 7.

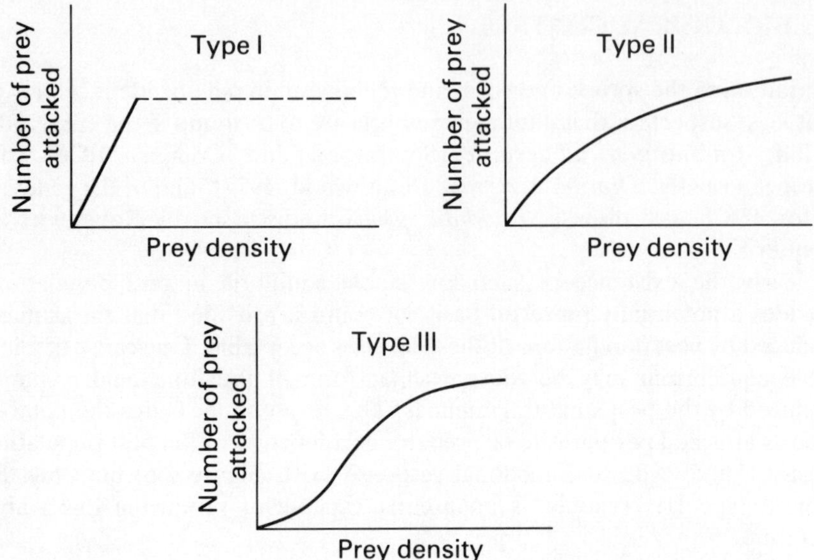

FIGURE 2.6 Functional responses of predators or parasites (after Hassell, 1978).

Quite clearly, these issues of the stabilizing potentialities of natural enemies are of great strategic importance. If low stable equilibria are possible then dramatic control of pests is feasible, as Hassell and Rabinovich demonstrate.

2.5. HOST OR TARGET POPULATION CHANGE, ΔH

The concentration on the dynamics of the pest or disease population in the past has tended to relegate the position of the host or target population to the status of a simple black box. This has been particularly true where crops are the host or target, and it is only recently that population genetics models have been used in the evaluation of crop resistance to pathogens or models of the dynamics of crop components used in damage assessment.

Rather more attention has been devoted to the human host populations of disease, largely because the critical role of population and subpopulation size was recognized early (see Anderson, Chapter 8). In the case of the often relatively simple microparasitic diseases such as the viruses, bacteria, and protozoa, the population N can be thought of as two changing subpopulations, the susceptibles (X), and the infected (Y), and the progress of the disease can be modeled in terms of their dynamics:

$$\mathrm{d}X/\mathrm{d}t = gX + \gamma Y - lX - \beta XY \tag{2.6}$$

$$\mathrm{d}Y/\mathrm{d}t = \beta XY - lY - (\alpha + \gamma)Y, \tag{2.7}$$

where g and l represent the natural gain and loss rates, α is the mortality rate from infection, γ is the recovery rate, and β is the transmission coefficient. Where immunity occurs, a third subclass has to be considered.

The important features of this system are that transmission of the disease depends on the sizes of subpopulations and, as Anderson shows for such diseases as rubella and measles, there is a critical threshold density T_D of the whole host population that permits the disease to be maintained. For macroparasites and for diseases transmitted by a vector the modeling becomes considerably more complex and in the latter situation disease transmission and maintenance thresholds become functions of both host and vector densities.

The existence of thresholds in human disease systems is clearly of strategic importance for control. For example, the existence of a threshold for hookworm suggests that a combined program of improved sanitation and education could bring about eradication of the disease. For gonorrhea, however, where no threshold exists, the only attack possible is a continual program of detection and chemotherapy to achieve acceptable control. There is also growing evidence of thresholds producing multiple equilibria in disease systems. For example, two stable states for schistosomiasis appear to exist dependent on a threshold in transmission efficiency. Such situations present opportunities for control analogous to those discussed in the previous section.

2.5.1. Resistance and Immunity, *R*

Where immunity exists, vaccination becomes a viable strategy of control and Anderson shows how calculations of the proportion of the population required to be vaccinated can be determined. A form of immunity may also exist in plant host–disease dynamics and Trenbath describes in Chapter 9 how this may work, but the principal interest of plant disease modelers, here, is in the dynamics of genetically inherited resistance where the host population consists of two or more subpopulations with differing degrees of susceptibility.

Trenbath looks at this situation for an idealized rust-type disease and cereal crop system where the populations are described in terms of complex exponential functions of the general form

$$\mathrm{d}H_i/\mathrm{d}t = r_{\mathrm{H}} H_i \tag{2.8}$$

$$\mathrm{d}P_j/\mathrm{d}t = r_{\mathrm{P}} P_j, \tag{2.9}$$

where H_i is the size of the ith variety in the host population, and P_j is the size of the jth race in the pathogen population. The growth rates r_{H} of the host varieties are modified by the degree of infection and the comparative vigor of competing varieties, while the growth rates r_{P} of the pathogen races are affected by a number of factors, including stabilizing selection and the degree of cross-protection.

Trenbath examines a number of strategies with this model, and his conclusions are rather surprising. It appears that there is very little to choose between control based on growing a mixture of varieties (multiline), a single variety possessing different resistance genes (superline), or using different varieties in some sequence or rotation. The result is surprising because there is a growing ecological wisdom that spatial heterogeneity has useful stabilizing properties (Hassell, for example, argues that it provides the key to natural enemy control of insect populations). On these grounds one might have expected the multiline to be a superior strategy.

2.5.2. Damage, *D*

The most significant work so far on the dynamics of damage relations has focused on cotton and cotton pests and these are the subject of Chapters 10 and 11. The stress on the dynamic aspects of pest–target interaction is apparent in the use by Curry and Cate in Chapter 10 of partial differential equations to describe the growth of crop components that are closely related to the Leslie matrix models (Leslie, 1945; Usher, 1972) used in human demography and insect population modelling (Conway, 1979). The sets of equations are of the form developed by von Foerster (1959):

$$\frac{\partial N}{\partial t} + \frac{\partial N}{\partial a} = -\mu(\cdot)N(ta), \qquad (2.10)$$

where N may be the population of leaves, skins, root, or fruits. $N(t,a)$ is the number density function (in mass or energy units), t is time, a is age, and $\mu(\cdot)$ is a multifactor death function (Gutierrez and Wang, 1978). These equations are then all linked by a carbohydrate pool submodel that determines photosynthate availability.

The critical feature of cotton crop dynamics that makes it so important as a case study is the existence of a carrying capacity in the population of fruits (bolls) that a cotton plant can support. Photosynthate availability limits the size of the fruit population and a process of natural shedding occurs. Any damage that insect pests cause to the fruit has to be quantitatively reconciled with this, and Gutierrez and his colleagues (Chapter 11) show how this affects both the understanding and detailed calculation of the economic threshold for determining when and how much spraying should be done. Both Gutierrez and his colleagues and Curry and Cate (Chapter 10) show that because of the pattern of development of both the boll and the boll weevil populations and, in particular, the timing of the natural shedding, there is a very strong argument for confining spraying to early in the season. This not only has the advantage of economy but also reduces the risk of upsetting the natural enemies of other cotton pests. Curry and Cate further demonstrate that the use of resistant varieties that reduce the development, reproduction and survival rates of the boll weevil, will, with a complementary low level of parasitism, provide very satisfactory control without the use of pesticides at all.

2.6. DETAIL AND ROBUSTNESS

Chapters 10 and 11 also raise a major issue in pest and disease modeling, concerning the degree of detail necessary, and hence the extent to which models require extensive laboratory and field data. As is apparent in this book, there is a gradation from the very intricate models of crop and pest population growth of Gutierrez and his colleagues and Curry and Cate, to some of the detailed plant disease simulation models (Kranz *et al.*, 1973; Waggoner and Horsfall, 1969; Waggoner *et al.*, 1972) with their strong base in empirical field data, to the conceptually simpler and often analytically tractable pest and disease models of Anderson, Hassell, and Trenbath, which are more deductively based. The contrast, however, is probably more apparent than real in terms of the validity of the modeling exercise. The difference stems primarily from the differing levels of generality at which the key questions are being asked. On the one hand, we have workers who enquire into what are likely to be the essential dynamic properties of a kind or class of pest or disease system, deducing from known features, and arriving at broad strategic conclusions; on the other, we have work, that focuses on a specific circumscribed system, where the broad dynamics are known but the particular details need to be incorporated to establish the most appropriate variants on the general strategy.

The critical issue is not simplification versus detail but, rather, whether each model is realistic in terms of its *chosen level of generality* and whether there is the information from either field or laboratory data or from accumulated experience to support this.

REFERENCES

Bellows, T. S. (1981) The descriptive properties of some models of density dependence. *J. Animal Ecol.* 50: 139–156

Conway, G. R. (1976). Man versus pests, in R. M. May (ed) *Theoretical Ecology: Principles and Applications* (Oxford: Blackwell) p257.

Conway, G. R. (1979) Case studies of pest control, in G. A. Norton and C. S. Holling (eds) *Pest Management*, Proc. Int. Conf., October 1976. IIASA Proceedings Series Vol. 4 (Oxford: Pergamon) pp177–200.

Conway, G. R. (1981) Man versus pests, in R. M. May (ed) *Theoretical Ecology: Principles and Applications* 2nd edn. (Oxford: Blackwell) pp356–386.

Conway, G. R., G. A. Norton, A. B. S. King, and N. J. Small. (1975) A systems approach to the control of the sugar cane froghopper, in G. E. Dalton (ed) *Study of Agricultural Systems* (London: Applied Science Publishers) pp193–229.

Gutierrez, A. P. and Y. Wang. (1979) Applied population ecology: models for crop production and pest management, in G. A. Norton and C. S. Holling (eds) *Pest Management*, Proc. Int. Conf. October 1976. IIASA Proceedings Series Vol. 4 (Oxford: Pergamon) pp255–80.

Hassell, M. P. (1978) *The Dynamics of Anthropod Predator–Prey Systems*. Monographs in Population Biology, 13. (Princeton: Princeton University Press).

Horn, H. S. (1978) Optimal tactics of reproduction and life history, in J. R. Krebs and N. B. Davies (eds) *Behavioural Ecology: An Evolutionary Approach* (Oxford: Blackwell) pp 411–429

Kranz, J., K. M. Mock, and A. Stumpf. (1973) EPIVEN—ein Simulator für Apfelschorf. *Z. Pflanzenkrankh. 80*:181–187.

Leslie, P. H. (1945) On the use of matrices in certain population mathematics. *Biometrika* 33:183–212.

Macarthur, R. H. and E. O. Wilson (1967) *The Theory of Island Biogeography* (Princeton, NJ: Princeton University Press).

May, R. M. (1976) Models for single populations, in R. M. May (ed) *Theoretical Ecology: Principles and Applications*. (Oxford: Blackwell) pp4–25.

May, R. M. (1977) Thresholds and breakpoints in ecosystems with a multiplicity of stable states. *Nature* 269:471–477.

May, R. M., G. R. Conway, M. P. Hassell, and T. R. E. Southwood. (1974) Time delays, density dependence and single species oscillations. *J. Animal Ecol.* 43:747–70.

Morris, R. F. (ed) (1963) The dynamics of epidemic spruce budworm populations. *Mem. Entomol. Soc. Canada* No. 31.

Peterman, R. M., W. C. Clark, and C. S. Holling. (1979) The dynamics of resilience: shifting stability domains in fish and insect systems, in R. M. Anderson B. D. Turner, and L. R. Taylor (eds) *Population Dynamics,* Proc. 20th Symp. Br. Ecol. Soc. (Oxford: Blackwell) pp321–41.

Perrin, R. M. (1980) The role of environmental diversity in crop protection. *Protection Ecol.* 2:77–114.

Southwood, T. R. E. (1977) The relevance of population dynamic theory to pest status, in J. M. Cherrett and G. R. Sagar (eds) *Origins of Pest, Parasite, Disease and Weed Problems,* Proc. 18th Symp. Br. Ecol. Soc. (Oxford: Blackwell) pp35–54.

Southwood, T. R. E. and H. N. Comins. (1976) A synoptic population model *J. Animal Ecol.* 45:949–65.

Stearns, S. C. (1976) Life history tactics: A review of the ideas. *Quart. Rev. Biol.* 51: 3–47

Stearns, S. C. (1977) The evolution of life history traits: a critique of the theory and a review of the data. *Ann. Rev. Ecol. Syst.* 8: 145–171

Sutherst, R. W., G. A. Norton, N. D. Barlow, G. R. Conway, M. Birley, and H. N. Comins. (1979) An analysis of management strategies for cattle tick (*Boophilus microplus*) control in Australia. *J. Appl. Ecol.* 16:354–382.

Usher, M. B. (1972) Developments in the Leslie matrix model, in J. N. R. Jeffers (ed) *Mathematical Models in Applied Ecology,* Proc. 12th Symp. Br. Ecol. Soc. (Oxford: Blackwell) pp29–60.

von Foerster, H. (1959) Some remarks on changing populations, in F. Stohlman (ed) *The Kinetics of Cellular Proliferation* (New York: Grune and Stratton).

Waggoner, P. E. and J. G. Horsfall. (1969) EPIDEM, a simulator of plant disease written for a computer. *Connecticut Agric. Exp. Station Bull.* 698.

Waggoner, P. E., J. G. Horsfall, and R. J. Lukens. (1972) EPIMAY, a simulator of southern corn leaf blight. *Connecticut Agric. Exp. Station Bull.* 729.

3 The Boomerang Effect in Models of Pest Population Control

A. S. Isaev, L. V. Nedorezov, and R. G. Khlebopros

3.1. INTRODUCTION

We can attempt to control phytophagous pests in a number of different ways. Chemical or bacterial pesticides usually produce a direct kill and a rapid reduction in the size of a pest population (often accompanied by a reduction in the population of the pests' natural enemies). Alternatively, control may be achieved indirectly, such as by increasing the natural enemy population, or by lowering the reproductive capacity of the pest population through the technique of male sterilization. However, these methods may be counterproductive, and the pest population may increase sharply at some time subsequent to the control attempt, to higher levels than before (Watt, 1968). Thus a serious problem may become much worse, necessitating a repetition and intensification of control measures.

It is often difficult to predict the outome of a particular control technique against a specific pest under given conditions. However, prior to the use of a control method, it may be helpful to attempt to identify a range of potential effects, particularly the boundaries of likely population change for both the pests and their natural enemies. In this chapter we present the "phase portraits" produced by several simple models of interacting pest and natural enemy populations as a way of illustrating the range of possible outcomes.

First, using a generalized model of coupled pest and natural enemy populations, we distinguish between the conditions that produce either an undercompensating or overcompensating response to control in the pest population. In the former situation the pest population returns smoothly to its precontrol equilibrium, while in the latter it overshoots, at least initially. In this chapter we refer to such overshooting as a "boomerang effect". We then examine a number of specific predator–prey models based on the Lotka–

Volterra equations and describe the potential range of population behavior following a pesticide application. Finally, we describe the conditions under which boomerang effects may be produced by biological control methods, i.e., by increasing the size of the predator or parasite populations.

3.2. THE GENERAL CASE

Two interacting populations may be coupled as follows:

$$dw/dt = F(w,v) \qquad dv/dt = G(w,v), \tag{3.1}$$

where w,v are, for example, the numbers of prey and predators (or of old or young individuals, or of individuals of different sex, etc.). It can be shown that all systems of this kind essentially have either a stable equilibrium point or a stable limit cycle (Kolmogorov 1936; May, 1972, 1975). In the latter case the populations oscillate in a cycle whose amplitude and period is determined uniquely by the parameters of the equations. If the system is perturbed it will return to the stable point or to the stable cyclic trajectory.

If a stable point exists it can be further characterized by the manner of the return to the point following perturbation. If every deviation from the stable point leads to an overcompensating rebound, or boomerang effect, then we define the point as a "focus" (Figure 3.1A). More precisely, the stable state (w_1,v_1) is a focus if at some time t following the disturbance from equilibrium $w(t) > w_1$, or $w(t) + v(t) > w_1 + v_1$. However, where the production of a boomerang effect is not inevitable and depends on the behaviour of trajectories near the stable state, we term the state a "knot". In this situation two domains in the phase plane can be distinguished. A perturbation producing an initial population shift into one of the domains produces a boomerang effect, while a shift into the other results in an undercompensating return (Figure 3.1B).

3.3. CONSEQUENCES OF PESTICIDE APPLICATION

3.3.1. Lotka–Volterra Models

The classic model for a deterministic interaction between a predator and a prey (or pest) population is that of Lotka (1925) and Volterra (1926):

$$dx/dt = x(\alpha - \gamma z) \qquad dz/dt = z(rx - \delta), \tag{3.2}$$

where $x(t)$ is the number of prey and $z(t)$ the number of predators at time t. The stability properties of this basic model and of many extensions and improvements to it have been investigated by a large number of workers (see, for example, Gilpin, 1975; Goh and Jennings, 1977; May 1972, 1975, 1981; Murdoch and Oaten, 1975; Rozenzweig and Macarthur, 1963; Tanner, 1975). The basic model exhibits a neutrally stable cycle whose amplitude is

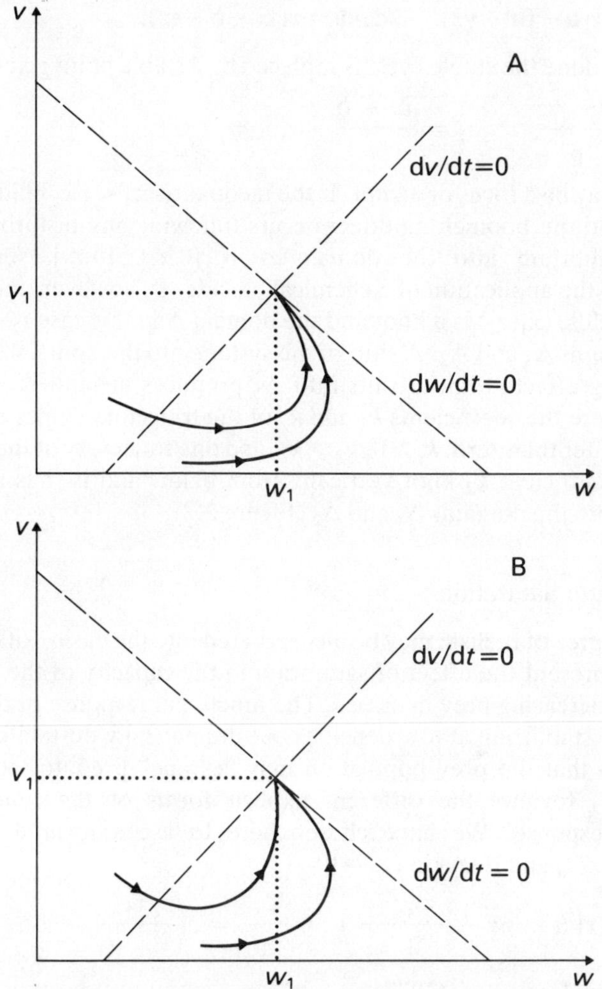

FIGURE 3.1 Behavior of populations perturbed from a stable equilibrium state (w_1,v_1). (A) The state is a focus and every perturbation into domain Δ results in an overshooting or boomerang effect in w. (B) The state is a knot and perturbations may or may not produce a boomerang effect. Broken curves are isoclines of stationary population w and v, and the dotted curve indicates the boundary of the domain Δ.

determined solely by the initial conditions. Any perturbation simply results in a new cycle with a new amplitude determined by the disturbance.

The basic model is highly unrealistic. The prey population, for example, is only limited by the predator population and in its absence shows unbounded growth. A more realistic modification is to include some degree of intraspecific competition in both equations, representing competition among the prey for their plant food resource or among the predators for the prey:

$$dx/dt = x(\alpha - \beta x - \gamma z) \qquad dz/dt = z(\tau x - \delta - \varepsilon z). \tag{3.3}$$

When this is done the stable cycle is replaced by a stable point given by

$$x_1 = \frac{\alpha - \gamma z}{\beta}, \qquad z_1 = \frac{\tau x - \delta}{\varepsilon}.$$

This point may be a focus or a knot. If the inequality $\varepsilon z_1 < \beta x_1$ holds, the point is a focus and the boomerang effect occurs following any disturbance of the system equilibrium into the domain $\Delta = [0,x_1]. [0,z_1]$, for example, produced by the application of a chemical biocide. If, however, the inequality $\varepsilon z_1 > \beta x_1$ holds, (x_1,z_1) is a knot and the domain Δ in this case is divided into two subdomains Δ_1 and Δ_2. A shift of the system into the domain Δ_2 produces a boomerang effect, while a shift into Δ_1 produces an undercompensating response. Here the coefficients k_1 and k_2 of the trajectory slopes entering the knot are greater than zero, $k_j > 0$, $k_2 > k_1$, and one trajectory of the system (l_1) enters the coefficient k_2 knot vertically from below and is thus a separatrix dividing Δ into the domains Δ_1 and Δ_2 (Figure 3.2).

3.3.2. Predator Saturation

A further degree of realism may be incorporated into the basic Lotka–Volterra model to represent the effect of saturation in the capacity of the predator to respond to increasing prey densities. The functional response of the predator may then be stabilizing at low densities but dramatically destabilizing at high densities, so that the prey population may "escape" predator control. May (1975, 1981) reviews the different explicit forms of the functional and numerical response. We have chosen here to use a form developed by Alexeyev (1973) and Bazykin (1974):

$$\frac{dx}{dt} = x\left(\alpha - \beta x - \frac{\gamma z}{1 + \pi x}\right) \qquad \frac{dz}{dt} = z\left(\frac{\tau x}{1 + \pi x} - \delta - \varepsilon z\right) \tag{3.4}$$

as modified by Isaev *et al.* (1979):

$$\frac{dx}{dt} = x\left(\alpha - \beta x - \frac{\gamma z}{1 + Ex^2 + Hz}\right)$$

$$\frac{dz}{dt} = z\left(\frac{\tau x}{1 + Rx + Fz^2} - \delta - \varepsilon z\right), \tag{3.5}$$

where E, H, R, and F are positive constants. This system now exhibits three equilibrium states, consisting of upper (x_2,z_2) and lower (x_1,z_1) stable states separated by an intermediate unstable state (x_r,z_r). The latter is also termed a saddle or breakpoint (see Chapter 2 and Figure 2.4). The domain Δ, in this case, may be divided into three subdomains Δ_1, Δ_2, and Δ_3 (Figure 3.3). A system shift into Δ produces an undercompensating response. A shift into Δ_2

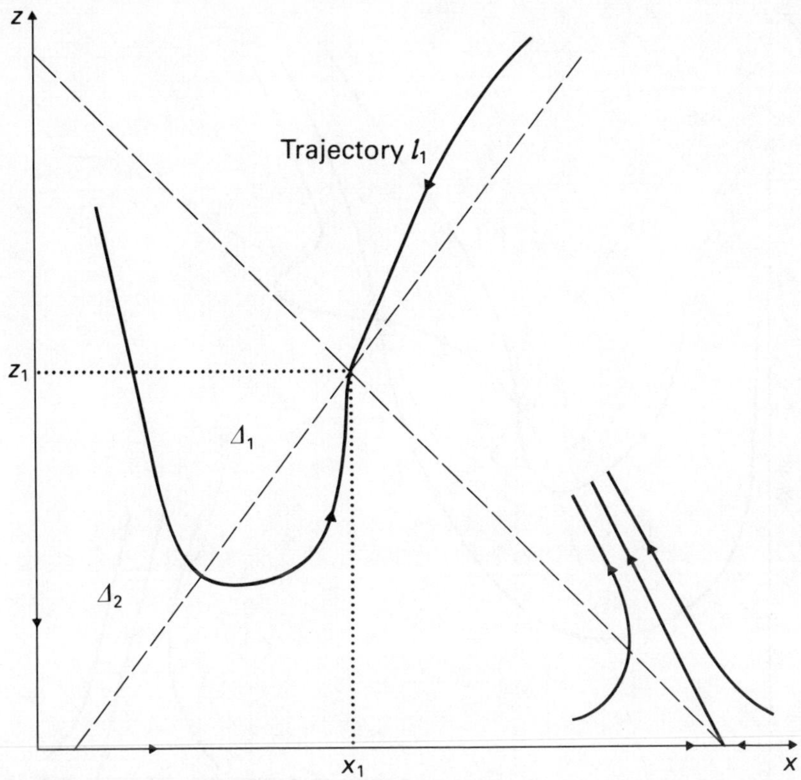

FIGURE 3.2 The structure of the domain Δ in model (3.3). The trajectory l_1 is the separatrix between Δ_1 and Δ_2.

produces a boomerang effect, but returning to the lower stable equilibrium. However, a shift into Δ_3 will produce a massive boomerang effect or mass outbreak, the pest population escaping to the upper stable point.

Once a pest population has moved to the upper stable point, further attempts at control may again produce a variety of responses. In the case depicted in Figure 3.4A there are two subdomains Δ_1 and Δ_2, a shift into the former producing an undercompensating response and into the latter a boomerang effect, but in both cases returning to the upper stable state. In Figure 3.4B, however, Δ_1 is further subdivided with a shift into Δ_1' producing either a boomerang return to the upper stable state or a movement of the population back to the lower stable state.

3.3.3. A General Problem

In a general system of ordinary differential equations modeling the predator–prey system:

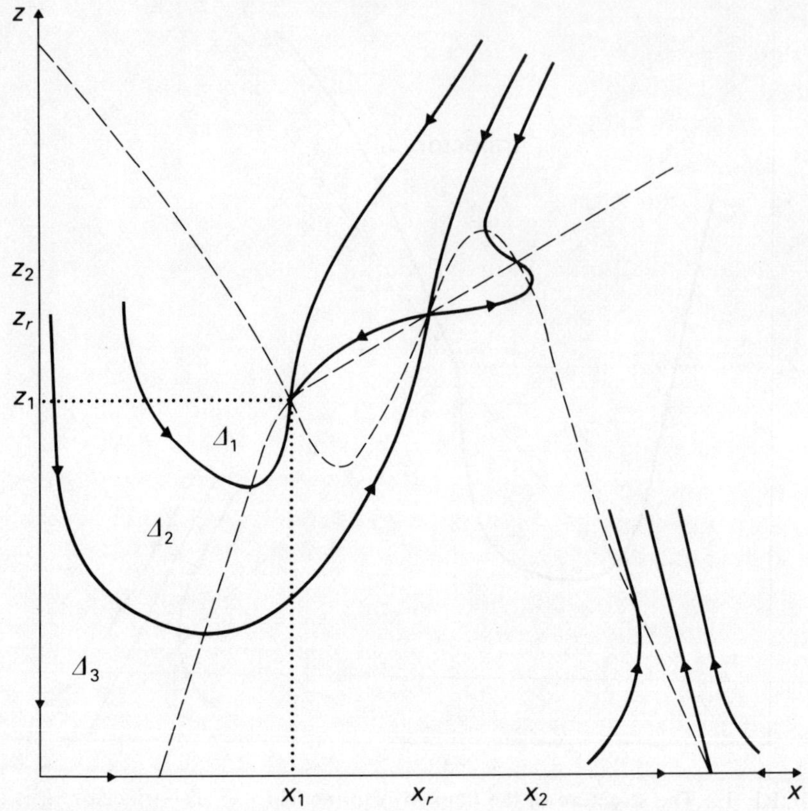

FIGURE 3.3 The structure of the domain Δ in model (3.5), where there are three equilibrium states.

$$\mathrm{d}x/\mathrm{d}t = xP(x,z) \qquad \mathrm{d}z/\mathrm{d}t = zQ(x,z). \tag{3.5}$$

It is usually supposed that functions $P(x,z)$ and $Q(x,z)$ satisfy the following conditions:

$$\frac{\partial P}{\partial z} < 0, \qquad \frac{\partial Q}{\partial x} > 0, \qquad \frac{\partial Q}{\partial z} < 0 \tag{3.6}$$

Then, if (x_1,z_1) is a stable knot in the phase space of system (3.5) and $P(x_1,z_1) = Q(x_1,z_1) = 0$, it follows from conditions (3.6), that the coefficients of the trajectory slopes k_1 and k_2 entering the knot are such that

$$k_1 k_2 = -z_1 \frac{\partial Q}{\partial x}(x_1,z_1) > 0, \tag{3.7}$$

i.e., they are of the same sign. If then the inequality $z_1\,(\partial Q/\partial z) < x_1\,(\partial P/\partial x)$,

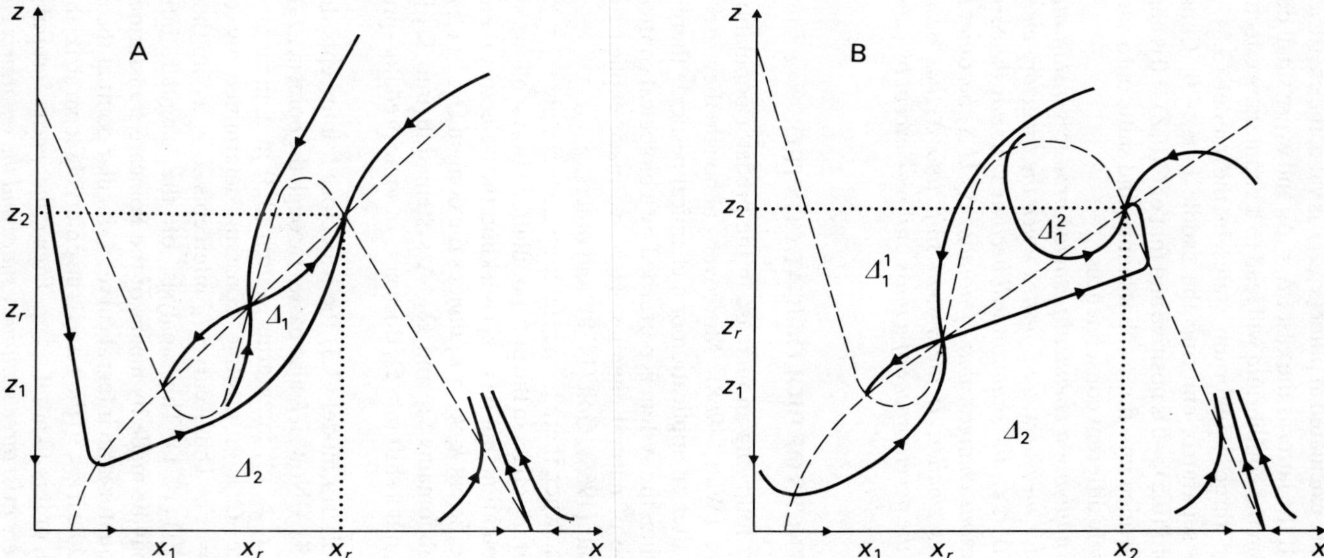

FIGURE 3.4 The structure of domain Δ when the system is at the upper stable state. (A) Δ_1 is an undivided domain; (B) Δ_1^1 is bounded by the separatrix entering the saddle and by the straight line $z = z_2$.

where derivatives are calculated at point (x_1,z_1), is valid, the coefficients $k_j > 0$; otherwise $k_j < 0$. If the latter is the case, $\Delta = \Delta_2$ and every small deviation of the population from the equilibrium will lead to a boomerang effect.

When there are three equilibrium states in the model (3.5), the slope coefficient of the separatrix entering the saddle is $\kappa > 0$. Consequently, because the root of $P(0,z) = 0$ is positive and that of $Q(0,Z) = 0$ is negative, the domain Δ is also divided into three subdomains, and shifts onto each of them lead to quantitatively different consequences.

For example, if applying a chemical pesticide produces a shift into domain Δ_2, *then if protective measures are repeated when the trajectory emerges from* Δ_2 *at* $x(t) = x_1$, $z(t) < z_1$, *the situation will become worse as the correlation of the domain dimensions changes, that of domain* $\Delta_2 \cup \Delta_3$ *becomes larger and that of* Δ_1 *becomes smaller. If the system shift into* Δ_3 *has occurred it is impossible to stop the development of a mass outbreak except by exterminating the population completely.*

3.4. CONSEQUENCES OF BIOLOGICAL CONTROL

Boomerang effects following an increase in the predator population number are extremely rare (Watt, 1968). Moreover, although they are produced relatively quickly after an application of a chemical pesticide there is a much slower response when the system is perturbed by a biological control method, such as the release of natural enemies. Here we make explicit the general observations of Watt (1968), Goh (1979), and others.

Consider the case when there is one stable stationary state (x_1,z_1). If (x_1,z_1) is a focus, a boomerang effect in the prey population always follows an increase in the predator population, but if (x_1,z_1) is a knot, the trajectory l_1 entering the knot with slope coefficient $k_2, k_2 > k_1$ divides the domain $\Omega = \{(x,z){:}x = x_1, z > z_1\}$ into two subdomains Ω_1 and Ω_2. A system shift into Ω_2 leads to a boomerang effect, but a shift into Ω_1 does not. In some predator–prey models $\Omega = \Omega_1$.

In the Lotka–Volterra model (3.3) the trajectory l_1 intercepts the straight line $x = x_1$ (Figure 3.6). Now, if l_1 also passes through the point $(x_3,z_3)\varepsilon\omega$ where $\omega = \{(x,z){:}z >> x_1 > x\}$, integrating system (3.3) in domain ω we obtain $z = Cx^{\varepsilon/\gamma}$, where C is a positive constant determined by conditions $x(0) = x_3$, $z(0) = z_3$. Thus, curve l_1 intercepts $x = x_1$ at the ordinate point $z = (z_3x_1^{\varepsilon/\gamma})/(x_3^{\varepsilon/\gamma})$. The analysis of the singular infinities of the system (3.3) may be made by means of the Poincaré transformation. The Poincaré sphere equator is an integral curve; a singular point at the end of the z axis is either a saddle if $\varepsilon < \gamma$ (the z axis is an emerging separatrix, the equator is an entering one), or it is a knot if $\varepsilon > \gamma$. *Then if the species is capable of mass outbreak, i.e., if there exist three stationary states and the separatrix entering the saddle intercepts domain* Ω, *an increase in the entomophagous insect population*

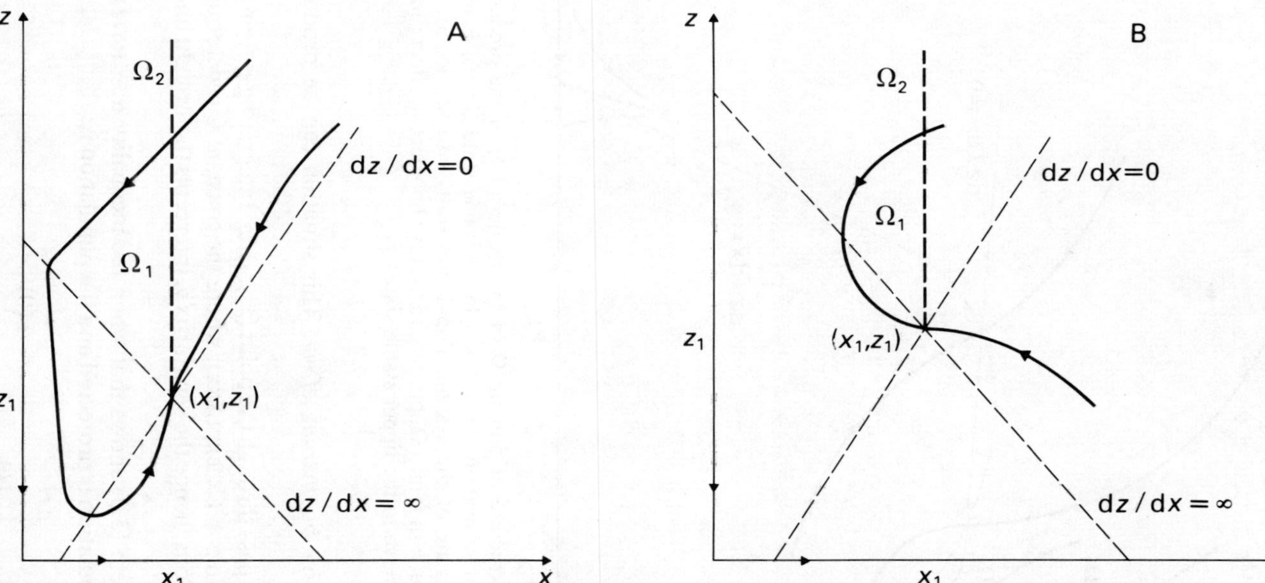

FIGURE 3.5 (A,B) The structure of domain Ω in the Lotka–Volterra model (3.3) at different values of the coefficients of trajectory slopes entering the knot (x_1,z_1).

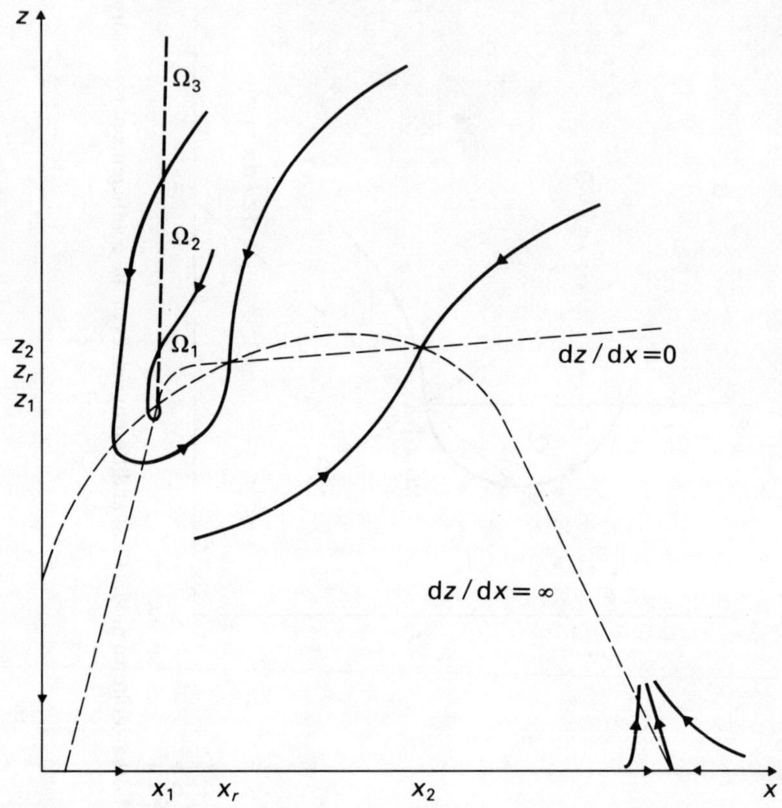

FIGURE 3.6 The structure of domain Ω in the model (3.4) of the predator–prey system with predator saturation. (x_1,z_1) is a lower stable state, (x_r,z_r) is a saddle, (x_2,z_2) is an upper state of the system. Population shifts into Ω_1 do not cause a boomerang effect, but a shift into Ω_3 produces a massive boomerang effect causing the prey population to move to the upper stable state.

may cause a massive boomerang effect. This situation can be produced by model (3.4).

Let separatrix y_r pass through the point $(x_3,z_3)\varepsilon\omega$. Integrating system (3.4) in domain ω we obtain $z = Cx^S\exp(S\pi x)$, where the constant C is determined by the original conditions; hence the separatrix y_r intercepts the straight line $x = x_1$ at $z > z_1$.

Analysis of system (3.5) shows that when the inequality $\alpha - (\gamma/H) > 0$ is valid, the prey population is protected and the population is:

$$x(t) > S, \qquad S = \frac{\alpha H - \gamma}{H\beta} \qquad \text{at} \quad x(0) > S.$$

If $S > 0$, and Y_r passes through point $(x_3,z_3)\varepsilon\omega$, $x_3 < S$, the increase in the entomophagous insect population does not cause a massive boomerang effect. But if there is no shelter, $S < 0$, y_r intercepts $x = x_1$. Thus, domain Ω may be divided into three subdomains Ω_j ($j = 1,2,3$).

Analogously, we can consider boomerang effects resulting from increases in the entomophagous insect population when the system of interacting populations is at an upper stable state. In this case a shift from equilibrium may cause a transition to the stable state (x_1,z_1) or to a return to the upper stable state with or without a boomerang effect. Domain Ω_3 is evidently absent here.

ACKNOWLEDGEMENTS

We would like to thank G. F. Kanevskaja, E. L. Krasova and G. B. Kofman for their helpful comments on the manuscript.

REFERENCES

Alexeyev, V. V. (1973) Effect of the saturation factor on the dynamics of the predator–prey system. *Biophysics* 18:922–6 (in Russian).

Bazykin, A. D. (1974) The Volterra system and the Michaelis–Menten system, in *Mathematical Genetics Problems* (Novosibirsk) pp103–43 (in Russian). (Available in English as *Structural and Dynamic Stability of Model Predator–Prey Systems* (1976) Research Memorandum, RM-76-8 (Laxenburg, Austria: International Institute for Applied Systems Analysis)).

Gilpin, M. E. (1975) Stability of feasible predator–prey systems. *Nature* 254:137–8.

Goh, B. S. and L. S. Jennings. (1977) Feasibility and stability in randomly assembled Lotka–Volterra models. *Ecol. Modelling* 3:63–71.

Goh, B. S. (1979) The usefulness of optimal control theory to ecological problems, in E. Halfon (ed) *Theoretical Systems Ecology* (New York: Academic Press) pp385–99.

Isaev, A. S., L. V. Nedorezov, and R. G. Khlebopros. (1979) The mathematical model of the escape effect in the predator–prey interactions, in *Mathematical Modeling of Forest Biogeocenosis Components* (Novosibirsk: Nauka) pp74–82 (in Russian).

Kolmogorov, A. N. (1936) Sulla teoria di Volterra della lotta per l'esistenza. *Gionale Instituto Ital. Attuari*, 7:74–80.

Lotka, A. J. (1925) *Elements of Physical Biology* (Baltimore: Williams and Wilkins).

May, R. M. (1972) Limit cycles in predator–prey communities. *Science* 177:900–2.

May, R. M. (1975) *Stability and Complexity in Model Ecosystmes* 2nd edn (Princeton, NJ: Princeton University Press).

May, R. M. (1981) Models for two interacting populations, in R. M. May (ed) *Theoretical Ecology: Principles and Applications* 2nd edn (Oxford: Blackwell) pp78–104.

Murdoch, W. W. and A. Oaten (1975) Predation and population stability. *Adv. Ecol. Res.* 9:2–131.

Rozenzweig, M. L. and R. H. Macarthur (1963) Graphical representation and stability condition of predator–prey interactions. *Am. Nat.* 97:209–23.

Tanner, J. T. (1975) The stability and the intrinsic growth rates of prey and predator populations. *Ecology* 56:855–67.

Volterra, V. (1926) Variations and fluctuations of the number of individuals in animal species living together. *J. Cons. Perm. Int. Entomol. Mer.* 3:3–5.

Watt, K. E. F. (1968) *Ecology and Resource Management* (New York: McGraw-Hill) (In Russian, published by Mir, Moscow).

4 Threshold Theory and its Applications to Pest Population Management

Alan A. Berryman

4.1. INTRODUCTION

Thresholds appear to be rather common phenomena in biology (e.g., see Chapter 2; May, 1977) and they have fascinated both theoreticians and experimentalists. Early medical epidemiologists puzzled over why certain diseases, which had remained at innocuous levels for long periods of time, suddenly exploded through human populations wreaking death and destruction. Likewise, modern psychologists try to understand why an apparently normal individual suddenly and inexplicably goes on a murderous rampage. Obviously, thresholds have been breached. For all this, however, the investigation of such phenomena remains difficult. By their definition thresholds are transient states, separating two distinct patterns of behavior, and thus presenting elusive targets for research. Because of the speed of change in a system, the threshold is rarely observed and empirical studies become difficult or impossible. For this reason, theoretical studies of threshold phenomena are especially important.

In this chapter I present some arguments for a biological theory of population thresholds that stem from a case study of the mountain pine beetle. I then discuss the relevance of threshold concepts to practical pest management—from the standpoint of the manager the most urgent need is for estimates of the likelihood of a pest outbreak in a particular location. I then show how the thresholds and the high-risk zones in the driving variables can be presented in a simple graphical form suitable for use as a basis for management.

4.2. THEORETICAL CONSIDERATIONS

Thresholds have been extensively investigated by biologists in many fields, most notably by disease epidemiologists (e.g., Kermack and McKendrick,

1927; Macdonald, 1965). Despite this, to my knowledge, there exists no formal and general theory of biological thresholds. On the other hand, there is a formidable area of mathematical theory that treats sudden and discontinuous patterns of behavior. Because of this there has been a temptation to adjust the biology to fit the mathematical theory. This is best exemplified by the current popularity of catastrophe theory (Thom, 1975; Zeeman, 1976; Jones and Walters, 1976; Harmsen *et al.*, 1976; Dodson, 1976) and the resulting controversy surrounding its application to the social and biological sciences (Kolata, 1977; Sussmann and Zahler, 1978). In this chapter I concern myself solely with some of the elements of a biological theory of population thresholds.

First, let us define a biological threshold as a boundary separating two distinct patterns of biological behavior. Although the two behaviors may be directly observable (e.g., endemic and epidemic population dynamics), the threshold itself can rarely be observed because it represents a transient state. *I further propose that threshold theory rests on the axiom that, if a biological system exhibits sudden and discontinuous changes in behavior, and if these changes cannot be* completely *explained by the action of exogenous driving variables, then a threshold must exist.*

Although the behavioral patterns exhibited on each side of a threshold may lead to distinct, permanent or long-term, equilibrium states (e.g., the catastrophe paradigm), this is purposefully not implied in the *biological* definition. In fact, many biological thresholds separate dramatically different behaviors that both lead, eventually, to the same equilibrium condition. Hence, a disease epidemic eventually returns to its endemic state, and the brain returns to its normal equilibrium following an epileptic seizure.

Perhaps a cybernetic perspective can help us clarify our concept of thresholds. Suppose we have a system with feedback S which produces an output O and is driven by an exogenous input E (Figure 4.1). For example, the input may be an environmental variable such as weather or food, and the output may be population density. If the output dynamics remain relatively

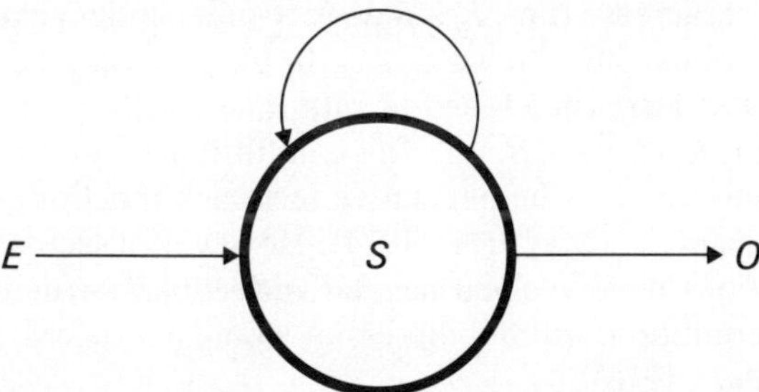

FIGURE 4.1 A simple feedback system S producing an output O from an environmental input E.

steady over a wide range of input values, then the system exhibits steady-state or equilibrium behavior that must be governed by internal negative feedback. However, suppose that the system, by design or circumstance, can only handle a certain range of input values so that parts of the negative-feedback structure are changed or disengaged outside of this range (i.e., the control mechanism(s) saturate as the input becomes large). What could be the possible outcome of exceeding such control tolerances? First, the system may go to a new equilibrium if some negative-feedback pathways are still open. Second, it may exhibit exponential growth or decay if only positive-feedback circuits remain open. Third, we may observe cyclic trajectories if time delays are introduced into the negative-feedback structure (Berryman, 1978a,b; Hutchinson, 1948; Wangersky and Cunningham, 1957; May, 1976). The last conclusion is perhaps the most interesting biologically because we often observe cyclical return to equilibrium in disease and insect epidemics, aggressive or emotional disturbances, etc.

This cybernetic viewpoint illustrates the point that threshold mechanisms are intimately associated with the feedback structure of biological systems. Thus, in order to understand threshold population phenomena we must delve into the theory of population dynamics and, in particular, into population regulation by negative feedback.

4.2.1. Population Theory

The classical Malthusian concepts of population regulation can be expressed in the elementary model:

$$R = Hf(N_g), \tag{4.1}$$

where R is the realized per capita replacement rate (or reproduction rate, see Chapter 2) of the gth generation ($R = N_{g+1}/N_g$), N_g is the density of the population in that generation, f is a density-dependent negative-feedback function, and H represents the external environment or habitat, which sets the level of "resources" available to the population and, therefore, the maximum individual rate of increase (i.e., H is a density-independent exogenous driving variable; see Berryman and Brown, 1981). In the classical models, f is assumed to be a monotonic decreasing function with $R \rightarrow 0$ as $N \rightarrow \infty$, so that a single equilibrium point K occurs at $R = 1$. This equilibrium may be stable or unstable depending on the slope of the negetative-feedback function in the neighborhood of equilibrium (see May *et al.*, 1974). However, it seems likely that most natural populations have evolved negetative-feedback structures that create globally stable equilibria; without these they would not survive in the long term (e.g., Hassell *et al.*, 1976).

For a given function f in this model, the equilibrium point is set by the habitat variable H. In other words, if H changes the system will self-adjust to a new

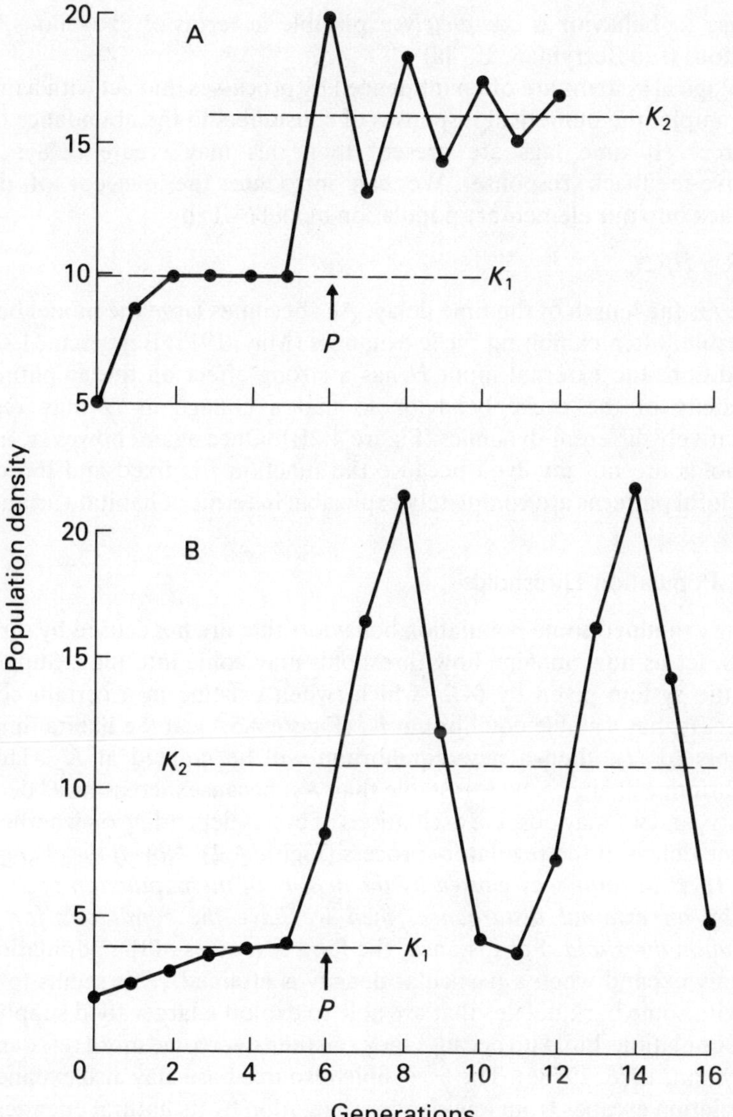

FIGURE 4.2 Dynamics of a simple population model, $R = H[\exp(-\alpha N_{g-t})]$ following an environmental change at P; $\alpha = 0.1$: (A) $H = 3$ prior to P and 6 afterwards, $t = 0$. (B) $H = 1.5$ prior to P and 3 afterwards, $t = 1$.

equilibrium level (Figure 4.2A). However, there exists a continuum of possible equilibria and although we may observe distinct equilibrium behaviors, the description of this system does not require notions of thresholds because its behavior is *completely* explicable in terms of exogenous habitat variations (see Berryman, 1978a).

Biological systems are often influenced by processes that act with a time lag; for example, the numerical responses of consumers to the abundance of their resources. If time lags are present then this may create delays in the negative-feedback response. We can introduce the concept of delayed feedback into our elementary population model (4.1) by

$$R = Hf(N_{g-t}), \tag{4.2}$$

where t is the length of the time delay. As t becomes large the model becomes less stable, often exhibiting cyclic dynamics (May, 1973; Berryman, 1978a,b). In addition, the external input H has a strong effect on the amplitude and periodicity of the cyclic behavior so that a change in H may result in qualitatively different dynamics (Figure 4.2B). Once again, however, internal thresholds are not involved because the function f is fixed and the distinct behavioral patterns are completely explicable in terms of habitat variations.

4.2.2. Population Thresholds

Having examined some population behaviors that are *not* caused by threshold effects, let us now imagine how thresholds may come into play. Suppose we have the system given by (4.2) which, when existing in a certain constant habitat H_1, has a stable equilibrium K_1 (Figure 4.3A). If the habitat improves suddenly to H_2, then a new equilibrium will be created at K_2. This new equilibrium will always be less stable than K_1, because increasing H decreases stability, and we may observe oscillations or cycles depending on whether there are time delays in the regulatory process (Figure 4.2). *Now if the change from H_1 to H_2 is in some way caused by the density of the population itself, rather than by an external disturbance, then we have the conditions for a true population threshold*. For instance, the food resources of the population may suddenly expand when a particular density is attained. This seems to be the case with some bark beetles that are able to exploit a larger food supply when their populations build up because they can then overcome more resistant trees (Berryman, 1976, 1978c). The accessible resource base may also expand when a population escapes from low-density regulation by its natural enemies, e.g., the eastern spruce budworm (Holling *et al.*, 1979) and the gypsy moth (Campbell and Sloan, 1978). Thus, although different mechanisms may be involved in setting the threshold, the basic process is similar. That is, an increase in population density above a certain level increases the quantity of resources available to the population, which in turn increases the reproduction and/or survival of the average individual. In this way the feedback structure of the system switches from negative to positive over a certain range of population densities (Figure 4.3B). This sinuous feedback function now has

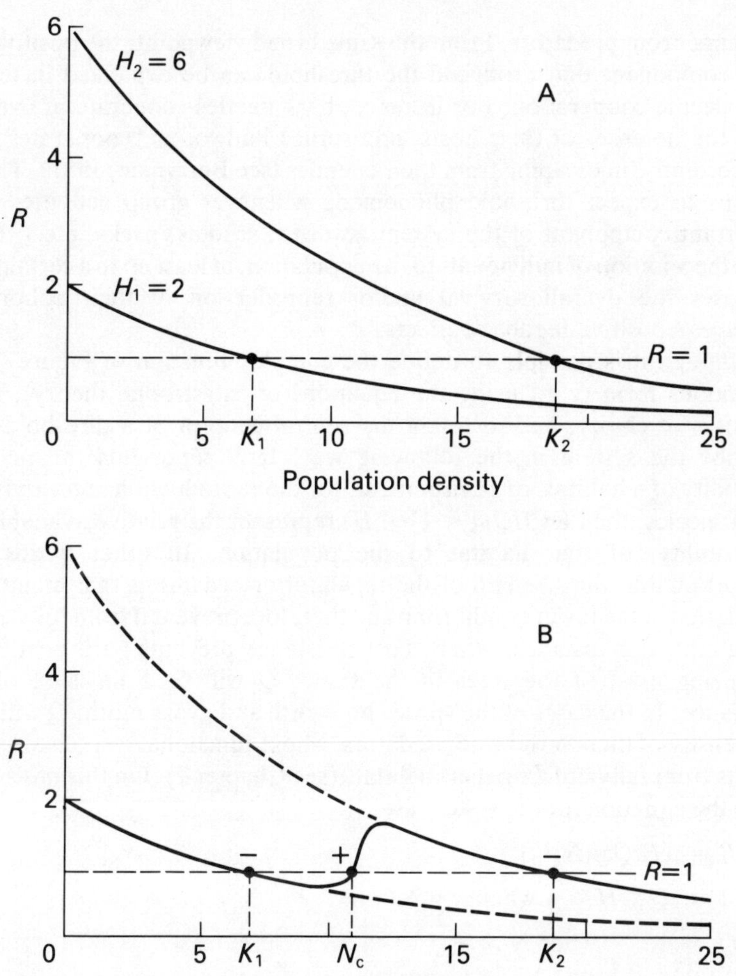

FIGURE 4.3 (A) Effect of a change in H on the model $R = H[\exp(-\alpha N_g)]$ showing equilibrium densities at K_1 and K_2. (B) Effect of positive feedback creating an unstable equilibrium at N_c.

three equilibrium points, two of which are potentially stable (K_1,K_2) and one is unstable (N_c). In fact it is easy to see that the unstable point N_c separates the system into two domains of behavior and, therefore, specifies a population threshold, i.e., the domain of attraction to K_1 is specified by N in $(0 \rightarrow N_c)$ and to K_2 by N in $(N_c \rightarrow \infty)$.

One more rather significant generalization emerges from this analysis. The negative-feedback component of the Malthusian model is often explained in terms of intraspecific competition for food and space, including space to hide, or escape from predators. From the same broad viewpoint, the positive-feedback component that produced the threshold can be explained in terms of intraspecific cooperation. For instance, bark beetles cooperate in overwhelming the defenses of their hosts, and spruce budworms "cooperate", albeit inadvertently, in escaping from their enemies (see Berryman, 1981). This may lead us to expect threshold phenomena whenever group activities are an important component of the system (swarms, schools, packs, etc.). In such cases the addition of individuals to the population, at least up to a certain point, increases the overall survival and/or reproduction of their cohorts and produces a positive-feedback effect.

Although it is possible to define the complex function of Figure 4.3B in continuous form (e.g., using the equations of catastrophe theory), for the purposes of clarity, and to retain the central concept of a threshold, let us describe the system in the following way. If H represents the maximum suitability of a habitat, or environment, for the reproduction and survival of a given species, then let $1/Q(1 < Q \leq H)$ represent the relative availability, or accessibility, of that habitat to the population. In other words, Q is proportional to the strength of the regulatory mechanisms that maintain the population at the lower equilibrium and therefore prevent it from fully utilizing its habitat. For instance, where host resistance prevents bark beetles from colonizing most of the trees in the stand, Q will be a measure of stand resistance. In the cases of the spruce budworm and gypsy moth, Q will reflect the density of their vertebrate predators, whose functional responses keep the insects from fully utilizing their habitats (see Chapter 2). On this basis we can write the equation for the system as

$$
\begin{aligned}
R &= (H/Q)f(N_{g-t}) \\
1 &< Q \leq H, \quad \text{when } N < N_c \\
Q &= 1, \quad \text{when } N > N_c.
\end{aligned}
\tag{4.3}
$$

Although this model assumes that low- and high-density equilibria are enforced by the same function f, which positive feedback maximizes the instant N_c is reached, and that N_c, Q, and H are constants, none of which are necessarily true in nature, it will suffice for illustrative purposes.

In a constant environment this model will equilibrate at its lower equilibrium provided the initial population density and the final equilibrium density are less than the threshold, i.e., $N_0 < N_c$ and $K_1 < N_c$. However, if a perturbation of N (say by immigration) carries the population density above N_c, then the system will move to its high-density equilibrium (Figure 4.4A). Of course the same is true in reverse.

4.2.3. Time Delays

From the above discussion it is apparent that H has an important influence on

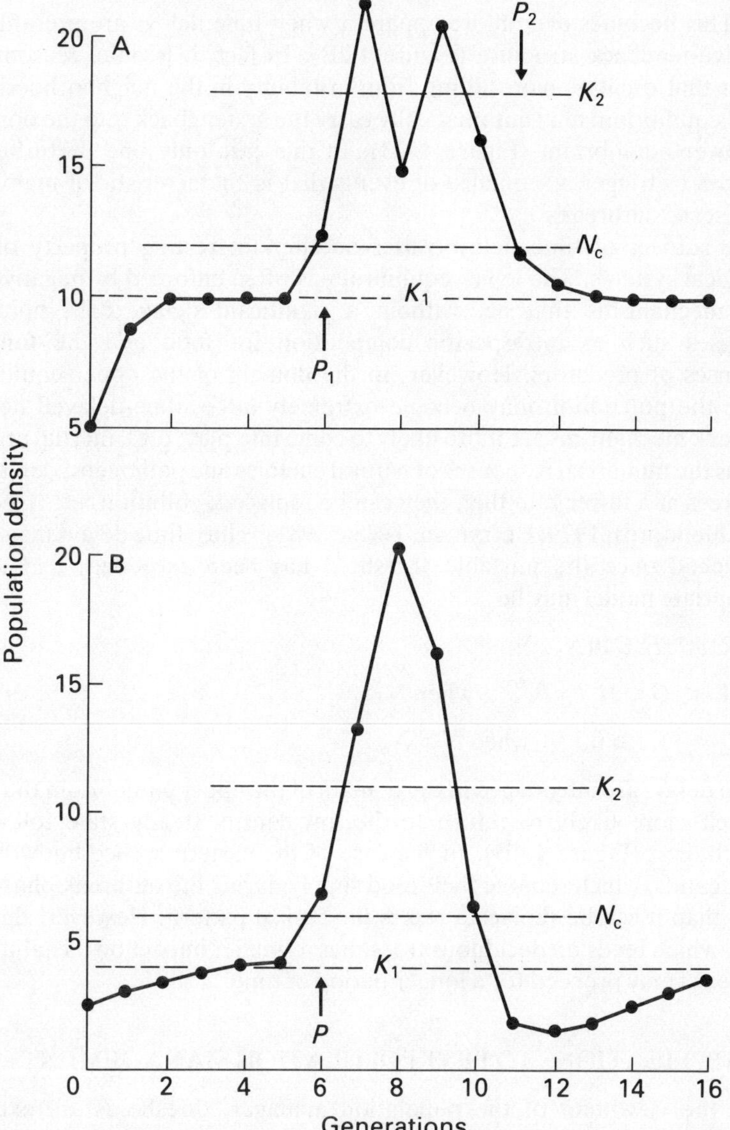

FIGURE 4.4 Dynamics of a simple population model, $R = (H/Q)[\exp(-\alpha N_{g-t})]$;$\alpha = 0.1$.
(A) $H = 6$; $Q = 2$; when $N < N_c$; $H = 12$ and $Q = 1$ when $N > N_c$; $t = 0$; and (B) $H = 3$; $Q = 2$ when $N < N_c$; $H = 6$ and $Q = 1$ when $N > N_c$; $t = 1$. P indicates perturbations to the system.

the *relative* stability of the simple population system (Figures 4.2 and 4.4A). Thus, the upper equilibrium will always be relatively less stable than the lower one. This becomes even more apparent when time delays are present in the negative-feedback structure (Figure 4.2B). In fact, it is quite reasonable to expect that oscillations resulting from instability in the neighborhood of the upper equilibrium may automatically carry the system back into the domain of the lower equilibrium (Figure 4.4B). In this case only one perturbation is required to trigger a sequence of events that is characteristic of many insect and disease outbreaks.

The subject of time delays raises another interesting property of some biological systems. The lower equilibrium is often enforced by negative-feedback mechanisms that act without a significant delay, e.g., noninertial processes such as intraspecific competition for food and the functional responses of predators. However, in the domain of the upper equilibrium, where the population may become extremely large, time-delayed negative-feedback mechanisms are more likely to come into play, e.g., inertial processes such as the numerical responses of natural enemies and pathogens, depletion of resources at a faster rate than they can be replaced, pollution, etc. (see Isaev and Khlebopros, 1979; Berryman, 1978a, 1981). Thus, time delays may also be introduced once the unstable threshold has been exceeded, and a more appropriate model may be

$$\begin{aligned} R &= (H/Q)f(N_{g-t}) \\ 1 &< Q \leqslant H, t=0, \qquad \text{when } N < N_c \\ Q &= 1, t \geqslant 0, \qquad \text{when } N > N_c. \end{aligned} \tag{4.4}$$

If time delays are introduced in the domain of the upper equilibrium the system is much more likely to return to the low-density steady state following a perturbation (Figure 4.4B). In the case of the eastern spruce budworm and bark beetles, which remove their food supply during the outbreak phase much faster than it can be replaced, this is the typical pattern. However, the gypsy moth, which feeds on deciduous trees, has a smaller impact on its habitat, and outbreaks may proceed for a longer period of time.

4.3 APPLICATIONS TO PEST POPULATION MANAGEMENT

From the viewpoint of the population manager, thresholds are extremely critical because they define the boundaries of resilience of the system (Holling, 1973; Berryman, 1983). But the patterns of behavior exhibited on each side of the threshold are usually known from historical records, or are of little concern. For example, the forest manager tends to be unconcerned with the behavior of endemic spruce budworm or bark beetle populations because they cause insignificant damage, and he usually has other more pressing problems on his mind. Moreover, he knows from experience the havoc that an outbreak can wreak, and does not require sophisticated models to prove it to him. However, he is extremely interested in the likelihood of a pest epidemic erupting in a *particular* forest stand. If he can predict this, given certain measureable system variables, then he can design a management plan and set priorities to minimize the risks of precipitating the epidemic.

The problem then is to define the epidemic threshold in terms of the system variables H and Q. For example, consider the density-dependent feedback function (Figure 4.4)

$$R = (H/Q)\exp(-\alpha N_{g-t}). \tag{4.5}$$

The system will be in equilibrium at K when $R = 1$, and thus

$$(H/Q)\exp(-\alpha K) = 1 \qquad K = \ln(H/Q)/\alpha. \tag{4.6}$$

From eqn. (4.3) we see that the two equilibria K_1 and K_2 are defined by

$$K_1 = \ln(H/Q)/\alpha \qquad K_2 = \ln(H)/\alpha. \tag{4.7}$$

In other words, K_1 decreases with Q but K_2 remains constant because it is independent of Q. The relationship between these two equilibria and the strength of the regulatory mechanism is shown in Figure 4.5, where we have relaxed the assumption that N_c is constant, allowing it to increase directly, and nonlinearly, with Q. This seems reasonable because, when the control mechanism is nonfunctional ($Q = 1$), the critical population density that overloads, or saturates, the control mechanism approaches zero ($N_c \rightarrow 0$). However, as the strength of the regulatory mechanisms increase, N_c would be expected to rise accordingly.

The dynamics of this system are such that $N \rightarrow K_2$ when $N_0 > N_c$ or $K_1 > N_c$, and $N \rightarrow K_1$ when $N_0 < N_c$ and $K_1 < N_c$. We can see from Figure 4.5 that there exists a point T, where $N_c = K_1$, which separates a zone ($Q < T$) where an epidemic is certain from one ($Q > T$) where the outcome is uncertain, depending on the initial population size N_0. In addition, the probability of the occurrence of an epidemic, given that $Q > T$, recedes as Q becomes larger because the magnitude of $N_c - K_1$ increases with Q (Figure 4.5). As Q becomes very large a point will be reached, where $N_c = K_2$, beyond which the probability of an epidemic is zero.

If we repeat this analysis in habitats of different suitability, which alters the level of K_2 (Figures 4.2 and 4.4), we obtain a series of points (T) that define the epidemic threshold in H,Q phase space (Figure 4.6). We now have a theoretical basis for identifying and defining epidemic threshold function.

4.3.1. Identifying Threshold Functions

Because thresholds describe transient states of a system they cannot be observed directly, so that precise empirical definitions are impossible to obtain. However, they can be approximated empirically if the system has been

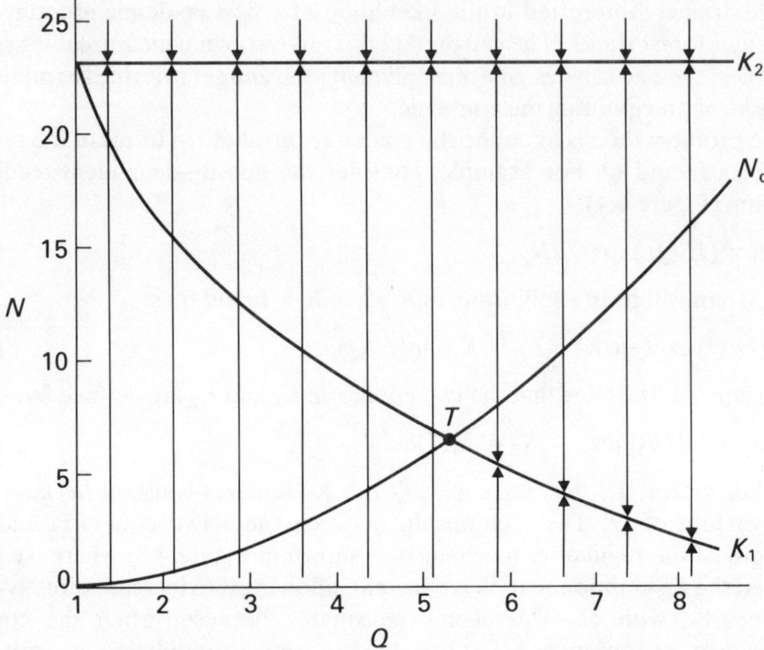

FIGURE 4.5 Effect of Q, the strength of the regulatory mechanism, on the population parameters: K_1 endemic equilibrium, K_2 epidemic equilibrium, and N_c population threshold, as defined by eqn. (4.7), with $\alpha = 0.1$, $H = 10$, and with the assumption that N_c varies directly with Q. Arrows show the qualitative dynamics of this system.

observed in both behavioral modes and under different values of H and Q, provided of course that these parameters can be expressed in terms of real system variables. In this way I have been able to obtain an approximate epidemic threshold function for the mountain pine beetle in lodgepole pine stands (Figure 4.7). The suitability of the habitat H for the beetle was defined as the percentage of trees in the stand with thick phloem, because these trees provide a more suitable environment for larval development and survival, and the strength of the regulatory mechanism Q was assumed proportional to stand resistance, where resistance was considered to be a function of stand density, age, and periodic growth rate (Figure 4.7).

Alternatively, epidemic thresholds can sometimes be identified by simulation modeling. For example, Holling *et al.* (1979) constructed a detailed model that simulated the population dynamics of the spruce budworm in eastern spruce–fir forests. In the process of simplifying and compressing the model's output, they displayed it as a catastrophe manifold, from which an approximate epidemic threshold function could be generated (Figure 4.8). In the case of the spruce budworm, habitat suitability H was measured by the density of balsam

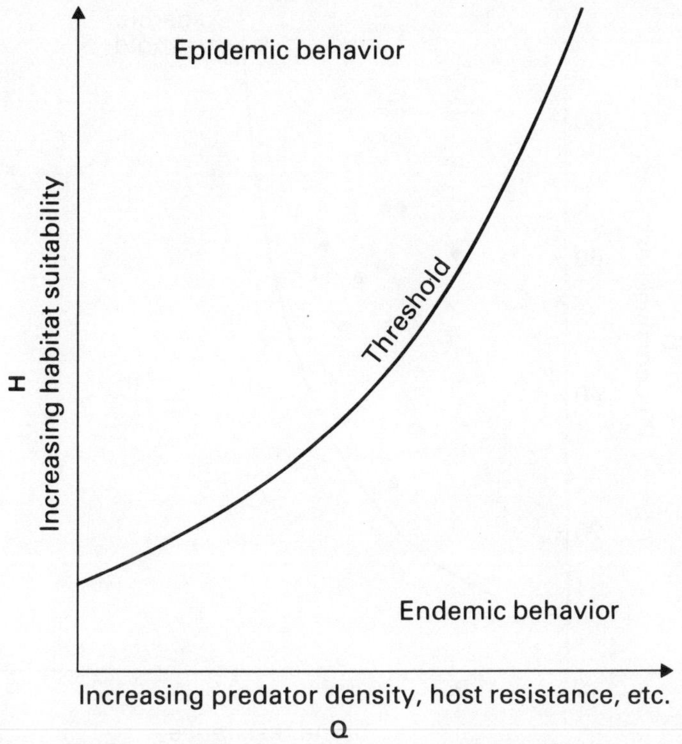

FIGURE 4.6 Theoretical derivation of the epidemic threshold function where Q, the strength of the regulating mechanism, may be measured in terms of host resistance or predator density and H is the suitability of the habitat.

fir branches, and the strength of the regulatory mechanism Q was assumed to be proportional to the density of insectivorous birds (Figure 4.8).

4.3.2. Dealing with Uncertainties

As we have seen, biological systems that are affected by thresholds are very susceptible to external perturbations. For example, drought and other environmental stresses may temporarily lower the resistance of lodgepole pine stands and precipitate mountain pine beetle epidemics, and spruce budworm outbreaks are often triggered by warm dry weather in spring, favoring larval survival. The system may also be perturbed by sudden changes in population density. In particular, immigration of large numbers of individuals from adjacent areas may be responsible for triggering outbreaks (Berryman, 1978a,c).

From the manager's standpoint, an environmental perturbation that cannot be anticipated is of little value for predicting when and where pest outbreaks

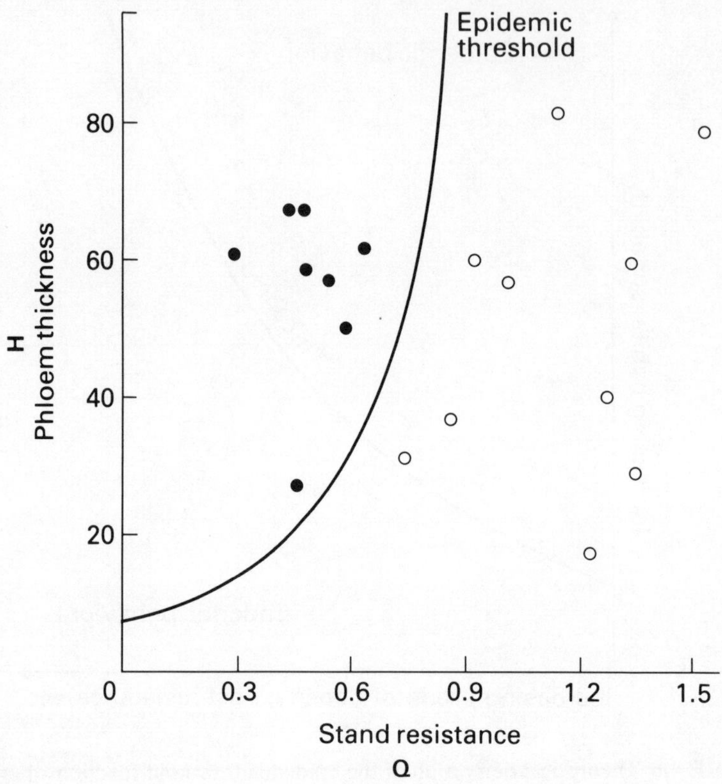

FIGURE 4.7 Approximate location of the epidemic threshold function for the mountain pine beetle in lodgepole pine stands: • epidemic; ○ endemic beetle populations; H = percentage of stand with phloem > 0.1 inch thick (a measure of food quantity and quality); Q = stand resistance measured by age, crown competition, and periodic growth rate (Mahoney, 1978) (After Berryman (1978c) with three abnormal data points omitted).

are likely to occur. However, we know that the probability of occurrence of an outbreak depends on how near the system is to the epidemic threshold, irrespective of the kind of perturbation involved. In addition, we know that once an outbreak has been triggered in a particular locality, the risk of its spreading to adjacent areas increases because migrating insects may raise the resident population density above the epidemic threshold (Berryman, 1978a). *Thus, the manager should treat compartments of his management unit that fall close to the threshold as high-risk areas, while those more distant can be assigned a lower risk. In this way he can produce a risk model, such as that in Figure 4.9, which he can then use to assign risk values to the compartments of his management unit. The risk zones should be drawn to the manager's specifications, and thus reflect his degree of risk aversion.*

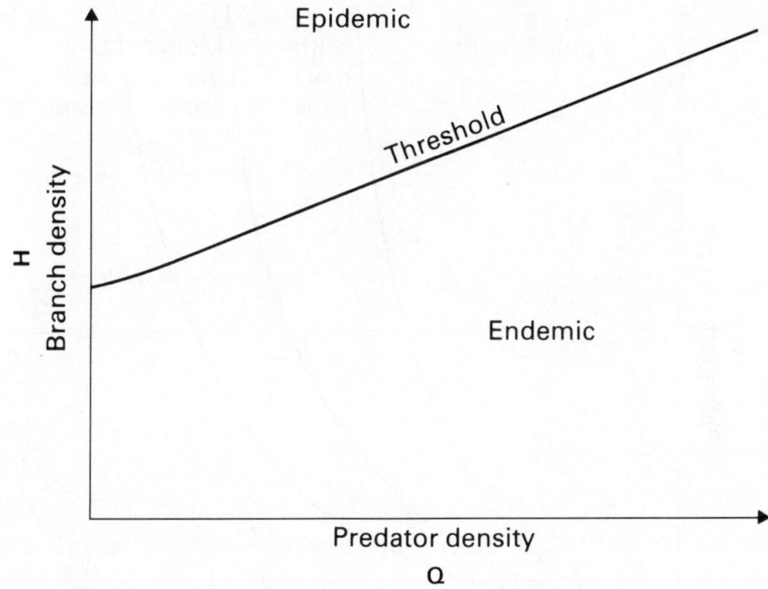

FIGURE 4.8 Epidemic threshold function for the spruce budworm in spruce-fir stands of eastern North America, where H is branch density and Q is density of avian predators (after Holling *et al.*, 1979, Figure 12).

Risk models of this kind are most valuable to the manager in setting priorities and in long-range planning. Obviously high-risk areas should receive high priority for management treatments, such as thinning to reduce branch area (spruce budworm) or to increase resistance (bark beetles), or perhaps manipulating bird populations using nesting boxes, etc. (spruce budworm). In addition, decision models can be used, in conjunction with complex prognosis models, to evaluate future risks and set long-range management goals. For example, Stage (1973) has implemented a stand prognosis model that is used by forest managers throughout the western United States. This model projects the structure of forest stands from inventory into the future, generating mensurational data at ten-year intervals. These data can be used to estimate lodgepole pine phloem thickness (Cole, 1973) and resistance to mountain pine beetle attack (Mahoney, 1978), and the trajectory of these crucial variables can be displayed on a risk decision model (Figure 4.10). Obviously, in this particular stand, the manager should be most concerned with the possibility of a beetle outbreak when the stand is between 80 and 110 years old.

Stage's stand prognosis model can also be used to evaluate various management practices on the growth and development of forest stands. For instance, a stand may be thinned at different times during the projection period and the trajectories of the critical variables displayed on the risk model. In this way the manager can choose the best thinning program to minimize the risk of

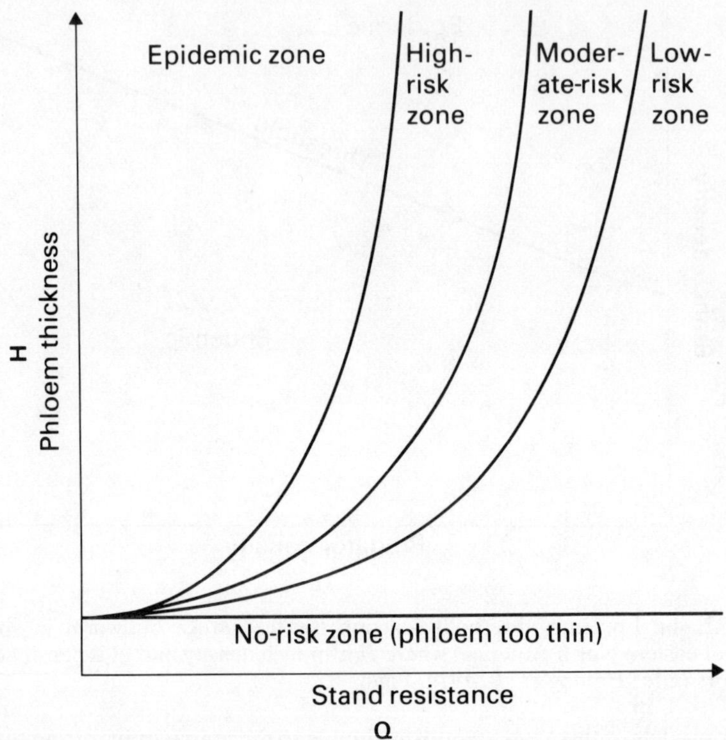

FIGURE 4.9 A model identifying the risk of mountain pine beetle outbreaks in lodgepole pine stands (after Berryman, 1978c).

bark beetle epidemics. *Thus, risk decision models can be used for evaluating the current danger of undesirable system behavior, predicting future risks, and formulating long-range management plans. Although empirical risk models have been developed for a number of forest pests (e.g., Keen, 1943; Morris and Bishop, 1951; Mahoney, 1978), threshold theory offers a formal and general framework upon which to formulate risk decisions (Berryman, 1980, 1982)*.

4.4. CONCLUSIONS

The study of thresholds is not only an interesting theoretical diversion, but also seems to offer a fundamentally sound framework for formulating resource management decisions. Although we have concentrated largely on pest management problems, the same principles can be applied to fisheries, wildlife, and other resource management fields. In these cases, however, a problem arises when a useful resource, normally at a high-density equilibrium, is harvested dangerously close to the unstable threshold. In this condition

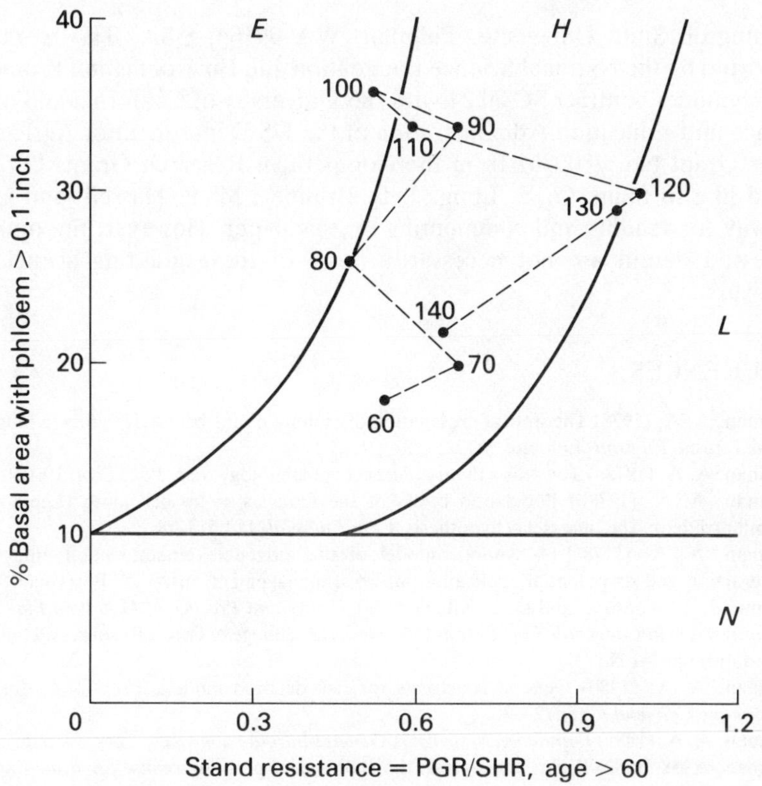

FIGURE 4.10 A lodgepole pine stand trajectory produced from a stand prognosis model (Stage, 1973) projected on the mountain pine beetle risk decision model (after Berryman, 1978c).

rather minor environmental perturbations may precipitate a collapse of the resource to its lower equilibrium (see effect of P_2 in Figure 4.4A). An interesting example is the case of the Atnarko River salmon run reported by Peterman (1977).

The mathematics used in this paper is straightforward and rather unsophisticated and there is a pressing need for a mathematical theory developed around the concept of biological thresholds. There seems to be no reason why mathematical tools cannot be developed for finding thresholds in *n*-dimensional space. Such developments would be of great value in the management of renewable resources and, perhaps, in psychology, medicine, and other biological disciplines as well.

ACKNOWLEDGMENTS

Scientific Paper No. 5729. College of Agriculture Research Center,

Washington State University, Pullman, WA 99164, USA. This work was supported by the National Science Foundation and Environmental Protection Agency under contract SC 00 24 with the University of California and by the Science and Education Administration of the US Department of Agriculture under Grant No. 7800861 from the Competitive Research Grants Office. I would like to thank G. E. Long, J. F. Brunner, M. P. Hassell, and G. R. Conway for reading and commenting on this paper. However, the opinions expressed herein are not necessarily those of these granting agencies or reviewers.

REFERENCES

Berryman, A. A. (1976) Theoretical explanation of mountain pine beetle dynamics in lodgepole pine forests. *Environ. Entomol.* 5:1225–33.

Berryman, A. A. (1978a) Towards a theory of insect epidemiology. *Res. Pop. Ecol.* 19:181–96.

Berryman, A. A. (1978b) Population cycles of the Douglas-fir tussock moth (Lepidoptera: Lymantriidae): The time-delay hypothesis. *Can. Entomol.* 110:513–18.

Berryman, A. A. (1978c) A synoptic model of the lodgepole pine/mountain pine beetle interaction and its potential application in forest management, in A. A. Berryman, G. D. Amman, R. W. Stark, and D. L. Kibbee (eds) *Theory and Practice of Mountain Pine Beetle Management in Lodgepole Pine Forests* (Moscow, ID: College of Forest Resources, University of Idaho) pp98–105.

Berryman, A. A. (1980) General constructs for risk decision models. *Proc. Soc. Am. For. (Spokane Convention)* pp123–8.

Berryman, A. A. (1981) *Population Systems: A General Introduction* (New York: Plenum).

Berryman, A. A (1982) Biological control, thresholds, and pest outbreaks. *Environ. Entomol.* 11: 544–549

Berryman, A. A. (1983) Defining the resilience thresholds of ecosystems, in W. K. Lavenroth, G. V. Skogerboe and M. Flug (eds) *Analysis of Ecological Systems: State-of-the-Art in Ecological Modelling* (Amsterdam: Elsevier)

Berryman, A. A. and G. C. Brown (1981) The habitat equation: A useful concept in population modeling, in D. G. Chapman and V. G. Gallucci (eds) *Quantitative Population Dynamics* Vol. 13, Stat. Ecol. Ser. (Burtonsville, MD: International Cooperative Publishing House).

Campbell, R. W. and R. J. Sloan (1978) Numerical bimodality among North American gypsy moth populations. *Environ. Entomol.* 7:641–46.

Cole, D. M. (1973) Estimation of phloem thickness in lodgepole pines. *USDA For. Serv. Res. Pap.* INT-148 (Ogden, UT: Intermtn. Forestry Range Expt. Station).

Dodson, M. M. (1976) Darwin's law of natural selection and Thom's theory of catastrophes. *Math. Biosci.* 28:243–74.

Harmsen, R., M. R. Rose, and B. Woodhouse (1976) A general model for insect outbreak. *Proc. Entomol. Soc. Ontario* 107:11–18.

Hassell, M. P., J. H. Lawton, and R. M. May (1976) Patterns of dynamical behaviour in single-species populations. *J. Animal Ecol.* 45:471–86.

Holling, C. S. (1973) Resilience and stability of ecological systems. *Ann. Rev. Ecol. Syst.* 4:1–23.

Holling, C. S., D. D Jones, and W. C. Clark (1979) Ecological policy design: A case study of forest and pest management, in G. A. Norton and C. S. Holling (eds) *Pest Management.* Proc. Int. Conf. October 1976 IIASA Proceedings Series Vol. 4 (Oxford: Pergamon) pp13–90.

Hutchinson, G. E. (1948) Circular causal systems in ecology. *Ann. N. Y. Acad. Sci.* 50:221–246.

Isaev, A. S. and R. G. Khlebopros (1979) Inertial and non-inertial factors regulating forest insect population density, in G. A. Norton and C. S. Holling (eds) *Pest Management*, Proc. Int. Conf., October 1976. IIASA Proceedings Series Vol. 4 (Oxford: Pergamon) pp317–30.

Jones, D. D. and C. J. Walters (1976) Catastrophe theory and fisheries regulation. *J. Fish. Res. Board Can.* 33:2829–33.

Keen, F. P. (1943) Ponderosa pine tree classes redefined. *J. Forestry* 38:597–8.

Kermack, W. O. and A. G. McKendrick (1927) Contributions to the mathematical theory of epidemics, Part I. *Proc. R. Soc.* A115:700–21.

Kolata, G. B. (1977) Catastrophe theory: The emperor has no clothes. *Science* 196:287.

Macdonald, G. (1965) The dynamics of helminth infections, with special reference to schistosomes. *Trans. R. Soc. Trop. Med. Hyg.* 59:489–506.

Mahoney, R. L. (1978) Lodgepole pine/mountain pine beetle risk classification methods and their application, in A. A. Berryman, G. D. Amman, R. W. Stark, and D. L. Kibbee (eds) *Theory and Practice of Mountain Pine Beetle Management in Lodgepole Pine Forests* (Moscow, ID: College of Forest Resources, University of Idaho) pp106–13.

May, R. M. (1973) *Stability and Complexity in Model Ecosystems* (Princeton, N. J: Princeton University Press)

May, R. M. (1976) (ed) *Theoretical Ecology: Principles and Applications*. (Philadelphia: Saunders; and Oxford: Blackwell).

May, R. M. (1977) Thresholds and breakpoints in ecosystems with a multiplicity of stable states. *Nature* 269:471–7.

May, R. M., G. R. Conway, M. P. Hassell, and T. R. E. Southwood (1974) Time-delays, density-dependence and single-species oscillations. *J. Animal Ecol.* 43:747–70.

Morris, R. F. and R. L. Bishop (1951) A method of rapid forest survey for mapping vulnerability to spruce budworm damage. *Forestry Chron.* 27:1–8.

Peterman, R. M. (1977) A simple mechanism that causes collapsing stability regions in exploited salmonid populations. *J. Fish. Res. Board Can.* 34:1130–42.

Stage, A. R. (1973) Prognosis model for stand development. *USDA For. Serv. Res. Pap.* INT-137. (Ogden, UT: Intermtn Forestry Range Expt Station).

Sussmann, H. J. and R. S. Zahler (1978) Critique of applied catastrophe theory in behavioral sciences. *Behavioral Sci.* 23:283–389.

Thom, R. (1975) *Structural Stability and Morphogenesis* (transl. D. H. Fowler) (New York: Addison-Wesley).

Wangersky, P. and Cunningham, W. J. (1957) Time lag in population models. *Cold Spring Harbor Symp. Quant. Biol.* 22:329–38.

Zeeman, E. C. (1976) Catastrophe theory. *Sci. Am.*, 234(4):65–83.

5 Chagas' Disease: Modeling Transmission

Jorge E. Rabinovich

5.1. INTRODUCTION

Chagas' disease is a trypanosomiasis that affects man in an extensive area of the tropics and subtropics of the New World. It predominates in rural areas wherever families live in simple huts or shacks, typically having mud-covered walls, palm leaves on the roof, and a soil floor. These conditions provide the ideal ecological niche for the insects that act as vectors of the disease (Pifano, 1971).

In Venezuela the most important vector species is the reduviid bug *Rhodnius prolixus*, which is widespread in both the foothills and plains. (Elsewhere in Latin America other species, including *Triatoma infestans, T. phyllosoma* and *Panstrongylus megistus* are important.) The causal agent of Chagas' disease is a trypanosome whose primary hosts are wild mammals and the incidence of the disease is a function of the abundance of the insect vector in rural dwellings and the degree of contact between man and domestic and wild animals.

The disease in man has a very short initial acute phase, but then stabilizes in a chronic and latent phase that evolves very slowly causing damage to the heart and to organs of the digestive and nervous systems. In Venezuela there are no recent reliable data on the seriousness of the situation, but a survey carried out in 1971 showed that 45.5% of the population, and 58.2% of the area of the country, are at risk (Gamboa, 1974). Some estimates (Rossell, 1976) show that the prevalence of infection in Venezuela is as high as 42.3% in the rural population, with 22.7% of these exhibiting heart problems. Of the 37% mortality due to cardiac causes in Venezuela, a substantial fraction can be attributed to Chagas' disease.

To date there has been no way of promoting active or passive immunity against infection by *Trypanosoma cruzi*, the immediate causal agent of the disease; neither is there a safe drug that eradicates the parasite. The campaign against the disease so far has focused on reducing the number of infested houses and the level of infestation, through reducing the size of the vector populations.

The object of the present study is a comparison of the standard technique of vector control using chemicals with more unorthodox means, namely biological control, using the parasitoid *Ooencyrtus trinidadensis*, and improvements in house construction. The model presented here represents an initial exercise to evaluate alternative methods of control and is primarily aimed at answering strategic questions. Because of the lack of adequate long-term quantitative information, no detailed verification of the model is possible and validation has been largely in terms of order of magnitude and direction of change, as indicators of the realism of the model. Furthermore, feedbacks from the field to the model are limited because several of the control actions considered have a relatively long-term effect.

In this chapter I first describe the model for the transmission of the disease from man to man via the vector. I next use the model to evaluate three control strategies (i) insecticide spraying at various frequencies; (ii) different grades of house improvement; and (iii) biological control using the hymenopterous parasitoid.

5.2. COMPONENTS OF THE MODEL

The incorporation of all the elements that have any relationship to the transmission of the Chagas' disease would necessitate a very complex model. Zeledón (1974) has provided an excellent graphical summary of all the possible connections between man, the insect, and both domestic and wild reservoirs. However, as indicated in Figure 5.1 some of these relationships are either not well known or occur very infrequently and thus play a minor role in the transmission of the disease. The model focuses on the cycle enclosed in the square in Figure 5.1, i.e., the cycle of domiciliary transmission, and concentrates only on insect–man transmission, ignoring the effect of animal reservoirs. Although in some situations these reservoirs are important they are less significant where *R. prolixus* is the vector since this species has a marked preference for human blood, even in the presence of domestic and wild animals (Pifano, 1973). The essential components of the domiciliary cycle are represented in Figure 5.2.

5.2.1. Assumptions

The simplifications assumed in the model are listed in Table 5.1.

5.2.2. Model Equations

The model consists of three differential equations describing the changes in insect and human populations and the relationship between the two.

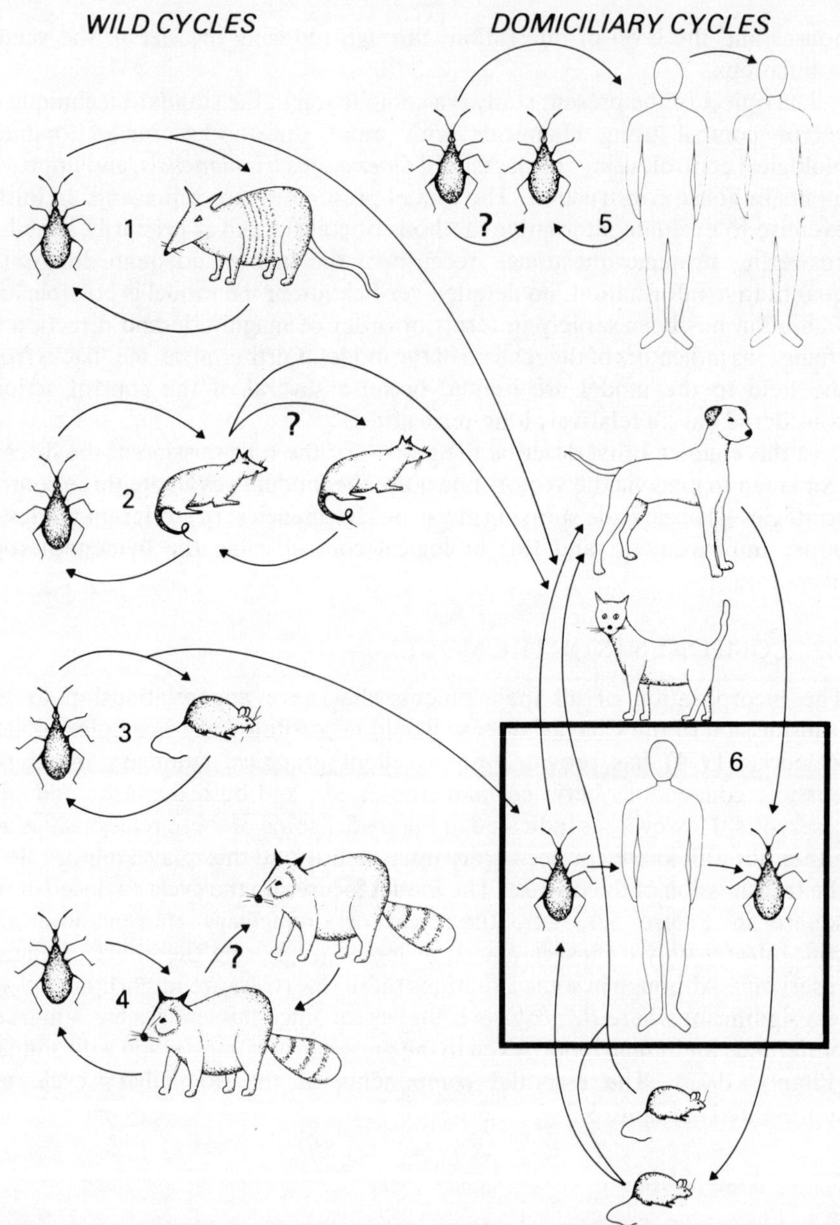

FIGURE 5.1 Wild and domestic cycles of Chagas' disease modalities and relationships. (1) *P. geniculatus* is often associated with armadillos. (2) Opossums (*Didelphis* spp.) are associated with several triatomines and both marsupials and insects visit human dwellings. (3) Rats and other rodents are associated with insects that also fly to houses. (4) Racoons may be associated with triatomines but, as with

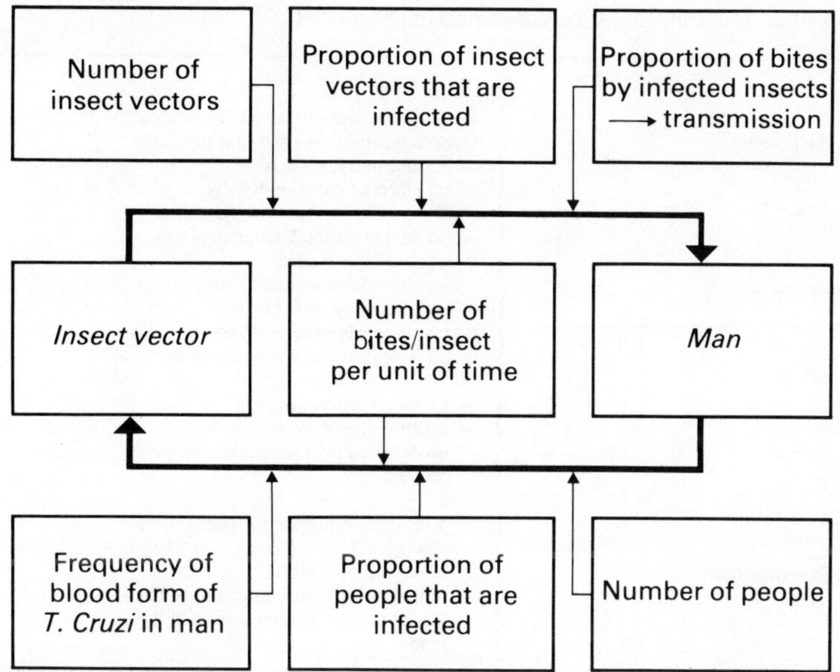

FIGURE 5.2 Block diagram of the components used in the model. The two main arrows connecting the box with insect vector and man represent the flux of transmission of *T. cruzi* in both directions; the rest of the boxes represent factors affecting this transmission that were considered in the simulation model.

Biting Acceptability Factor

This factor is introduced to describe the observed reduction in feeding on the host that occurs at high vector densities. When the number of bites per person exceeds a certain threshold the resulting nuisance and irritation results in increased movement of the host that in turn interrupts the bug's normal feeding. The net effect is a reduction in fecundity, together with an increase in mortality and a delay in molting in the vector. This phenomenon is treated by defining a biting acceptability factor B which is set equal to 1 at low vector densities and declines as densities increase, depending on the threshold T_B (Figure 5.3):

---

opossums, they may transmit *T. cruzi* among themselves without participation of the insect. (5) In the absence of domestic animals there is a cycle between the insect and man. Transmission from one insect to another has been suggested and transmission from man to man is a fact, both transplacentally and transfusionally. (6) Other domestic animals might participate in the cycle and some of them may become infected by eating small rodents (after Zeledón, 1974). The rectangle indicates the components included in the simulation model.

TABLE 5.1 Simplifications assumed in the model.

In the people	— no age structure — static population (no births, no deaths, no migration) in a six-year period — no immunity, no cure — no effect of parasite density
In the insect	— no developmental instar structure — no seasonality effect — no alternative animal hosts — logistic population growth — no effect of parasite infection on molting, development rate or survival
In the parasite	— no strain differences — no animal reservoirs — no density effects on transmission of disease
In the parasitoid	— no explicit equation for population change — no search behavior — functional response as evaluated in the laboratory assumed valid in the field

$$\frac{dB}{dt} = r_B B\left(1 - \frac{B}{K_B}\right) - \left(C\frac{V_t}{P}\right)\left(\frac{B^2}{T_B^2 + B^2}\right), \tag{5.1}$$

where V_t is the number of vectors, P is the number of human hosts, K_B is the maximum and r_B the instantaneous rate of change of the biting acceptability factor, and C is a proportionality constant related to the increase in irritation in the host population. The net effect is then summarized as a reduction of the vector's carrying capacity per house through the expression $B^2/(T_B^2 + B^2)$ (Figure 5.4), so that the vector population grows according to the logistic

$$\frac{dV_h}{dt} = r_v V_t\left[\left(1 - \frac{V_T}{K_v P}\right)\left(\frac{T_B^2 + B^2}{B^2}\right)\right], \tag{5.2}$$

where V_h is the number of noninfected bugs, r_v is the vector intrinsic rate of increase, and K_v is the carrying capacity for an average house.

The Parasitoid Attack

Laboratory studies to investigate the pattern of attack by the parasitoid *Ooencyrtus trinidadensis* on the vector have been conducted (Feliciangeli and Rabinovich, 1977). The important descriptor of attack is the functional response that measures the relation between the number of hosts attacked per

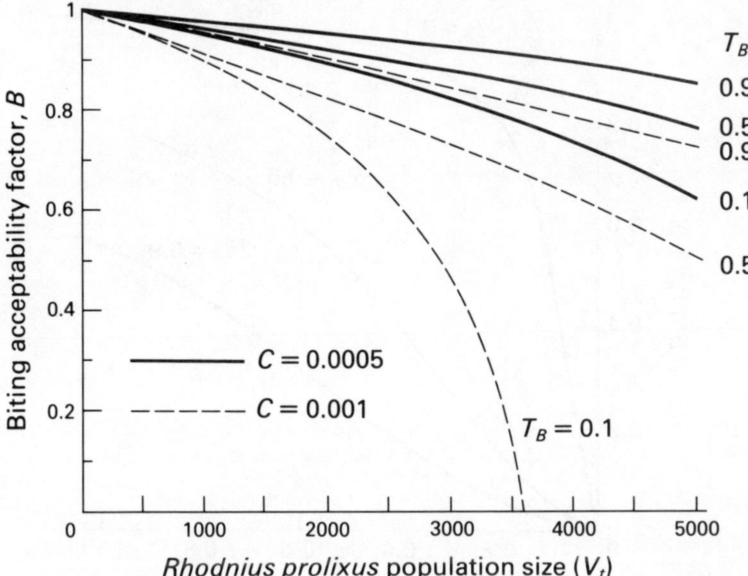

FIGURE 5.3 Changes in the biting acceptability factor B as a function of the density of the insect vector. The irritation response of the human host depends on the coefficient C and the threshold value T_B. The smaller the value of C, the stronger is the irritation response of the human host; similarly with respect to T_B. This figure was constructed assuming that the insect vector density increases from a lower to a higher density for the computation of B.

parasitoid as a function of the host density (Holling, 1959). Its form is critical to the dynamics of the relationship, and depends on the searching efficiency of the parasitoid and various saturation effects related to handling time, limits on the number of eggs available, and so on. In a type II functional response the number of hosts attacked increases at a constantly decreasing rate to the upper asymptote, while a type III response is sigmoid in form with a point of inflexion in the curve (see Chapter 2 and Hassell, 1978).

The experimental data were obtained by varying the number of female parasitoids and the number of *Rhodnius* eggs offered. However, the results do not clearly distinguish between a possible type II or type III response (Figure 5.5). In general it is common for type II responses to be obtained under laboratory conditions where the host and parasitoid are closely confined. Under natural field conditions, however, it is more likely that parasitoids will search less efficiently at low host densities and sigmoid response will occur. For these reasons I have chosen to fit a type III functional response of the form

$$F_p = \beta \frac{V_t^2}{\alpha^2 + V_t^2}, \tag{5.3}$$

where α determines the scale of densities of the vector population at which

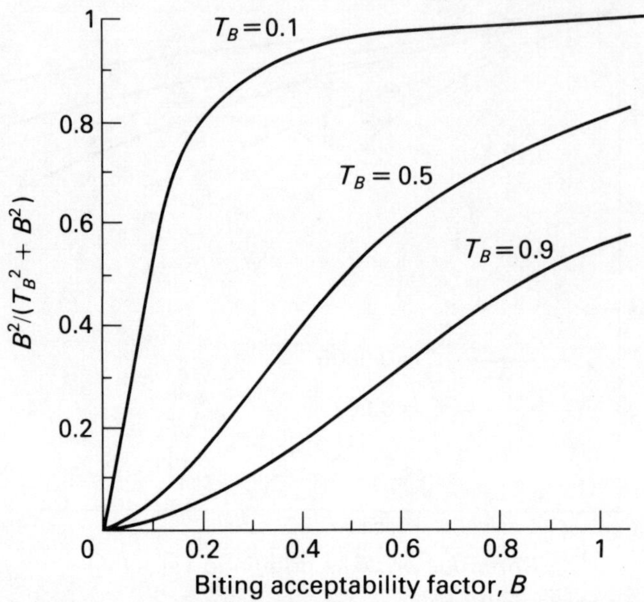

FIGURE 5.4 Effects of T_B and B on the function $B^2/(T_B^2 + B^2)$, which alters the carrying capacity of the insect population.

saturation begins to take place, and β is the saturation value. The best fit of this model to the data (Figure 5.5) gives estimates of $\alpha = 150$ and $\beta = 200$. If now, following the stability analysis of Ludwig *et al.* (1978), we plot the per capita rate of growth of the vector (5.2) and the per capita death rate due to parasitism (5.3) as a function of the vector density, three equilibrium points emerge: a lower-density stable equilibrium point that occurs at a vector density of 20; an intermediate density, which corresponds to a point of unstable equilibrium at 1890 vectors; and finally, an upper-density stable equilibrium point that occurs at 3090 vectors. An interesting peculiarity of this graph (Figure 5.6) is that the unstable and the upper stable equilibrium points are relatively close together and very near to the minimum tangent point of the concave part of the curve. This means that as soon as the per capita rate of growth of the hosts drops only slightly, these two points of equilibria are lost, and only the lower stable equilibrium point remains.

Vector Infection

The number of vectors that become infected each week N_{vi} is a product of the size of the vector V_t, and infected human populations P_i, and the proportion of infected people carrying blood forms of *T. cruzi* S_p, divided by the overall human population size and biting interval for the vector in weeks R_B, i.e.,

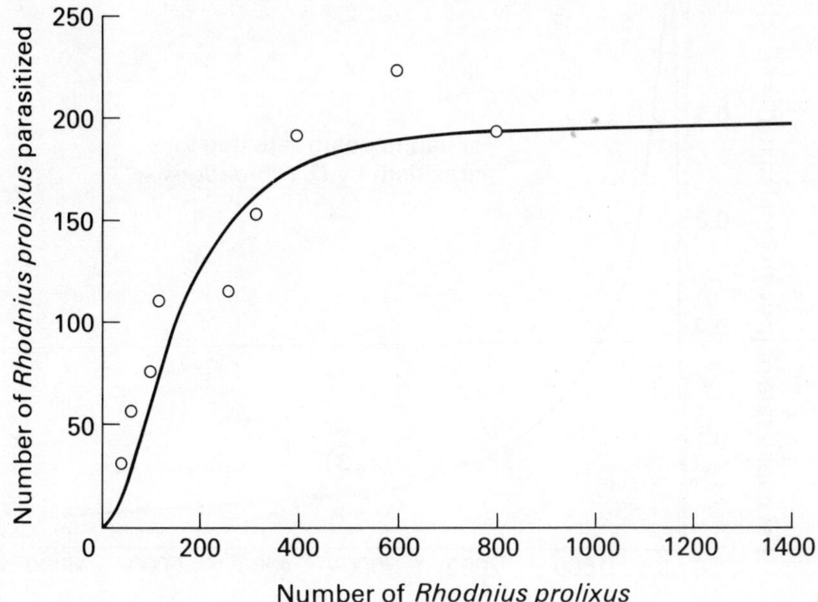

FIGURE 5.5 Parasite's functional response $F_p = \beta[V_t^2/(\alpha^2 + V_t^2)]$. ○ = from laboratory information; — expected.

$$N_{vi} = (V_t P_i S_p)/(PR_B) \tag{5.4}$$

The final equations for vector population change thus become:

$$\frac{dV_h}{dt} = r_v V_t \left[\left(1 - \frac{V_t}{K_v P}\right) \left(\frac{T_B^2 + B^2}{B^2}\right) \right] - N_{vi} - F_p \tag{5.5}$$

$$\frac{dV_i}{dt} = N_{vi} - \left[\left(\frac{r_v V_i^2}{K_v p}\right) \left(\frac{T_B^2 + B^2}{B^2}\right) \right] - F_p, \tag{5.6}$$

where V_h and V_i are the number of healthy and infected vectors, respectively.

Host (human) Population

The size of the human population in a house is considered to be constant; that is, no deaths, births or migration occur in the simulation time period. The rate of change of infected people P_i is thus the simple product of the number of healthy people times the risk of infectionR_t:

$$dP_i/dt = P_h R_t, \tag{5.7}$$

where

$$R_t = [(V_i/R_B)P_t]/P. \tag{5.8}$$

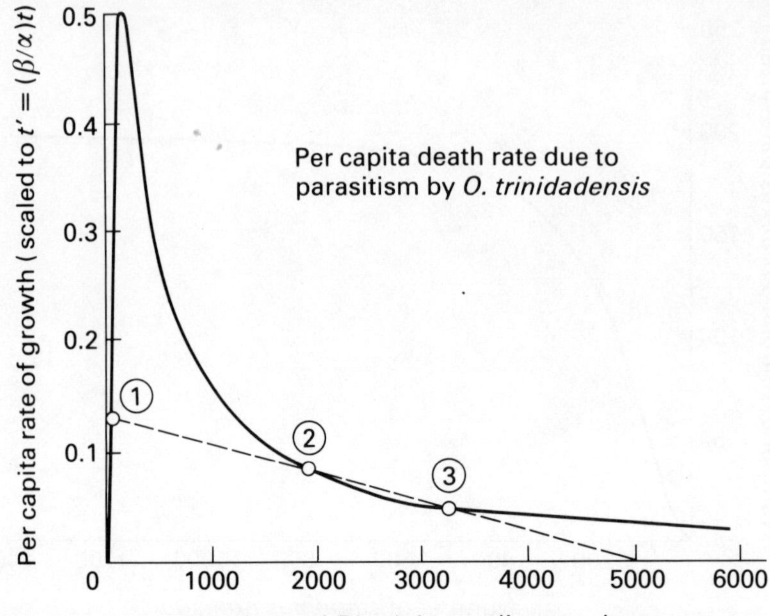

FIGURE 5.6 Analysis of the equilibrium of the *Rhodnius prolixus–Ooencyrtus trinidadensis* system assuming constant number of residents of a house (α = 150, β = 200, r_v = 0.17, K = 5000). (1) Density of lower stable equilibrium = 20. (2) Density of unstable equilibrium = 1890. (3) Density of upper stable equilibrium = 3090.

Hence the rate of change of healthy people, P_h:

$$dP_h/dt = -\bar{P}_i. \tag{5.9}$$

5.3. CONTROL STRATEGIES

Table 5.2 lists the control strategies considered in the simulation studies. Conventional insecticide control is assumed to produce a 85% kill of the total vector population but with no residual action. Household participation is aimed at rendering the dwellings less favorable for reproduction and development of the vector population and its effectiveness is measured in terms of a reduction in carrying capacity (Figure 5.7A). Where control is through collecting or catching the bugs by members of the household it is assumed that only visible insects are caught and the percentage of the population visible is assumed to be a linear function of density (Figure 5.7B). Finally, parasitoid control is modeled by using the sigmoid type III response curve, discussed above (Figure 5.7C).

5.3.1. Parameter Values

The parameter values and initial state variables used in the simulations are listed in Table 5.3.

TABLE 5.2 Description of control strategies.

Conventional control

Level 0 = No spraying
Level 1 = Spray every four years
Level 2 = Spray every three years
Level 3 = Spray every two years
Level 4 = Spray every year

Nonconventional control

(I) *Household participation*
Level 0 = No participation
Level 1 = House improvement grade 1 (= wall plastering)
Level 2 = House improvement grade 2 (= grade 1 + replacement of thatched roof by tin)
Level 3 = House improvement grade 3 (= grade 2 + cleaning of house and hygiene)
Level 4 = Insect collection by household

(II) *Biological control*
Release of hymenopterous parasites, *Ooencyrtus trinidadensis*.

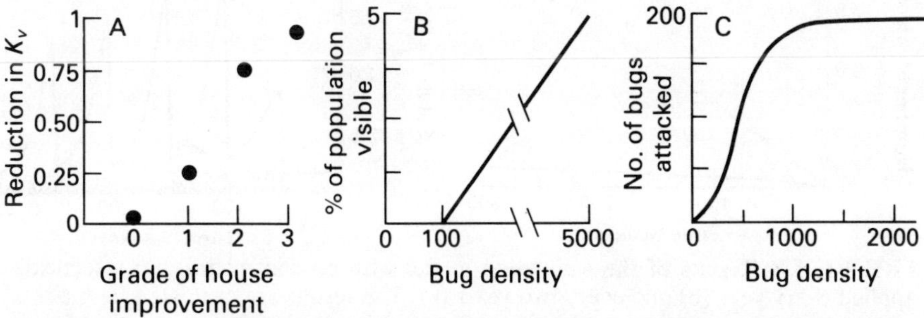

FIGURE 5.7 Graphs of the basic functions for the nonconventional control actions. A, Reduction factor affecting the carrying capacity of the insect vector population as a function of the degree of house improvement. B, Percentage of insect vector population that becomes visible and is collected as a function of its density. C, Parasitism as a function of bug density.

5.3.2. Results of the Simulations

The effectiveness of the various control strategies were evaluated in terms of: (1) the number of bugs per household; (2) the proportion of infected bugs; (3) the number of infected people; and (4) a safety factor to measure the number of weeks between infected bites ($1/R_t$)

In the absence of control (Figure 5.8) the density of bugs per house rapidly rises to its carrying capacity of almost 5000 individuals. The proportion of

TABLE 5.3 Values of the parameters and initial state variables used in the simulations.

State variables (initial value)
V_t = 1000 insects; V_h = 600 insects, V_i = 400 insects
P = 10 people; P_h = 10 people; P_i = 0 people

Parameter values
K_v = 500 insects/person
T_B = 0.1–0.5; $K_B = 1$; $C = 0.0005$; $r_B = 1$ per week
R_B = 1.2 weeks; $P_t = 0.0005$; $S_p = 0.001$
α = 150 insects; $\beta = 200$ insects

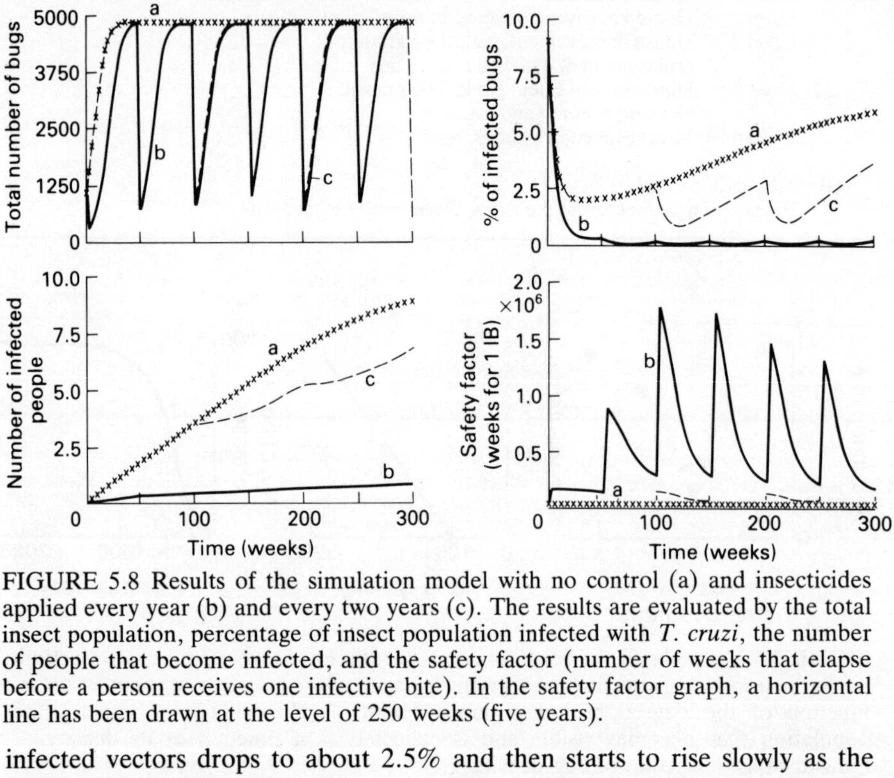

FIGURE 5.8 Results of the simulation model with no control (a) and insecticides applied every year (b) and every two years (c). The results are evaluated by the total insect population, percentage of insect population infected with *T. cruzi*, the number of people that become infected, and the safety factor (number of weeks that elapse before a person receives one infective bite). In the safety factor graph, a horizontal line has been drawn at the level of 250 weeks (five years).

infected vectors drops to about 2.5% and then starts to rise slowly as the number of infected people also rises. The safety factor (the time between infections) remains permanently below an arbitrary line of 250 weeks (approximately five years).

Conventional insecticide control (Figure 5.8) produces a fairly constant and low proportion of infected bugs, of the order of less than 1% when sprayed every year, as compared with about 2–3% when sprayed every two years. However, this produces a considerable difference in terms of the number of infected people which rises steeply to about seven infected persons in six years'

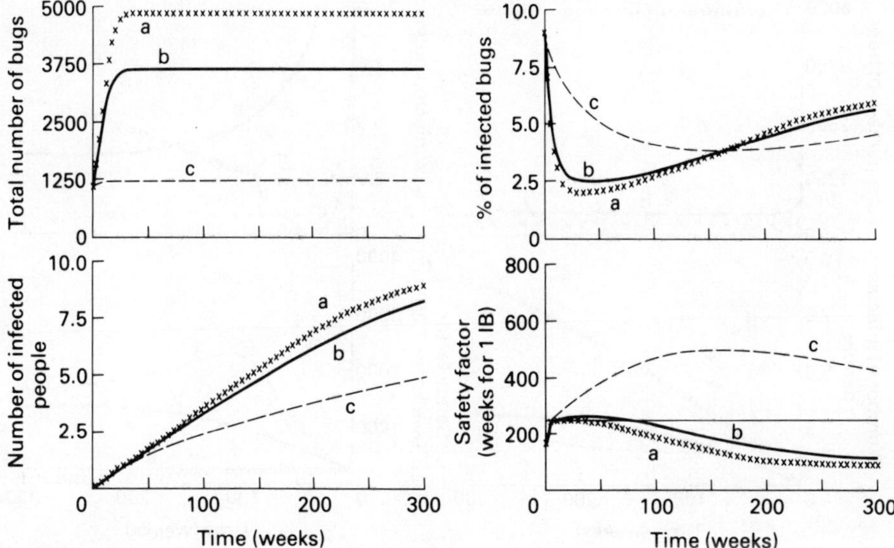

FIGURE 5.9 Results of the simulation model with house improvement to level 1 (wall plastering), (b); and level 2 (plus a tin roof), (c). No control (a).

time if spraying is biannual, but only reaches one per infected person in the case of spraying every year. Biannual spraying also permits the safety factor to drop below the five-year line, while it is always above the line when spraying is annual.

House improvement (Figures 5.9 and 5.10) shows a progressive reduction in percentage of vectors and numbers of people infected. A combination of wall plastering, a tin roof, cleaning of the house and hygiene reduces the proportion of infected bugs to about 3% and results in approximately 2.5 infected people after six years. However, household collection of the bugs produces a rather similar result. Both approaches keep the safety factor above the five-year line.

If the assumptions about the form of the parasitoid attack are correct, biological control (Figure 5.11) produces very dramatic results. The bug population is driven to a low stable equilibrium of 20 per house (see Figure 5.6), *the proportion of infected bugs rapidly goes to zero and the household human population remains free of the disease.*

5.3.3. Compression of Results

Nomograms of the results have also been produced (Figure 5.12), which express the effectiveness of the control strategies in terms of the safety factor, i.e., the number of years between infections.

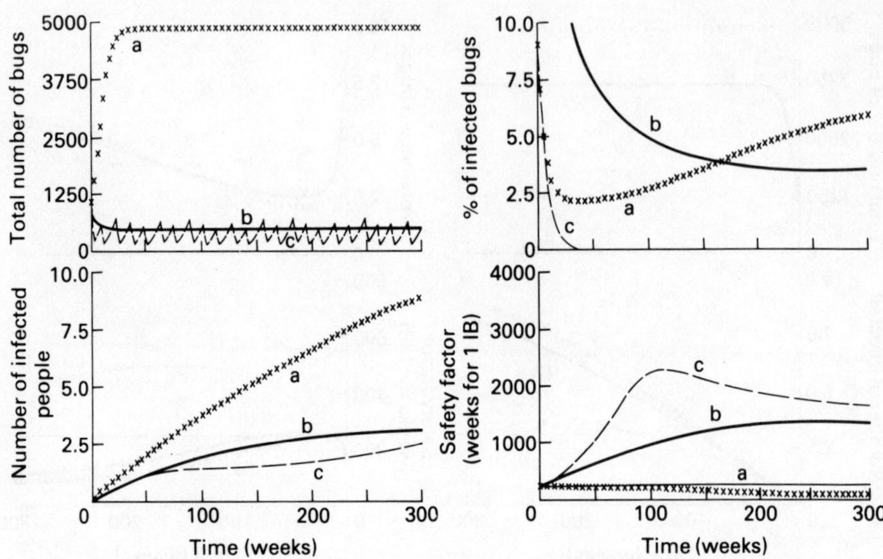

FIGURE 5.10 Results of the simulation model with house improvement level 3 (level 2 plus house cleaning and improved hygiene) (b); and insect collection by the household (c). No control (a).

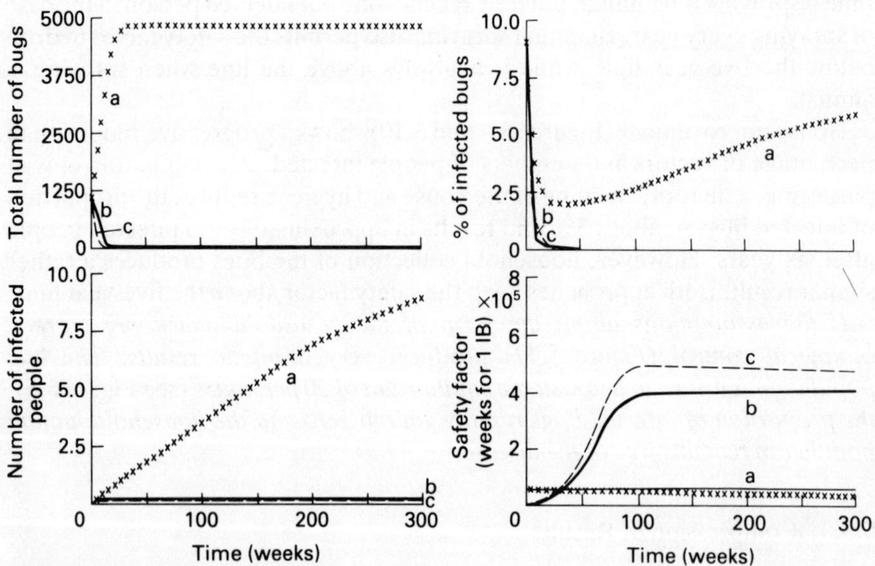

FIGURE 5.11 Results of the simulation model with biological control with microhymenopterous parasites (b), and the combination of house improvement level 2 plus biological control (c). No control (a).

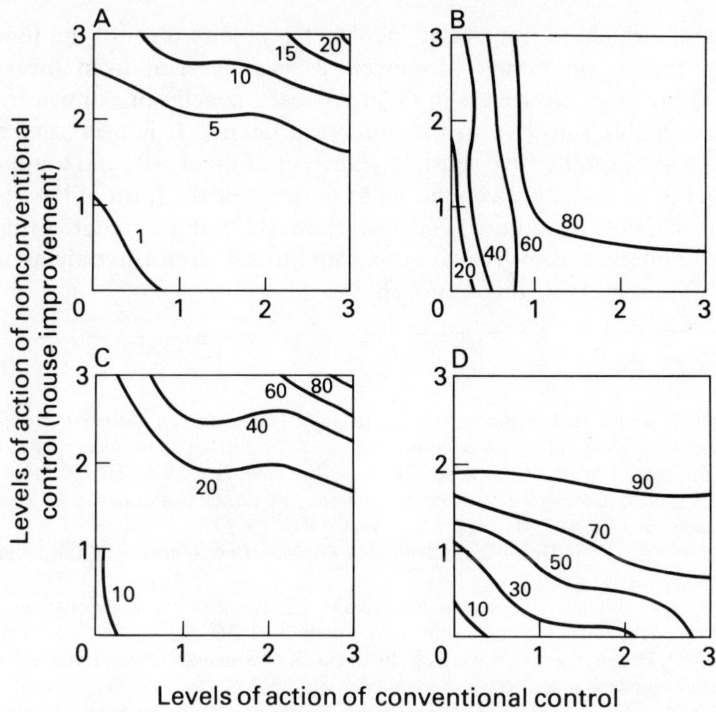

FIGURE 5.12 Nomograms for a global evaluation of action control alternatives. $P_T = 0.001$, initial number of insects = 1000. The lines represent points in the plane with the same numerical values for four different indicators: A, the average safety factor (i.e., the number of years for a healthy person to become infected); B, coefficient of variation of the safety factor (%); C, the maximal value of the safety factor, and D, the simulation weeks with safety factor above the threshold value of five years (%). The plane is determined by four different values of levels of action of nonconventional control by house improvement. On the *x*-axis, four different levels of actions of conventional control by insecticide spray. Relatively horizontal lines indicate small sensitivity to changes in the insecticide level and a strong effect as a result of the nonconventional control.

5.4. CONCLUSIONS

The results suggest that satisfactory control can be obtained either by annual spraying or by household collection of the bugs or by a combination of house improvements. But it would appear that dramatically better control can be achieved by using the hymenopterous parasite. *The effectiveness of biological control, however, is critically dependent on the assumptions made about the form of the attack by the parasitoid on the bug. I have here assumed a type III functional response but if in the field the attack is of the type II kind, then control will not be as effective*. So far, our assumptions are based on limited laboratory experiments which do not clearly identify the precise form of the functional response that is operating. Moreover, as can be seen from the shallow intersection of the two curves in Figure 5.6, the conclusions drawn from the model are highly sensitive to the numerical details. It is necessary now to undertake a comprehensive sensitivity analysis of the model and to repeat the experiments on an extensive scale so as to discover the form of the response under normal field conditions. The potential of the biological control suggested by these simulations is so great as to warrant further careful investigation of the assumptions underlying this approach.

REFERENCES

Feliciangeli, D. and J. E. Rabinovich (1977) Effecto de la densidad en *Ooencyrtus trinidadensis* (Chalcidoidea, Encyrtidae), un parásito endófago de los huevos de *Rhodnius prolixus* vector de la enfermedad de Chagas en Venezuela. *Rev. Inst. Med. Trop. Sao Paulo* 19:21–34.

Gamboa, J. (1974) Ecología de la Tripanosomiasis Americana (Enfermedad de Chagas) en Venezuela. *Bol. Inf. Dir. Malariol. y San. Amb.* 14(1–2):3–20.

Hassell, M. P. (1978) *The Dynamics of Arthropod Predator–Prey Systems*. (Princeton: Princeton University Press).

Holling, C. S. (1959) The components of predation as revealed by a study of small mammal predation on the European pine sawfly. *Can. Entomol.* 91:293–320.

Ludwig, D., D. D. Jones, and C. S. Holling (1978) Qualitative analysis of insect outbreak systems: the spruce budworm and forest. *J. Animal Ecol.* 47: 315–332.

Pifano, F. (1971) *La enfermedad de Chagas como entidad nosológica problema del medio rural venezolano* (mimeo) Central University of Venezuela.

Pifano, F. (1973) La dinámica epidemiológica de la enfermedad de Chagas en el Valle de los Naranjos, Estado Carabobo, Venezuela. II. Infección chagásica en la población rural del área. *Arch. Venez. Med. Trop. y Paras. Méd.* 5(2):31–45.

Rossell, O. J. (1976) *Evaluación de la Transmisión de la enfermedad de Chagas en dos Caseríos del Estado Guárico (Venezuela) sometidos a Rociamientos*. Promotion Work (Mérida: Faculty of Sciences, University of the Andes).

Zeledón, R. (1974) Epidemiology, modes of transmission and reservoir hosts of Chagas' disease, in *Trypanosomiasis and Leshmaniasis with Special Reference to Chagas' disease* (Amsterdam: Elsevier-Exerpta Medica-North Holland) pp51–85.

6 Host–Parasitoid Models and Biological Control

Michael P. Hassell

6.1. INTRODUCTION

Insect parasitoids have been of considerable interest to many ecologists involved in very different studies. Theoretical ecologists have delighted in their characteristic life cycles that make them such ideal subjects for the development of relatively simple predator–prey population models; experimental ecologists have found them very convenient for laboratory studies on the factors affecting host finding; field workers have been impressed by their abundance and the high levels of host mortality that they can inflict; and biological control workers have often found them to provide a most efficient means of pest control. However, despite this common thread linking theory with pest control, there has been little meeting ground, to date, between those developing general population models for parasitoid–host interactions and those involved in classical biological control programs.

This chapter moves toward bridging the gap by exploring just one type of parasitoid–host model and its contribution to the understanding of some biological control practices. These practices are primarily those of classical biological control used against exotic pests of perennial standing crops. Natural enemies are imported, released and then, hopefully, become established and cause a marked decline in the pest population, with both natural enemy and pest populations persisting thereafter at relatively low densities. Perennial crop systems are singled out since they come closer to providing the continuity of interaction comparable to that in a simple, coupled population model, in contrast to systems in annual crops where any continuous interaction between natural enemies and their hosts are regularly disrupted by the upheavals of harvesting, plowing, etc.

The chapter is divided into four parts. First, the basic model structure is outlined, indicating why this is well suited to parasitoids and especially to some biological control programs. Second, the question is then posed: "What accounts for the persistence (= stability) of host and parasitoid populations at very low average levels of abundance (= equilibrium levels) following a successful program?" Third, a further question is then asked: "Does theory support the practice of multiple parasitoid introductions?". Finally, a case study is briefly examined, showing how a simple model can contribute to an understanding of an observed biological control outcome.

6.2. STABILITY AND EQUILIBRIUM LEVELS

Let us consider a simple difference equation model of the form:

$$\begin{aligned} N_{t+1} &= \lambda N_t f(N_t,P_t) \\ P_{t+1} &= N_a = N_t[1 - f(N_t,P_t)], \end{aligned} \tag{6.1}$$

where N_t,N_{t+1} and P_t,P_{t+1} are the host and parasitoid populations in successive generations, respectively, N_a is the number of hosts parasitized, and λ is the net rate of increase of the hosts per generation. The function f defines the survival of hosts from parasitism and hence contains all assumptions made about the searching efficiency of the parasitoids. A full discussion of this model with a variety of forms for f is given in Hassell and May (1973) and Hassell (1978).

By being framed in difference rather than differential equations, the model is clearly more appropriate to populations with fairly discrete generations, as are frequently found amongst insects in temperate regions. In addition, the model is particularly appropriate to insect parasitoids rather than to predators in general, for two reasons. First, parasitoids are, in effect, a special kind of predator with only a single stage (the adult female) searching for a host. The lack of age structure in the model is therefore a less important omission. Second, parasitoids normally oviposit in or on their hosts so that reproduction is directly dependent on host mortality. The assumption that the next generation of parasitoids P_{t+1} stems directly from the number of hosts parasitized N_a in eqn. (6.1) is therefore acceptable for parasitoids, but inadequate for most predators where reproduction bears a more complex relationship to prey consumption (Lawton *et al.*, 1975; Beddington *et al.*, 1976a,b). A true predator–prey model could not get away with these simplifications and would need to include at least some provision for age-dependent differences in searching efficiency and a more realistic treatment of predator reproduction.

Apart from being more appropriate to parasitoids than to predators in general, eqn. (6.1) also leans well towards classical biological control programs since these, of all parasitoid–host systems, often come closest to being coupled two- or three-species systems, rather than involving the more intricate guilds of natural enemies typical of natural systems (e.g., Rejmanek and Stary, 1979; Zwölfer, 1979).

6.2.1. The Critical Factors

With this background in mind we now consider the two examples in Figure 6.1: that of the red scale (De Bach *et al.*, 1971) and the winter moth (Embree, 1965). In both cases the populations were reduced from outbreak levels to a very low equilibrium level by the action of parasitoids, and then persisted at this level without further outbreaks. These evocative pictures, which must resemble that from many of the classical biological control successes (e.g., Doutt, 1958; Huffaker and Kennet, 1966), prompt two obvious questions: What determines the degree of depression in the host equilibrium caused by the parasitoids?, and what is responsible for the stability of the equilibrium?

There is certainly little mystery about the first question. The level of the host equilibrium N^* depends upon the balance between the effective rate of increase of the host (the fecundity corrected for all mortalities other than parasitism) and the parasitoid's efficiency (searching ability, tempered by any mortalities acting upon the parasitoid population; Hassell and Moran, 1976; Hassell, 1977). Such a simple statement, however, belies the huge task in measuring these for natural populations. Apart from the difficulties in measuring parasitoid searching efficiency under field conditions (one needs

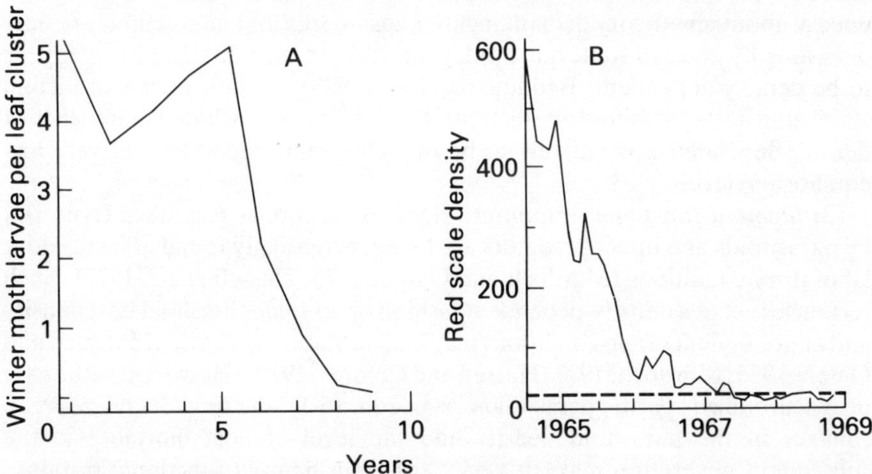

FIGURE 6.1 The decline in two insect populations due to insect parasitoids. A, Winter moth (*Operophtera brumata*) in Nova Scotia following the establishment of two parasitoid species, *Cyzenis albicans* and *Agrypon flaveolatum* (after Embree, 1965). B, California red scale (*Aonidiella aurantii*) after a DDT-induced outbreak. The decline in density is due to the reestablishment of *Aphytis melinus* and *Comperiella bifasciata* (after De Bach *et al.*, 1971).

estimates of the density of searching adults in addition to the level of parasitism), life table information for both host and parasitoid is also needed to establish the average magnitude of the different mortalities affecting each population (Varley, 1970a). Such information, combining estimates of parasitoid efficiency and life tables, has rarely been collected (e.g., Broadhead and Wapshere, 1951; Klomp, 1966; Varley and Gradwell, 1968).

Even harder to assess from field populations alone is the identity of the factors contributing to the stability of parasitoid–host equilibria, especially when at the very low levels associated with a completely successful biological control project. However, there are a number of theoretical possibilities.

(1) *Host density dependence*. It is possible that the persistence of such parasitoid–host interactions as shown in Figure 6.1 is not due to the parasitoid at all, but due to some other density dependence acting upon the host population. It is well known that an otherwise quite unstable interaction can be rendered stable simply by introducing a density-dependent host rate of increase (Beddington *et al.*, 1975; May and Hassell, 1980) or some density-dependent host mortality, such as that caused by generalist predators. Recent work has also suggested that interactions may be quite sensitive to the particular point at which this density dependence enters the host life cycle (Wang and Gutierrez, 1980; May *et al.*, 1981). This, however, is unlikely to be the solution that we seek. In the first place, we cannot invoke host resource limitation at these very low population levels, which immediately removes one of the possible causes of host density dependence, and second, one should be uneasy about any theory demanding that host–parasitoid interactions are only persisting by grace of some quite independent host mortality that just happens to be density dependent. Beddington *et al.* (1978), in their review of factors stabilizing host–parasitoid interactions, have similarly concluded that such host density dependence is unlikely to be of widespread importance at very low equilibrium levels.

(2) *Sigmoid functional responses*. Sigmoid functional responses (type III) by parasitoids and insect predators are being increasingly found at least under laboratory conditions (Murdoch and Oaten, 1975; Hassell *et al.*, 1977). Such responses act in a density-dependent fashion up to some threshold host density and hence under certain conditions may lead to stable equilibria (Murdoch and Oaten, 1975; Murdoch, 1977; Hassell and Comins, 1978). However, in the case of synchronized specific parasitoids, as in eqn. (6.1), the time delays between changes in the parasitoid density and the level of host mortality in the subsequent generation make it very hard for a sigmoid functional response alone to stabilize the populations (Beddington *et al.*, 1978; Hassell and Comins, 1978).

(3) *Mutual interference*. There are plentiful data showing that, under laboratory conditions, increasing parasitoid density leads to increasing interference between individuals and thus to a decline in the per capita

searching efficiency (e.g., Hassell, 1978). Any model including such effects can be made very stable if interference is marked, simply because of the powerful density dependence now acting upon the parasitoids themselves. However, this too we must discard as the "universal stabilizer" since it is hard to see how such interference could be significant at the extremely low population densities of parasitoids and hosts evident in successful biological control projects, a point forcibly made by Free *et al.* (1977) and Beddington *et al.* (1978).

(4) *Nonrandom search in a patchy environment.* In discussing the responses under (2) and (3) above, only the total host population has been considered, without any regard to spatial heterogenities in parasitoid and host distributions. That such heterogenities are crucial components in population models is now widely appreciated (e.g., Hassell and May 1974; Levin, 1976; Hastings, 1977; Ziegler, 1977; Gurney and Nisbet, 1978). While random search for a prey dwelling in an homogeneous world has the merit of mathematical simplicity, it is clearly a gross simplification of the real world, where prey are usually clumped in their distribution, and predators and parasitoids tend to spend more of their searching time where prey are plentiful rather than scarce. Some examples of such behavior amongst insects are shown in Figure 6.2.

The means of describing such responses within the model framework of eqn. (6.1) has been considered elsewhere (e.g., Hassell and May 1973, 1974; Beddington *et al.*, 1978; Hassell, 1979; Comins and Hassell, 1979) but, however they are described, the general conclusion remains the same: aggregation of parasitoids in high host density patches can be a most powerful stabilizing mechanism. The extent of this stability hinges on at least three things:

(1) It is enhanced by increasing amounts of parasitoid aggregation, provided the host distribution is sufficiently uneven.
(2) It is also enhanced by increased contagion in the host distribution for a given amount of parasitoid aggregation. Indeed, if there is too little contrast in the host densities per patch, no amount of parasitoid aggregation will be sufficient for stability.
(3) Any tendency towards stability is counteracted by an increased host rate of increase (λ).

These conclusions are independent of the detailed way that the parasitoid aggregative responses are modeled, and therefore hold for naive parasitoids that can only respond to the proportion of hosts per patch available at the start of each generation (Hassell and May, 1973), as well as to omniscient beasts that forage optimally and whose distribution of searching time per patch maximizes the rate of discovering healthy hosts (Comins and Hassell, 1979).

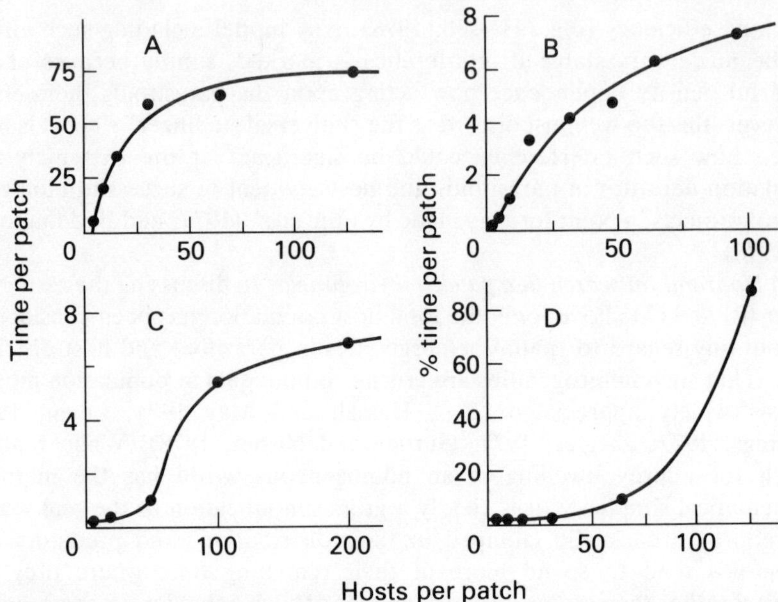

FIGURE 6.2 Some examples of aggregative responses by insect parasitoids searching in a patchy environment (all curves fitted by eye). A, Searching time per leaf of the braconid wasp, *Diaeretiella rapae*, at different densities per cabbage leaf of its aphid host, *Brevicoryne brassicae* (Akinlosotu, 1973). B, Percentage searching time of the ichneumonid wasp, *Diadromus pulchellus*, at different densities per unit area of leak moth pupae, *Acrolepia assectella* (Noyes, 1974). C, Number of parasitoid hours spent on plants of different *Pieris brassicae* density by the braconid, *Apanteles glomeratus* (Hubbard, 1977). D, Percentage of searching time by the ichneumonid, *Nemeritis canescens*, at different densities per container of its flour moth host, *Ephestia cautella* (Hassell, 1971).

Essentially the same picture emerges from much simpler models that include no details of parasitoid behavior but just describe the outcome of nonrandom search. Such is the case for the parasitoid–host model proposed by May (1978) where host survival, $f(N_t,P_t)$ in eqn. (6.1) is given by

$$f(N_t,P_t) = [1 + (aP_t)/k]^{-k}. \tag{6.2}$$

This expression is merely the zero term of the negative binomial distribution, where the constant k governs the degree of contagion in parasitoid attacks (small k = more contagion), a is the searching efficiency, and aP_t is the mean of the distribution (i.e., the average number of encounters per host). The model succeeds in capturing the essence of the examples in Figure 6.2, because increasing aggregation of searching parasitoids in patches of high host density will inevitably lead to a more clumped overall distribution of parasitism and so to a reduced value for the parameter k. The extent of stability in this model is governed solely by the degree of contagion in the

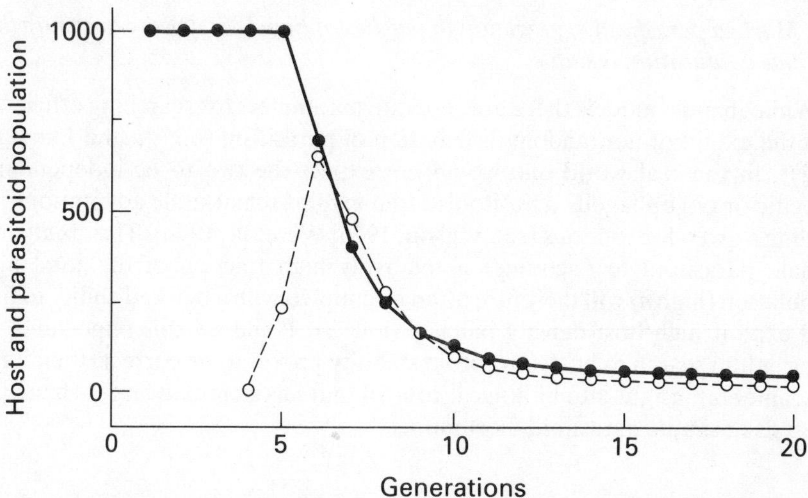

FIGURE 6.3 A numerical simulation using model (6.1) with f defined from eqn. (6.2). ● host population; ○ parasitoid population. The parasitoid is introduced in generation 4 and the host is assumed to have a maximum size of 1000, above which it cannot rise. Assumed parameter values: $\lambda = 2$; $a = 0.75$; $k = 0.15$ (after Hassell, 1978).

parasitoid attacks k, with the model always being stable if and only if $k < 1$ (Figure 6.3).

The underlying explanation for the stability arising from these models rests solely on the differential exploitation of the patches of different host density. By aggregating in the most profitable patches, the parasitoids are leaving the other, lower host density patches, as partial refuges from parasitism, and it is this refuge effect that will contribute to stability. Conclusions such as this emerge most forcibly from simple analytical models. They are not dependent on the detailed form of the models and thus provide a counter-example to Wang and Gutierrez's (1980) claim that simple models yield only specific conclusions depending upon the particular form of the model.

From the stance of classical biological control, this nonrandom search, however described, is important because, for the first time, we have a plausible mechanism by which the parasitoids can stabilize interactions *irrespective* of the level of the equilibrium populations. Thus, very low equilibria are just as easily rendered stable as higher ones. This has led Beddington *et al.* (1978) to conclude that aggregative parasitoid behavior is the most likely factor responsible for the stability of the very low equilibria arising from successful biological control projects. Other things being equal, two crucial requirements therefore emerge for a successful biological control agent:

(1) A high effective searching efficiency relative to the host's rate of increase to achieve a sufficient depression in the host equilibrium.

(2) *Marked parasitoid aggregation in patches of high host density to ensure that this equilibrium is stable.*

While in most models there are separate parameters for searching efficiency and the extent of nonrandom distribution of parasitism (e.g., a and k in eqn. (6.2)), in the real world one would not expect the two to be independent. Specific or oligophagous parasitoids often exhibit remarkable adaptations for finding a given host species (see Vinson, 1976; Weseloh, 1976). The ability of a female parasitoid to encounter a relatively high fraction of the total host population (high a) will therefore often be coupled with a marked ability to find and exploit high host density patches (low k). If indeed this importance of nonrandom search to host–parasitoid stability proves to be correct, then it is a fundamental insight into biological control that has stemmed largely from the analysis of simple parasitoid–host models.

6.3. MULTIPLE INTRODUCTIONS

Most biological control programs have involved attempts to establish more than one parasitoid species, a practice that has led to controversy over whether or not such multiple introductions are the most effective strategy. Some workers, such as Turnbull and Chant (1961), Turnbull (1967, and Watt (1965), have argued on *a priori* grounds that the most effective parasitoid species will always cause a greater depression of the host equilibrium on its own than in competition with other species. Opposing this view, van den Bosch (1968), Huffaker (1971), and others take a more pragmatic stance and argue that the identification of the "best" species is impractical. They point to the many cases where multiple introductions have seemingly improved pest control, and to the absence of examples where the situation has been worsened.

This problem has recently been explored by May and Hassell (1981), using an extension of eqn. (6.1) to include two parasitoid species P and Q:

$$\begin{aligned} N_{t+1} &= \lambda N_t f_1(N_t,P_t)\, f_2(N_t,Q_t) \\ P_{t+1} &= N_t[1-f_1(N_t,P_t)] \\ Q_{t+1} &= N_t f_1(N_t,P_t)[1-f_2(N_t,Q_t)]. \end{aligned} \tag{6.3}$$

The functions f_1 and f_2, describing host survival from species P and Q respectively, are of the same form given in eqn. (6.2), namely

$$f_1 = \left[1 + \frac{a_1 P_t}{k_1}\right]^{-k_1} \qquad f_2 = \left[1 + \frac{a_2 P_t}{k_2}\right]^{-k_2} \tag{6.4}$$

Equation (6.3) can therefore be made stable by the parasitoid's nonrandom searching behavior (k_1 and k_2 small) or by making the host rate of increase λ density dependent.

The full analysis of this model is given in May and Hassell (1981), from which several conclusions emerge. Of these, the most emphatic is that the two parasitoid species are most likely to coexist as long as each parasitoid–host link is contributing significantly to stability (i.e., k_1 and k_2 are small, corresponding to marked nonrandom search by both species). The possibilities for coexistence are greatly reduced if only one of the parasitoids contributes to stability while the other searches more or less randomly. In particular, it proves impossible to have a demarcation of roles within the system such that one species is the sole cause of stability while the other has the higher searching efficiency, and is thus the major cause of depression in the host equilibrium.

From the viewpoint of biological control, establishment and coexistence are only the first requirements. It is also crucial that the host equilibrium be sufficiently depressed below an economic injury level. This, as in the single parasitoid–host systems, will be enhanced by high effective searching efficiencies of the parasitoids. In addition, the models indicate that the establishment of *both* parasitoid species will normally lead to a greater depression of the host equilibrium than that caused by one species on its own. There *are* conditions in which this does not hold, but the difference between the effects of the "best" species and both together is slight and, in any event, given the difficulties in accurately estimating parameters of searching efficiency and degree of aggregation in different patches, it would be impractical to identify these special cases.

I conclude, therefore, that these models point firmly towards multiple introductions being a sound biological control strategy. In addition, it is interesting that the models also point to the same optimal criteria for parasitoid searching behavior as found from the single parasitoid–host system already described. Thus, the selection of parasitoids with high searching efficiency and a marked ability to seek out patches of high host density remains the procedure most likely to lead to low and stable equilibrium host populations, with coexisting parasitoids.

6.4. A CASE STUDY

The models discussed above are completely general and relate only to biological control in providing some theoretical basis for biological control practices, or in suggesting useful attributes for a successful biological control agent. In this section an attempt is made to show how such simple analytical models can be applied to a specific biological control situation. The example chosen is that of the winter moth in Nova Scotia, where the introduction and establishment of two parasitoid species, a tachinid fly (*Cyzenis albicans*) and an ichneumonid wasp (*Agrypon flaveolatum*), led to a spectacularly successful outcome. It is unique example in that, apart from the detailed studies on the winter moth in Nova Scotia (Embree, 1965, 1966), the winter moth was also studied intensively between 1950 and 1968 in a native habitat at Wytham Wood, Berks, UK. Throughout this period the winter moth at Wytham Wood, together with some associated natural enemy species (including *Cyzenis*) was regularly sampled on or under five oak trees; so providing detailed life table information. Details of this study have been given elsewhere (e.g., Varley, 1970a,b; Varley and Gradwell, 1963, 1971; Varley *et al.*, 1973) and a full treatment of the analysis reported here is to be found in Hassell (1980).

6.4.1. Winter Moth at Wytham Wood

The winter moth study is of particular interest in relation to eqn. (6.2), since the sampling program each year (= generation) provides the information necessary to obtain estimates of the parameters a and k for *Cyzenis*. Winter moth larvae are freely superparasitized by *Cyzenis* so that dissection of the sampled hosts provides a complete frequency distribution of parasitoid "attacks" per host, from which a and k can be obtained, as shown in Table 6.1. Such data are available for eight years of the study at Wytham Wood when parasitism by *Cyzenis* reached appreciable levels (> 4%), thus providing eight values for a and k. While the values for a show no discernible trends ($\bar{a} = 0.18 \pm 0.09$ (s.e.) m^2 per generation), this is not true for the eight values of k that exhibit a clear positive relationship with overall winter moth abundance per year, as shown in Figure 6.4. In other words, the distribution of *Cyzenis* attacks per host is more clumped in years of low winter moth density than when the winter moth density is high. The implication of this is that *Cyzenis* searches nonrandomly, causing higher levels of parasitism in patches (trees) of high host density, but this effect is most marked in years of low winter moth density. This is exactly the situation observed at Wytham Wood (Hassell, 1980).

TABLE 6.1 The distribution of *Cyzenis* larvae within winter moth hosts from samples taken in 1965 from Wytham Wood. k has been estimated using the approximate method of Bliss and Fisher (1953).

	Frequency of hosts with the following numbers of *Cyzenis* larvae						
Distribution	0	1	2	3	4	> 4	χ^2
Observed	1066	176	48	8	5	0	
Negative binomial ($k = 0.547$)	1066.0	179.1	2.6	11.1	3.0	1.2	4.14 ($0.5>p>0.3$)
Poisson	1022.4	247.9	30.1	(> 2 = 2.6)			74.96 ($p < 0.001$)

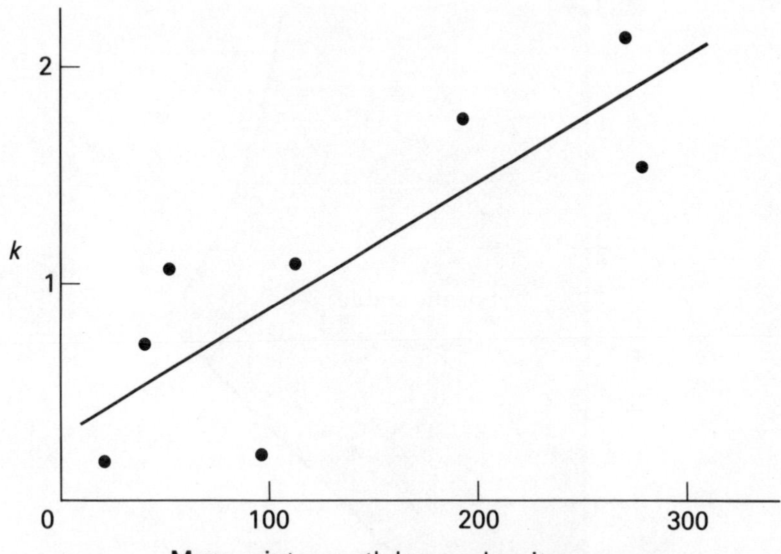

FIGURE 6.4 Relationship between the degree of contagion of *Cyzenis* larvae per winter moth host (expressed by k from the negative binomial distribution) and the mean winter moth density per tree for eight different years. $Y = 0.28 + 0.006X$; $b_{yx} \pm 0.004$ (95% c.l.) (after Hassell, 1980).

With Figure 6.4 in mind, eqn. (6.2) can be rewritten as

$$f(N_t, P_t) = \left[1 + \frac{aP_t}{\alpha + bN_t}\right]^{-\alpha + bN_t}, \tag{6.5}$$

where N_t is the density of winter moth hosts and α and b are respectively the intercept and slope of the relationship. Equation (6.5) within the general model framework of eqn. (6.1) yields interactions that may be locally stable, as shown in Figure 6.5. This merely tells us, however, that parasitism by *Cyzenis* has the potential to stabilize a parasitoid–host interaction. The full assessment of the actual role of *Cyzenis* in winter moth population dynamics is more complex and involves a model incorporating other aspects of the life tables of both species. The detailed development of this model is given in Hassell (1980), and culminates in:

$$N_{t+1} = \lambda N_t s_1 s_2 s_3 s_4 s_5 \tag{6.6a}$$

$$P_{t+1} = 0.16[N_t(1 - s_2)]^{0.63}, \tag{6.6b}$$

where s_1 to s_5 are the survivals from different mortalities, as given in Table 6.2. Of these different factors, s_2 is given by eqn. (6.5) and s_3 and s_4 represent minor mortalities treated here as constants. The remaining submodels, s_1 and s_5, are both crucially important in modeling the winter moth at Wytham Wood.

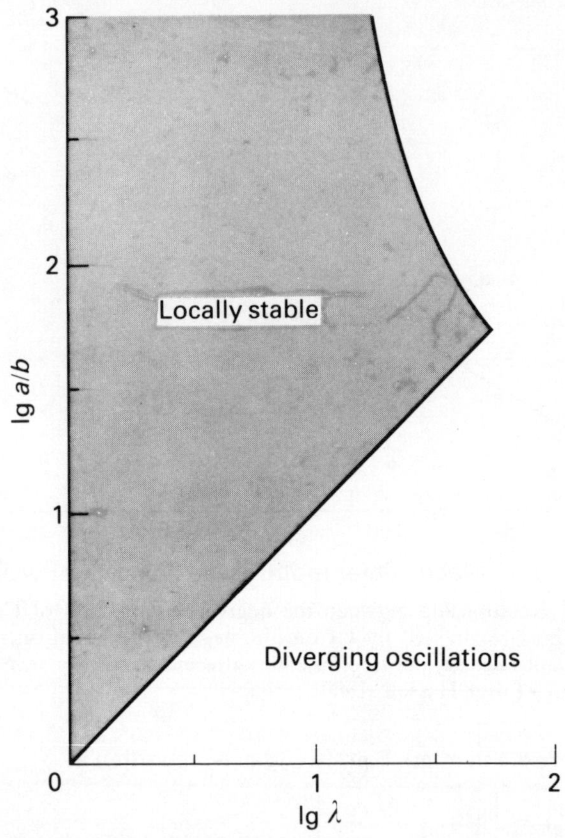

FIGURE 6.5 Local stability boundaries of model (6.1) with *f* defined from eqn. (6.5) and $\alpha = 0$, expressed in terms of the parameter combination *a/b* from eqn. (6.5) and the host rate of increase λ (after Hassell, 1980).

Submodel s_5 represents the survival of winter moth prepupae and pupae in the soil, mainly from beetle predators and an ichneumonid parasitoid (*Cratichneumon culex*). It is a major mortality that is also density dependent over the 19 years of the study, as shown in Figure 6.6. Density-dependent soil mortality also explains the form of eqn. (6.6b) for *Cyzenis*. The *Cyzenis* puparia are in the soil for approximately ten months and exposed to the same pupal predators as the winter moth, accounting for the density-dependent relationship in Figure 6.7.

Some numerical simulations from eqns. (6.6a,b) corresponding to the Wytham Wood populations are shown in Figure 6.8. They commence with the observed populations for 1950 and are then run in two ways: (*a*) using the observed s_1-values between 1950 and 1968 which, not unexpectedly, gives a very good correspondence between observed and calculated populations, since

TABLE 6.2 Submodels used in the winter moth–*Cyzenis* models for Wytham Wood.

1. Mortality between egg and mature larval stages	s_1 = observed values or $s_1 = 0.11$
2. Parasitism by *Cyzenis*	$s_2 = [1 + (aP_t/k)]^{-k}$ where $a = 0.18$ m²/generation and $k = 0.28 + 0.006\, N_t$
3. Parasitism by other larval parasitoids	$s_3 = 0.79C^{0.03}$ where C = winter moth surviving *Cyzenis*
4. Parasitism by a microsporidian	$s_4 = 0.93$
5. Winter moth soil mortality	$s_5 = 0.60\, N_t^{-0.31}$
6. Reproductive rate per adult moth	$\lambda = 75$
7. Soil mortality of *Cyzenis*	$s_{cyz} = 0.16\, N_a^{-0.31}$ where N_a = winter moth parasitized by *Cyzenis*.

the key factor is being introduced as observed; and (*b*) using the average value of $s_1 = 0.11$, which reveals the predicted equilibrium populations of $N^* = 64.7$ m^{-2} and $P^* = 0.4$ m^{-2}.

Two conclusions are emphasized by these models. (1) *Cyzenis* plays a very minor part in the winter moth population dynamics at Wytham Wood, and indeed could be completely omitted from the models with only a relatively small change in the winter moth populations ($N^* = 80.3$ m^{-2}, $P^* = 0$). (2) The equilibrium is the result of the density-dependent soil mortality, as emphasized by Varley and Gradwell (1963). This, therefore, is the primary regulating factor in the system.

The winter moth population dynamics at Wytham Wood are thus largely governed by the balance between the destabilizing effects of the key factor s_1 and the stabilizing effects of the density-dependent soil mortality s_5. With this picture in mind, we now turn to the winter moth—*Cyzenis* interaction in Nova Scotia.

6.4.2. Winter Moth in Nova Scotia

The winter moth was accidentally introduced to Nova Scotia in the early 1930s where it became a pest by defoliating shade and apple trees (Embree, 1971). A biological control program was initiated in 1954 in which six parasitoid species

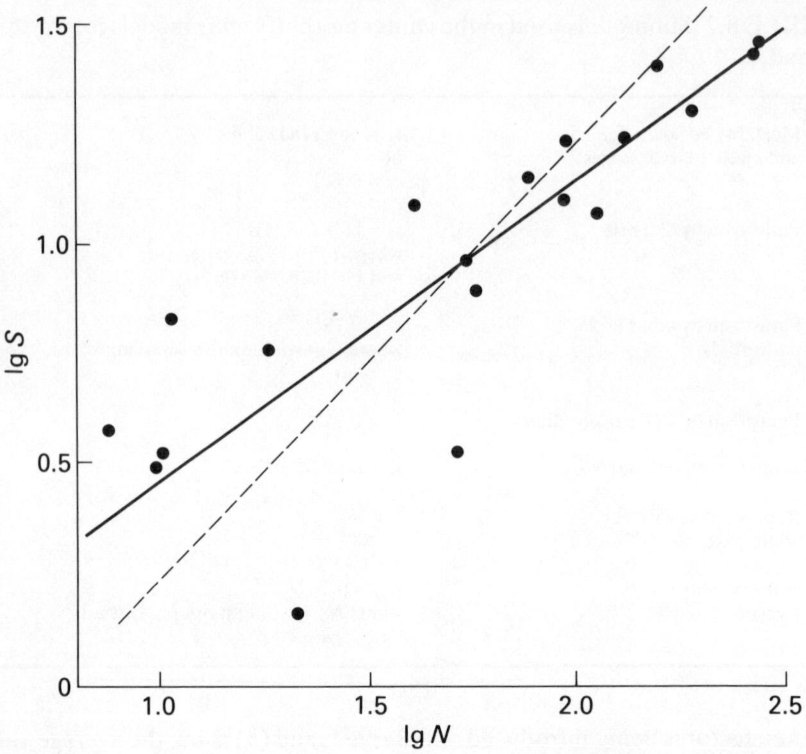

FIGURE 6.6 Density-dependent relationship between the density of winter moth entering the soil to pupate, N, and the subsequent emerging adults, S. Full line, $Y = 0.22 + 0.69X$; $b_{yx} \pm 0.21$ (95% c.l.); broken line, slope of unity.

were released (Graham, 1958). Of these, *Cyzenis* and an ichneumonid, *Agrypon flaveolatum*, became established and soon caused a dramatic decline in winter moth populations (see Figure 6.1A) that have remained at these low levels up to the present time. The populations dynamics of the winter moth in Nova Scotia were studied between 1954 and 1962 by Embree (1965), and when this information is combined with the data from Wytham Wood, we have a unique opportunity to analyze a biological control program in depth. It is an especially interesting exercise since we now know that *Cyzenis* has a negligible effect on the winter moth populations at Wytham Wood and yet it has produced very high levels of parasitism in Nova Scotia.

The components of the Nova Scotia model presented below are all taken from the data in Embree (1965) with the exception of the submodel for *Cyzenis*, which follows that for Wytham Wood. Several differences from Wytham Wood are immediately apparent, but the most important difference lies in the soil mortality. While at Wytham there is a heavy density-dependent mortality, in Nova Scotia there is no sign of density dependence and the

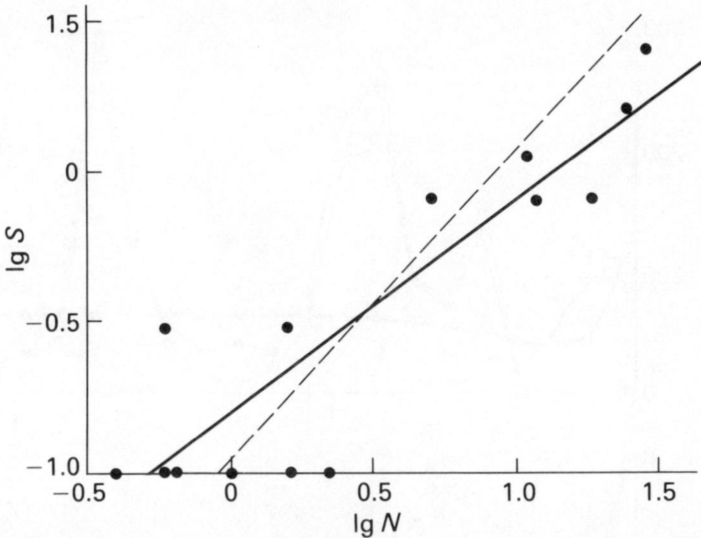

FIGURE 6.7 As Figure 6.6, but showing the comparable relationship for the survival of *Cyzenis* in the soil (N and S are densities of parasitized winter moth and subsequently emerging *Cyzenis* adults, respectively). $Y = 0.80 + 0.69X$; $b_{yx} \pm 0.29$ (95% c.l.) (after Hassell, 1980).

average level of mortality is much lower ($\bar{x} = 35\%$). We will assume that this is a constant mortality that affects both winter moth and *Cyzenis* while in the soil. On putting these components together, the full winter moth–*Cyzenis* model for Nova Scotia becomes:

$$N_{t+1} = \lambda N_t s_1 s_2 s_5 \tag{6.7a}$$

$$P_{t+1} = 0.65\, N_a, \tag{6.7b}$$

where $\lambda = 89$, $s_1 = 0.02$, s_2 is as for the Wytham Wood model, and $s_5 = 0.65$ (see Hassell, 1980, for further details). A numerical simulation of this system is shown in Figure 6.9. It commences with the winter moth population at the Wytham Wood equilibrium ($N^* = 64.7$ m^{-2}). When the parameters are changed to those for Nova Scotia, the winter moth population increases exponentially in the absence of any density dependence, until cut off at an assumed upper limit of 300 m^{-2}. This was chosen as a density slightly higher than the 1957 level at Wytham ($N_t = 277.4$ m^{-2}) when extensive defoliation occurred. *Cyzenis* is then introduced at a very low density, corresponding to the Nova Scotia introductions made in 1954, and increases rapidly causing a marked decline in the winter moth population. The system then converges on the locally stable equilibrium $N^* = 9.4$ m^{-2}, $P^* = 0.9$ m^{-2}.

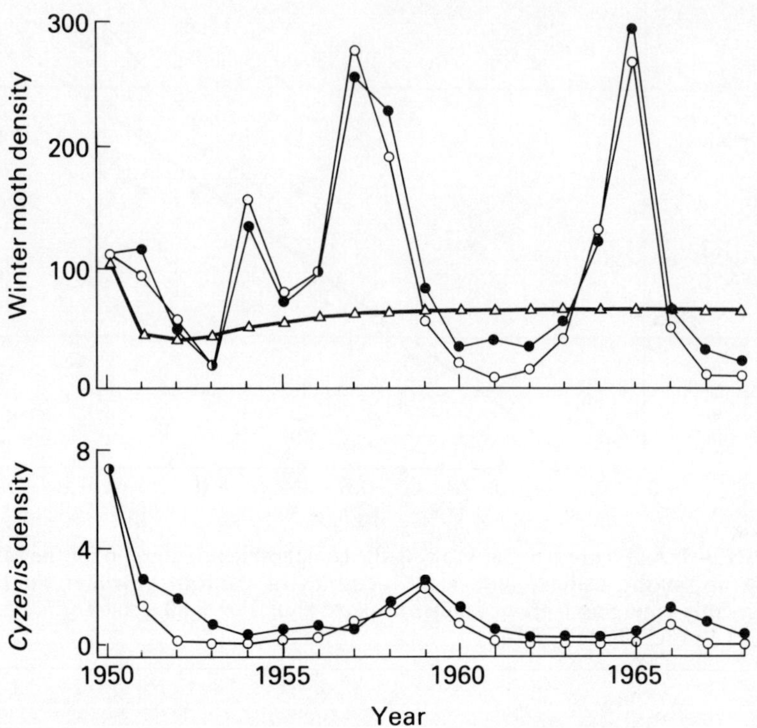

FIGURE 6.8 Observed winter moth and adult *Cyzenis* populations per m^2 canopy area. ● Wytham Wood, 1950–68, compared with the predicted populations using eqn. (6.6a,b); ○ predicted populations using observed s_1 values; △ predicted winter moth populations using average $s_1 = 0.11$ (after Hassell, 1980).

The picture at Nova Scotia is therefore one where Cyzenis not only causes the depression of the winter moth equilibrium but is also the sole cause of the observed stability. The reason that *Cyzenis* can achieve this depends upon three things: (1) its high searching efficiency (a = 0.18 m^{-2}/generation); (2) the absence of a heavy soil mortality, making the effective *overall* efficiency of *Cyzenis* heavily dependent upon the actual *searching* efficiency; and (3) its nonrandom search causing higher levels of parasitism in patches of high winter moth density. The latter is entirely responsible for the stability in the Nova Scotia model. Were *Cyzenis* to search at random ($k \rightarrow \infty$ in eqn. (6.5)), the model populations would exhibit unstable oscillations.

We have therefore a situation very much in accord with that explored by Beddington *et al.* (1978) and discussed above, where nonrandom search in a patchy environment is the most likely cause of stability in parasitoid–host systems at very low equilibrium levels.

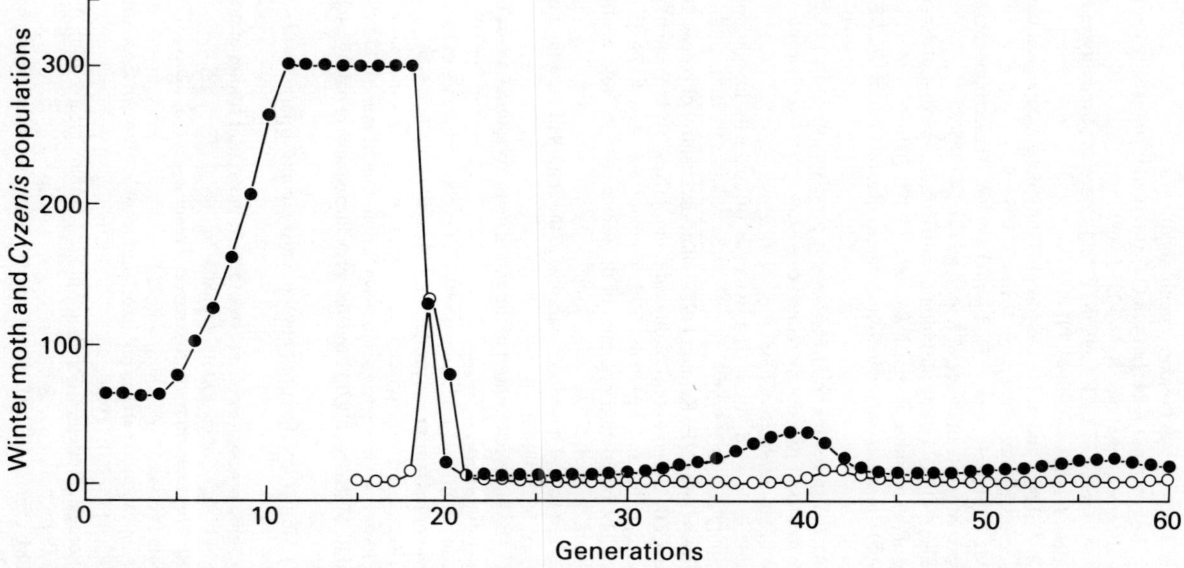

FIGURE 6.9 Numerical simulation of the Nova Scotia system using eqns. (6.7a,b). ● Winter moth larvae; ○ adult *Cyzenis* (after Hassell, 1980).

REFERENCES

Akinlosotu, T. A. (1973) *The Role of Diaeretiella rapae (McIntosh) in the Control of the Cabbage Aphid*. PhD Thesis, University of London (unpublished).

Beddington, J. R., C. A. Free, and J. H. Lawton (1975) Dynamic complexity in predator–prey models framed in difference equations. *Nature* 225:58–60.

Beddington, J. R., C. A. Free, and J. H. Lawton (1976a) Concepts of stability and resilience in predator–prey models. *J. Animal Ecol.* 45:791–816.

Beddington, J. R., C. A. Free, and J. H. Lawton (1978) Modelling biological control: on the characteristics of successful natural enemies. *Nature* 273:573–9.

Beddington, J. R., M. P. Hassell, and J. H. Lawton (1976b) The components of arthropod predation II: The predator rate of increase. *J. Animal Ecol.* 45:165–85.

Bliss, C. I. and R. A. Fisher (1953) Fitting the binomial distribution to biological data and a note on the efficient fitting of the negative binomial. *Biometrics* 9:176–200.

van den Bosch, R. (1968) Comments on population dynamics of exotic insects. *Bull. Entomol. Soc. Am.* 14:112–15.

Broadhead, E. and A. J. Wapshere (1951) *Mesopsocus* populations on larch in England—the distribution and dynamics of two closely related coexisting species of *Psocoptera* sharing the same food resource. *Ecol. Mon.* 36:327–88.

Comins, H. N. and M. P. Hassell (1979) The dynamics of optimally foraging predators and parasitoids. *J. Animal Ecol.* 48:335–51.

De Bach, P., D. G. Rosen and C. E. Kennett (1971) Biological control of coccids by introduced natural enemies, in C. B. Huffaker (ed) *Biological Control* (New York: Plenum) pp165–94.

Doutt, R. L. (1958) Vice, virtue and the redalia. *Bull. Entomol. Soc. Am.* 4:119–23.

Embree, D. G. (1965) The population dynamics of the winter moth in Nova Scotia 1954–1962. *Mem. Entomol. Soc. Can.* 46:1–57.

Embree, D. G. (1966) The role of introduced parasites in the control of the winter moth in Nova Scotia. *Can. Entomol.* 98:1159–68.

Embree, D. G. (1971) The biological control of the winter moth in eastern Canada by introduced parasites, in C. B. Huffaker (ed) *Biological Control* (New York: Plenum) pp217–26.

Free, C. A., J. R. Beddington, and J. H. Lawton (1977) On the inadequacy of simple models of mutual interference for parasitism and predation. *J. Animal Ecol.* 46:543–54.

Graham, A. R. (1958) Recoveries of introduced species of parasites of the winter moth, *Operophtera brumata* (L.) (Lepidoptera: Geometridae) in Nova Scotia. *Can. Entomol.* 90:595–6.

Gurney, W. S. C. and R. M. Nisbet (1978) Predator–prey fluctuations in patchy environments. *J. Animal Ecol.* 47:85–102.

Hassell, M. P. (1971) Mutual interference between searching insect parasites. *J. Animal Ecol.* 40:473–86.

Hassell, M. P. (1977) Some practical implications of recent theoretical studies of host–parasitoid interactions, *Proc. XV. Int. Congr. Entomol.* pp808–16.

Hassell, M. P. (1978) *The Dynamics of Arthropod Predator–Prey Systems*. (Princeton, NJ: Princeton University Press).

Hassell, M. P. (1979) Non-random search in predator–prey models. *Fortschr. Zool.* (Stuttgart, New York: Gustave Fischer) 25(2/3):311–30.

Hassell, M. P. (1980) Foraging strategies, population models and biological control: A case study. *J. Animal Ecol.* 49:603–28.

Hassell, M. P. and H. M. Comins (1976) Discrete time models for two-species competition. *Theor. Pop. Biol.* 9:202–21.

Hassell, M. P. and H. N. Comins (1978) Sigmoid functional responses and population stability. *Theor. Pop. Biol.* 14:62–7.

Hassell, M. P., J. H. Lawton, and J. R. Beddington (1977) Sigmoid functional responses in invertebrate predators and parasitoids. *J. Animal Ecol.* 46:249–62.

Hassell, M. P. and R. M. May (1973) Stability in insect host–parasite models. *J. Animal Ecol.* 42:693–736.

Hassell, M. P. and R. M. May (1974) Aggregation in predators and insect parasites and its effect on stability. *J. Animal Ecol.* 43:567–94.

Hassell, M. P. and V. C. Moran (1976) Equilibrium levels and biological control. *J. Entomol. Soc. S. Africa* 39:357–66.

Hastings, A. (1977) Spatial heterogeneity and the stability of predator–prey systems. *Theor. Pop. Biol.* 12:37–48.

Hubbard, S. F. (1977) *Studies on the Natural Control of Pieris brassicae with Particular Reference to Parasitism by Apanteles glomeratus*. DPhil Thesis, University of Oxford (unpublished).

Huffaker, C. B. (ed) (1971) *Biological Control* (New York: Plenum).

Huffaker, C. B. and C. E. Kennett (1966) Studies of two parasites of olive scale, *Palatoria oleae* (Colvee): IV. Biological control of *Parlatoria oleae* through the compensatory action of two introduced parasites. *Hilgardia* 37:283–335.

Klomp, H. (1966) The dynamics of a field population of the pine looper, *Bupalus piniarius* L. (Lep., Geom.). *Adv. Ecol. Res.*, 3:207–305.

Lawton, J. H., M. P. Hassell, and J. R. Beddington (1975) Prey death rates and rate of increase of arthropod predator populations. *Nature* 255:60–2.

Levin, S. A. (1976) Population dynamics in heterogeneous environments. *Ann. Rev. Ecol. Syst.* 7:287–310.

May, R. M. (1978) Host–parasitoid systems in patchy environments: a phenomenological model. *J. Animal Ecol.* 47:833–44.

May, R. M. and M. P. Hassell (1981) The dynamics of multi-parasitoid–host interactions. *Am. Nat.* 117, 234–61

May, R. M., M. P. Hassell, R. M. Anderson, and D. W. Tonkyn (1981) Density dependence in host–parasitoid models. *J. Animal Ecol.* 50, 855–65

Murdoch, W. W. (1977) Stabilizing effects of spatial heterogeneity in predator–prey systems. *Theor. Pop. Biol.* 11:252–73.

Murdoch, W. W. and A. Oaten (1975) Predation and population stability. *Adv. Ecol. Res.* 9:2–131.

Noyes, J. S. (1974) *The Biology of the Leek Moth, Acrolepia assectella (Zeller)*. PhD Thesis, University of London (unpublished).

Rejmanek, M. and P. Stary (1979) Connectance in real biotic communities and critical values for stability of model ecosystems. *Nature* 280:311–13.

Turnbull, A. L. (1967) Population dynamics of exotic insects. *Bull. Entomol. Soc. Am.* 13:333–7.

Turnbull, A. L. and D. A. Chant (1961) The practice and theory of biological control of insects in Canada. *Can. J. Zool.* 39:697–753.

Varley, G. C. (1970a) The need for life tables for parasites and predators, in R. L. Rabb and F. E. Guthrie (eds) *Concepts of Pest Management* (Raleigh, NC: North Carolina University Press) pp59–68.

Varley, G. C. (1970b) The effects of natural predators and parasites on winter moth populations in England. *Proc. Tall Timbers Conf. on Ecol. Animal Control by Habitat Management* 2:103–16.

Varley, G. C. and G. R. Gradwell (1963) The interpretation of insect population changes. *Proc. Ceylon Ass. for the Advancement of Science* 18:142–56.

Varley, G. C. and G. R. Gradwell (1968) Population models for the winter moth. *Symp. R. Entomol. Soc. Lond.* 4:132–42.

Varley, G. C. and G. R. Gradwell (1971) The use of models and life tables in assessing the role of natural enemies, in C. B. Huffaker (ed) *Biological Control* (New York: Plenum) pp93–112.

Varley, G. C., G. R. Gradwell, and M. P. Hassell (1973) *Insect Population Ecology* (Oxford: Blackwell).

Vinson, C. G. (1976) Host selection by insect parasitoids. *Ann. Rev. Entomol.* 21:109–33.

Wang, Y. and A. P. Gutierrez (1980) An assessment of the use of stability analyses in population ecology. *J. Animal Ecol.* 49:435–52.

Watt, K. E. F. (1965) Community stability and the strategy of biological control. *Can. Entomol.* 97:887–95.

Weseloh, R. M. (1976) Behavior of forest insect parasitoids, in J. F. Anderson and H. K. Kaya (eds) *Perspectives in Forest Entomology* (New York: Academic Press).

Zeigler, B. P. (1977) Persistence and patchiness of predator–prey systems induced by discrete event population exchange mechanisms. *J. Theor. Biol.* 67:687–713.

Zwölfer, H. (1979) Strategies and counter-strategies in insect population systems competing for space and food in flower heads and plant galls. *Fortschr. Zool.* 25(2/3):331–53 (Stuttgart, New York: Gustave Fischer).

7 Modeling Predation

A. P. Gutierrez and J. U. Baumgaertner

7.1. INTRODUCTION

Many studies have been conducted on the feeding habits of predaceous insects (e.g., Holling, 1966; Hagen and Sluss, 1966; Fraser and Gilbert, 1976; Tamaki *et al.*, 1974), but little effort has been made toward developing a model for the allocation of nutrients. Two exceptions are Mukerji and LeRoux (1969) on *Podisus*, and Rogers and Randolph (1978) on *Glossina*. Beddington *et al.* (1976a; see Hassell, 1978 for a review) focused theoretical attention on this component of predator population growth.

In this chapter we summarize a model for predator growth and development proposed by Gutierrez *et al.* (1981), which is derived from the metabolic pool model developed by Gutierrez *et al.* (1975), Wang *et al.* (1977) for cotton and by Gutierrez *et al.* (1976) for the Egyptian alfalfa weevil (*Hypera brunneipennis* Boheman). In the present study the experimental animal is the female ladybird beetle, *Hippodamia convergens* Guerin.

7.1.1. Biology of the Beetle

The larval and adult stages of the ladybird beetle are important predators of aphids in western North America (Hagen, 1974; Neuenschwander *et al.*, 1975; Fraser and Gilbert, 1976). Adult beetles cease reproduction and begin to migrate out of fields at low aphid densities, but larvae must remain in the field until they die or find sufficient food to complete their development. The aphid biomass eaten must be divided into components of respiration (i.e., maintenance and searching costs, and prey tissue conversion costs), waste, growth, and reproduction by adults. The food supply also affects ageing, survivorship, and net immigration rates. This model links all of these aspects of predator biology.

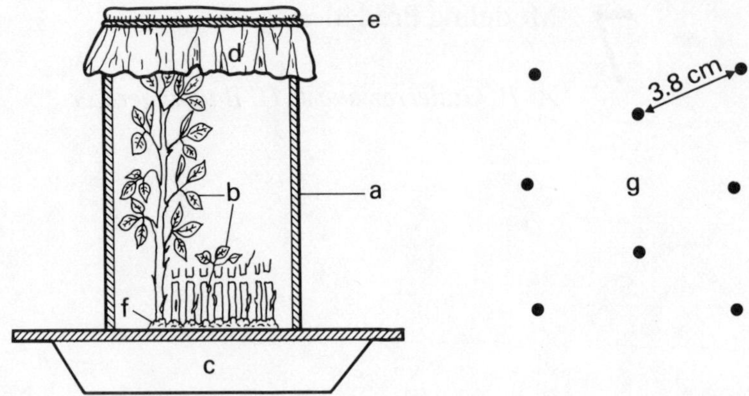

FIGURE 7.1 The experimental arena: a, plexiglass tube and plate; b, eight cut alfalfa stems; c, shallow water filled pan; d, cheesecloth top; e, rubber band; f, putty; and g, the spacing pattern of the stems.

7.1.2. The Experimental Approach

The experiments to formulate the model consisted of feeding beetles in an arena (Figure 7.1) with specified quantities of aphids and measuring weight changes and/or oviposition by the beetles. Time and age are expressed here in day degrees (D°) above the estimated developmental threshold of 8.8 °C (Butler and Dickerson, 1972). All experiments were conducted at 24.8 °C and long day lengths (14–16 h). The time step in our model Δt is 16 day degrees (D°) or one day under our conditions. This latter fact greatly influences the form of our models. For a complete description of methods and review of the literature, see Gutierrez *et al.* (1981).

7.2. THE MODEL

The model originally proposed by Gutierrez and Wang (1979) proved to be inadequate because it combined gross food intake with the process of assimilation, and a better model has been produced that separates the two (Figure 7.2). Here hunger and behavior determine food demand, while metabolic needs determine assimilate requirements. It appears that the hunger submodel described in Section 7.2.1 can be ignored, and it is only necessary to incorporate the behavioral component of prey consumption. In Section 7.2.2 we present the assimilation submodel and estimate the two critical parameters of respiration and excretion. We then turn to the behavioral component, showing how the prey demand and search rates can be estimated and incorporated in the functional response equation. Finally, we describe the priority scheme for allocating assimilates to growth and reproduction.

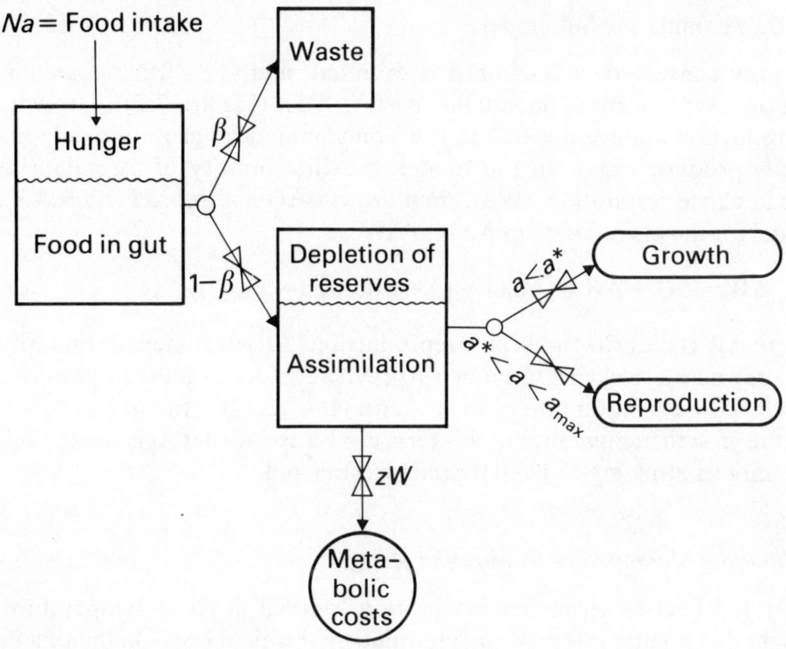

FIGURE 7.2 Metabolic pool model for a generalized predator separating hunger from assimilate allocation.

7.2.1. Hunger Submodel

The quantity of prey Na eaten by a predator during some period of time Δt is determined by its searching success, which is conditioned by behavior and physiological state. The rate of search is greatly influenced by beetle size and hunger H_t. Hunger is defined here as

$$0 \leq H_{t+\Delta t} = \Omega_t - Na_t - F_t + E_t + \Delta\Omega_t < \Omega(W)_{\max}, \tag{7.1}$$

where Ω_t is the gut capacity at time t; F is the food in gut; E is the food leaving gut via excretion and assimilation (i.e., β is the proportion of Na excreted); $\Delta\Omega$ is the change in gut capacity during Δt only if the beetle is immature; $\Omega(W)_{\max}$ is the maximum gut capacity of the beetle $= f(W)$; and W is the mass of the beetle. All of these variables are expressed as mg of biomass.

Laboratory experiments indicate that food eaten is processed within one day, hence the daily flow of Na into the gut is balanced by the outflows (i.e., excretion, assimilation, and metabolic costs). Hunger is thus a transitory state and this submodel can be ignored with little apparent effect. The influence of behavior on Na can be incorporated in the search A and prey consumption demand b rates, which are related to predator size, as discussed below.

7.2.2. Assimilation Submodel

Of prey consumed, a fraction β is excreted, and (1 − β)*Na* is assimilated and/or used to meet metabolic costs $zW\Delta t$ (Figure 7.2). Growth and reproduction are age dependent (i.e., only immatures grow and only mature adults produce eggs). In the model, the first priority of assimilate use is maintenance respiration $zW\Delta t$, then growth ΔG and reproduction ΔR. The model for the assimilation of *Na* into ΔW is:

$$\Delta W = \Delta G + \Delta R = [Na_t(1-\beta) - zW_t\Delta t](1 - z\Delta t/2) \tag{7.2}$$

where ΔW is the growth (ΔG) or reproduction (ΔR) increment during Δt; *Na* is the prey mass attacked and eaten = *f*(prey, predators); *z* is the respiration rate; and β is the excretion rate. The last term $(1 - z\Delta t/2)$ is the maintenance cost for the growth increment, but this term can be excluded if Δt is small. Only the parameters *z*(mg/mg/D°) and β need be estimated.

Estimating Maintenance Respiration Rate z

Fully fed beetles given free water were starved at room temperature and weighed at regular intervals to determine their weight loss. On the first day t_1, the weight changes rapidly as the gut empties, while on subsequent days there is a more gradual loss of weight $\hat{z}$ per day due to the utilization of the body fat. The model does not define the body tissues used, but we recognize that fat and protein have different caloric values. As we shall see, the model remains reasonably robust despite this shortcoming.

Average maintenance cost of beetles was estimated from t_2 to t_4 as $\hat{z} = 1 - (\bar{W}_{t+1}/\bar{W}_t) = 0.09$ mg/day. $\hat{z}$ is the maximum rate because the beetles were starved, and were searching at their maximum rate. All beetles died by time $t = 3.5$, so we estimate that they utilized a maximum of 21% of their total body weight during the 2½-day period. We cannot include the first day in the computation because the beetles still had food in their guts. The maximum rate z_{max} per D° is 0.006. The rate decreases as $Na/bW\Delta t \rightarrow 1$. The coefficient (0.2) in eqn. (7.3) is estimated from unpublished searching behavior data:

$$0.005 \leqslant z(1 - 0.2\,Na/bW\Delta t) \leqslant 0.006. \tag{7.3}$$

A second independent estimate for z_{max} is given in the next section, while the prey demand rate *b* and *Na* are derived in subsequent sections.

Estimating Excretion Rate β

If *Na* is the prey biomass eaten during Δt, $zW\Delta t$ is the maintenance cost and ΔW is the weight gain, then we can solve eqn. (7.2) for β:

$$\beta = [Na - zW\Delta t - \Delta W/(1 - z\Delta t/2)]/Na \quad (7.4)$$

$$\beta = \text{wastes}/Na.$$

All of the variables were measured in an independent experiment, and z is estimated above. Ovipositing adult beetles were given eight 1.5-mg aphids per day for a 9.25-day period at 24.8 °C, and the changes in weight and numbers of eggs deposited were determined. The average daily values (mg) were substituted in eqn. (7.2)

$$\overline{\Delta G} + \overline{\Delta R} = (\bar{N}a - \beta\bar{N}a - \hat{z}W_t)(1 - \hat{z})$$

$$\beta = [Na - \hat{z}W_t - (\Delta G + \Delta R)/(1 - \hat{z}/2)]/Na = 0.77.$$

In fact, food quality also influences $\bar{\beta}$; hence β is prey specific (Hagen and Sluss, 1966).

The model (7.2) was then tested against a larger, independent set of data in Figure 7.3. The broken curve excludes maintenance costs for the growth increment, while the full curve includes them and yields a reasonable fit to the data. The x intercept estimates the compensation point for the predators ($Na(1 - \beta) = zW\Delta t$), and the y intercept estimates total maintenance costs.

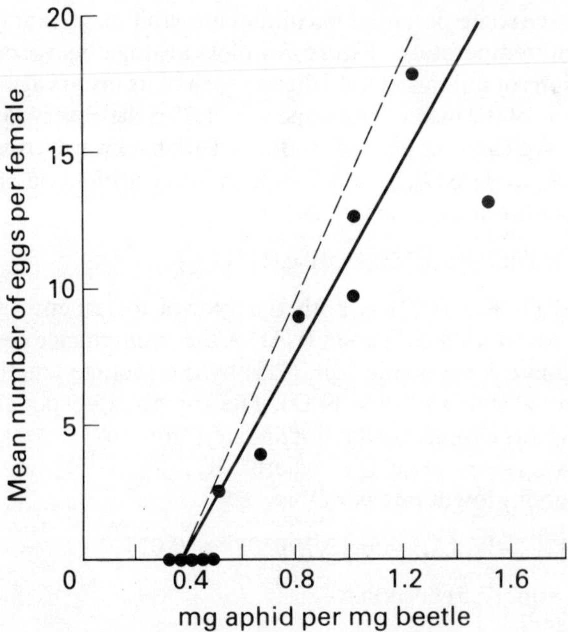

FIGURE 7.3 Mean number of eggs produced per 16 D° > 8.8 °C by female ladybird beetles fed varying mg of aphids per mg of body weight.

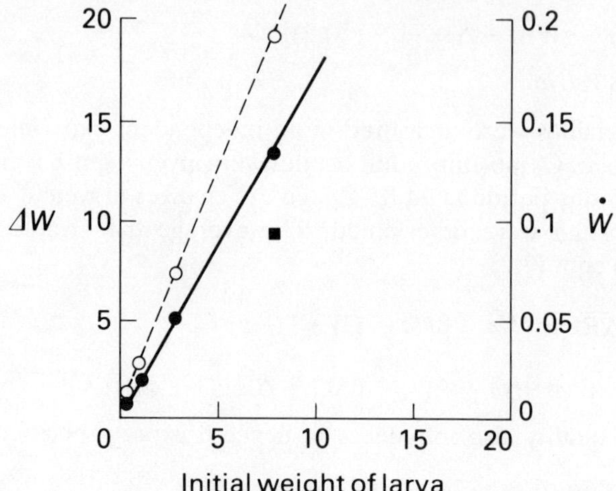

FIGURE 7.4 Relationship between weight gain (ΔW) plotted as ●, and the rate of gain ($\dot{W}$) plotted as ○, per life stage on the initial weight at the beginning of each life stage. $\dot{W}$ for fourth instar larvae was corrected for the fact that the length of the stage as measured also included a quiescent period (i.e., ■).

Prey Demand Rate b per Predator

The beetles have some potential maximum growth rate v^* that is a function of body mass and temperature. Figure 7.4 plots average beetle net weight gain under conditions of unlimited food during each of its instars against the weight at the beginning of the instar. The slope $v^* = 1.75$ is the observed rate of growth per life stage. We can now use the model to work backward to estimate b (note $\Delta W_{\max} = \Delta R_{\max} + \Delta G_{\max} = 1.75W$). A little arithmetic shows that the demand $\Delta\hat{b}$ per life stage of length Δt^* is

$$\Delta\hat{b} = [1.75W_t(1 + z\Delta t^*/2) + zW_t\Delta t^*], \tag{7.5}$$

where $1.75\ W_t(1 + z\Delta t^*/2)$ is both the growth increment and its average maintenance cost during Δt^*, and $zW_t\Delta t^*$ is the maintenance cost of the initial mass. To estimate $\hat{b}$ we divide eqn. (7.5) by the average length of the active parts of the larval stages ($\Delta t^* = 39°$D). The growth rate v per D° is 0.045, but the predator grows exponentially, not linearly, from t to $t + \Delta t^*$; hence a better model is $W_{t+\Delta t^*} = W_t e^{v\cdot\Delta t^*}$, where $W_{t+\Delta t^*} = W_t + 1.75W_t$ and the instantaneous growth rate per D° is:

$$v = \ln(W_{t+\Delta t^*}/W_t)/\Delta t^* = \ln(2.75)/39\text{D}° = 0.025.$$

If Δt is small, eqn. (7.5) becomes

$$b = \gamma[ve^v W(1 + z/2) + zW],$$

where $\gamma = 1 + \beta/(1 - \beta)$ corrects for excretion. The estimate b reconciles the observed food intake and growth. Note that demand depends upon weight, rather than age.

7.2.3. The Functional Response

Prey Consumption, Na

All of the above discussion presupposes that we know the searching success *Na* of beetles eqn. (7.2). This is clearly not the case. Many predation submodels for estimating *Na* have been proposed (*see* Lotka, 1925; Volterra, 1926; Nicholson, 1933; Holling, 1966; and Royama, 1971), but only that proposed by Fraser and Gilbert (1977) was developed from laboratory and field experiments and was field tested. Gutierrez and Wang (1979) analyzed some of the theoretical properties of the Fraser–Gilbert model (1976):

$$\begin{aligned} Na &= N_0 f(N,M,W,P) \\ &= N_0 M(1 - \exp\{[-bWPt/(N_0 M)][1 - \exp(-AN_0 M/b)]\}) \end{aligned} \tag{7.6}$$

where $N_0 = X_c$ is the corrected number of hosts available (*see* Gutierrez *et al.*, 1980); P is the number of predators ($0 < P < \infty$); $T = \Delta t$ is an increment of time in physiological units (D°); M = mg per prey; and A is the predator search rate $g(b)$. All other variables were defined above.

The Lotka–Volterra and Nicholson–Bailey models are special cases of the Fraser–Gilbert model. This deterministic model (7.6) describes the searching success of the individual or of the population, and this fact marks an important transition point, as all prior discussion concerned invididual predators. In the model, the quantity of prey captured (*Na*) may be expressed in terms of mass or numbers; in our study both are of interest. Only the predator search rate *A* has not been explained.

Rate of Search, A

Laboratory studies indicate that *A* is greatly influenced by weight and hunger. If we ignore hunger *per se*, and use *b* as a surrogate, we define the search rate as follows:

$$A_t = A_0 b_t / b_0, \tag{7.7}$$

where A_0 is the search rate of a newly hatched larva [0.008 in our experiment, i.e., (number of stems searched)/(number of stems × Δt) = 1/(8 × 16)], and b_0 and b_t are the prey demand rates of a newly hatched larva and a larva at its current growth state (see derivation of *b*).

7.2.4. The Metabolic Pool Submodel

Metabolic Pool Priority Scheme

The priority scheme for allocating assimilate to growth and reproduction is similar to that proposed for cotton by Gutierrez and Wang (1979).

$$\text{(1)}\quad \text{If } \Delta W > 0 \quad \begin{matrix} \Delta G > 0 \text{ if } a \leqslant a^* \\ \Delta R > 0 \text{ if } a > a^* \text{ and } W_t \geqslant W_{\max} \end{matrix} \qquad (7.8)$$

(see eqn. 7.2), where a = age (D°), a^* = 337 D° is the minimum age at reproduction, $W_{\max}$ is the maximum prior mass achieved, and the number of eggs produced = ΔR/(mass of one egg = 0.22 mg).

(2) If $\Delta W = 0$, then $\Delta G = \Delta R = 0$.

(3) If $\Delta W < 0$, then $\Delta G = \Delta W$, $\Delta R = 0$ and $0.8\ W_{\max} < W_{t+\Delta t} = W_t + \Delta W$. In the model the beetles are assumed to use all of their reserves and die when the weight loss is greater than $0.2\ W_{\max}$. The threshold comes from our experiment to estimate z (see above). The costs of converting prey tissues (Na) to predator tissues (G,R) are included in β and not in z, which measures only search and maintenance respiration. For simplicity, we assume ΔG and ΔR accrue via the same process, but the allocation is age-dependent as immatures increase in body size, while reproductive adults produce eggs. In general, nondiapause beetles do not have large fat reserves.

7.3. MODELING OTHER EFFECTS

It is well known that partially starved insects require longer to complete each life stage and their mortality rates increase accordingly.

Ageing

The influence of different levels of constant food (mg/mg) on the adult preoviposition period is seen to be a decreasing function of mg prey/mg predator (Figure 7.5A), while Figure 7.5B shows the effect of increased variability of daily food intake ($S/\bar{X}$ of mg/mg) on the preoviposition period. Only experiments with three or more observations were used to compute $S/\bar{X}$ values, which explains the difference in the number of data points in Figure 7.5B. Much of the scatter in the data results from the highly variable daily consumption rates caused by the uneven searching success of the beetles.

In the model, beetle eggs hatch at $a = 29$ D°, and during each Δt they age in increments of Δa. The maximum value of Δa is determined by temperature, but $\Delta t \neq \Delta a$ because the actual accrued age must also be modified by some measure

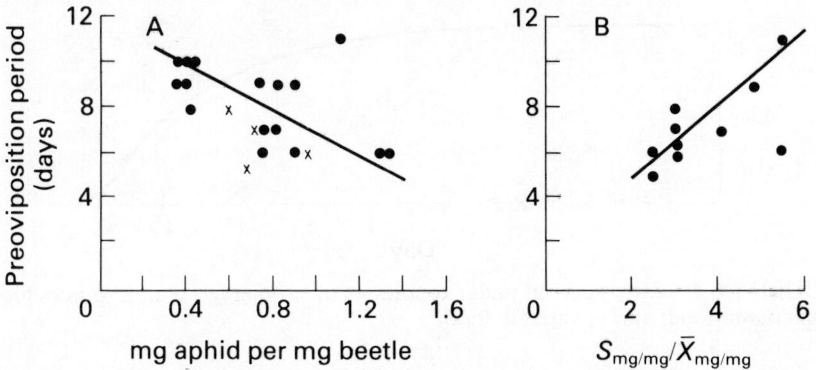

FIGURE 7.5 Influence of the ratio mg aphid per mg beetle on preoviposition time in days (16 D°/day) of females (A), and the influence of variability of prey consumption ($S_{mg/mg}/\bar{X}_{mg/mg}$) on preoviposition time (B).

of the beetles' nutritional state ϕ, which in this model is the ratio of food eaten Na to the prey mass required to meet maximum demands for growth, reproduction, and metabolic costs (i.e., $bW\Delta t$). The following formulates this notion:

$$A_{t+\Delta t} = a_t + \Delta a\phi, \qquad \text{where } 0 \leqslant \phi = Na/(bPW\Delta t) < 1. \tag{7.9}$$

Mortality

The probability of survival S during Δt for individuals or for the population can be modeled as

$$0 \leqslant S_t = \alpha + (1-\alpha)\, Na/(0.8bPW\Delta t) \leqslant 1 \tag{7.10}$$

(the coefficient 0.8 was estimated from computer simulations).
The starvation experiments show that most fully fed beetles die within four days when food is withheld because they have depleted their reserves (i.e., $W_t = 0.8W_{\max,t-4}$). If beetles are starved (i.e., $Na = 0$), $\alpha = 0.5$ is the estimated probability of survival during Δt(16 D°); hence the probability (7.11) of surviving four days equals 0.06, while $\dot{\alpha} = \alpha/\Delta t$.

$$\prod_{i=1}^{4} S_i = 0.5^4 = 0.06. \tag{7.11}$$

In view of the estimation error involved in the experiment, α appears to be quite reasonable. Furthermore, as $Na \rightarrow bPW\Delta t$, $S \rightarrow 1$ but, as we shall see in the next section, Na is always less than $bPW\Delta t$ because searching success is imperfect. Figure 7.6 (curves a and c) shows the two extremes of survivorship predicted by eqn. (7.11) (starved and fully fed), while curve b shows an intermediate pattern.

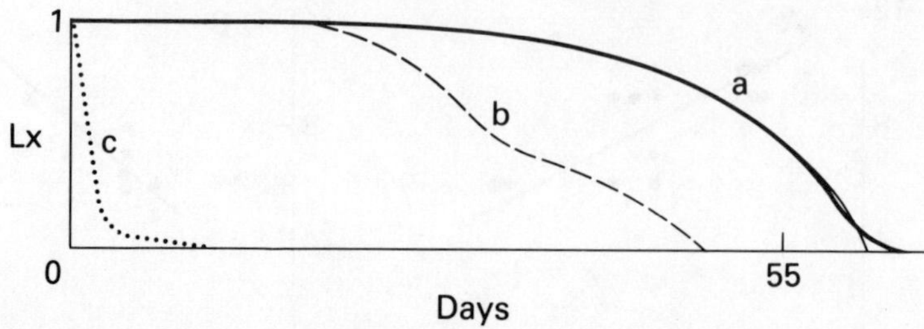

FIGURE 7.6 Predator survival under conditions of varying food. a, maximum food; b, minimum food; and c, variable food.

Emigration

The probability (E) that an adult beetle emigrates from the field is assumed to be a function of prey supply and demand.

$$0 < E_t = Na/bPW\Delta t < 1. \tag{7.12}$$

This fits well with field observations of Hagen and Sluss (1966), who noted that on average 1 mg of aphids per alfalfa stem was required to keep adult beetles in the field and to begin reproduction. This quantity of aphids is approximately equal to the 0.38 mg prey per mg predator predicted by the model as the metabolic compensation point (i.e., $Na(1 - \beta) = zW\Delta t$).

Immigration

Estimating immigration (IM) is a more difficult problem because the population (P^*) surrounding the study field is unknown, and the attraction rate of individual fields is likewise unknown. The model for immigration should be some function ($0 \leqslant IM_t = g(Na/bPW\Delta t, P^*) \leqslant 1$) of food availability, but further work is required before it can be formulated.

7.4. SIMULATION OF INDIVIDUAL PREDATOR GROWTH AND DEVELOPMENT

Only the resuts for individual beetles are presented here, while those describing the population dynamics of the alfalfa plant, pea aphid (*Acyrthosiphum pisum* Harris), and blue alfalfa aphid (*A. kondoi* Shinji), and *H. convergens* are described elsewhere (Gutierrez *et al.*, 1976; Gutierrez and Baumgaertner, unpublished).

Figure 7.7 shows the simulated growth and development of a newly hatched larva in our experimental arena under the conditions of 1, 3, or 5 aphids per stem at a constant temperature of 24.8 °C. The results (7A) show that ageing rate (as time to reproduction) slows, while growth and fecundity decrease as food

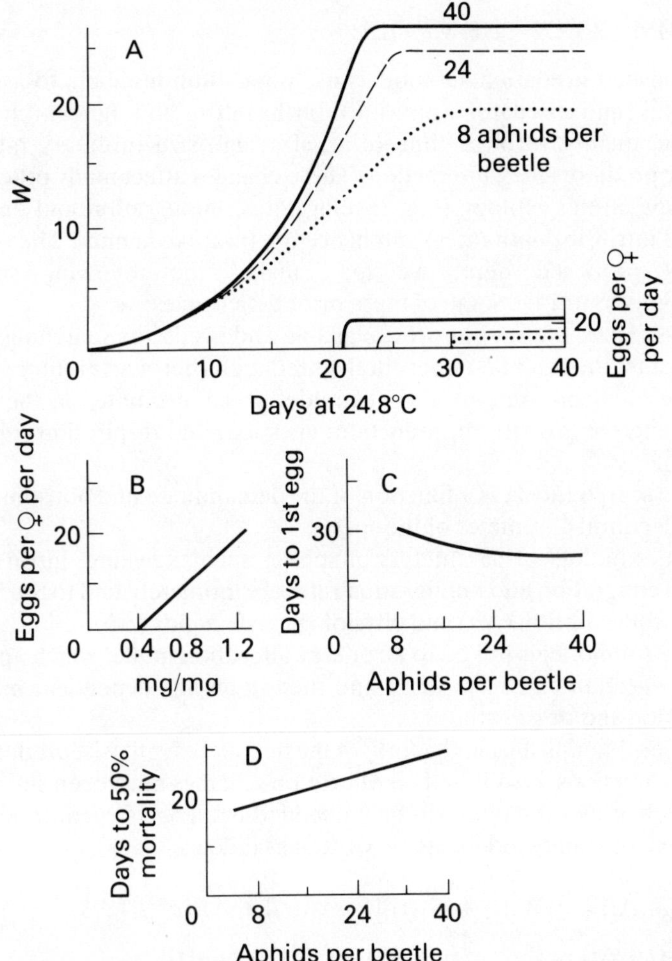

FIGURE 7.7 Simulated effects on A, growth; A, B, fecundity; C, developmental time; and, survivorship of ladybird beetle larvae exposed from hatching for a 50-day period at 24.8 °C (Δt = 16 D°) to 2, 3, and 5 aphids on each of eight stems (8, 24, and 40 aphids).

availability decreases. Figure 7.7 (B, C, D) shows the effects of beetle nutrition on fecundity, developmental time, and estimated 50% survivorship respectively. The effects of nutrition on the preoviposition of adults can be estimated in Figure 7.7A, and they compare reasonably well with laboratory results (Figure 7.5A). The model, while relatively simple, yields a very rich dynamic behavior which is also reasonably realistic. We have resisted the temptation to add the complexities of caloric content of tissues and more subtle aspects of beetle behavior.

7.5. SUMMARY OF THE MODEL

Most published predation submodels use population numbers to assess host death rates and predator (parasite) birth rates, and ignore changes in population quality through time (e.g., size, age structure, sex ratio, etc.) caused by predator–prey interaction. These changes affect many other aspects of the population's biology (e.g., ageing rates, immigration and emigration rates, and intrinsic death rates) and hence the future dynamics. The predation submodel proposed here, we feel, makes the following significant contributions to resolve some of these prior deficiencies:

(1) Mass (size, energy) of prey available and predators searching replaces (or augments) the use of numbers in the predation (herbivore) submodel.

(2) The demand rate in the submodel is an estimate of the species' requirements for growth, reproduction, egestion, and respiration (i.e., Δb = demands).

(3) The search rate A is a function of the demand rate and not some loosely linked behavioral parameter of the species.

(4) Other biological parameters of species such as ageing, intrinsic death rates, and emigration and immigration rates are intimately tied to the interplay of energy consumption (Na) and physiological demands (Δb).

(5) The submodel is linked to an energy allocation model which apportions it first to egestion and respiration and then in an age-dependent manner to reproduction and/or growth.

(6) Several similarities in the form of the net photosynthesis, predation, and herbivore functions as well as their allocation schemes have been shown.

(7) Lastly, this submodel can be embedded in an age structure population model. One of many models can be written as follows:

$$\begin{aligned} W_1(t+\Delta t) &= \{[W_1(t) + G_1(t)][1 - \phi_1(t)] + \Delta R(a^*,t)\}S_1\hat{S}_1 \\ W_{i+1}(t+\Delta t) &= \{[W_{i+1}(t) + \Delta G_{i+1}(t)][1 - \phi_{i+1}(t)]S_{i+1} \\ &\quad + \phi_i(t)S_iW_i(t)\}\hat{S}_{i+1} \\ W_n(t+\Delta t) &= \{[W_n(t) + \Delta G_n(t)][1 - \phi_n(t)]S_n + \phi_{n-1}(t) \\ &\quad S_{n-1}W_{n-1}(t)\}\hat{S}_n + I_n - \mathrm{E}_n. \end{aligned} \tag{7.13}$$

For the sake of clarity all variables will be redefined, hence:

$W_1,\ldots,W_{i+1},\ldots,W_n$ are the massess of all n cohorts in the population.
$i = 1,\ldots,n$ are ages or stage categories.
$\phi_i = \phi_i(t,Na/\Delta b)$ is the ageing rate of the ith cohort as a function of the ratio of food supply Na and maximum food demand Δb at time t.
$\Delta R(a^*_,t)$ is the reproduction by adults of reproductive age a^* as in eqn. 7.2.
ΔG_i is growth by the ith cohort, where ΔG_i can be + or − as in eqn. 7.2.

$S_i = S_i(t, Na_i/\Delta b_i)$ is the intrinsic survivorship as a function of the ratio of food supply–demand.
$\hat{S}_i = S(t,\boldsymbol{e})$ is the survivorship from all other biotic or abiotic factors ($\boldsymbol{e}$).
$\mathrm{E}_i = \mathrm{E}_i(\mathrm{Na}_i/\Delta b_i)$ is the mass of age i immigrating into the population. Note that some stages cannot emigrate.
$I_i = I_i(\mathrm{Na}/\Delta b, W_i^*)$ is the mass of age i immigrating from a population of unknown size W_i^*.

A similar model can be written for population numbers and linked to the appropriate equation above, wherein little or no modifications are required for various parameters. Thus we see some natural linkages in this model between cohorts and between populations. This model form is a modification of one proposed by Wang and Gutierrez (1980) where ageing and growth and reproduction precedes predation.

7.5. CONCLUSIONS

At the level of the individual, the major facets of predator growth and development have been incorporated into a simple model as eqn. (7.2), similar to that described by Schoener (1973). The model is an extension of the metabolic pool model developed by Gutierrez *et al.* (1975), Gutierrez and Wang (1979), and Wang *et al.* (1977) for plant growth and development, which showed how the interplay of photosynthate supply- and demand-controlled growth and development processes in single plants or plant populations. The cotton model, while relatively simple, has proven to be very useful and is currently used in pest management programs in the Central Valley of California and in Nicaragua (see Chapter 11). The metabolic pool model (7.2) can be rewritten for completeness, ignoring the metabolic costs of the growth increment as

$$\Delta W = N_0 M f(N,M,P,W)(1-\beta) - zPW\Delta t, \qquad \text{where } 0 < P < \infty. \qquad (7.14)$$

The function $f(\cdot)$ through the origin is the Fraser–Gilbert model (Figure 7.8A, full curve), which is a type II functional response (Holling, 1966). Subtracting the metabolic costs (y intercept $= zPW\Delta t$) shifts the curve to the right and the correction for excretion $(1-\beta)$ lowers it. The curves saturate at high prey density, because predator demands ($bPW\Delta t$) are met asymptotically. Similar but less analytical models have been developed by Gutierrez *et al*, (1976), Gutierrez and Wang (1979), Beddington *et al.* (1976a), and later by Readshaw and Cuff (1982), who modeled Nicholson's blowfly data from a more realistic perspective. Note that both the demand b and the search A rates in $f(\cdot)$ are variables. If $N_0 M \rightarrow N_a$, then dividing x by WP and y by 0.22 mg yields the function shown in Figure 7.8B. Hunger increases searching costs (z_h) and these must be added to basic maintenance costs, z_m, (i.e., $z_t = z_m + z_h$). These costs are incorporated in $f(\cdot)$ via the parameters b and A. These added

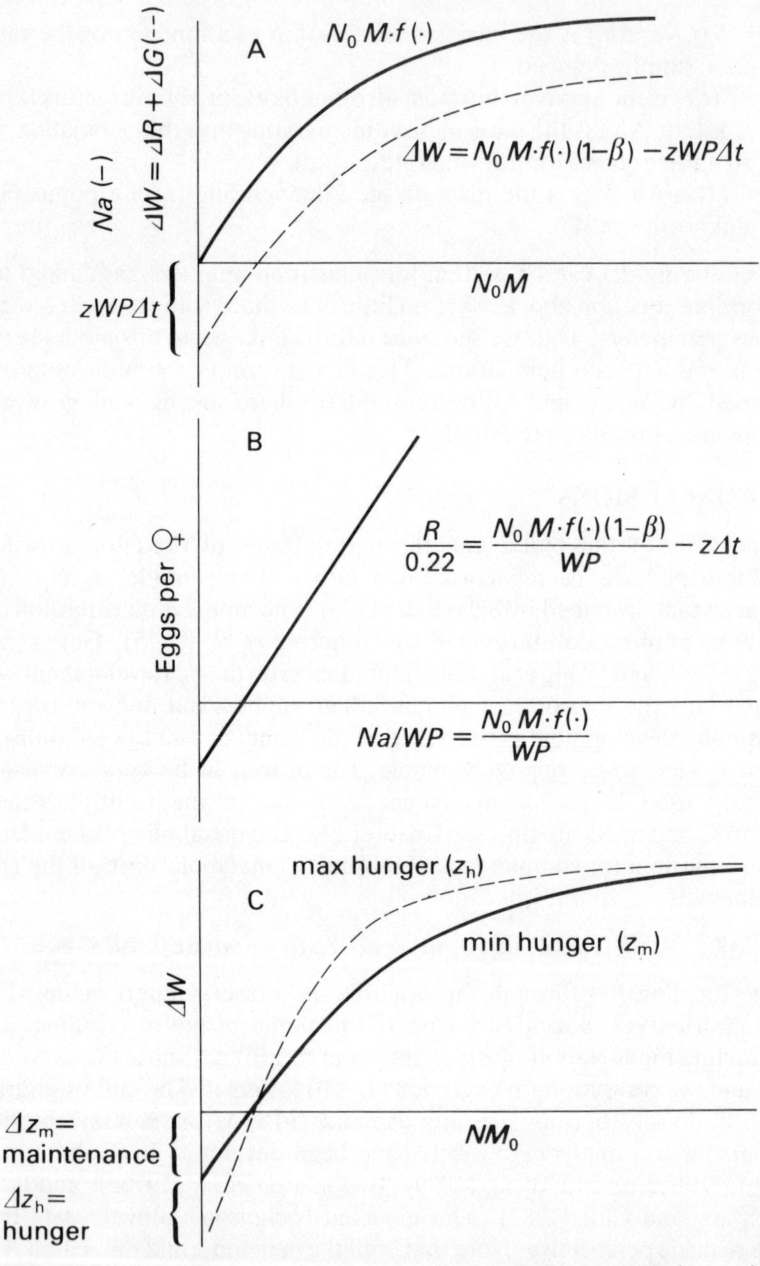

FIGURE 7.8 Generalized net growth function for predators: A, with (– – – –) and without (____) incorporating metabolic costs due to maintenance respiration (z) and losses due to excretion (β). B, the production of eggs per female as a function of mg prey per mg predator. C, As Figure 7.8A, with the inclusion of hunger effects on increased maintenance respiration. All variables are defined in the text.

costs are reflected in Figure 7.8C as an increase in the slope of the function.

The net predator growth function (Figure 7.8A) is analogous to a standard net photosynthesis function (Thornley, 1977, Chapter 4), which brings the argument full circle: *animal demographic equations have been used to model plant population, and a heretofore plant metabolite pool model has been modified to describe changes in predator growth and development, adult fecundity, ageing, survivorship, and net immigration rates at both the individual ($P = 1$) and the population level ($P > 1$)* (see Gutierrez *et al.*, 1981).

ACKNOWLEDGMENTS

This study profited greatly from discussions with K. S. Hagen and P. E. Kenmore, and the technical assistance of D. W. Zimpfer and A. Wu.

The publication was supported in part by the National Science Foundation and the Environmental Protection Agency, through a grant (NSF DEB-77-25260) to the University of California. The findings, opinions and recommendations expressed herein are those of the authors and not necessarily those of the University of California, the National Science Foundation, or the Environmental Protection Agency. Support was also obtained from grant NSF DEB-75-04223.

REFERENCES

Baumgaertner, J. U., A. P. Gutierrez, and C. G. Summers (1981) The influence of aphid prey consumption on searching behavior, weight increase, developmental time, and mortality of *Chrysopa carnea* (Neuroptera: Chrysopidae) and *Hippodamia convergens* (Coleoptera: Coccinellidae) larvae. *Can. Entomol.* 113:1007–14.

Beddington, J. R., M. P. Hassell, and J. H. Lawton (1976a) The components of arthropod predation II: The predation rate of increase. *J. Animal Ecol.* 45:165–86.

Beddington, J. R., C. A. Free, and J. H. Lawton (1976b) Concepts of stability and resilience in predator–prey models. *J. Animal Ecol.* 45:791–816.

Butler, G. D. and W. A. Dickerson (1972) Life cycle of the convergent lady beetle in relation to temperature. *J. Econ. Entomol.* 65:1508–9.

Fraser, B. D. and N. E. Gilbert (1976) Coccinellids and aphids: A qualitative study of the impact of adult lady birds (Coleoptera: Coccinellidae) preying on field populations of pea aphids (Homoptera: Aphididae). *J. Entomol. Soc. Brit. Col.* 73:33–56.

Gutierrez, A. P., J. U. Baumgaertner, and K. S. Hagen (1981) A conceptual model of predator growth, development and reproduction in the lady beetle *Hippodamia convergens* Guerin (Coleoptera: Coccinellidae) *Can. Entomol.* 113:21–31.

Gutierrez, A. P., J. B. Christensen, C. M. Merritt, W. B. Loew, C. G. Summers, and W. R. Cothran (1976) Alfalfa and the Egyptian alfalfa weevil (Coleoptera: Curculionidae). *Can. Entomol.* 108:635–48.

Gutierrez, A. P., L. A. Falcon, W. B. Loew, P. A. Leipzig, and R. van den Bosch (1975) An analysis of cotton production in California: A model for Acala cotton and the effects of defoliators on its yields. *Environ. Entomol.* 4:125–36.

Gutierrez, A. P., C. G. Summers, and J. U. Baumgaertner (1980) The phenology and within field distribution of aphids in California alfalfa. *Can. Entomol.* 112:489–495.

Gutierrez, A. P. and Y. Wang (1979) Applied population ecology: Models for crop production and pest management, in G. Norton and C. S. Holling (eds) *Pest Management*, Proc. Int. Conf., October 1976. IIASA Proceedings Series Vol. 4 (Oxford: Pergamon) pp255–80.

Hagen, K. S. (1974) The significance of predaceous Coccinellidae in biological and integrated control of insects. *Entomophaga Mem.* 7:25–44.

Hagen, K. S. (1976) Role of insect nutrition in insect management. *Proc. Tall Timbers Conf. on Ecol. Animal Control by Habitat Management* 6:221–261

Hagen, K. S. and R. R. Sluss (1966) Quantity of aphids required for reproduction by *Hippodamia* spp. in the laboratory, in I. Hodek (ed), *Ecology of Aphidophagous Insects*. pp47–59

Hassell, M. P. (1978) *The Dynamics of Arthropod Predator–Prey Systems*. (Princeton: Princeton University Press).

Holling, C. S. (1966) The functional response of invertebrate predators to prey density. *Mem. Entomol. Soc. Can.* 48:1–86.

Lotka, A. J. (1925) *Elements of Physical Biology* (Baltimore: Williams and Wilkins).

Mukerji, M. K. and E. J. LeRoux (1969) A study on energetics of *Podisus maculiventris* (Hemiptera: Pentatomidae). *Can. Entomol.* 101:449–60.

Neuenschwander, P., K. S. Hagen, and R. F. Smith (1975) Predation on aphids in California's alfalfa fields. *Hilgardia* 43(2):53–78.

Nicholson, A. J. (1933) The balance of animal populations. *J. Animal Ecol.* 2:131–78.

Readshaw, J. L. and W. R. Cuff (1982) A model of Nicholson's blowfly cycles applied to predation theory. *J. Animal Ecol.* 49:1005–1010

Rogers, D. J. and S. E. Randolph (1978) Metabolic strategies of male and female tsetse (Diptera: Glossinidae) in the field. *Bull. Entomol. Res.*, 68:639–54.

Royama, T. (1971) A comparative study of models for predation and parasitism. *Res. Pop. Ecol.* Suppl. 1. September 1971.

Schoener, T. (1973) Population growth regulation by intraspecific competition for energy or time: Some simple presentations. *Theor. Pop. Biol.* 4:56–84.

Streifer, W. (1974) Realistic models in population ecology. *Adv. Ecol. Res.* 8: 199–266.

Tamaki, G., J. V. McGuire, and J. E. Turner (1974) Predator power and efficacy: A model to evaluate their impact. *Environ. Entomol.* 3:625–30.

Thornley, J. H. L. (1977) *Mathematical Models in Plant Physiology*. (New York: Academic Press).

Volterra, W. (1926) Variazioni e fluttuazioni del numero d'individui in specie animali conviventi. *Mem. R. Accad. Naz. dei Lincei* 6(2):31–113.

von Foerster, H. (1959) Some remarks on changing populations, in F. Stohlman, Jr. (ed) *The Kinetics of Cellular Proliferation* (New York: Grune and Stratton).

Wang, Y. H. and A. P. Gutierrez (1980) An assessment of the use of stability analyses in population ecology. *J. Animal Ecol.* 49:435–52.

Wang, Y., A. P. Gutierrez, G. Oster, and R. Daxl (1977) A population model for cotton growth and development: Coupling cotton–herbivore interactions *Can. Entomol.* 109:1359–74.

8 Strategies for the Control of Infectious Disease Agents

Roy M. Anderson

8.1. INTRODUCTION

Our current understanding of the population dynamics of infectious disease agents is surprisingly limited when considered in the light of their enormous veterinary and medical significance. Such knowledge is clearly of central importance to the control or eradication of disease agents since, before attempting to disrupt their population stability, we ideally need to understand the mechanisms that normally enable disease agents to persist (Yorke *et al.*, 1979).

Some progress has been made towards developing models of the population dynamics of directly transmitted viral and bacterial parasites of man with a view to designing optimal control policies (*see* Bailey, 1975; Wickwire, 1979). More generally, however, there is an urgent need for a theoretical framework that relates the biology and structure of the disease agent's life cycle to the likely success of various control options.

This chapter examines some simple mathematical models that mimic the infection dynamics of a variety of disease agents, ranging from viruses to helminths, which are transmitted either directly or indirectly between hosts. The specific aim is to establish criteria for disease persistence based on the biological parameters that control transmission success. Once defined, these criteria may be used to assess the relative effectiveness of different control methods.

The chapter is organized as follows. First, two general concepts of disease dynamics are examined, namely, those concerning the basic reproductive rate of infection and the threshold host density required for parasite persistence. Next, models are developed for three types of directly transmitted agents (viral infections, gonorrhea, and hookworm). Criteria are derived for disease maintenance and the impact of various control methods examined. Two indirectly transmitted infections, malaria and schistosomiasis, are then examined and similar criteria derived. The final sections briefly consider the significance of seasonal variation in transmission and discuss the relationship between parasite dynamics and control strategy.

8.2. CONCEPTS OF DISEASE DYNAMICS

Basic Reproductive Rate of an Infectious Disease, R

Over the past few decades one of the most important concepts to have emerged from theoretical studies of the epidemiology of infectious disease agents is that of the *basic reproductive rate* of a disease, R, which is determined by a series of biological processes. These include such factors as the density of the host population N, transmission success between hosts β, the duration of acquired immunity to reinfection $1/r$, the length of the period during which an infected host is infectious $1/\gamma$, and the rate of turnover of the host population (Macdonald, 1957, 1965; Bailey, 1975; Dietz, 1974, 1976; Anderson, in press a, b, 1982).

The number of population parameters that influence R is determined by the structure and form of the disease agent's life cycle. Malarial parasites, for example, are transmitted indirectly between humans by means of an arthropod vector, and hence R is determined by a complex of parameters describing the population characteristics of the parasite within both the human and mosquito host. Life cycle structure also determines the manner in which specific parameters act to influence R. The transmission success of directly transmitted viral agents (e.g., measles and rubella) is related to the density of the host population and the frequency of contact between hosts (Bailey, 1975). In contrast, the transmission efficiency of vector-borne agents is dependent on both the man-biting habits of the vector and the density of vectors relative to human density (Macdonald, 1957; Anderson, in press a).

In the case of *microparasitic* disease agents such as viruses, bacteria, and protozoa, R represents the average number of secondary cases that one infected host gives rise to during its infectious lifespan. Clearly, for the disease to persist within a given population of hosts this quantity must be greater than unity, irrespective of whether the parasite is transmitted directly or indirectly (Dietz, 1974, 1976). The definition of R for *macroparasitic* infections, caused by helminths and arthropods, is somewhat different, representing the average number of offspring produced by a single adult parasite throughout its reproductive lifespan (or female adult parasite in the case of dioecious species) that successfully complete their life cycles to reach sexual maturity. Again, R must be greater than unity for parasite persistence.

This distinction between micro- and macroparasites is somewhat arbitrary but is a consequence of the fact that microparasites multiply within their host to directly increase population size (direct reproduction) while the reproduction of macroparasites in their final host is invariably geared to the intermediate step of producing transmission stages (transmission reproduction) (Anderson and May, 1979a; May and Anderson, 1979; Anderson, 1979a). The number of individual microparasites within an infected host is usually very large and rarely measurable, and thus we conventionally define the host as the basic unit of study for the dynamics of these diseases. In contrast, the number of macroparasites within a host is usually quantifiable (and has direct relevance to the impact of the parasite on host survival and reproduction) and it is thus customary to regard the parasite as the basic unit of study for helminth and arthropod infections. Both units (host and parasite) are sometimes used to describe the dynamics of specific disease agents. The schistosome flukes, for example, which are important pathogens of man, have a phase of direct reproduction in the intermediate snail host. Models of the dynamics of schistosomiasis conventionally consider the mean number of parasites per human host but describe transmission via the molluscan host by reference to the number or proportion of infected snails (Macdonald, 1965, May, 1977a).

Irrespective of these distinctions, the concept of a basic reproductive rate is a unifying one that enables the population dynamics of different disease agents to be compared. A knowledge of the components that determine R is central to our understanding of the factors that enable infectious disease agents to persist within host populations. Eradication is the converse of persistence and represents the ultimate goal of disease control. In the design of optimum control or eradication policies measures of R thus provide a quantitative framework within which the efficiency of various strategies can be assessed.

It is important to note that although R is commonly referred to as a *rate* (see Macdonald, 1965; Dietz, 1976), it is in fact a dimensionless quantity, measuring the reproductive success of an infection (or parasite).

Threshold Host Density for Disease Persistence, N_T

The second major concept of disease dynamics concerns the relationship between host density and the ability of a parasite to persist within its host population. This concept was first outlined by Kermack and McKendrick (1927) in the context of directly transmitted microparasitic agents.

There exists for virtually all parasitic organisms, whether viral protozoan or helminth, a critical host density N_T below which the parasite is unable to persist within the host population (Anderson and May, 1979a; May and Anderson, 1979). This concept is clearly associated with transmission success, which is itself a component of R. For a disease agent to persist, the number of secondary cases arising from one infected host (microparasites) or one mature adult parasite (macroparasites) must be greater than unity. Irrespective of the way in which a parasite is transmitted between hosts, host density clearly influences the likelihood of successful transmission. In the case of indirectly transmitted agents, transmission will be dependent on both the density of final hosts and the density of vectors. If host density is such that $R < 1$ the disease will die out and thus we can define a threshold density N_T for disease maintenance. As in the case of R, many population parameters determine N_T. It is intuitively obvious that diseases with high transmission efficiency will be potentially able to persist in low-density host populations, while those with low transmissability can only survive in high-density populations. Section 8.3 outlines the derivation of expressions for R and N_T for a range of disease agents, including viral, bacterial, protozoan, and helminth parasites.

8.3. DIRECTLY TRANSMITTED DISEASE AGENTS

8.3.1. Viral Diseases

A large and mathematically sophisticated literature has evolved to describe the dynamics of directly transmitted viral diseases such as measles and rubella. This literature has been reviewed by Bailey (1975).

The host is invariably treated as the basic unit in such studies, and its population is conventionally divided into a number of categories containing susceptible, infected, and immune individuals numbering $X(t)$, $Y(t)$, and $Z(t)$ at time t respectively. In such models attention is focused on the rate at which individual hosts flow between compartments. For disease persistence, $Y(t)$ must obviously be greater than zero.

To derive an expression for R, we consider a model in which the host population is of constant size N, where the net rate of input of susceptibles (new births) $(bN + \alpha Y)$ exactly balances the net rate of loss of hosts due to natural bN and disease-induced mortalities αY. It is assumed that infected hosts are immediately infectious (i.e., no disease incubation period) and that the net rate of transmission is βXY, where β is a transmission coefficient. Hosts are assumed to suffer mortalities due to infection at a rate α and to recover from infection at a rate γ where $1/\gamma$ represents the average duration of infection. Recovered hosts are assumed to be immune to reinfection for the rest of their lives. These assumptions lead to the following differential equations:

$$\mathrm{d}X/\mathrm{d}t = bN + \alpha Y - \beta XY - bX \tag{8.1}$$

$$\mathrm{d}Y/\mathrm{d}t = \beta XY - (\alpha + b + \gamma)Y \tag{8.2}$$

$$\mathrm{d}Z/\mathrm{d}t = \gamma Y - bZ. \tag{8.3}$$

The net rate at which susceptibles are infected, βXY, is formed from the total number of contacts they have per unit of time βNX, times the proportion of those contacts which are infective, Y/N. Where the proportion of susceptible, infected, and immune hosts are denoted by x, y, and z, respectively, then the model has two possible equilibrium solutions. If

$$\beta N/(\alpha + b + \gamma) > 1 \tag{8.4}$$

$$x^* = (\alpha + b + \gamma)/\beta N \tag{8.5}$$

$$y^* = \left(1 - \frac{1}{x^*}\right)\left(\frac{b}{\gamma + b}\right) \tag{8.6}$$

$$z^* = (\gamma y^*)/b. \tag{8.7}$$

If eqn. (8.4) is not satisfied the system settles to the state $x^* = 1, y^* = z^* = 0$; in other words, the disease is unable to persist. These two patterns of behavior are illustrated graphically in Figure 8.1.

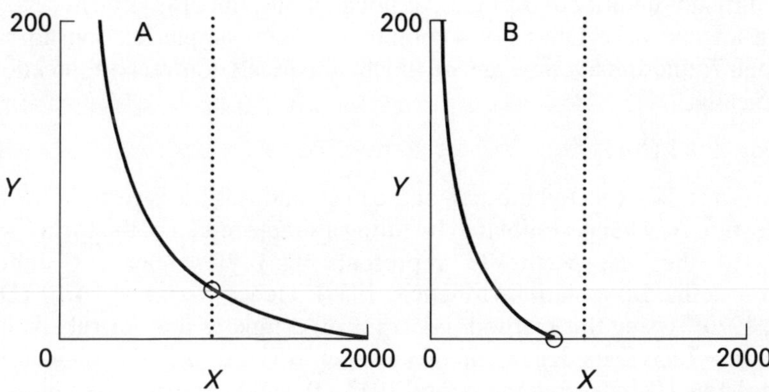

FIGURE 8.1 Phase planes of $bN + \alpha Y - \beta XY - bX = 0$ (full curve), and $\beta XY - (\alpha + b + \gamma)Y = 0$ (dotted line). A, disease persistence where $Y^* > 0$ ($\alpha = 0, \beta = 0.33 \times 10^{-4}, \gamma = 0.033, N = 2000, b = 0$). B, disease extinction where $Y^* = 0$ ($\alpha = 0, \beta = 0.33 \times 10^{-4}, \gamma = 0.333, N = 750, b = 0$). Threshold density for disease maintenance N_T is represented by the vertical dotted lines.

Basic Reproductive Rate, R

The basic rate R of the infection is

$$R = \beta N/(\alpha + b + \gamma), \tag{8.8}$$

which must be greater than unity for disease maintenance. It clearly represents the number of secondary cases produced by one infective throughout its infectious period, if introduced into a population of N susceptibles. Put another way, R represents the potential rate of transmission βN, times the expected lifespan of the infectious host $[1/(\alpha + b + \gamma)]$. As pointed out by Dietz (1976), it is easy to understand why the proportion of susceptibles at equilibrium $x^* = 1/R$, since at the steady state each infected host should produce on average one secondary case, i.e., $Rx^* = 1$, where Rx^* is the actual reproductive rate of the disease in a population where only the proportion x^* is susceptible, and the proportion $1 - x^*$ of contacts is wasted on nonsusceptibles.

If we explicitly take into account a disease incubation period in infected hosts where for a time they are infected but not infectious, R is multiplied by a fraction f representing the proportion of infected hosts that survive to become infectious. In the case of human diseases such as measles and rubella, where this incubation period is short (a matter of days) in relation to the expected lifespan of the host, this proportion is effectively one and can thus be ignored.

For a defined population, R can be estimated from a knowledge of the biological characteristics of a disease. For example, in the case of rubella virus in European populations, $1/\gamma$ is roughly 14 days, while α is effectively zero and the human lifespan $1/b$ is in the region of 70 years. Invariably, β is the most difficult quantity to estimate, although it may not always be necessary to obtain a value since Dietz (1974) points out that a simple relationship exists between R and the average age at which individuals contract an infection A. Specifically,

$$R = 1 + 1/(bA), \tag{8.9}$$

where $1/b$ is the expected lifespan of the host, and A is the inverse of the force of infection Λ, which is estimated by fitting a simple model of the form $\bar{z} = 1 - e^{-\Lambda a}$, to the age-specific ($a$ represents age) prevalence of immunes ($\bar{z}$) in a defined population (Muench, 1959). Hence $R = 1 + (\Lambda/b)$ (Dietz, 1974, 1976). Using this method, Dietz (1976) estimates that for rubella in the FRG, $R \approx 7.65$. Data documented in a paper by Griffiths (1974) on measles in Cirencester, UK covering the period 1947–50, yield $R \approx 13.7$. Recent papers by Anderson and May (1982, 1983) provide extensive reviews of available estimates of A and R for a variety of infections.

Control Strategies

It is intuitively obvious that diseases with high R will be the most difficult to eradicate. For directly transmitted viral agents no effective chemotherapy exists and hence control has centered on vaccination programs. Effective vaccines exist for both measles and rubella, for example (Smith, 1970). The effect of a vaccination program can be examined by assessing its impact on R. If we denote the proportion of susceptibles vaccinated shortly after birth as p, then

$$\hat{R} = R(1 - p). \tag{8.10}$$

As noted by Dietz (1976), this yields the condition

$$p > [1 - (1/R)] \tag{8.11}$$

for disease eradication. *For rubella in the FRG, this requires that 87% of the population be vaccinated shortly after birth, while in the case of measles in Cirencester, 92% vaccination is required.* Both figures are very high and provide an explanation of why both rubella and measles are still endemic in Western Europe. The methods outlined above, however, clearly provide a firm quantitative footing for the design of control measures if the eradication of these diseases is felt to be desirable in the future. It should be noted, however, that these methods are based on the assumption that the force of infection, Λ, is independent of host age. Recent work suggests that age-dependency in contact with infections acts to slightly reduce the critical levelof vaccination coverage required for eradication.

One final point to note concerns the threshold host density for disease persistence N_T. From eqn. (8.4) it can be seen that

$$N_T = (\alpha + b + \gamma)/\beta. \tag{8.12}$$

In other words, the introduction of a few infected individuals into a community will only result in an epidemic provided host density is greater than N_T. The value of this concept to the design of vaccination programs, concerns the identification of communities with densities greater than N_T, and hence potentially able to support endemic disease. Specifically, vaccination efforts should be directed at epidemiologically important population groupings such as schools (Yorke *et al.*, 1979).

8.3.2. Veneral Diseases: Gonorrhea

Gonorrhea is a human disease caused by the bacterium *Neisseria gonorrheae*, which is directly transmitted between hosts by sexual contact. Compared with viral diseases such as measles and rubella, gonorrhea has a number of distinctive epidemiological characteristics (Yorke *et al.*, 1978) that are related to certain biological properties of the infection: the disease only occurs within the sexually active proportion of the population; the duration of infectiousness is long and differs substantially between sexes; and acquired immunity to reinfection is essentially nonexistent.

In recent years a number of models have been developed to describe the dynamics of gonorrhea (Cooke and Yorke, 1973; Bailey, 1975; Reynolds and Chan, 1975; Hethcote, 1976; Lajmanovich and Yorke, 1976). The following model is based on components from a series of these published papers. We consider a freely mixing community of N_1 males and N_2 females with at any time Y_1 and Y_2 infected males and females and X_1 and X_2 susceptibles. It is assumed that in a unit of time, X_2 susceptible females contact S_2X_2 males, of whom the proportion (Y_1/N_1) are infected. S_2 thus represents the average number of sexual partners per female per unit of time. We assume that the probability of infection passing from an infected male to a susceptible female during a single "partner contact" is g_2. The net rate at which new female infections occur is thus $(S_2g_2X_2Y_1)/N_1$. For convenience we define $\beta_1 = (S_2g_2/N_1)$. In a similar manner the net rate at which male infections occur is $(S_1g_1X_1Y_2)/N_2$, and we define $\beta_2 = (S_1g_1/N_2)$. Here S_1 is the average number of sexual partners per male per unit of time, and g_1 is the probability of infection passing from an infected female to a susceptible male during a single "partner contact". We further assume that the average duration of male and female infectiousness is $1/\gamma_1$ and $1/\gamma_2$, respectively. Where the populations of sexually active males and females are constant, these assumptions lead to the following differential equations.

$$dX_1/dt = -\beta_2X_1Y_2 + \gamma_1Y_1 \tag{8.13}$$

$$dY_1/dt = \beta_2X_1Y_2 - \gamma_1Y_1 \tag{8.14}$$

$$dX_2/dt = -\beta_1X_2Y_1 + \gamma_2Y_2 \tag{8.15}$$

$$dY_2/dt = \beta_1X_2Y_1 - \gamma_2Y_2. \tag{8.16}$$

This model has two equilibrium solutions where the disease may either persist or not be maintained within the host population. Positive equilibria for Y_1 and Y_2 only occur when

$$S_1S_2g_1g_2 > \gamma_1\gamma_2. \tag{8.17}$$

Basic Reproductive Rate, R

R is thus given by

$$R = \frac{S_1S_2g_1g_2}{\gamma_1\gamma_2}, \tag{8.18}$$

where R is defined as the product of the rates of effective transmission from male to female and vice versa $[(S_1g_1)\ (S_2g_2)]$, times the product of the expected durations of infectiousness in males and females $[(1/\gamma_1)\ (1/\gamma_2)]$. Note that in contrast to the expression for R derived in eqn (8.8), *for gonorrhea, R is independent of host population size. In other words, there is no critical threshold density for disease maintenance*. Disease persistence is simply dependent on the prevailing degree of sexual promiscuity within the population (i.e., S_1 and S_2).

Control Strategies

Methods of control of gonorrhea are basically of two types, namely: the identification of infections by screening programs and chemotherapeutic treatment; and education aimed to reduce the likelihood of transmission either by reducing promiscuity or by males using condoms during intercourse. In the latter case control aims to reduce S and g. Due to the absence of acquired immunity the development of a vaccine appears unlikely, so that control measures in the USA and Europe center on screening programs and chemotherapy. Unfortunately, however, chemotherapy does not result in any degree of protection to reinfection, so that treated individuals immediately rejoin the pool of susceptible hosts.

We can examine the impact of chemotherapy by modifying our model in the following manner. We assume that our screening program detects a constant proportion q of the infected population per unit of time and that all these individuals are cured by chemotherapy (where, for the sake of simplicity, 50% are males and 50% females). This will lead to additional loss terms in eqns. (8.14) and (8.15) of the form cY_1 and cY_2, respectively, where c is defined as

$$c = -\ln(1 - q). \tag{8.19}$$

The new reproductive rate $\hat{R}$ becomes:

$$\hat{R} = \frac{S_1 S_2 g_1 g_2}{(\gamma_1 + c)(\gamma_2 + c)} . \tag{8.20}$$

For disease eradication we need to treat a sufficient proportion of the infected population such that $\hat{R} < 1$. The estimation of R is fraught with problems because it requires behavioral information on average sexual habits (Yorke *et al.*, 1978). Darrow (1975), on the basis of a large number of interviews in a city community, estimated that the average patient with gonococcal infection has 1.46 partners per 30-day period ($S_1 = S_2 = 1.46/30$ days). Constable (1975) and Reynolds and Chan (1975) suggest that the average duration of infectiousness ($1/\gamma$) is 10 days and 100 days for males and females, respectively. Assuming that each "sexual partner contact" always results in infection ($g_1 = g_2 = 1.0$), this gives, from eqn. (8.18), $R = 2.4$. The use of these parameter estimates in eqn. (8.20) gives the depressing result that 22% of the infected individuals must be treated each week! This estimate is certainly too large since each sexual contact between an infected and susceptible individual will not always result in transmission, but it certainly illustrates the magnitude of the problem. Gonorrhea is endemic in Western societies, and screening programs have had little effect as yet on reducing the prevalence of the disease (although in the US they appear to be preventing increases at present).

The above analysis is obviously crude and does not take into account the statistical distribution of sexual promiscuity within a population. Certain segments of a population are more at risk than others and available evidence from any given community suggests that transmission is essentially maintained by a "core" of individuals who are very sexually active (S large). As suggested by Yorke *et al.* (1978), control measures directed at this core could be very effective. In fact, in the "noncore" segment of the population, R may be less than unity so that the maintenance of endemic disease is entirely dependent on the activities of the core who continually reintroduce the infection into the remainder of the population. The noncore group thus effectively detracts from the overall efficiency of disease transmission. A much more detailed treatment of this problem, and others associated with "nonapparent" infections, is given in Yorke *et al.* (1978). *Finally, it is important to note that any degree of control will act to reduce R and hence the prevalence of infection.*

8.3.3. Helminth Infections: Hookworm

Two hookworm species, *Ancylostoma duodenale* and *Necator americanus*, are important nematode parasites of man. Stoll (1947) estimated in 1947 that 456 million people were infected with hookworm throughout the world; today, this figure is probably closer to 700 million. In many regions of the world the disease is endemic, with the prevalence of infection amongst adults approaching 100%. The effect of the parasite on the host is related to the worm burden, and heavy infections cause severe anemia and damage to the gut wall (the adult worm lives in the gut of man and feeds on blood obtained from the capillaries in the gut mucosa). The parasite has a direct life cycle (Figure 8.2) and mature female worms produce eggs in the human host, which pass into the external habitat in the feces. Larvae hatch from the eggs and eventually develop into an infective stage (L3 larva), which gains entry to the host by consumption of contaminated food, or by direct penetration of the skin (Chandler, 1929; Muller, 1975).

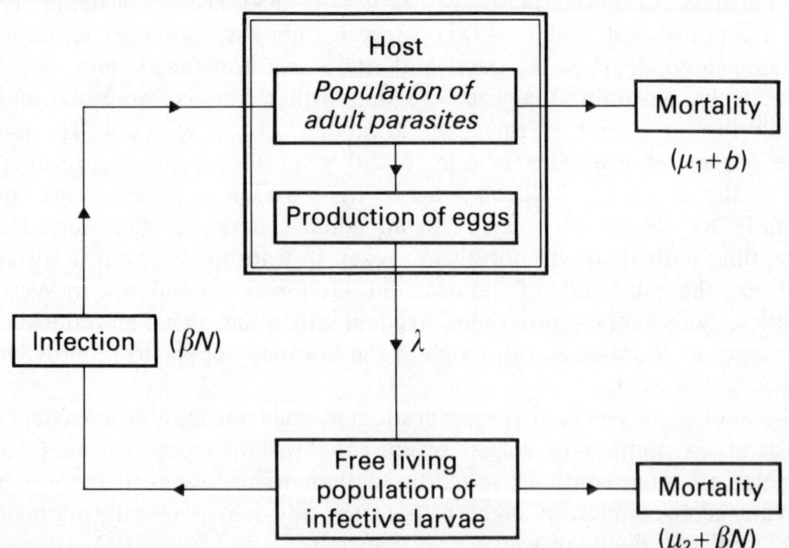

FIGURE 8.2 Populations and parameters involved in the life cycles of human hookworm parasites.

Despite the medical importance of these parasites, little work has been done on the dynamics of transmission, both from a practical and theoretical point of view. The following model attempts to capture the major features of the parasite's dynamics and is based on a framework that utilizes the parasite as a basic unit. (Thus, in contrast with the previous models of microparasites, this model considers the mean number of parasites per host, rather than the proportion of infected individuals.)

We assume that the host population is of constant size N and denote the number of mature adult worms and free-living infective larvae by P and L, respectively. The death rates of the human host, adult worms, and larvae are represented by b, μ_1 and μ_2, respectively, and the net rate of transmission into the total population of adult worms is defined as βLND_1, where D_1 represents the proportion of larvae infecting hosts that survive to maturity. The rate of pick-up of larvae is assumed to be proportional to the density of hosts times the density of larvae, i.e., βLN (β is a transmission coefficient). The net rate of input into the larval population is defined as $\frac{1}{2}\lambda PD_2$, where λ represents the rate of egg production per female worm (the ½ enters as a result of the 1:1 sex ratio of male to female worms), and D_2 represents the proportion of those eggs produced that hatch to yield larvae and survive to the L3 infective state. The death rate of the adult worms is assumed to be density dependent and of the form $\hat{\mu}(i) = \mu_1 + \alpha_i$, where i is the number of parasites in an individual host and α is a constant measuring the severity of density dependence. (See Anderson and May, 1978; May and Anderson, 1978, for a more detailed discussion of this assumption). Density dependence acts on both worm mortality and egg production but, for simplicity, in this model its action is restricted to mortality. The biological evidence for density dependence is discussed in Nawalinski *et al*. 1978a,b; Hoagland and Schad, 1978.)

The net rate of density-dependent mortality is dependent on the statistical distribution of parasite numbers per host, which is assumed to be negative binomial in form, with a dispersion parameter k. The net rate of worm loss is thus

$$\left[\mu_1 + \alpha \frac{(k+1)}{k} \frac{P}{N}\right] P$$

(see Anderson and May, 1978). These assumptions lead to the following differential equations for P and L:

$$\frac{dP}{dt} = \beta LND_1 - (\mu_1 + b)P - \alpha \frac{(k+1)}{k} \frac{P^2}{N} \tag{8.21}$$

$$\frac{dL}{dt} = \tfrac{1}{2}\lambda PD_2 - \mu_2 L - \beta LN. \tag{8.22}$$

This model has two possible equilibrium solutions. If

$$\tfrac{1}{2}\beta\lambda N D_1 D_2 > (\mu_1 + b)(\mu_2 + \beta N), \tag{8.23}$$

the disease persists and positive equilibrium states for P and L occur. If eqn. (8.23) is not satisfied the parasite is unable to maintain itself and $P^* = L^* = 0$.

Basic reproductive rate, R

From eqn. (8.23) it can be seen that R is given by

$$R = \frac{\tfrac{1}{2}\beta\lambda N D_1 D_2}{(\mu_1 + b)(\mu_2 + \beta N)} \tag{8.24}$$

R is formed from the overall rate of transmission through each segment of the parasite's life cycle ($\beta N D_1 D_2$), times the rate of worm reproduction ($\tfrac{1}{2}\lambda$), divided by the product of the net rates of adult worm ($\mu_1 + b$) and infective larval ($\mu_2 + \beta N$) mortality. In contrast to the previous models, R represents the basic rate of parasite reproduction (8.24) rather than the basic rate of infected host or infection reproduction eqns. (8.8) and (8.18).

The threshold density for disease persistence can be easily derived from eqn. (8.24):

$$N_T \simeq \frac{(\mu_1 + b)\mu_2}{\tfrac{1}{2}\beta\lambda D_1 D_2}\,. \tag{8.25}$$

Further insights into the dynamics of the model defined in eqns. (8.21) and (8.22) can be made by noting that the expected lifespan of the infective larval stage (a few weeks) is many orders of magnitude less than that of the adult worm (a few years). The two differential equations can thus be decoupled, by assuming the "short-lived" infective stages are adjusted essentially instantaneously to their equilibrium level ($\mathrm{d}L/\mathrm{d}t = 0$) for any given vale of P (see May and Anderson, 1979). This gives

$$\frac{\mathrm{d}P}{\mathrm{d}t} = (R + 1)(\mu_1 + b)P - \alpha\,\frac{(k + 1)}{k}\,\frac{P^2}{N}\,. \tag{8.26}$$

A further simplification can be achieved by noting that the mean parasite burden per host $M = P/N$ and thus:

$$\frac{\mathrm{d}M}{\mathrm{d}t} = M\left[(R - 1)(\mu_1 + b) - \frac{\alpha(k + 1)M}{k}\right]. \tag{8.27}$$

The equilibrium intensity of infection per host M^* is therefore

$$M^* = [(R - 1)(\mu_1 + b)k]/[\alpha(k + 1)], \tag{8.28}$$

from which it can be seen that R must be greater than unity for disease persistence. It is also apparent that eqn. (8.27) is essentially of logistic form (Verhulst, 1838); thus, where $A = (R - 1)(\mu_1 + b)$ and $B = \alpha(k + 1)/k$, eqn. (8.26) becomes

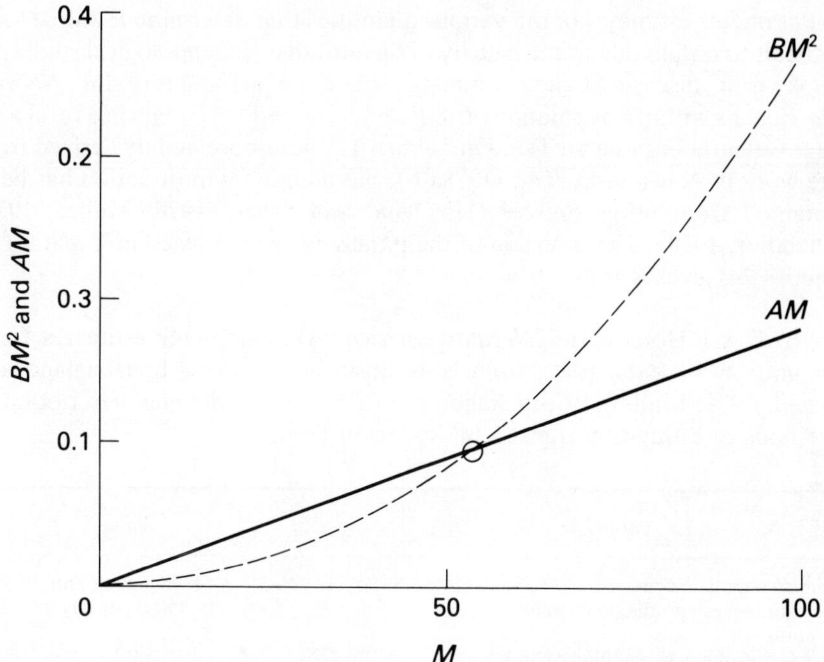

FIGURE 8.3 Phase plane representation of the gain (AM) and loss (BM^2) terms of eqn. (8.29). Full line, gain term; broken line, loss term. The intersection of the two lines defines the globally stable equilibrium mean worm burden M^*. Chemotherapy would have to make the slope of the full line $\leqslant 0$ for disease eradication. Selective chemotherapy aimed at the most heavily infected hosts would increase B and hence the rate at which the broken line rises.

$$\frac{\mathrm{d}M}{\mathrm{d}t} = AM\left(1 - \frac{B}{A}M\right), \tag{8.29}$$

where AM and BM^2 represent the "gain" and "loss" terms. A graphical representation of the dynamics of this equation is presented in Figure 8.3. In a human population of constant size and stable age distribution, under conditions of stable endemic disease, the intensity of infection at age a, $M(a)$ is given by (Anderson and May, 1982):

$$M(a) = M^*\left[\frac{1 - \mathrm{e}^{-\bar{\gamma}}}{1 + (1 - 1/R)\mathrm{e}^{-\gamma}}\right], \tag{8.30}$$

where $\bar{\gamma} = (2R - 1)(b + \mu)a$. Furthermore, the prevalence of infection in each age group (proportion of infected hosts), $p(a)$, is given by

$$p(a) = 1 - \left[1 + \frac{M(a)}{k}\right]^{-k}. \tag{8.31}$$

Parameter estimates of the various quantities that determine R and M^* are difficult to obtain due to the paucity of quantitative epidemiological studies of hookworm disease. Crude estimates of these parameters for *Necator americanus* within a population of 1803 children aged 1–11 years in a rural area near Calcutta in India are listed in Table 8.1. These were mainly derived from the work of Nawalinski *et al.* (1978a,b), but additional information has been obtained from other sources (Hoagland and Schad, 1978; Muller, 1975; Chandler, 1929). The insertion of the parameter values listed in Table 8.1 in eqn. (8.24) gives $R = 2.7$.

TABLE 8.1 Hookworm (*Necator americanus*). Parameter estimates for a community in India (data from Nawalinski *et al.*, 1978a,b; Hoagland and Schad, 1978; Muller, 1975; Chandler, 1923). A more detailed description of methods of estimation is given in Anderson (1980).

Parameter	*Symbol*	*Value*
Human population density	N	1895 per sq. mile
Daily rate of egg production/female worm	λ	15000 per day
Proportion of larvae that infect a host which survive to reach maturity	D_1	0.1
Proportion of eggs that hatch to produce larvae that survive to the L3 stage	D_2	0.4(?)
Death rate of mature adult worms (expected lifespan)	μ_1 $(1/\mu_1)$	0.78×10^{-3} per day (3.5 years)
Death rate of human host (expected lifespan)	b $(1/b)$	0.55×10^{-4} per day (50 years)
Death rate of L3 larvae (expected lifespan)	μ_2 $(1/\mu_2)$	0.1955 per day (5 days)
Transmission coefficient (per larvae per host per day per 2 sq. miles)	β	2.9×10^{-10}
Density-dependent mortality parameter	α	7.0×10^{-6}
Inverse measure of the degree of aggregation of adult worms in the host population	k	0.34

Control Strategies

Control of hookworm disease is at present based on two approaches, often used concomitantly in endemic areas. A range of chemotherapeutic agents are available (see Muller, 1975), but a single course of chemotherapy aimed at eradication is unlikely to be successful in endemic areas since, once treated, patients are immediately susceptible to reinfection. The aim of such treatment is to reduce the average worm burden below the level causing clinical symptoms. This level varies with endemic region as a result of differences in the level of iron intake and general nutritional status. A second approach combines good sanitation, the disposal or chemical treatment of feces containing hookworm eggs and larvae, combined with education, for example encouraging the wearing of shoes helps to diminish the rate of contact with infective larvae.

Within our model framework chemotherapy acts to increase the death rate of adult worms μ_1. For example, if we treat a proportion g of the human population per day and this treatment kills a proportion h of their individual worm burdens. Then, assuming that the proportion g harbor the average worm burden of the population M, the extra death rate applied to the worm population by chemotherapy, c (defined as per day/worm) is given by:

$$c = -\ln(1 - gh) \tag{8.32}$$

Equation (8.27) thus becomes

$$\frac{\mathrm{d}M}{\mathrm{d}t} = M\left[(R-1)(\mu_1 + b) - c - \frac{\alpha(k+1)}{k}M\right], \tag{8.33}$$

and the new basic reproductive rate $\hat{R}$ is

$$\hat{R} = \frac{\tfrac{1}{2}\beta\lambda N D_1 D_2}{(\mu_1 + b + c)(\mu_2 + N)}. \tag{8.34}$$

For hookworm eradication ($\hat{R} < 1$) the death rate c required by chemotherapeutic treatment is

$$c > (R-1)(\mu_1 + b). \tag{8.35}$$

For the population of Indian children near Calcutta (where $R = 2.7$) this gives $c > 0.00142$ (per day). Thus if our chemotherapeutic agent is 80% effective over a period of seven days ($h = 0.8$/week) then 1.2% of the population must be treated per week. Unfortunately, however, since the equilibrium state $M^* > 0$ is globally stable, this rate of treatment must continue unabated for a number of years to reach the state $M^* = 0$ particularly as the lifespan of adult worms is long (3.5+ years). For this reason disease eradication by chemotherapy alone is likely to be extremely difficult. However, it is important to note that chemotherapy acts to reduce the average burden of worms per host and is thus beneficial irrespective of the rate of treatment, if applied continuously over long periods of time within a given population. For example, eqn. (8.29) indicates that under long-term chemotherapy the equilibrium mean worm burden M^* becomes

$$M^* = (A - c)/B, \tag{8.36}$$

where the additional worm death rate c reduces the average worm load. As indicated above, the reduction of M^* below a certain level in a defined region can alleviate clinical symptoms (note, however, that the statistical distribution of worms per host is contagious and hence a few individuals will always have very high worm burdens).

The observed overdispersed distribution of worm numbers per host (see Anderson, 1980c) suggests that an optimal strategy for lowering the average worm burden is simply to treat those individuals with high fecal egg counts. This form of action will operate to increase the severity of density-dependent losses (i.e., will increase B in eqns. (8.29) and (8.36)). Note, however, that although such action will certainly be very effective in reducing M^*, it will not necessarily result in disease eradication (i.e., *see* eqn. (8.36)). Finally, it is clear from the preceding discussion and analysis that since the equilibrium state M^* is globally stable (Figure 8.3), chemotherapy will be of little use either in reducing M^* or in attempting to achieve eradication unless it is applied continuously. *Short programs of chemotherapy will produce short-term benefits, but once they cease, the equilibrium worm burden will return to its original pre-treatment level.*

Both sanitation and the wearing of shoes to protect individuals from larval penetration through the feet act to reduce β in eqn. (8.24). For example, in the Indian community ($R = 2.7$) a 30% reduction in β would lead to disease eradication. *Clearly, therefore, sanitation and education could ultimately achieve eradication, particularly if used in conjunction with chemotherapy.*

Vaccination against human hookworm is a potential method of control for the future since effective vaccines have recently been developed for dog hookworms (*Anculostoma caninum*) based on γ-irradiated L3 larvae (Miller, 1978; Clegg and Smith, 1978). Although effective, these canine vaccines have a short shelf-life and thus at present they have not been accepted by pharmaceutical companies as a viable economic proposition (Miller, 1978). A vaccine could in theory be produced for human hookworm infections, but development costs and safety regulations concerning the use of prophylactic agents make this unlikely in the near future.

Within our model framework the likely effectiveness of a vaccine may be considered as follows. If we vaccinate a proportion g of the susceptibles within the population per day, and our vaccine gives effective protection for $1/V$ years then $\hat{R}$ is:

$$\hat{R} = R[1 - (g/V)], \tag{8.37}$$

where g/V is the proportion of the population that must be protected at any one time. *Clearly, if the duration of protection provided by the vaccine is short, the vaccination rate must be high.* The larger R of the disease in a defined region, the greater g/V must be to ensure disease eradication. Figure 8.4 illustrates this point and shows that *in the Indian community, 63% of the population must be protected if hookworm is to be eliminated.* A more detailed discussion of the hookworm model, including the complications arising from seasonal factors and the probability of worm pairing, is given in Anderson (1980c).

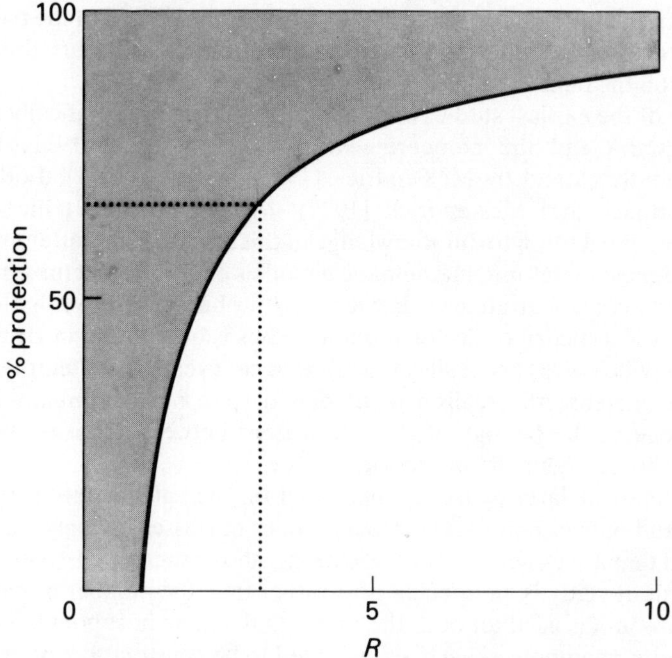

FIGURE 8.4 Relationship between the percentage of the population that must be protected (vaccinated) at any one time for disease eradication, and R. In the case of hookworm this is hypothetical, since there is no effective human vaccine. The shaded region denotes disease eradication, the unshaded region infection maintenance. The dotted line indicates R estimates from the epidemiological survey of children in Calcutta.

8.4. INDIRECTLY TRANSMITTED DISEASE AGENTS

Many tropical disease agents of man have complex life cycles involving two or more hosts and the dynamic properties of such diseases are considerably more complex than those discussed so far. The following treatment of two such indirectly transmitted agents, malaria and schistosomiasis, is necessarily brief and centers on R.

8.4.1. Malaria

In many regions of the world, particularly in India and Africa, malaria is one of the most widespread human diseases. Roughly 1472 million people live in malarious areas of the world, resulting in more than 200 million cases of the disease annually, of which several million are fatal (Garnham, 1966; Bruce-Chwatt, 1968; Cahill, 1972). The most pathogenic of the protozoan parasites that cause malaria in man is *Plasmodium falciparum*, and the following treatment is restricted to this species. The vectors of the parasites are mosquitoes of the genus *Anopheles* and transmission occurs during blood feeding on the human host.

Some of the earliest studies of disease dynamics were specifically associated with malaria and the pioneering work of Ross (1904, 1911, 1916) was subsequently extend by McKendrick (1912), Martini (1921), Lotka (1923), and Kermack and McKendrick (1927). Macdonald (1957) made an outstanding contribution to our knowledge of this disease by his attempts to insert parameter estimates into mathematical models and to predict the effectiveness of various control strategies. Reviews of the historical development of the mathematical theory of vector-borne diseases can be found in Bruce-Chwatt and Glanville (1973) and Bailey (1975). In recent years few attempts have been made to increase the realism of models despite our continually improving knowledge of the biology of the interactions between parasite, vector, and primary host (Anderson, in press a).

Models of malaria dynamics are based on the categorization of both the vector and human population into a number of classes, namely, susceptibles, infected that are either latent or infectious, and immunes. In other words, the basic unit of study is an infected host rather than the number of parasites per host. This unit is used for both the vector and human host populations, where the absolute numbers of each are assumed to be constant, denoted by N_2 and N_1, respectively. A diagrammatic representation of the flow of vectors and humans between various compartments of their respective populations is shown in Figure 8.5.

The following definition of R for malaria derives from a model described in Anderson (in press a), which is itself based on the work of Ross (1911), Macdonald (1957) and Dietz (1974), but includes refinements.

Basic Reproductive Rate, R

R is of the form

$$R = \frac{\beta^2(N_2/N_1)D_1D_2}{(b_1 + \alpha_1 + \gamma)(b_2 + \alpha_2)}, \tag{8.38}$$

and, as usual, must be greater than unity for disease persistence. If $R < 1$, the prevalence of infection in man and mosquitoes declines to zero. In eqn. (8.38) β represents the man-biting rate of the mosquito vector (per vector/unit of time), D_1 is the proportion of infected humans that survive to become infectious, and D_2 is the proportion of mosquitoes that survive to become infectious. Note that β^2 appears in the numerator of eqn. (8.37), since the mosquito is responsible for transmission from man to vector and vector to man. The denominator is formed from the total rate of mortality of infectious humans $(b_1 + \alpha_1 + \gamma)$, consisting of natural mortality b_1, disease-induced

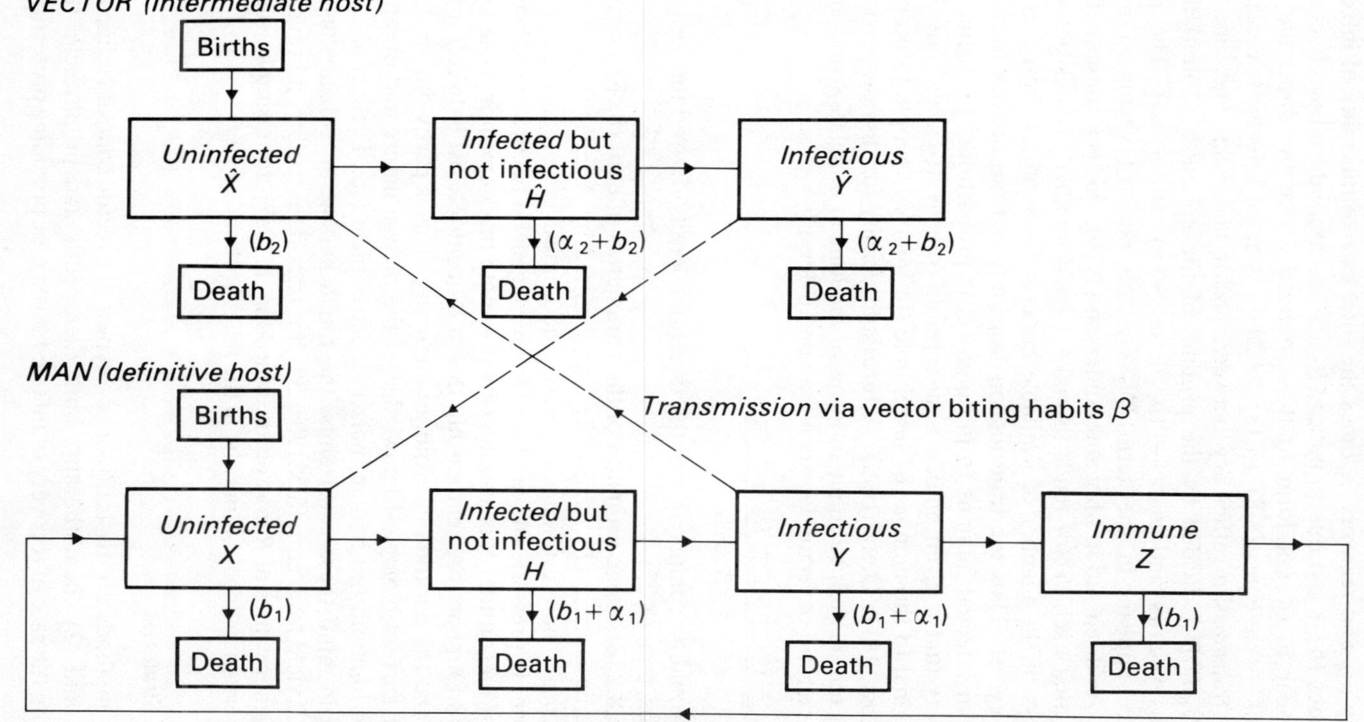

FIGURE 8.5 Flow of humans and vectors through the various compartments of the populations defined in mathematical models of malara dynamics.

mortality α_1, and recovery γ, times the total rate of mortality of infectious mosquitoes $(b_2 + \alpha_2)$ due to natural b_2 and disease-induced α_2 effects. R is thus similar in overall form to those derived above for direct life cycle infections — *see* eqns. (8.8), (8.18), (8.24) — since it effectively consists of the net transmission efficiency between vector and man, and vice versa $(\beta_2(N_2/N_1)D_1D_2)$ divided by the product of the net rates of mortality of infectious vectors and man — $(b_1 + \alpha_1 + \gamma)$, $(b_2 + \alpha_2)$. The major difference appears in the term (N_2/N_1), the ratio of vector to human density. As discussed in May and Anderson (1979), this is a consequence of the fact that each vector tends to make a fixed number of bites per week, regardless of the number of available humans (mosquitoes show a regular feeding cycle). Thus the transmission from infected mosquitoes to humans (and from infected humans to mosquitoes) is proportional to β, times the probability that a given human is susceptible (or infected), and not simply proportional to the number of susceptible (or infected) people (Macdonald, 1957; Ross, 1911; Dietz, 1974). A threshold density for disease persistence can thus not be defined solely in terms of the human population and instead becomes a ratio of vector density N_2 to human density N_1:

$$\frac{N_2}{N_1} = \frac{(b_1 + \alpha_1 + \gamma)(b_2 + \alpha_2)}{\beta^2 D_1 D_2}, \tag{8.39}$$

which should be compared with the threshold density derived for hookworm infections, eqn. (8.25).

Table 8.2 lists some estimates of the parameters of eqn. (8.38) for regions of endemic malaria (*P. falciparum*), transmitted by *Anopheles gambiae*, in East Africa. These estimates are very crude, however (sources are given in Anderson, in press a), and are listed to give a general picture of comparative orders of magnitude. It is also important to note that current malaria models (Figure 8.5) ignore the relationship that is thought to exist between superinfection and the duration of acquired immunity. This is partly due to the fact that our current biological knowledge of the precise nature and duration of acquired immunity to human malaria is incomplete (see Dietz *et al.*, 1974). Bearing in mind these constraints, the rough parameter estimates listed in Table 8.2 yield from eqn. (8.38) $R \approx 39$. It is important to note, however, that much higher values of R may occur. Macdonald (1957), for example, suggests that in certain instances R may lie in the neighborhood of 1000.

Control Strategies

Control methods are basically of two types: (1) vector control (reduction in N_2/N_1), and (2) chemotherapy, both therapeutic and prophylactic (either increasing the rate of recovery of infected hosts r, or providing protection to a

TABLE 8.2 Malaria (*Plasmodium falciparum*). Data sources are given in Anderson (in press a).

Parameter	*Symbol*	*Value*
Vector density/human density (vector is *Anopheles gambiae*)	N_2/N_1	10.0 (highly variable)
Man biting rate per mosquito per day	β	0.25
Proportion of infected humans that survive to become infectious (incubation period 10 days)	D_1	0.90 (?)
Proportion of infected mosquitoes that survive to become infectious (incubation period 22 days at 20 °C)	D_2	0.11
Natural human mortality rate (expected lifespan $1/b_1$ = 50 years)	b_1	0.55×10^{-4} per day
Disease-induced human mortality rate (30% case mortality rate)	α_1	0.0048 per day
Rate of human recovery from infection (average duration of infection 90 days)	γ	0.0111 per day (highly variable)
Natural mosquito mortality rate (expected lifespan $1/b_2$ = 10 days)	b_2	0.10 per day
Disease-induced mosquito mortality rate	α_2	0.00 per day (?)

proportion of the population). Other methods such as the use of mosquito nets or mosquito repellants reduce β.

A reduction in N_2/N_1 also reduces R, and hence decreases the prevalence and intensity of infection. For disease eradication where $R = 39$, this ratio must be less than 0.26 mosquitoes per human (Figure 8.6A). Prophylactic chemotherapy in essence operates in the same manner as a vaccine, acting to protect a proportion of the population against infection. For example, if a proportion p are protected at any one time, under equilibrium conditions the basic reproductive rate becomes:

$$\hat{R} = R(1-p), \tag{8.40}$$

which is the same expression as that derived for the control of directly transmitted viral agents by means of vaccination, eqn. (8.10). High R values imply that a large proportion of the population must be protected if infection is to be eradicated (for R = 39, where control measures aim for $\hat{R} > 1$, eqn. (8.40) gives p = 0.974) (Figure 8.6B). *In endemic areas of malaria, where R values are high, infection eradication is clearly difficult to achieve with either prophylactic chemotherapy or vector control. An integrated approach to control would appear to be optimum where prophylactic and therapeutic chemotherapy are used in conjunction with programs aimed at reducing vector density.*

8.4.2. Schistosomiasis

The term schistosomiasis is used to describe the diseases caused by four species

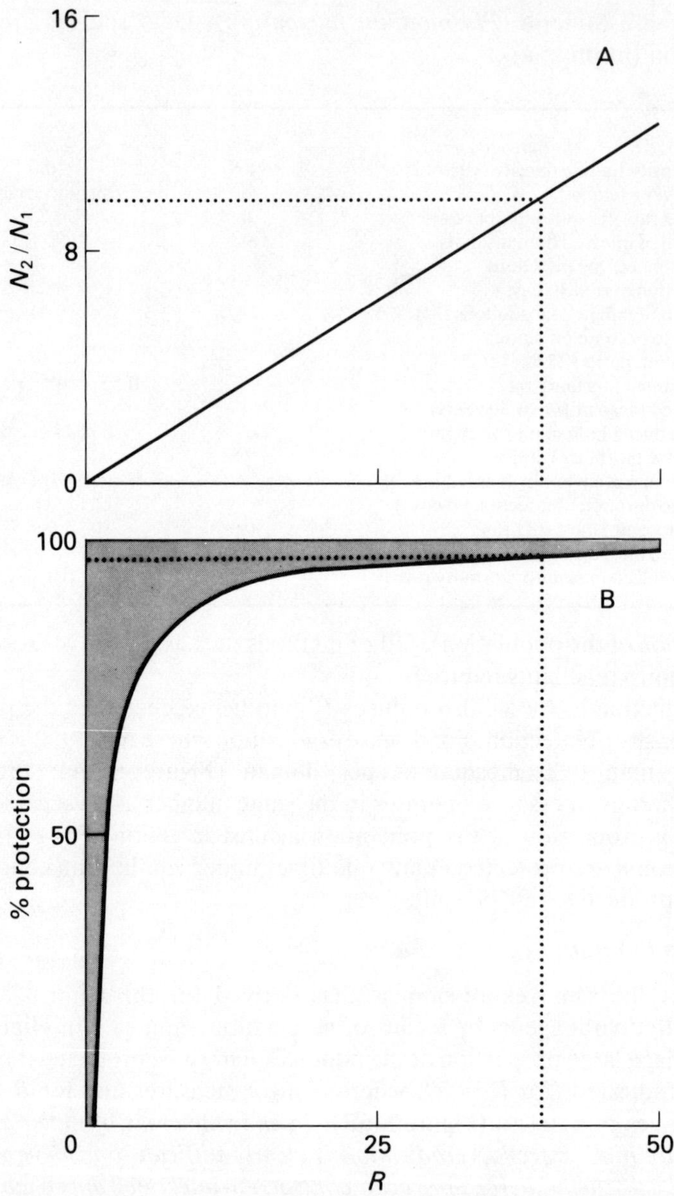

FIGURE 8.6 A, relationship of mosquito density per human host N_2/N_1 with R. For disease eradication the ratio N_2/N_1 must produce $R < 1$. The dotted line indicates the crude estimate of R for *P. falciparum* in East Africa. B, relationship between the population that must be protected by prophylactic chemotherapy for disease eradication (%) and R. Shaded and unshaded areas are defined in Figure 8.4; the dotted line represents the estimate of R.

of trematodes: *Schistosoma mansoni*, *S. intercalatum*, *S. japonicum* and *S. mansoni*, which are important human parasites in many regions of the world. In 1973 it was estimated that at least 150 million people were infected (Ansari, 1973) and today this figure is probably an underestimate (Muller, 1975). As in the case of hookworm the pathogenesis of infection is related to the worm burden carried by an individual and light infections may not give rise to clinical symptoms. Heavy infections, however, are extremely debilitating and are thought to reduce life expectancy (Wright, 1972; Cohen, 1976).

The trematode parasites have indirect life cycles utilizing a molluscan intermediate host (Figure 8.7). Transmission between man and snail, and vice versa, is achieved by means of two free-living aquatic larvae, the miracidium and cercaria, respectively, Both larvae gain entry to their respective hosts by penetration. The parasites have a phase of sexual transmission reproduction in the human host, the eggs passing to the exterior either in the urine (*S. haematobium*) or in the feces (*S. mansoni* and *S. japonicum*). On contact with water the miracidia hatch from these eggs, direct asexual reproduction occurs in the molluscan host, and cercariae are eventually released from infected snails.

Macdonald (1965) was the first to use mathematical models to study the transmission dynamics of schistosomiasis, and recent work has built on his efforts (Nasell and Hirsch, 1973; Lewis, 1975; Cohen, 1976; May, 1977a; Bradley and May, 1978; Barbour, 1978; Goddard, 1978). All these models are based on a framework that describes the dynamics of the mean worm burden per human host M, and the proportion of infected snails y, where both snail and human densities are assumed to be constant (N_1 and N_2, respectively). The basic unit for the study of infection dynamics therefore differs within the two hosts (i.e., the parasite is the basic unit for the human phase of the life cycle and the host is the basic unit for the molluscan segment of the cycle).

Basic Reproductive Rate, R

Without going into detail concerning its derivation (see May, 1977a; Anderson, 1980b), R is of the form:

$$R = \frac{\tfrac{1}{2}\beta_1\beta_2\lambda_1\lambda_2 D_1 D_2 N_1 N_2}{[(b_1 + \alpha_1\ \mu_1)(\mu_2 + \beta_2 N_2)(b_2 + \alpha_2 + \mu_3)(\mu_4 + \beta_1 N_1)]}. \tag{8.41}$$

The numerator represents overall transmission and reproductive success, where β_1 is the coefficient of transmission from cercariae to man, β_2 is the coefficient of transmission from miracidia to snail, $\tfrac{1}{2}\lambda_1$ is the rate of egg production per mature adult fluke (the ½ appears as a consequence of the assumption of a 1:1 sex ratio), λ_2 is the rate of cercarial production per cercarial-shedding snail, D_1 is the proportion of cercariae that penetrate human hosts and survive to maturity, and D_2 is the proportion of infected

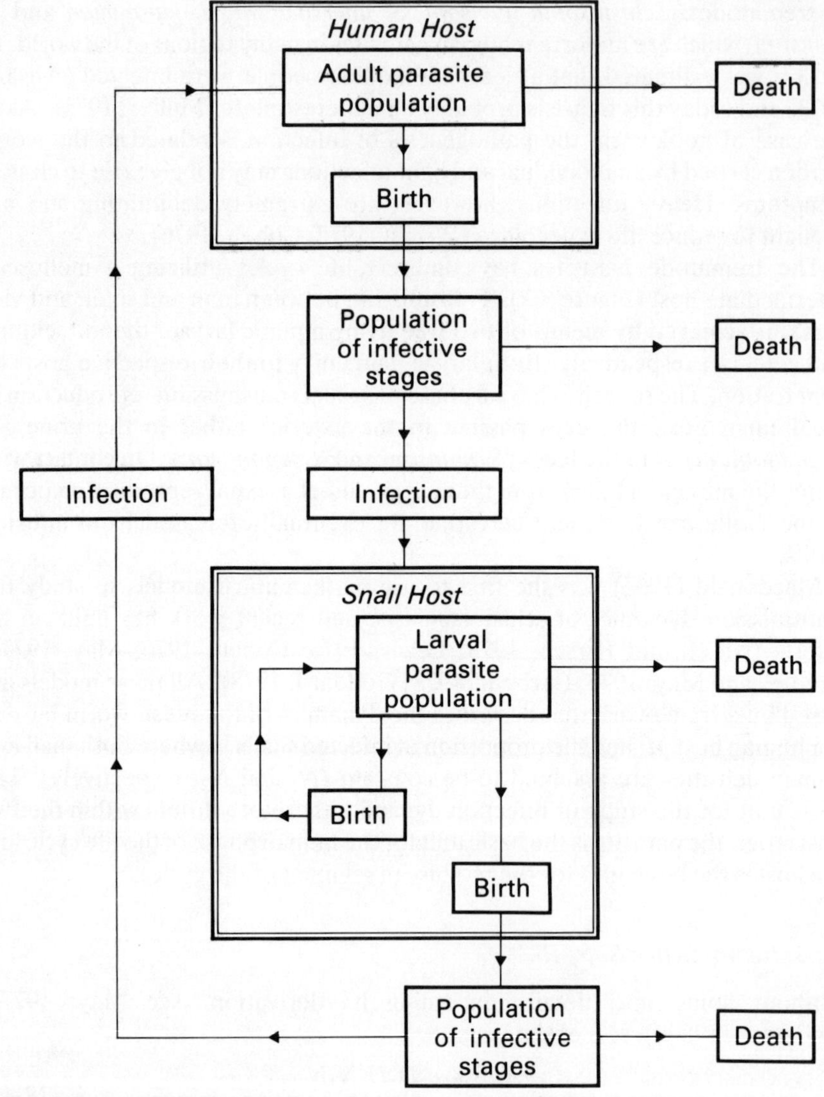

FIGURE 8.7 Populations and rate processes involved in the life cycle of *Schistosoma mansoni*.

snails that survive to release cercariae (see Anderson, in press b). The denominator represents the products of the total death rates in each segment of the life cycle. Loss of mature worms is due to natural host mortality (b_1); disease-induced host mortality (α_1); and natural parasite mortality (μ_1). Mortality of miracidia is due to natural mortality (μ_2) and losses due to infection ($\beta_2 N_2$), while losses of infected snails are due to natural (b_2) and parasite-induced (α_2) mortalities, plus recovery of snails from infection (μ_3). Cercarial mortalities are, as in the case of the miracidia, due to natural (μ_4) and infection losses ($\beta_1 N_1$). The derivation of eqn. (8.41) is based on a simplified model that ignores the effects of worm mating at low parasite densities, the statistical distribution of worm numbers per host (see May, 1977a), the effects of acquired immunity to reinfection (see Jordan, 1972; Warren, 1973), and the density-dependent effects of immunological responses by the human host on worm survival and egg production (Smithers and Terry, 1969; Smithers, 1972). The inclusion of these factors in models of schistosomiasis will clearly influence equilibrium and stability properties, but will not generally affect the basic form of R (eqn. 8.41).

Two important general points emerge from the parameter structure of eqn. (8.41). First, note the basic similarities and differences of this expression with those for malaria (8.38), hookworm (8.24), gonorrhea (8.18), and directly transmitted viral infections (8.8). Second, the threshold host density for parasite persistence N_T is now formed from the product of the densities of man and snail $N_1 N_2$, where

$$N_1 N_2 > \frac{[(b_1 + \alpha_1 + \mu_1)(\mu_2 + \beta_2 N_2)(b_2 + \alpha_2 + \mu_3)(\mu_4 + \beta_1 N_1)]}{\frac{1}{2}\beta_1 \beta_2 \lambda_1 \lambda_2 D_1 D_2} \tag{8.42}$$

Control Strategies

The methods currently employed for the control of schistosomiasis are basically of four types: (1) reduction in snail density N_2 by the use of molluscides or alteration of snail habitats; (2) chemotherapeutic treatment of the human host (which acts to increase the death rate μ_1 of adult worms); (3) education to reduce the contact of the human population with aquatic habitats containing infected snails (reduction in β_1); and (4) improvement of sanitation to reduce the release of fecal material, containing eggs, into habitats harboring the snail intermediate host (reduction in β_2). For each of these methods, either used separately or in conjunction in an integrated approach to control, eqn. (8.41) clearly provides a quantitative framework within which an assessment can be made of the probable effectiveness of various strategies.

Unfortunately, our quantitative knowledge of the many parameters that influence R is extremely limited. Table 8.3, for example, provides a rough guide to the relative orders of magnitude of some of these parameters for *S. mansoni* transmitted by the snail *Biomphalaria glabrata* on the island of St Lucia (West Indies). We have at present little information on β_1 and β_2, although attempts have been made to obtain estimates from age prevalence and age intensity curves of infection in human and snail populations (see Cohen, 1976). Such estimates, although an important step in the right

TABLE 8.3 Schistosomiasis (*Schistosoma mansoni*). Data sources are given in May and Anderson (1979), Anderson and May (1979b), and Anderson (in press b).

Parameter	*Symbol*	*Value*
Natural human mortality rate (expected lifespan $1/b_1 = 50$ years)	b_1	0.55×10^{-4} per day
Natural parasite mortality rate in man (expected lifespan $1/\mu_1 = 5$ years)	μ_1	0.55×10^{-3} per day
Rate of egg production by mature female worms	λ_1	50–60 per day
Rate of natural miracidial mortality (expected lifespan $1/\mu_2 = 0.26$ days)	μ_2	3.8 per day
Natural rate of snail mortality (expected lifespan $1/b_2 = 26$ days)	b_2	0.04 per day
Rate of parasite-induced snail mortality (expected lifespan of infected snails $1/(b_2 + \alpha_2) = 19$ days)	α_2	0.011 per day
Rate of snail recovery from infection	μ_3	0.0 per day (?)
Rate of cercarial production per infected snail	λ_2	1000–3000 per day
Proportion of infected snails that survive to release cercariae (at 25 °C)	D_2	0.26
Natural rate of cercarial mortality (expected lifespan $1/\mu_4 = 0.54$ days)	μ_4	1.84 per day

direction, are at present extremely crude and are obtained by the use of models that are very poor reflections of biological reality.

Control programs aiming to reduce snail density, improve sanitation, and apply chemotherapy have had a degree of success in reducing the prevalence and intensity of human infection (Rosenfield, 1978). But if precise predictions are to be made concerning the level of control required for eradication in a defined region, we clearly require a much better database for the many parameters that determine R. Encouragingly, past theoretical work suggests that the equilibrium mean worm load in endemic areas is not globally stable (due to the nonlinearities introduced by the probability of worm pairing and the distribution of worm numbers per host) and that lowering the intensity of infection below a threshold level may move the dynamics into an area where the stable state is parasite extinction (see May, 1977a). If this theoretical prediction is correct, then control measures need not necessarily aim to reduce R to less than unity but simply to reduce the intensity of infection below this threshold state. The threshold level, however, is likely to be very low as a consequence of the aggregated distribution of worm numbers per host. As demonstrated in the hookworm model, eqns. (8.33) and (8.36), R is one determinant of the equilibrium intensity and prevalence of infection, the precise interelationship depending on model structure.

8.5. SEASONAL CHANGES IN *R*

Many of the parameters that determine *R* for a given disease agent will clearly be time dependent due to seasonal changes in climatic and host behavioral factors. Time dependence in any of the population parameters, may, in principle, produce seasonal cyclic variations in the levels of intensity and prevalence of infection. Such patterns are observed in the epidemiology of *all* the disease agents discussed in this chapter. In the case of directly transmitted microparasitic agents, β_1 often exhibits annual periodicity due to both climatic effects influencing survival and dispersal of infective stages, and to seasonal changes in social behavior (children returning to school after a long vacation). Annual periodicity in transmission rates can produce complicated nonseasonal cycles in the prevalence of viral infections (Yorke *et al.*, 1979; Dietz, 1974). A mechanism of this kind is thought to be responsible for the regular biennial cycle, alternating between years of high and low incidence, for measles in New York City between 1948 and 1964 (Yorke and London, 1973; London and Yorke, 1973). In contrast, gonorrhea tends to exhibit simple annual cycles in prevalence although these differ in periodicity in the male and female segments of the population. This is thought to be due to seasonal changes in sexual activity acting in conjunction with the differing durations of infectiousness in males and females (Yorke *et al.*, 1978).

Seasonal factors influence the dynamics of hookworm infections principally via their impact on the expected lifespan of the L3 infective larva. In many regions of endemic disease, transmission may cease during dry seasons since the larvae die extremely rapidly in dry soil habitats (Nawalinski et al., 1978a,b; Muller, 1975). Low temperatures also influence transmission since the rate of larval development to the L3 stage is maximal at 25–35 °C. The intensity and prevalence of hookworm infections, and hence *R*, are highest in tropical regions with more than 100 cm of rain per year (Muller, 1975).

Seasonal changes in the prevalence of malaria are principally due to changes in vector density, correlated with the prevailing patterns of annual rainfall (Garrett-Jones and Shidrawi, 1969), and the impact of temperature on the incubation period of the parasite in the mosquito host. Similar factors operate in the population dynamics of schistosome parasites, where temperature and rainfall affect snail density, infective stage survival, and the latent period of infection in the mollusc (Anderson and May, 1979b). In summary, therefore, seasonal factors influence *R* by their action on host behavior, infective stage survival, vector density, and the rate of parasite development within the intermediate (or vector) host. *For a given disease, R will thus alter on a temporal and spatial basis in response to prevailing climatic and social conditions.*

With respect to disease control, the significance of seasonal changes in *R* may be summarized as follows. First, changes in *R* will be reflected by changes in the prevalence and intensity of infection. Secondly, *R* may fall below unity during certain periods of the year, but infection may persist on a year to year basis provided the period during which $R < 1$ is shorter than the maximum lifespan of the longest stage in the parasite's life cycle (i.e., the duration of the host infection in the case of directly transmitted microparasites, or the lifespan of the adult parasites in the case of hookworm or schistosomiasis) (Figure 8.8). Third, estimates of parameters that determine R must be made at a series of intervals in order to quantify the impact of seasonal factors on disease dynamics. Finally, the application of control measures should be intensified during periods when R is at a minimum.

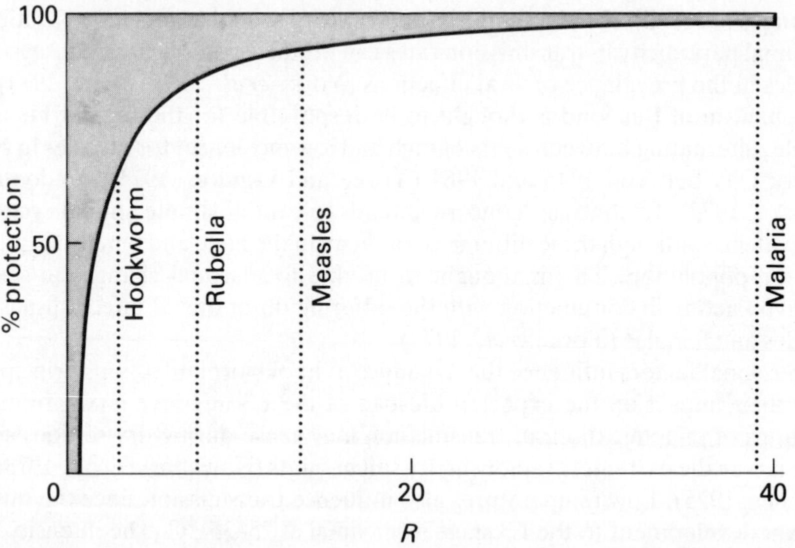

FIGURE 8.8 Relationship between the percentage protection of a population against disease required for infection eradication and R. The dotted lines represent estimates of R for hookworm, rubella, measles, and malaria.

8.6. CONCLUSIONS

Theoretical studies of the dynamics of infectious disease agents highlight two general points that are of central importance to the design of control programs. First, an understanding of the stability properties of equilibrium levels of parasite intensity and prevalence in endemic areas provides basic information on the robustness of the system to various forms of perturbation induced by human intervention. For example, if the equilibrium intensity of infection is globally stable, as indicated for hookworm disease, control measures must be applied continuously to sustain reductions in disease prevalence and intensity. Cessation of control measures will result in the system returning to its original (precontrol) endemic level. In contrast, if the system is locally stable and multiple stable states exist, each with its own region of attraction, the lowering of disease intensity below some critical level may result in the system settling to a new equilibrium state, ideally that of parasite extinction (May, 1977b; Anderson, 1979b; May and Anderson, 1979). In such cases short sharp burst of control measures may produce the desired outcome of infection eradication.

The second point concerns the significance of R on our understanding of disease dynamics and persistence. For the five disease agents discussed in this chapter the expressions that determine R are of similar form, although interesting differences exist (Table 8.4). Enmeshed in the framework of these expressions is also the concept of a critical host density, below which a parasite is unable to maintain itself within a host population.

For a given disease agent, R will vary according to locality and season due to the prevailing physical conditions and habits of the human population. In a defined habitat, the range over which R varies provides a rough guide to the likely success of infection control. For example, if $R \leqslant 1$ at certain times of the year, disease eradication would appear to be a realistic aim. However, it is important to note that the methods available to public health workers and the socioeconomic conditions of the disease environment are central to the success of control programs. For example, it may be extremely difficult to reduce what already appear to be low R values for hookworm in endemic regions, since individuals who have received chemotherapy are immediately susceptible to reinfection. Hookworm transmission may be high because of the use of human feces as soil fertilizers. If a vaccine is available, however, eradication, or the reduction of prevalence to a low level, may be feasible for disease agents with high R values, such as measles and rubella in Europe. The proportion of the population that must be protected at any one time for disease eradication is directly related to R (see Figure 8.8).

We have seen how precise description of the parameters that determine R, helps to focus attention on the information required to assess the impact of various control strategies. *Field epidemiological studies should be designed not only to measure the prevalence and intensity of infection in a given community, but also to provide quantitative measures of these parameters*. At present, as is abundantly clear, our knowledge of these parameters is extremely limited, except in the case of a few directly transmitted viral agents. It is hoped that theoretical studies delineating the significance of such parameters to the overall dynamics and control of infection will stimulate the acquisition of this essential quantitative information.

ACKNOWLEDGMENTS

Thanks are due to the UNDP/World Bank/WHO Special Programme for Research and Training in Tropical Diseases, and the UK Department of the Environment for financial support for this work.

TABLE 8.4 Basic reproductive rate R for major infectious disease types.

Disease	*No. of hosts in life cycle*	*Basic reproductive rate, R*
Measles virus Rubella virus	1	$\frac{\beta N}{(b + \alpha + \gamma)}$
Gonorrhea (*Neisseria gonorrhaeae*)	1	$\frac{g_1 g_2 S_1 S_2}{\gamma_1 \gamma_2}$
Hookworm (*Necator americanus*)	1	$\frac{\frac{1}{2}\lambda\beta N D_1 D_2}{[(\mu_1 + b)(\mu_2 + \beta N)]}$
Malaria (*Plasmodium falciparum*)	2	$\frac{\beta^2 (N_2/N_1) D_1 D_2}{[(b_1 + \alpha_1 + \gamma_1)(b_2 + \alpha_2)]}$
Schistosomiasis (*Schistosoma mansoni*)	2	$\frac{\frac{1}{2}\beta_1\beta_2 N_1 N_2 D_1 D_2 \lambda_1 \lambda_2}{[(b_1 + \alpha_1 + \mu_1)(\mu_2 + \beta_2 N_2)(b_2 + \alpha_2 + \mu_3)(\mu_4 + \beta_1 N_1)]}$

REFERENCES

Anderson, R. M. (1979a) Depression of host population abundance by direct life cycle macroparasites. *J. Theor. Biol.* 82:283–311.

Anderson, R. M. (1979b) The influence of parasitic infection on the dynamics of host population growth, in R. M. Anderson, B. D. Turner, and L. R. Taylor (eds) *Population Dynamics* (Oxford: Blackwell) pp245–81.

Anderson, R. M. (in press a) Population dynamics of indirectly transmitted disease agents: the vector component, in R. Harwood and C. Koehler (eds) *Comparative Aspects of Animal and Plant Pathogen Vectors*. (New York: Praeger).

Anderson, R. M. (in press b) Reproductive strategies of trematode species, in K. G. Adiyodi and R. G. Adiyodi (eds) *Reproductive Biology of Invertebrates* VI (New York: Wiley).

Anderson, R. M. (1980) The dynamics and control of direct life cycle helminth infections, in G. Salvini (ed) *Mathematical Models in Biology: Lecture Notes in Biomathematics* (Berlin: Springer-Verlag) pp278–322.

Anderson, R. M. (1983) (ed) *Population dynamics of infectious disease agents: theory and application*. (London, Chapman and Hall)

Anderson, R. M. and R. M. May (1978) Regulation and stability of host–parasite population interactions I: Regulatory processes. *J. Animal Ecol.* 47:219–49.

Anderson, R. M. and R. M. May (1979a) Population biology of infectious diseases, Part I. *Nature* 280:361–7.

Anderson, R. M. and R. M. May (1979b) The prevalence of schistosome infections within molluscan populations: Observed patterns and theoretical predictions. *Parasitology* 79:63–94.

Anderson, R. M. and R. M. May (1982) Directly transmitted infectious diseases: control by vaccination. *Science* 215:1053–60

Anderson, R. M. and R. M. May (1983). Vaccination against rubella and measles: quantitative investigations of different policies. *J. Hyg*. 90, 259–325

Ansari, N. (ed) (1973) *Epidemiology and Control of Schistosomiasis* (*bilharziasis*) (Basel: Karger).

Bailey, N. T. J. (1975) *The Mathematical Theory of Infectious Diseases* 2nd ed (London: Griffin).

Barbour, A. D. (1978) Macdonald's model and the transmission of bilharzia. *Trans. R. Soc. Trop. Med. Hyg*. 72:6–15.

Bradley, D. J. and R. M. May (1978) Consequences of helminth aggregation for the dynamics of schistosomiasis *Trans. R. Soc. Trop. Med. Hyg*. 72:262–73.

Bruce-Chwatt, L. J. (1968) Malaria zoonosis in relation to malaria eradication. *Trop. Geog. Med.* 20:50–87.

Bruce-Chwatt, L. J. and V. J. Glanville (1973) *Dynamics of Tropical Disease* (London: Oxford University Press).

Cahill, K. M. (ed) (1972) Malaria *Clinical Tropical Medicine* Vol. II (Baltimore, MD: University Park Press). pp1–109.

Chandler, A. C. (1929) *Hookworm Disease* (London: Macmillan).

Clegg, J. A. and M. A. Smith (1978) Prospects for the development of dead vaccines against helminths. *Adv. Parasitol.* 16:165–218.

Cohen, J. E. (1976) Schistosomiasis, a human host–parasite system, in R. M. May (ed) *Theoretical Ecology: Principles and Applications* (Oxford: Blackwell) pp237–56.

Constable, G. M. (1975) The problems of VD Modeling. *Bull. Inst. Int. Statist.* 106:256–63.

Cooke, K. L. and J. A. Yorke (1973) Some equations modeling growth processes and gonorrhea epidemics. *Math. Biosci.* 16:75–101.

Darrow, W. W. (1975) Changes in sexual behaviour and veneral disease. *Clin. Obstet. Gynecol.* 18:255–67.

Dietz, K. (1974) Transmission and control of arbovirus diseases. *Epidemiology* (Philadelphia: SIAMS) pp104–21.

Dietz, K. (1976) The incidence of infectious diseases under the influence of seasonal fluctuations, in J. Berger, W. Bühler, R. Repges, and P. Tautu (eds) *Mathematical Models in Medicine; Lecture Notes in Biomathematics* Vol. II (Berlin: Springer) pp1–15.

Dietz, K., L. Molineaux, and A. Thomas (1974) A malaria model tested in the African Savannah. *WHO Bull.* 50:347–57.

Garnham, P. C. C. (1966) *Malaria Parasites and Other Haemosporidea* (Oxford: Blackwell).

Garrett-Jones, C. and G. R. Shidrawi (1969) Malaria vectorial capacity of a population of *Anopheles gambiae. WHO Bull.* 40:531–45.

Goddard, M. J. (1978) On Macdonald's model for schistosomiasis. *Trans. R. Soc. Trop. Med. Hyg.* 72:123–31.

Griffiths, D. A. (1974) A catalytic model of infection for measles. *Appl. Statist.* 23:330–9.

Hethcote, H. W. (1976) Qualitative analyses of communicable disease models. *Math. Biosci.* 28:335–56.

Hoagland, K. E. and G. A. Schad (1978) *Necator americanus* and *Ancylostoma duodenale*: Life history parameters and epidemiological implications of two sympatric hookworms of humans. *Exptl. Parasitol.* 44:36–49.

Jordan, P. (1972) Epidemiology and control of schistosomiasis. *Br. Med. Bull.* 28:55–9.

Kermack, W. O. and A. G. McKendrick (1927) Contributions to the mathematical theory of epidemics. *Proc. R. Soc.* A115:700–21.

Lajmanovich, A. and J. A. Yorke (1976) A deterministic model for gonorrhea in a nonhomogeneous population. *Math. Biosci.* 28:221–36.

Lewis, T. (1975) A model for the parasitic disease bilharziasis. *Adv. Appl. Prob.* 7:673–704.

London, W. P. and J. A. Yorke (1973) Recurrent outbreaks of measles, chickenpox and mumps: I. Seasonal variation in contact rates. *Am. J. Epidemiol.*, 98:453–68.

Lotka, A. J. (1923) Contributions to the analysis of malaria epidemiology: I. General. *Am. J. Hyg.*, 3:1–121.

Macdonald, G. (1957) *The Epidemiology and Control of Malaria.* (London: Oxford University Press).

Macdonald, G. (1965) The dynamics of helminth infections, with special reference to schistosomes. *Trans. R. Soc. Trop. Med. Hyg.* 59:489–506.

Martini, E. (1921) *Berechnungen und Beobachtungen zur Epidemiologie und Bekampfung der Malaria.* (Hamburg: Gente).

May, R. M. (1977a) Togetherness among schistosomes: its effects on the dynamics of the infection. *Math. Biosci.* 35:301–43.

May, R. M. (1977b) Thresholds and breakpoints in ecosystems with a multiplicity of stable states. *Nature* 269:471–7.

May, R. M. and R. M. Anderson (1978) Regulation and stability of host–parasite population interactions: II. *J. Animal Ecol.* 47: 249–67.

May, R. M. and R. M. Anderson (1979) Population biology of infectious diseases: Part II. *Nature* 280: 455–61.

McKendrick, A. G. (1912) On certain mathematical aspects of malaria (Proc. Imperial Malarial Comm., Bombay, 1911). *Paludism* 4:54–63.

Miller, T. A. (1978) Industrial development and field use of the canine hookworm vacinne. *Adv. Parasitol.* 16: 333–42.

Muench, H. (1959) *Catalytic Models in Epidemiology* (Cambridge, MA: Harvard University Press).

Muller, R. (1975) *Worms and Disease* (London: Heinemann Medical).

Nasell, I. and W. M. Hirsch (1973) The transmission dynamics of schistosomiasis. *Can. J. Pure Appl. Maths.* 26:395–453.

Nawalinski, T., G. A. Schad, and A. B. Chowdhury (1978a) Population biology of hookworms in children in rural west Bengal I: General parasitological observations. *Am. J. Trop. Med. Hyg.* 27:1152–61.

Nawalinski, T., G. A. Schad, and A. B. Chowdhury (1978b) Population biology of hookworms in children in rural west Bengal: II. Acquisition and loss of hookworms. *Am. J. Trop. Med. Hyg.* 27:1162–73.

Reynolds, G. H. and Y. K. Chan (1975) A control model for gonorrhea. *Bull. Inst. Int. Statist.* 106:264–79.

Rosenfield, P. (1975) *Development and Verification of a Schistosomiasis Transmission Model*, PhD Dissertation. John Hopkins University, Baltimore.

Ross, R. (1904) The anti-malaria experiment at Mian Mir. *Br. Med. J.* 2:632–5.

Ross, R. (1911) *The Prevention of Malaria* 2nd edn (London: Murray).

Ross, R. (1916) An application of the theory of probabilities to the study of *a priori* pathometry I. *Proc. R. Soc.* A92:204–30.

Smith, C. E. G. (1970) Prospects for the control of infectious disease. *Proc. R. Soc. Med.* 63:1181–90.

Smithers, S. R. (1972) Recent advances in the immunology of schistosomiasis. *Br. Med. Bull.* 28:49–54.

Smithers, S. R. and R. J. Terry (1969) The immunology of schistosomiasis. *Adv. Parasitol.* 7:41–93.

Stoll, N. R. (1947) This wormy world. *J. Parasitol.* 33:1–18.

Verhulst, P. F. (1838) Notice sur le loi que la population suit dans son accroissement. *Corresp. Math. Phys.* 10:113–21.

Warren, K. S. (1973) Regulation of the prevalence and intensity of schistosomiasis in man immunology or ecology. *J. Infectious Diseases* 127:595–609.

Wickwire, K. (1979) Mathematical models for the control of pests and infectious diseases: a survey. *Theor. Pop. Biol.* 11:182–238.

Wright, W. H. (1972) A consideration of the economic impact of schistosomiasis. *WHO Bull.* 47:559–66.

Yorke, J. A., H. W. Hethcote, and A. Nold (1978) Dynamics and control of the transmission of gonorrhea. *Sex. Trans. Diseases* 5:51–6.

Yorke, J. A. and W. P. London (1973) Recurrent outbreaks of measles, chickenpox and mumps: II. Systematic differences in contact rates and stochastic effects. *Am. J. Epidemiol.* 98:469–82.

Yorke, J. A., N. Nalhanson, G. Pianigiani, and J. Marlin (1979) Seasonality and the requirements for perpetuation and eradication of viruses in populations. *Am. J. Epidemiol.* 109: 103–23.

9 Gene Introduction Strategies for the Control of Crop Diseases

B. R. Trenbath

9.1. INTRODUCTION

The development of disease resistant varieties has in many cases allowed farmers to continue growing crop species that would otherwise have been abandoned. After finding one or more plants resistant to the current population of disease organisms, the breeder has usually transferred the gene or genes responsible for the resistance to agronomically desirable plants, multiplied the type, and released it as a new "resistant" variety. Since major genes conferring high levels of resistance are relatively easily identified and transferred, it is such genes that have provided most of the success stories in this field.

The history of plant breeding is, however, full of examples where a new variety controls some crucial disease extremely well for a few years, but thereafter the disease rapidly returns to its original destructive level (van der Plank, 1963). A new variety carrying a different resistance gene is then produced by plant breeders and this, in turn, is exploited until destroyed. Being subject to this boom-and-bust cycle, the productivity of the crop is characterized by instability. If the rate of breakdown of the current variety is underestimated, catastrophic losses may occur before a new variety can be released.

With the decrease in genetic diversity of agricultural landscapes and the wider use of irrigation and high nutrient levels, pathogen growth rates are tending to rise, and the rate of passage round the cycle tending to accelerate. At the same time, the supply of disease-resistance genes is in some crops seen to be limited (Watson, 1970; Frey *et al.*, 1973; MacKenzie, 1979). Aware of the threat of being overtaken by the pathogen, plant breeders are urgently exploring the possibility of using more durable, usually polygenic, types of resistance (Robinson, 1977). However, until durable resistance is successfully incorporated into new varieties, it seems likely that the vulnerable major genes will still be used to protect many crops (Person *et al.*, 1976).

Among methods of gene utilization proposed for extending their useful life, the strongest claims have been made for the use of multicomponent mixtures of varieties, i.e., multilines (Browning and Frey, 1969), and varieties carrying multiple resistance genes (Watson, 1970). Indeed, multilines have been in successful commercial use for ten years in Iowa (J. A. Browning, personal communication), and multigene varieties have been used extensively in Australia and elsewhere, also with success (Watson, 1970; Ou, 1977). Rotation of single-gene varieties has occasionally been considered as a theoretical possibility (van der Plank, 1968; Kiyosawa, 1972), but is likely to present logistic problems of seed supply. However, owing to the vast scale of the necessary experiments, no controlled comparison of these methods with the traditional sequence of single-gene varieties is ever likely to be possible. Such a situation invites the use of mathematical models to evaluate the various approaches.

The aim of this chapter is to report the use of a simulation model to compare the rate of breakdown of three resistance genes, *A*,*B*, and *C* under four gene utilization strategies:

(1) Cultivation every year of a multiline mixture of single-gene varieties ($A + B + C$).
(2) Cultivation every year of a multiple-gene "superline" variety (*ABC*).
(3) Sequential introduction in single-gene varieties (*A*,*A*,*A*,*A*,*B*,*B*,*B*,*B*,*C*,*C*,..).
(4) Rotation of single-gene varieties (*A*,*B*,*C*,*A*,*B*,*C*,*A*,*B*,*C*,.....).

After first outlining the target host–pathogen system, I describe the model; I then report the results of some simple experiments with it, and finally present the comparison of the gene utilization strategies.

9.2. THE HOST–PATHOGEN SYSTEM

The system modeled consists of a land area uniformly planted every year with a cereal species that is potentially attacked by a destructive fungal rust. After infection occurs, the rust reproduces swiftly on susceptible plants by means of small, ideally well dispersed, asexual spores. Ultimately, the pathogen population may grow to the level where it reduces crop growth and leads to yield loss. Disease development is taken to be uniform over an area assumed to be large enough for edge effects, immigration, and emigration to be ignored. This idealized system is rather similar to black stem rust (*Puccinia graminis*) attacking wheat in North America.

For the purpose of this current comparison, it is assumed that the three disease-resistance genes *A*, *B*, and *C*, are available for incorporation into pure-line varieties of the self-fertilizing crop. As mentioned above, the comparison will involve single-gene varieties that can be identified by the major dominant gene they carry, *A*, *B*, or *C*, and also a multigene variety *ABC*. Varieties of the crop, or other susceptible plants such as related weeds that have none of these three resistance genes, are designated *O*. Apart from their response to disease, which is determined only by their complement of major resistance genes, all plants are taken to have potentially identical behavior.

The rust is taken to be always in a vegetative, diploid state, with a gene-for-gene correspondence between its own genes for virulence and those for resistance in the host (Flor, 1956). The pathogen genes giving virulence on varieties with genes *A*, *B*, and *C* are here denoted *a*, *b*, and *c*, respectively. Since the virulence genes are taken to be fully recessive (Flor, 1956), they will only confer virulence on a race when present in a homozygous state. To allow races, like varieties, to be identified by their phenotypes, a race carrying for example *a* in double dose will be referred to as race (*a*), whereas races carrying the gene either in single dose or not at all will be collectively referred to as race (*o*). Mutation and/or reassortment is assumed to take place so that factors for phenotypic virulence can be added to, or lost from, a race one at a time. To simplify description, the production of new phenotypes will be considered to be only through "mutation" (defined in a broad sense). Apart from determining differences in host range, a race's complement of virulence genes is assumed to act in either of two key ways. Under the first option, "stabilizing selection" (van der Plank, 1963) is supposed to be present such that each additional virulence gene carried imposes a burden on the race's fitness. Potential multiplication rates diminish as virulence genes accumulate. Under the second option, stabilizing selection is taken to be absent so that all races have the same potential multiplication rate when growing on susceptible plants.

9.2.1. The Attack Matrix

The host ranges of the various races are summarized in the "attack" matrix (Table 9.1). Thus, variety *A* can only be attacked by the races (*a*), (*ab*), (*ac*), and (*abc*) in the present system; variety *O* can be attacked by the race (*o*), which is confined to such plants, and also by all the other races containing genes *a*, *b*, and *c*; variety *ABC* can only be attacked by race (*abc*). The more complex races, i.e., those containing more virulence genes, have a wider host range. The most complex race in the system, (*abc*), can attack all varieties considered and is called the "super-race". Its dynamics will be seen to be crucial in determining the duration of protection under the various gene utilization strategies.

Where a mixture of hosts or pathogens is present, several effects are expected that are not found when only one of each is present. Because these relatively little-studied effects have often been referred to in the debate on the

TABLE 9.1 Attack matrices of two of the simulated systems.

(i) Multiline of three single-gene varieties*

		Rust race							
		(*o*)	(*a*)	(*b*)	(*c*)	(*ab*)	(*ac*)	(*bc*)	(*abc*)
Variety	*O*	1	1	1	1	1	1	1	1
	A	0	1	0	0	1	1	0	1
	B	0	0	1	0	1	0	1	1
	C	0	0	0	1	0	1	1	1

(ii) Superline variety*

		(*o*)	(*a*)	(*b*)	(*c*)	(*ab*)	(*ac*)	(*bc*)	(*abc*)
Variety	*O*	1	1	1	1	1	1	1	1
	ABC	0	0	0	0	0	0	0	1

*10% of all crops is taken to consist of fully susceptible variety of type *O*.

long-term efficacy of multilines (Browning and Frey, 1969; Johnson and Allen, 1975; Burdon, 1978), they are outlined below. (A more detailed description is given in Trenbath, 1977.)

(1) The multiplication rate of a pathogen race is lower in a mixture of host and nonhost plants than in a pure stand of host plants because a larger proportion of dispersing spores is lost, either on the ground, or deposited on nonhost surfaces (the "fly-paper" effect).

(2) Lightly attacked components may grow more vigorously and not only compensate to some extent for yield loss in the heavily attacked components of the mixture, but also enhance the fly-paper effect.

(3) If stabilizing selection is present, simpler races of the pathogen have a reproductive advantage over more complex races when growing together on a common host.

(4) In the case of rusts and some mildews, spores deposited on nonhost leaves may reduce the probability of infection by virulent spores. One effect is that the avirulent spores physically block the stomata and prevent some of the infective germ tubes from entering otherwise susceptible leaves. There is, however, also an induced physiological resistance. The overall effect is "cross-protection" (Yarwood, 1956), which is synonymous with "induced resistance" (Littlefield, 1969).

These four types of effect are included in the model described in the section 9.3.

9.2.2. Further Assumptions

While all growing seasons are assumed to be uniformly favorable for the growth of both crop and pathogen, these seasons are taken to be separated by off-seasons that are uniformly unfavorable for the pathogen and in which all races suffer equal percentage mortality. The pathogen growth season is preceded by a pathogen-free period during which the crops grow to a third of their potential final biomass. Inoculum in the form of vegetative spores (urediospores) is then assumed to be dispersed on the crop and to form lesions, the abundance of which, compared with crop biomass, determines initial "percent cover". Simultaneous growth of crop and pathogen races occurs till harvest, 50 days later.

The four gene utilization strategies to be compared are assumed to start at a point where the old varieties, all designated O, are to be replaced because they are heavily attacked by races of type (o). It is assumed that the strategies can only be used over 90% of the area because there are some uncooperative farmers who continue to grow the old varieties. Patches of these are assumed to be sufficiently interspersed with cultures of the new varieties for there to be free exchange of spores.

9.3. THE HOST–PATHOGEN MODEL

The model used in the simulations is very similar to that described elsewhere (Trenbath, 1977) and only the main features and changes will be given here.*

9.3.1. Host Growth Rate

The model assumes a system of m cereal varieties and n rust races with variety–race interactions defined by an $m \times n$ "attack" matrix as in Table 9.1. The growth rate $\mathrm{d}v_i/\mathrm{d}t$ of the ith variety during the rust growth season ($0 \geqslant t \geqslant 1$) is given by:

$$\mathrm{d}v_i/\mathrm{d}t = v_i a F_{Gi} F_{Ci}(1-t) \qquad (i=1,\ldots,m), \tag{9.1}$$

where v_i is the biomass of the variety per unit area. When $t=0$, $\sum_i v_i = 1$; a is the maximum possible relative growth rate of all varieties attained at the start of the rust growth season under rust-free conditions; F_{Gi} is a "vigor" factor, depending on percent rust cover; F_{Ci} is a "competition" factor depending on the vigor of the other varieties in a mixture; and t is physiological time. When $t = 0$, spores are dispersed, population growth begins, and crop relative growth rate is maximal with a value of a. When $t = 1$, all growth ceases.

*An error in eqn. (13) of the original publication is corrected (eqn. 9.2). This error and one in the original eqn. (21) were absent from the program used to produce the published results.

TABLE 9.2 Values of parameters used in the model.

a	Maximum possible initial relative growth rate of the crop varieties (per season)	2.2
b	Maximum possible relative growth rate of the rust races (per season)	10
c	Mortality rate of the rust races during the rust growing season (per season)	0.9
z	Mutation rate between races differing in one effective virulence gene (mutants per unit of new rust per season)	10^{-6}
k_1	Parameter determining the gradient of the response of host relative growth rate ("vigor") to rust cover	14.2
k_2	Rust cover at which host growth becomes negligible	0.75
k_3	Parameter determining the strength of the stabilizing selection	
	none	0
	standard strength	0.34
	high	1
k_4	Parameter determining the strength of cross-protection	
	none	0
	standard strength	69
	high	200
k_5	Parameter determining the proportion of dispersing rust spores lost on the ground or blown out of the system.	0.2

The vigor factor F_{Gi} is calculated using a negative logistic expression, the parameter values of which are chosen to make it vary effectively from unity at a rust cover up to 10% to zero at cover over 75%:

$$F_{Gi} = 1/\left\{1 + 100/\exp\left[k_1\left(k_2 - \frac{1}{100}\sum_j \frac{A_{ij}r_j}{\sum_k A_{kj}v_k}\right)\right]\right\}. \tag{9.2}$$

Parameter values are summarized in Table 9.2. The competition factor F_{Ci} varies about unity, the value in a pure stand. It is calculated using de Wit's (1960) competition equation with the "crowding coefficients" replaced by the vigor factors F_{Gi}:

$$F_{Ci} = \frac{F_{Gi}\sum_j v_j}{\sum_j F_{Gj}v_j}. \tag{9.3}$$

Grain yield per unit area is estimated as that part of the shoot biomass that is accumulated in excess of 1.5 units. With rust absent, a unit area of crop produces 3 units of biomass so that the maximum yield is also 1.5 units. The maximum harvest index is therefore 50%.

9.3.2. Pathogen Growth Rate

The growth rate $\mathrm{d}r_j/\mathrm{d}t$ of the jth rust race $(0 \geqslant t \geqslant 1)$ is given by:

$$\mathrm{d}r_j/\mathrm{d}t = r_j[bF_{Sj}(\sum_i A_{ij}F_{Ai}F_{Pi}F_{Di} - c)] + Z_j \tag{9.4}$$

$(j = 1,\ldots,n)$, where r_j is the amount of rust race j per unit area in arbitrary units. The percentage cover of the race j in a pure stand of a susceptible variety i is given by r_j/v_i, while in a mixture of varieties, the percentage cover on the susceptible varieties is $r_j/(\sum_k A_{kj}v_k)$; and b is the maximum possible relative growth rate of the simplest rust race (o). This could only be achieved at very low cover on an infinitely abundant host in the absence of any cross protection, mortality, and mutation. F_{Sj} is a parameter determining the strength of stabilizing selection; and A_{ij} is an attack matrix of ones and zeros (*see* Table 9.1). If $A_{ij} = 1$, race j attacks variety i. If $A_{ij} = 0$, it is avirulent on variety i. F_{Ai} and F_{Pi} are availability and cross-protection factors, respectively, for all races attacking variety i. F_{Di} is a deposition factor representing the probability of a spore landing on variety i. c is the relative mortality rate of all races on all hosts during the pathogen growing season. Z_j is the net rate of appearance of race j by mutation.

When stabilizing selection is required in the simulation, the factor F_{Sj} reduces b to an extent depending on the number of virulence genes N_j carried by race j:

$$F_{Sj} = 1/(1 + k_3N_j). \tag{9.5}$$

Here the model differs from the previous one (Trenbath, 1977) in that it accommodates the possibility (Ogle and Brown, 1970) that survival ability depends primarily on reproductive capacity, which is summed up in the value of b. In principle, therefore, differences in survival ability should be predictable from performance in pure culture.

The availability factor F_{Ai} is the proportion of the surface area of shoot of variety i that is available for the production of further lesions. It is calculated as:

$$F_{Ai} = 1 - \frac{1}{100}\left(\sum_j \frac{A_{ij}r_j}{\sum_k A_{kj}v_k}\right). \tag{9.6}$$

In the previous model, F_{Ai} was used also to represent the action of stabilizing selection on each race on variety i. This was achieved by setting it equal to the "physiological" space available on this host calculated using rust covers weighted according to the number of virulence genes that they carried. In the present version, however, F_{Ai} represents directly the physical space on variety i available for new lesions.

The cross-protection factor F_{Pi} is reduced as the amount of rust avirulent on variety i, namely $\sum_j(1 - A_{ij})r_j$, increases; it is increased as the area effectively receiving these spores, namely, $\sum_j v_j + k_5$, increases. F_{Pi} is given by

$$F_{Pi} = \left\{1 + \frac{k_4}{100}[\sum_j(1 - A_{ij})r_j]/(\sum_j v_j + k_5)\right\}^{-1}, \tag{9.7}$$

where the term with $k_4/100$ as coefficient is a proportional cover term analogous to those in eqns. (9.2) and (9.6), and k_5 represents the area of ground surface over which rust spores are distributed at the same density as on the plant surfaces.

The deposition factor F_{Di} gives the shoot surface area of variety i as a fraction of the whole receiving area. Thus

$$F_{Di} = v_i/(\sum_j v_j + k_5). \tag{9.8}$$

Calculation of the rate of appearance of mutant phenotypes of race j from race k is based in the new version of the model on the "birth" term in eqn. (9.4) since it is in new lesions derived from mutant spores that mutations enter the modeled populations. The birth term for the kth race can be written as $B_k = r_k b F_{Sk}(\sum_i A_{ik} F_{Ai} F_{Pi} F_{Di})$. The net increase Z_j in the jth pathogen race due to mutations is then:

$$Z_j = z(\sum_k M_{kj} B_k - B_j \sum_k M_{jk}), \tag{9.9}$$

where z is the assumed common mutation rate and M_{kj} is an element of the $m \times m$ mutation matrix M containing ones and zeros (see Table 9.3). In eqn. (9.9) the first term represents gain of mutant rust from other races and the second term represents loss.

TABLE 9.3 Mutation matrix for rust races present in a three-gene system.

		Receiving race							
		(o)	*(a)*	*(b)*	*(c)*	*(ab)*	*(ac)*	*(bc)*	*(abc)*
Donor race	*(o)*	0	1	1	1	0	0	0	0
	(a)	1	0	0	0	1	1	0	0
	(b)	1	0	0	0	1	0	1	0
	(c)	1	0	0	0	0	1	1	0
	(ab)	0	1	1	0	0	0	0	1
	(ac)	0	1	0	1	0	0	0	1
	(bc)	0	0	1	1	0	0	0	1
	(abc)	0	0	0	0	1	1	1	0

9.4. ESTIMATION OF PARAMETER VALUES

At the present stage of development of the model no strong claims can be made for the accuracy of the parameter values. "Reasonable" estimates have been used in order to imitate the main features of a wheat–rust system. Some justification of the values chosen (Table 9.2) is given in Trenbath (1977) and some brief further comments are given below.

The value taken for *a* (Table 9.2) allows the simulated growth of an unattacked crop stand to follow closely that of a healthy field crop of wheat (Puckridge and Donald, 1967). A unit of biomass in the simulations is equivalent to about 3000 kg/ha.

With F_{Di} varying within the season between 1/1.2 and 3/3.2 in a pure, insignificantly damaged crop, the values of *b* and *c* given in Table 9.2 will lead the calculated relative growth rate of the rust to vary between 0.15 and 0.17 per day. These rates are near the fastest in a stripe rust epidemic observed by Zadoks (1961, p199), although they fall well below rates in the stem rust epidemic observed by Asai (1960).

The within-season rate of $c = 0.9$ represents a convenient guess introduced to make populations without planted hosts continue to decrease during the crop-growing season. The rust mortality rate of 0.99 between seasons is based on an average value of 0.996 for rice blast (Kiyosawa and Shiyomi 1976). Use of Zadoks' (1961) much higher estimates for stripe rust mortality would have necessitated the simulation of much longer growing seasons. Given the parameter values of Table 9.2, the maximum rate of loss of a race is a little more than 2 orders of magnitude per simulated year while the maximum annual multiplication rate is about 36 (3600 within a season; see Figure 9.9A).

The sigmoid damage function defined by eqn. (9.2) and the assumed values for k_1 and k_2 approximate the generalized function given by Jones (1966) for crop yield responding to initial pest population. The damage function of the model is however applied to the instantaneous growth rate and the resulting final biomass is generated dynamically.

As described in Trenbath (1977), the parameter k_3 determining the strength of stabilizing selection was chosen by finding the value that caused the model to reproduce the experimental results of Watson and Singh (1952). These authors cultured 1:1 mixtures of rust races differing in one virulence gene, as do (*o*) and (*a*), on susceptible wheat plants in a glasshouse. To imitate Watson and Singh's five experimental generations, a whole season of simulated time was allowed and an initial lesion cover of 1% was arbitrarily chosen to represent the outcome at the spore density used by them. At this level of inoculum, a k_3 value of 0.34 produced a final race ratio (*a*)/(*o*) of 0.18, the average of Watson and Singh's results in Figure 9.1A, the responses given at different inoculum levels show the error likely in the estimation of k_3 if the assumed level of 1% is too high or too low by an order of magnitude.

The value of the cross-protection parameter k_4 was estimated in a similar way using data of Johnson and Allen (1975). In this case, the authors measured a 32% reduction in rust growth in one generation when a wheat leaf was preinoculated with an equal quantity of avirulent race. Using the model, the value of k_4 was found which gave a 32% reduction in growth of (*a*) on variety *A* when (*o*) had been added. Again, an initial lesion cover of 1% of (*a*) was assumed and an equal amount of (*o*) was specified. Only one-fifth of a season was allowed in order to represent the single generation. The growth response of (*a*) to the k_4 value used is shown in Figure 9.1B. Here, the curves of responses at different inoculum levels diverge greatly; this suggests that slight errors in the assumed inoculum level could much more seriously affect the accuracy of the estimate.

To provide some contrasting levels of stabilizing selection and cross-protection, in addition to the standard strength values already considered, "high" and "none" values (Table 9.2) are also used later. Their effects can be gauged by referring to Figure 9.1A,B. The parameter k_5 determines the proportion of dispersing rust spores that are lost. As a first approximation it may be ignored (van der Plank, 1963, p110), but here it is set to 0.2.

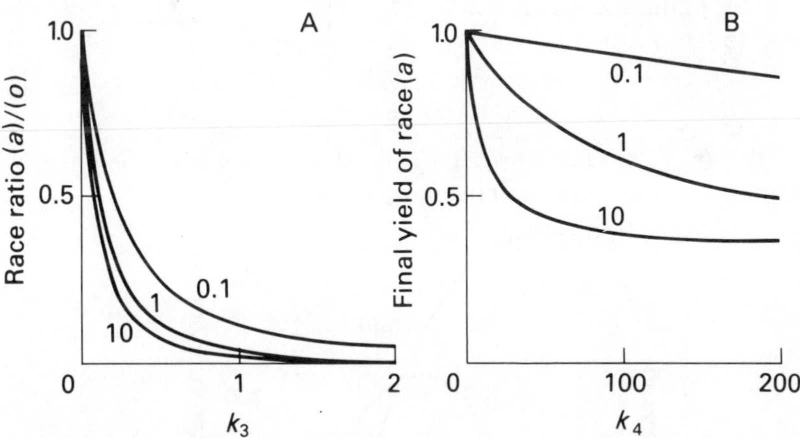

FIGURE 9.1 Response of the model to values of parameters k_3 (strength of stabilizing selection) and k_4 (strength of cross-protection). The three curves are for initial inoculum levels giving 0.1, 1, and 10% rust cover; the 1% lines were arbitrarily chosen to correspond to the levels used in the experiments of Watson and Singh (1952) and Johnson and Allen (1975).

If the 50-day season is seen as five 10-day cycles of reproduction, the mutation rate of 10^{-6} per season represents a value of 2×10^{-7} per generation. This is bigger than the estimate by Person *et al.* (1976) of 10^{-7} for powdery mildew, but much smaller than Kiyosawa's (1977) estimate of 10^{-2} for a particularly mutable avirulence gene in rice blast.

9.5. SIMULATIONS OF SIMPLE CROP–PATHOGEN SYSTEMS

To show how the simulated crops and pathogens grow and interact, the model was first used to simulate the behavior of a series of relatively simple systems.

One Crop Variety and One Rust Race

The patterns of growth of the crop and pathogen through the season depend on the initial level of inoculum (Figure 9.2A,B). The damage function (Figure 9.2C) drawn from Figure 9.2A shows that yield of grain is zero for initial

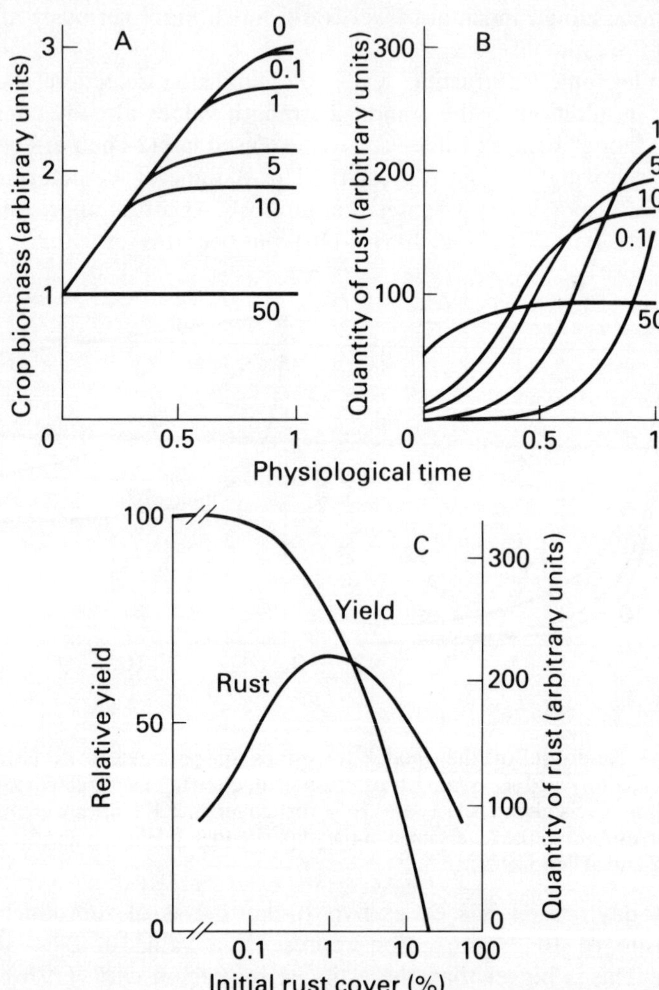

FIGURE 9.2 Patterns of simulated growth of A, crop and B, rust within a season with six levels of initial rust cover (%). C, Crop yield and final quantity of rust in relation to initial rust cover.

infections giving greater than about 20% cover. Pathogen output is greatest from an initial cover of 1% (Figure 9.2C).

One Crop Variety, O, and Two Pathogen Races (o) and (a)

Because the races differ in the number of virulence genes they carry, any stabilizing selection usually acts to disadvantage race (*a*) with its unnecessary virulence gene. To show the effect of stabilizing selection ($k_3 = 0.34$) in both pure and mixed cultures, replacement series have been simulated with proportions of the two races ranging from 1:0 to 0:1 (Figure 9.3). At the medium level of inoculum (Figure 9.3B), the ratio of race outputs (*a*)/(*o*) from a mixture starting with a 50:50 proportion is 0.18, since it was to this level of inoculum that Watson and Singh's (1952) value was taken to apply. At the low and high levels of inoculum, corresponding values of this ratio are 0.12 and 0.35, respectively. Selection against race (*a*) clearly intensifies with

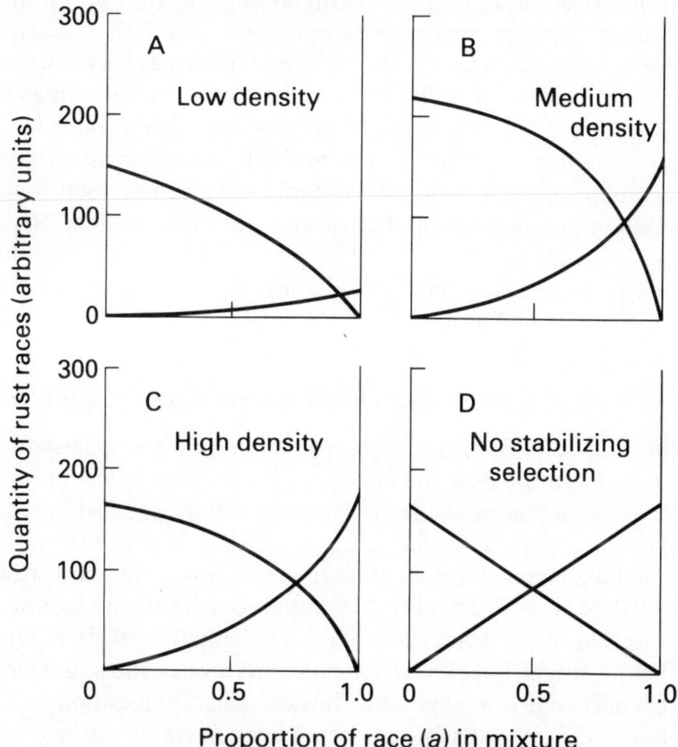

FIGURE 9.3 A–C, Final quantities of rust races produced in mixtures of races (*o*) and (*a*) growing on variety *O*. Proportions of the races are varied from 1:0 to 0:1 at three levels of initial rust cover (0.1, 1, and 10%). Stabilizing selection is of standard strength. D, As in C, but with no stabilizing selection.

increase of inoculum level; such a dependence of "competitive ability" on density has been observed in laboratory rust cultures (Katsuya and Green, 1967; Rastegar, 1976). Interestingly, in pure culture the much more slowly growing (*a*) may actually be more productive than (*o*) because it does not depress host growth so much (Figure 9.3C). Selection coefficients thus depend on both inoculum density and race frequency. If stabilizing selection is absent, the graphs will be of the type shown in Figure 9.3D.

Two Crop Varieties, A and B, and Two Pathogen Races (a) and (b)

Here only one sort of race will be growing on each variety so there can be no stabilizing selection. However, spores will be falling on plants on which they are nonvirulent so that as proportions of *A* and *B* are varied between 1:0 and 0:1, varying intensities of cross-protection will occur. To show that the effect of cross-protection depends on pathogen density, replacement series varying variety proportions have been simulated (Figure 9.4A,B) for a low level and a high level of inoculum, in each case with cross-protection set at three levels, none, standard strength, and high (k_4 = 0, 69, and 200, respectively). To provide the two inoculum levels, the races had initial covers of either 0.1% or 10%, each on its own host. In Figure 9.4, the "none" and the "high" lines are further apart when the pathogen densities are higher implying that, as expected, a higher density of nonvirulent spores cross-protects more effectively (Johnson and Taylor, 1976). With low inoculum, there is no effect of intraspecific competition on multiplication rate *M* so that ln *M* is a linear function of proportion of susceptible plants. With high inoculum, high proportions of susceptible plants lead to competition for space on the hosts and there is departure from linearity.

Two Crop Varieties, A and B, and Three Pathogen Races, (a), (b), and (ab)

Both stabilizing selection and cross-protection will act against (*ab*) in this system. Demonstrating how the strength of stabilizing selection depends on other processes, replacement series varying variety proportions have been simulated (Figure 9.4C,D) for the same low and high inoculum levels, as before each with stabilizing selection set at three levels, none, standard strength, and high (k = 0, 0.34, 1, respectively). The initial cover of the additional race (*ab*) was equal to that of the other races at each inoculum level. In Figure 9.4, the "none" and the "high" lines are now further apart when the pathogen densities are less because there is very little intraspecific competition to lower the "none" line. The depression of multiplication rates around the 50:50 proportion in Figure 9.4D is due to a fairly strong cross-protection of both components rather than the very strong cross-protection of just the minority component in other mixtures.

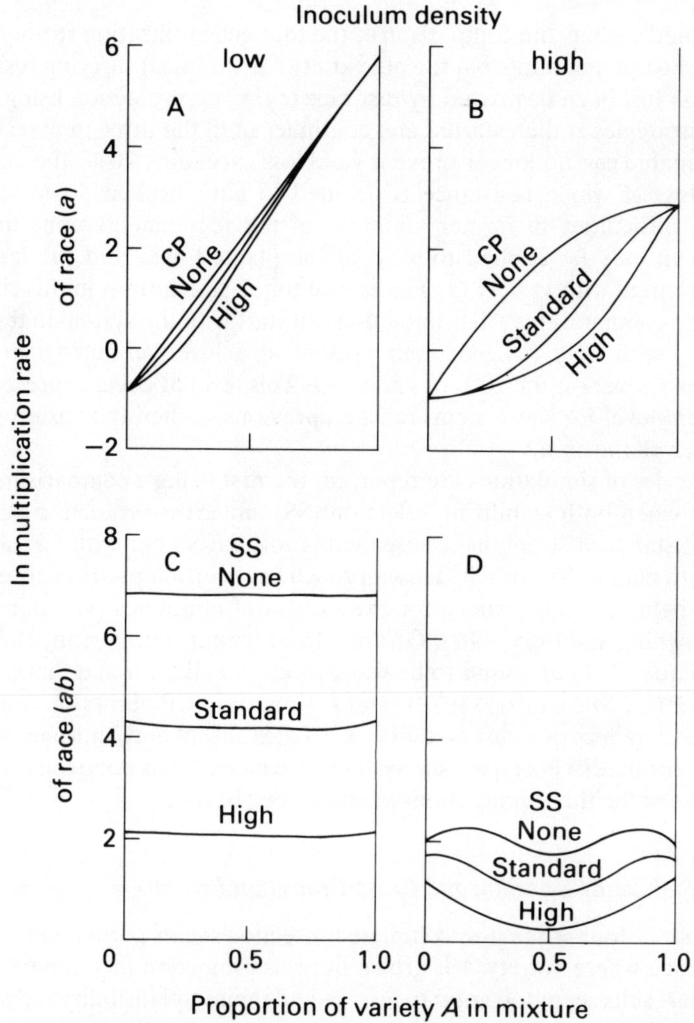

FIGURE 9.4 A,B, Dependence of multiplication rate of rust race (*a*) on proportion of susceptible hosts (variety *A*) in a mixture. In the simulations, races (*a*) and (*b*) are together attacking a mixture of varieties *A* and *B*; proportions of the varieties are varied from 0:1 to 1:0. Three strengths of cross-protection (CP) are assumed, none, standard, and high. C,D, Dependence of multiplication rate of rust race (*ab*) on host proportions when growing mixed with races (*a*) and (*b*) on a range of mixtures of varieties *A* and *B*. Three strengths of stabilizing selection (SS) are assumed: none, standard, and high. Two levels of initial rust cover are used: A,C, low, 0.1% B,D, high, 10%.

9.6. SIMULATION OF COMPLEX CROP–PATHOGEN SYSTEMS

As explained earlier, the comparison of the four gene utilization strategies has been carried out assuming that the old variety (or varieties) carrying resistance gene *O* has just been destroyed by rust race (*o*). Crop protection using one of the four strategies is then started and continues until the three new resistance genes available can no longer prevent yield loss exceeding 20%, the arbitrary level of loss at which resistance is deemed to have broken down. Due to supposed difficulties in farmer adoption of the recommendation, the new strategy can only be applied to 90% of the planted area and the last 10% remains planted with variety *O*. This is true for all simulations in this chapter. To provide comparability, the simulations all start with the system in the same state, i.e., with only (*o*) inoculum present at a level producing an initial 2.125% rust cover on the 10% of variety *O*. This level of cover represents the equilibrium level for the system, reached previously when the variety *O* was grown over all the area.

Three series of simulations are reported, the first being a comparison of the strategies when both stabilizing selection (SS) and cross-protection (CP) are present at standard strengths, the second is without SS but with CP, and the third is with neither SS nor CP. This approach has been adopted because (1) SS is widely believed to be crucial for the success of multilines (van der Plank, 1963; Browning and Frey, 1969; Groth, 1976; Fleming and Person, 1978); (2) SS has frequently been found to be weak or absent (Brown and Sharp, 1970; Martens, 1973; MacKenzie, 1979); and (3) omitting CP allows the model to mimic the dynamics of a rust system where CP is absent and, more important, of the wider range of host–parasite systems in which CP has not been reported. The results of the three simulations are described below.

Series 1 (Stabilizing Selection and Cross-Protection Present)

In three out of four strategies, complete protection seems permanent. Only in the sequence where variety *A* is grown pure, is protection incomplete (Table 9.4). Such results seem too good to be true and their implausibility is discussed later. In all four strategies, the initial abundance of race (*o*) gives rise to a burst of mutant forms of all races. Only in the sequence do the races not all decline to negligible levels; here, race (*a*) comes into equilibrium at a level giving slight yield loss.

Series 2 (Cross-Protection But No Stabilizing Selection)

In the absence of SS, all strategies show slight yield losses after 12–14 years with destruction in the following year (Table 9.4). The sequence also suffers small periodic losses before this. In the multiline, after year 1, races increase or decrease steadily according to the proportion of the plants that each can attack

TABLE 9.4 Yield losses (%) and durations of protection (years) in three series of simulations of four gene-introduction strategies with stabilizing selection (SS) and cross-protection (CP) either present (+) or not (−). Yield losses are in the final year of protection unless otherwise stated. Protection is considered effective until a loss of more than 20% is suffered.

	Series		
Strategy	1 SS^+, CP^+	2 SS^-, CP^+	3 SS^-, CP^-
Multiline *A+B+C*	0% (∞ years)	15% (13 years)	4% (12 years)
Superline *ABC*	0% (∞ years)	8% (14 years)	1% (13 years)
Sequence *A,A,A,A,B,B,...*	7%[a] (∞ years)	17%, 4%, 8%[b] (14 years)	7%, 4%, 8%[c] (13 years)
Rotation *A,B,C,A,B,...*	0% (∞ years)	5% (12 years)	1% (11 years)

[a] Yield loss increases to 7% over 20 years and stabilizes at this level.
[b] Losses occur in years 6, 10, and 14.
[c] Losses occur in years 5, 9, and 13.

(Figure 9.5). As the super-race becomes more common it feeds the populations of the other races by reverse mutation.

In the superline, after year 1, all races with two virulence genes and less share the same proportion of susceptible plants. Therefore they all decrease initially at the same rate (Figure 9.6). The pathogen population as a whole is less fit on the superline than on the multiline (Person *et al.*, 1976). As the super-race becomes commoner, back mutation reverses the declines of the populations, first of "two-gene" races, then of "one-gene" races, and then of race (*o*).

In the sequence, the races show a complex pattern of increase and decrease depending on whether the variety being planted is a host or not (Figure 9.7). The successive rises of races (*a*), (*ab*), and (*abc*) force the replacement of all varieties until *C* is destroyed. The 4–6 year lifespan of a single rust-resistant variety is consistent with expectations (Stevens and Scott, 1950) and field experience (Frey and Browning, 1971).

The slowing of the rates of increase of (*ab*) and (*abc*) (Figure 9.7) when the current varieties are heavily infected either by (*a*) or (*ab*) is apparently due to competition for space on their common hosts. The destruction of varieties *B* and *C* is thereby delayed about three years. However, because large populations of (*a*) and (*ab*) produce numerous mutants, the growth rates of (*ab*) and (*abc*) are higher than under the other strategies. There is clearly no

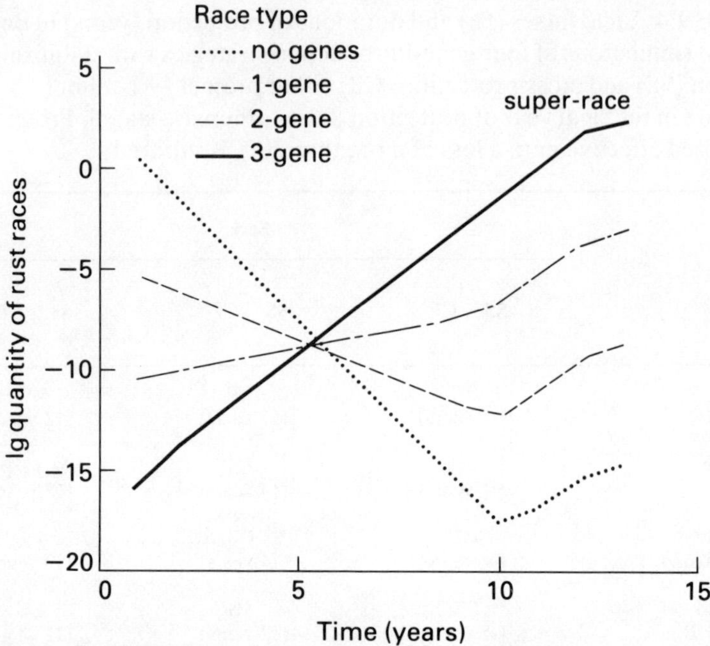

FIGURE 9.5 Changes in populations of rust races over 15 years of simulated time when 90% of the land area is planted to a multiline of varieties *A*, *B*, and *C*. Rust races are of four types according to how many virulence genes they carry: none (*o*); one (*a*), (*b*), and (*c*); two (*ab*), (*ac*), and (*bc*); three, the super-race *abc*.

chance of "recycling" old varieties (e.g., *A*) in this system (Stevens, 1949); without SS, virulence genes accumulate (Person *et al.*, 1976).

In the rotation, the race populations show a pattern of changes similar (Figure 9.8) to that in the multiline (Figure 9.5). The super-race, which is again responsible for ending the useful life of this strategy, shows uninterrupted growth at the same rate as under the superline and multiline strategies.

The yield losses that destroy the varieties are usually 37%, which corresponds to an equilibrium population level to which the system always tends. It conforms with Suneson's (1960) suggestion that where water is not limiting crop growth, natural rust epidemics in wheat should rarely cause losses greater than 50%.

Series 3 (No Stabilizing Selection or Cross-Protection)

The suppression of CP does not alter the pattern of yield losses, but causes protection periods to end just one year earlier than in the previous series (Table 9.4). As in earlier runs, a short-lived initial burst of growth of race (*o*) on variety *O* produces avirulent spores on plants of the new variety. With CP absent in this series, these plants are now not cross-protected. The shorter

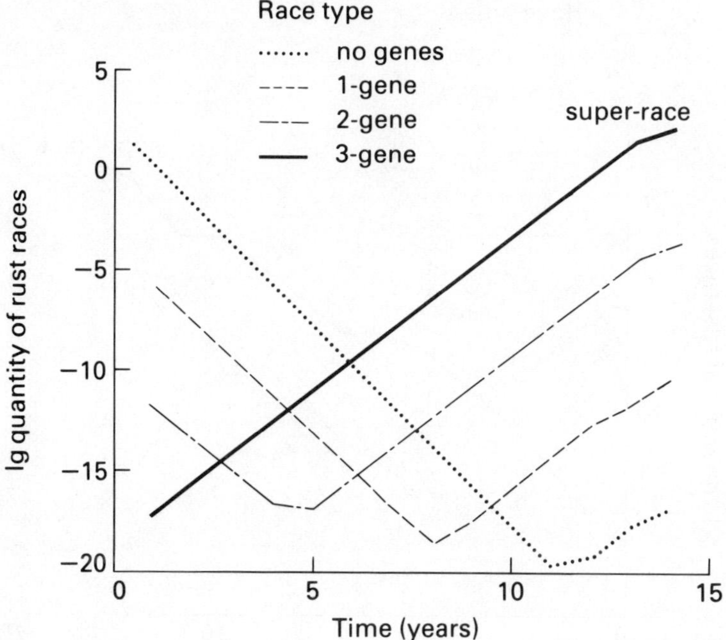

FIGURE 9.6 Changes in populations of rust races over 15 years of simulated time when 90% of the land area is planted to a superline variety (*ABC*). For race types, see Figure 9.5.

duration of protection in this series is the consequence of faster rust growth on the new variety that occurs at this stage.

9.7. COMMENTS ON THE OPERATION OF THE MODEL

To show why the presence of SS (as modeled) led to decreases of all rust races in both multiline and superline, the dependence of within-season multiplication rate M on the proportion of plants susceptible is plotted in Figure 9.9A. The lines are based on simulations using very low inoculum levels of races with zero, one, two, and three virulence genes attacking crops with varying proportions of susceptible plants. The action of SS in reducing the "birth" term in the model (9.4, 9.5) shows up as a reduction of gradient in the lines of Figure 9.9.

Because between-season mortality reduces populations by a factor of 100, for a race population to show net growth in long-term simulations, its M value must exceed 100. In terms of Figure 9.9A, ln M (equal to the within-season mean relative growth rate) must exceed 4.6. In Figure 9.9, the ln M values of the various race categories are shown as filled points above the proportions of the crop mixture that are susceptible to them. With SS active, no race has a value high enough to give net growth in a multiline. In a superline crop, the

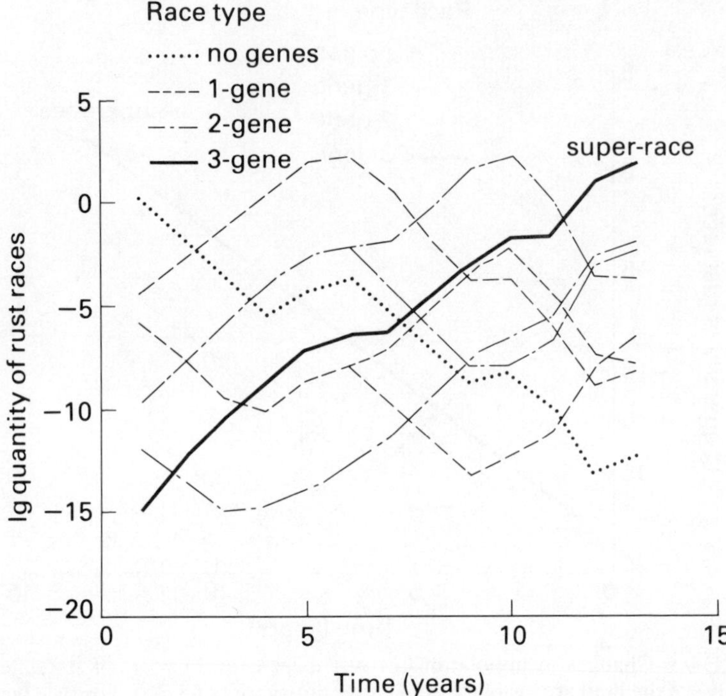

FIGURE 9.7 Changes in populations of rust races over 15 years of simulated time when 90% of the land area is planted to a sequence of varieties carrying single resistance genes (varieties *A*, *B*, and *C*). The current variety is changed for the next in the sequence when it is likely to suffer more than 20% yield loss. For race types, see Figure 9.5.

situation is similar although the one- and two-gene races will here be like race (*o*) in being able to attack only 10% of the crop. Accordingly, with this level of SS, either a multiline or a superline strategy provides indefinite protection.

Without SS in the system, all races have the same response of ln *M* to proportion of crop that is susceptible. All behave as race (*o*) except that their host ranges differ. The appropriate values of ln *M* for the four categories of race in the multiline crop are shown in Figure 9.9A by open symbols. The super-race (*abc*) and the two-gene races have values above the threshold and will increase exponentially, the super-race at a rate nearly four times faster than the two-gene races. The simpler races, on the other hand, have values below the threshold and will decrease exponentially, the simplest race at a rate about 2.5 times faster than the one-gene races. These relationships can be seen directly in Figure 9.5, where populations are shown on a log scale. In a superline crop without SS, only the super-race has a value of ln *M* above the threshold (Figure 9.9A). This race will therefore grow at the full maximum rate (Figure 9.6), as in the multiline. All other races can attack only 10% of the crop and so will

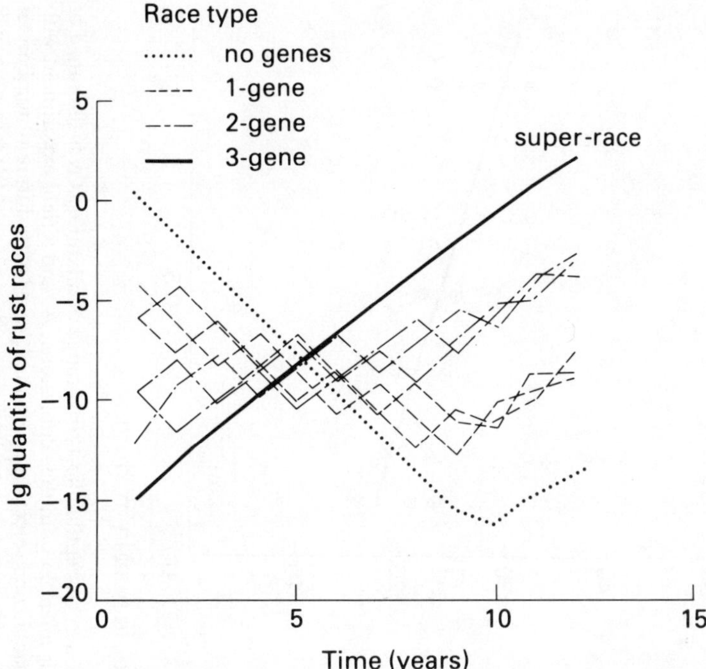

FIGURE 9.8 Changes in populations of rust races over 15 years of simulated time when, on 90% of the land area, single-gene varieties (*A*, *B*, and *C*) are annually rotated. For race types, see Figure 9.5.

decrease at the same rate as (*o*) did in the multiline. In both multiline and superline, the effect of mutation between races finally halts the population declines; equilibrium levels are reached as the super-race ends the protection by overcoming all three resistance genes.

The form of the dependence of ln *M* on the proportion of susceptible plants (Figure 9.9A) is of interest since it contrasts with expectations according to the model of Leonard (1969a). This latter model assumes nonoverlapping generations of pathogen, and relates the multiplication rates in mixed and pure stands of host plants:

$$M_m = f^n M_p, \tag{9.10}$$

where M_m and M_p are multiplication rates, taken over n generations of the pathogen in the mixed and pure stand respectively, and f is the proportion of susceptible plants in the mixture. When log-transformed,

$$\ln M_m = n \ln f + \ln M_p, \tag{9.11}$$

and this implies that the relationship between ln M and proportion of susceptible plants f will only be linear when f is plotted on a logarithmic scale.

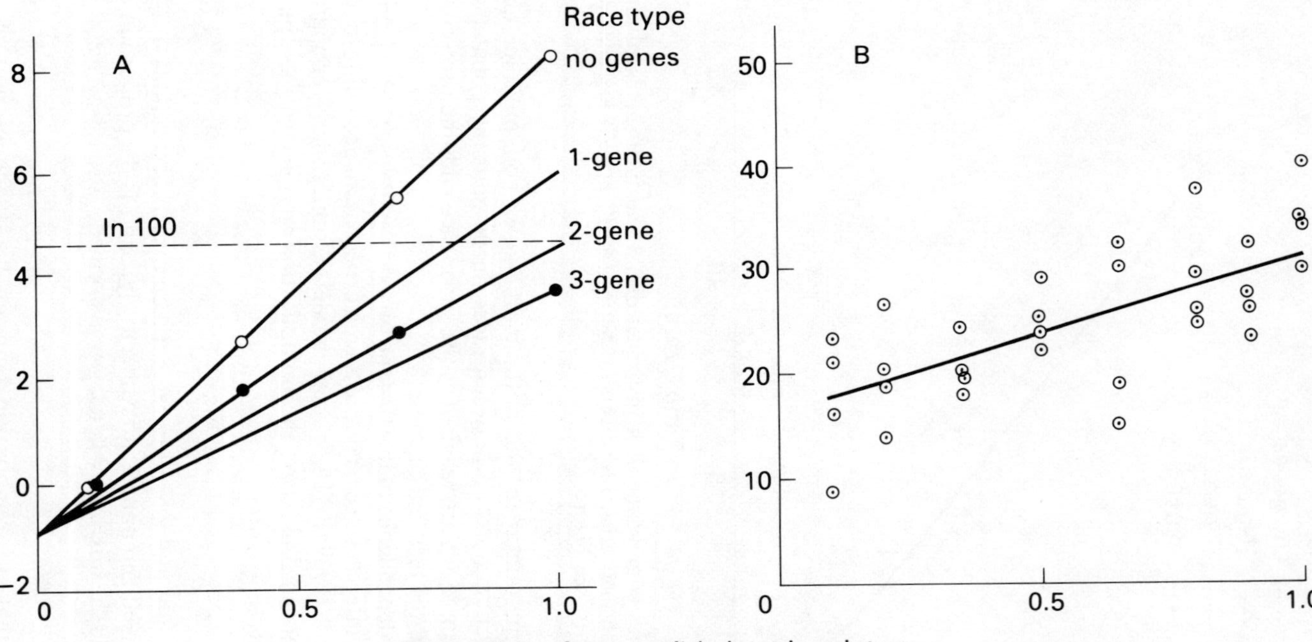

FIGURE 9.9 Responses of pathogen multiplication rate to proportion of susceptible hosts in a range of mixtures. Over the range of proportions considered in the simulations (0.1 to 1.0), the response is linear in both the simulations, A, and a field experiment with barley mildew, B (after Burdon and Whitbread, 1979). In A, when stabilizing selection is assumed, races with different numbers of virulence genes (see Figure 9.5) will differ in their potential maximum multiplication rate. In the multiline, races of the four types will be able to attack 0.1, 0.4, 0.7, and all of the plants, respectively; rates are shown by full circles. When stabilizing selection is absent, all races will have the same potential maximum multiplication rate. Hence, their rates in the multiline will lie on the same line (open circles). The multiplication rate ($\log_e 100 = 4.605$) that has to be exceeded for a race population to increase is shown as a broken line.

The relationships of Figure 9.9A showing the linear dependence on f plotted on an arithmetic scale is, however, a clear consequence of the present model. If the model (9.4) is written in simplified form applicable to a system of one rust growing in a mixture of a single susceptible component with a resistant one, we have

$$dr/dt = r(bf - c), \tag{9.12}$$

where $f \approx v_1/(v_1 + v_2 + k_5)$, the deposition factor of eqn. (9.8), and mutation is ignored. The integrated form is

$$r(t) = r(o) \exp (bf - c), \tag{9.13}$$

whence

$$\ln \frac{r(t)}{r(o)} = \ln M = bf - c, \tag{9.14}$$

where the multiplication rate is calculated over the pathogen growth season. This is the linear form seen in the arithmetic plot of Figure 9.9A. As expected, the intercept with the ordinate gives the value of the mortality rate of pustules, $c = 0.9$, that was input for all the simulations (Table 9.2).

Although experimental and simulated data of ln M (the relative growth rate, or “infection rate” of van der Plank, 1963) have on several occasions been plotted with ln f as the abscissae (Leonard, 1969a; Kiyosawa, 1976; Burdon and Chilvers, 1977), the linearity in Figure 9.9A suggests that ln M may sometimes be just as appropriately plotted with f on an arithmetic scale. The introduction of a dispersal gradient and a consideration of total disease instead of only new lesions in the simulations of Kiyosawa (1976) both led to curvature upward in a plot using ln f as the abscissae; on an arithmetic f scale, Kiyosawa’s results consequently come near to linearity. Possibly because of dispersal gradients and their counting of total disease, the extensive data of Burdon and Whitbread (1979) are also linear on an arithmetic f scale (Figure 9.9B). Regressions using two less extensive sets of data (Leonard, 1969a; Burdon and Chilvers, 1977) show almost as good a fit with an arithmetic scale as with a logarithmic one, possibly for the same reasons. The similarity of Figure 9.9A and B shows that the present model adequately mimics the field response to admixture of resistant plants over the range of f used.

9.8. CONCLUSIONS

The earlier study with a similar model (Trenbath, 1977) showed that there might be little difference between one superline and two multiline strategies in the amount of pathogen super-race produced in seven years of simulated time. The present work has extended these results, with some improvements in the model, to the point where the gene introduction strategies have either finally broken down or reached a state where crop protection will continue indefinitely. For the crop–pathogen systems considered, we now have evidence bearing on the key question: what strategy will best prolong the useful life-span of a single resistance gene (Person *et al.*, 1976)? *In agreement with the earlier work, the new results show that there is little to choose between four gene introduction strategies, given any of three systems (see Table 9.4). In the system where stabilizing selection (SS) and cross-protection (CP) are present, all strategies give more or less complete protection over a indefinite period. Where SS is absent, all strategies give about the same finite duration of more or less complete protection (12–14 years). Without CP, protection lasts in all cases one year less (11–13 years).*

To ensure that these results are directly relevant to the current debate on the usefulness of multilines (Barrett and Wolfe, 1978; Fleming and Person, 1978; Macpherson, 1978; Chin, 1979), the parameter values and model structure were chosen to represent the main features of a cereal–rust system. This is the system in which disease control by multilines has been the most strongly advocated (Browning and Frey, 1969). Although there are gross simplifications in the modeling procedure, significant similarities between the model's output and observations on rust epidemics in cereals have been remarked on above. In the absence of suitable data for rusts, a critical part of the model's behavior is validated by comparison with results for mildew (Figure 9.9B). This seems acceptable because of the fundamental similarity of the population dynamics of a wide range of *r*-type airborne pathogens (van der Plank, 1963).

Widely differing intensities of SS are used in the simulations because SS varies greatly in strength among real-world rust systems (Thurston, 1961; Katsuya and Green, 1967; Leonard, 1969b; Martens *et al.*, 1970; Watson, 1970). Widely differing intensities of CP are also allowed for, so that the results will span the range of cases where CP is present (as in rusts and powdery mildews; Johnson and Taylor, 1976; Chin, 1979) and where CP is absent.

An effort has been made to make the assumed conditions realistic: contrasting with the 100% farmer acceptance assumed by most other authors of models (Kiyosawa *et al.*, 1982 for an exception), here only 90% of the cropped area is taken to consist of the new variety or varieties. The remainder of the area is taken to be cropped by "conservative" farmers possibly preferring the old varieties for reasons of quality or familiarity, and hoping that large areas of new resistant material will alleviate the previous disease problem. Actually, in this latter case, Figures 9.5–9.8 suggest that their hopes were fulfilled; Painter (1958) gives an example of host rarity giving protection from Hessian fly. The 10% of fully susceptible plants may alternatively represent the usually present, related weeds in or near the crop (e.g., Suneson, 1960). The introduction of the new material over a large proportion of the area follows recommended practice for multilines (Browning and Frey, 1969; van der Plank, 1968, p171), and the rapidity of the introduction is in line with Wolfe and Barrett's (1976) suggestion for resistant genotypes that are to be used over a large area.

In the first series of simulations (with SS and CP), the total success of three out of the four strategies was unexpected. The reason became clear when the behavior of the third strategy was examined. In this sequence strategy, a stable equilibrium develops with variety *A* on 90% of the area suffering a constant low level of attack by race (*a*). There is a yearly 7% loss of yield. Whereas race (*a*) can just maintain a population on a totally susceptible crop, reference to Figure 9.9A shows that, with SS present, the more complex races only decrease. The strategies involving all three genes simultaneously or in quick succession are therefore permanently safe from attack.

The implausibility of this result suggests that the modeled strength of SS is too high and/or that selection for fitness (aggressiveness) within races cannot be ignored. There are reasons to believe that the SS assumed is indeed too strong. In the absence of any information on the pustule density used in Watson and Singh's (1952) experiment, I assumed arbitrarily that it was 1%. To give the resource competition between races that van der Plank (1963, p231) believes was probable in this experiment, the pustule density would have to be higher. Hence the value of k_3 probably needs to be reduced (*see* Figure 9.1A), so weakening the SS effect in the simulations. In addition to this, it can be argued that the known variation in reproductive fitness within mutant races (Watson, 1970), combined with intense *r*-selection at the beginning of an epiphytotic, would lead to races rapidly overcoming handicaps associated with extra virulence. If this were not so, as varieties are bred with more and more resistance genes, newly virulent races would be progressively less damaging. Because there is no evidence of this, and also because of the evidence cited earlier that in many cases SS is weak or absent, the assumption of "no SS" made in the second series of simulations may be more realistic.

The second series of simulations showed that without SS the average breakdown rate of resistance genes is about the same under all strategies. Leonard's (1969c) recommendation of a superline variety in a situation where SS is absent is upheld by the simulations, although the total protection that it affords is the longest by only one year. If an ideal 100% cover of superline had been assumed, its advantage would certainly have been much greater.

The third series of simulations showed that CP in this system extended the protection relatively very little. As with SS, there is reason to suppose that the strength of the CP has been greatly exaggerated. The rust cover used in Johnson and Allen's (1975) experiment was considerably higher than the assumed 1% (R. Johnson, personal communication). Reference to Figure 9.1B shows that k_4 should therefore have a smaller value, and hence a smaller effect on the results. However, more recent work has confirmed the importance of cross-protection in the field (Johnson and Taylor, 1976; Chin, 1979). Also, it has been found that avirulent rust spores can trigger a degree of systemic immunity (Hoes and Dorrell, 1979). Although CP at the assumed strength has influenced the present results very little, a study using the model has identified a multiline configuration where CP at this strength but without SS leads to a stable equilibrium with some disease, but with negligible loss of yield. Is this, perhaps the "optiline"? It seems that more attention should be paid to CP; it has rarely been considered in models.

Although a number of modeling studies have evaluated gene introduction strategies using race frequencies at equilibrium, such a criterion seems only distantly related to the concerns of farmers and research institutes. To calculate the more relevant yield criteria, it is necessary to have estimates of the absolute population size through time and a damage function. If stable equilibria are to be exploited in real-world climates, other criteria to be considered may be their sensitivity to environmental variation and the rate of the systems' return to equilibrium after pertubation. Although it has been claimed that equilibria among pathogen races will be either neutral or unstable (Barrett *et al.*, quoted in Barrett and Wolfe 1978), the equilibria encountered in the present study seem at least locally stable; their properties will be reported elsewhere.

To make the most useful evaluation of gene introduction strategies, the target farming enterprise and all its costs must be considered. This has been attempted by Macpherson (1977) for a rich Western agrobusiness and for a poor subsistence farmer. Optimizations were carried out using either discounted yields or the farmer's nutritional status as the criterion of utility. However, the differing costs of breeding the plant material for the various strategies have not yet been considered. Although the traditional sequence of single-gene varieties involves costs of monitoring disease and costs from occasional lost yield, postponment of the early need for numerous new resistance genes may lead to considerable savings. If disease is to be rationally managed, the managers should avoid too narrow a focus on the disease.

REFERENCES

Asai, G. N. (1960) Intra-and inter-regional movement of urediospores of black stem rust in the upper Mississippi River Valley. *Phytopathology* 50:535–41.

Barrett, J. A. and M. S. Wolfe (1978) Multilines and super-races—a reply. *Phytopathology* 68:1535–7.

Brown, J. F. and E. L. Sharp (1970) The relative survival ability of pathogenic types of *Puccinia striiformis* in mixtures. *Phytopathology* 60:529–33.

Browning, J. A. and K. J. Frey (1969) Multiline cultivars as a means of disease control. *Ann. Rev. Phytopathol.* 7:355–82.

Burdon, J. J. (1978) Mechanisms of disease control in heterogeneous plant populations—an ecologist's view, in P. R. Scott and A. Bainbridge (eds) *Plant Disease Epidemiology* (Oxford: Blackwell) pp193–200.

Burdon, J. J. and G. A. Chilvers (1977) Controlled environment experiments on epidemic rates of barley mildew in different mixtures of barley and wheat. *Oecologia (Berlin)* 28:141–6.

Burdon, J. J. and R. Whitbread (1979) Rates of increase of barley mildew in mixed stands of barley and wheat. *J. Appl. Ecol.* 16:253–8.

Chin, K. M. (1979) *Aspects of the Epidemiology and Genetics of the Foliar Pathogen, Erysiphe graminis f. sp. Hordei, in Relation to Infection of Homogeneous and Heterogeneous Populations of the Barley Host (Hordeum vulgare)*. PhD Thesis, Cambridge University.

Fleming, R. A. and C. O. Person (1978) Disease control through use of multilines: a theoretical contribution. *Phytopathology* 68: 1230–3.

Flor, H. H. (1956) The complementary genic systems in flax and flux rust. *Adv. Genet.* 8:29–59.

Frey, K. J. and J. A. Browning (1971) Breeding crop plants for disease resistance, in *Mutation Breeding for Disease Resistance* IAEA Publ. No. STIPUB 271 (Vienna: IAEA) pp45–54.

Frey, K. J., J. A. Browning and M. D. Simons (1973) Management of host resistance genes to control diseases. *Z. für Pflanzenkrankh. Pflanzenschutz* 80:160–80.

Groth, J. V. (1976) Multilines and "super-races": a simple model. *Phytopathology* 66:937–9.

Hoes, J. A. and D. G. Dorrell (1979) Detrimental and protective effects of rust in flax plants of varying age. *Phytopathology* 69:695–8.

Johnson, R. and D. J. Allen (1975) Induced resistance to rust diseases and its possible role in the resistance of multiline varieties. *Ann. Appl. Biol.* 80:359–63.

Johnson, R. and A. J. Taylor (1976) Effects of resistance induced by non-virulent races of *Puccinia striiformis. Proc. 4th European and Mediterranean Cereal Rusts Conf., September 1976, Interlaken*, pp49–51.

Jones, F. G. W. (1966) *Rothamsted Report for 1965* pp301–316.

Katsuya, K. and G. J. Green (1967) Reproductive potentials of races 15B and 56 of wheat stem rust. *Can. J. Bot.* 45:1077–91.

Kiyosawa, S. (1972) Theoretical comparison between mixture and rotation cultivations of disease-resistant varieties. *Ann. Phytopathol. Soc. Japan* 38:52–9.

Kiyosawa, S. (1976) A comparison by simulation of disease dispersal in pure and mixed stands of susceptible and resistant plants. *Jap. J. Breeding* 26:137–45.

Kiyosawa, S. (1977) Some examples of pest and disease epidemics in Japan and their causes, in P. R. Day (ed) *The Genetic Basis of Epidemics in Agriculture Ann. NY Acad. Sci.* 287:35–44.

Kiyosawa, S., N. Matsumoto, and M. Inoue (1982) The use of two resistance genes to control plant diseases *SABRAO J.*

Kiyosawa, S. and M. Shiyomi (1972) A theoretical evaluation of the effect of mixing resistant variety with susceptible variety for controlling plant diseases. *Ann. Phytopathol. Soc. Japan* 38: 41–51.

Kiyosawa, S. and M. Shiyomi (1976) Simulation of the process of breakdown of disease-resistant varieties. *Jap. J. Breeding* 26:339–52.

Leonard, K. J. (1969a) Factors affecting rates of stem rust increase in mixed plantings of susceptible and resistant oat varieties. *Phytopathology* 59:1845–50.

Leonard, K. J. (1969b) Selection in heterogeneous populations of *Puccinia graminis* f. sp. avenae. *Phytopathology* 59: 1851–6.

Leonard, K. J. (1969c) Genetic equilibria in host–pathogen systems. *Phytopathology* 59:1858–63.

Littlefield, L. J. (1969) Flax rust resistance induced by prior inoculation with an avirulent race of *Melampsora lini. Phytopathology* 59: 1323–8.

MacKenzie, D. R. (1979) The multiline approach to the control of some cereal diseases. *Proc. Rice Blast Workshop* (Los Banos, Philippines: International Rice Research Institute) pp199–216.

Macpherson, D. K. (1977) *Optimal Use of Crop Disease-Resistance Genes*, PhD Thesis, Australian National University, Canberra.

Martens, J. W. R. (1973) Competitive ability of oat stem rust races in mixtures *Can. J. Bot.* 51:2233–6.

Martens, J. W. R., R. I. H. McKenzie and G. J. Green (1970) Gene-for-gene relationships in the *Avena:Puccinia graminis* host–parasite system in Canada. *Can. J. Bot.* 48:969–75.

Ogle, H. J. and J. F. Brown (1970) Relative ability of two strains of *Puccinia graminis tritici* to survive when mixed. *Ann. Appl. Biol.* 66:273–9.

Ou, S. H. (1977) Genetic defence of rice against disease, in P. R. Day *The Genetic Basis of Epidemics in Agriculture Ann. NY Acad. Sci.*, 287:275–286.

Painter, R. H. (1958) Resistance of plants to insects. *Ann. Rev. Entomol.* 3: 267–90.

Person, C. O., J. V. Groth and O. M. Mylyk (1976) Genetic change at the population level in host–parasite systems. *Ann. Rev. Phytopathol.* 14:177–88.

van der Plank, J. E. (1963) *Plant Diseases: Epidemics and Control* (New York: Academic Press).

van der Plank, J. E. (1968) *Disease Resistance in Plants* (New York: Academic Press).

Puckridge, D. W. and C. M. Donald (1967) Competition among wheat plants sown at a wide range of densities. *Aust. J. Agric.* 18:193–211.

Rastegar, M. F. (1976) Competitive ability of an induced mutant race of *Puccinia hordei* Otth in mixtures. *Proc. 4th European and Mediterranean Cereal Rusts Conf., Interlaken, September 1976*, pp58–59.

Robinson, R. A. (1977) Plant pathosystems, in P. R. Day *The Genetic Basis of Epidemics in Agriculture. Ann. NY Acad. Sci.* 287:238–42.

Stevens, N. E. and W. O. Scott (1950) How long will present spring oat varieties last in the Central Corn Belt? *Agron. J.* 42:307–9.

Stevens, R. B. (1949) Replanting "discarded" varieties as a means of disease control. *Science* 110:49.

Suneson, C. A. (1960) Genetic diversity—a protection against plant diseases and insects. *Agron. J.* 52:319–21.

Thurston, H. D. (1961) The relative survival ability of races of *Phytophthora infestans* in mixtures. *Phytopathology* 51:748–55.

Trenbath, B. R. (1977) Interactions among diverse hosts and diverse parasites. *Ann. NY Acad. Sci.* 287: 124–50.

Watson, I. A. (1970) Changes in virulence and population shifts in plant pathogens. *Ann. Rev. Phytopathol.* 8:209–30.

Watson, I. A. and D. Singh (1952) *J. Aust. Inst. Agric. Sci.*, 18:190–7.

Wolfe, M. S. and J. A. Barrett (1976) The influence and management of host resistance on control of powdery mildew on barley. *Proc. 3rd Int. Barley Genetics Symp.*, pp433–9.

de Wit, C. T. (1960) On competition. *Versl. landbouwk. Onderz.*, 668:1–82.

Yarwood, C. E. (1956) Cross-protection with two rust fungi. *Phytopathology* 46:540–4.

Zadoks, J. C. (1961) Yellow rust on wheat: Studies in epidemiology and physiological specialization. *Neth. J. Plant Pathol.* 67:69–256.

10 Strategies for Cotton—Boll Weevil Management in Texas

Guy L. Curry and *James R. Cate*

10.1. INTRODUCTION

Cotton (*Gossypium hirsutum* L.) is grown commercially across the subtropical regions of the southern USA; in the 1978 growing season it produced a yield of US $1.5 billion. One of the key pests, from Texas eastward, is the boll weevil, *Anthonomus grandis* Boheman. The damage attributed directly to boll weevil in Texas alone was $4.6 million in 1978. Other pests are estimated to have contributed $14.5 million in additional losses, a large proportion from secondary pest outbreaks of *Heliothis* spp. Historically, 10–15 spray applications annually were used in boll weevil and subsequent *Heliothis* spp. control. There are several natural enemies of boll weevil, but under present production systems they alone do not appear to provide economic control. However, predators and parasites generally maintain the *Heliothis* spp. under economic control, unless the system is upset by spraying for the boll weevil.

In recent years growers in Texas have begun to adopt a variety of approaches to boll weevil control that are less reliant on heavy pesticide use. One approach has been to abandon the traditional production policy of producing a top crop, a second growth phase after the first set of bolls has matured. The top crop greatly extends the length of the growing season, allowing the insect population to increase to economic numbers, as well as contributing to the number of late-season migrating and diapausing weevils. Growing cotton with low inputs of water and nitrogen, in conjunction with varieties that retain a higher percentage of early fruit, also serves to reduce the development and reproduction rates of the weevil populations. Growers practicing these production methods obtain yields that are equal or only slightly reduced, but profit from reduced production expenditures and escape serious pest problems. In addition, early-season sprays for the boll weevil appear to be less likely to disrupt the predator–parasite complex that controls the *Heliothis* spp.

With these approaches the average number of sprays per season has been reduced to 4.4 (Walker *et al.*, 1978), with 67% of the fields never being sprayed.

The general reduction in pesticide usage and the adoption of a sound integrated control philosophy by cotton producers has further provided the opportunity to develop and implement additional management tactics. It is highly likely that additional crop and pest management strategies can be employed to reduce further pesticide, fertilizer, and irrigation usage in cotton. Further advancement is likely to utilize many strategies in concert, each one contributing relatively low levels of population suppression. It is also possible that, due to the density-responsive resilience of the biotic factors, many of these tactics may affect the role of other strategies either detrimentally or beneficially.

Such complex, low-level population suppressive strategies in the cotton crop are difficult, if not impossible, to evaluate in the field, so that a boll weevil–cotton crop model has been developed by Curry *et al.* (1980) to evaluate potential control practices and decision processes. It is hoped that by careful selection of the most important parameters to be measured, subsequent field experimentation can determine the validity of results from the model analyses. So far, the model has been utilized to evaluate strategies of host plant resistance, and biological and insecticidal control. It has also been used to examine the economic threshold for insecticide application decisions as it currently exists, and as it might be developed on a seasonal basis.

In this chapter first we give more details of the major features of the weevil–cotton crop system. The model is then briefly described and the results are presented of the comparative evaluations we have carried out of insecticidal and biological control and the use of host plant resistance.

10.2. THE BOLL WEEVIL–COTTON SYSTEM

10.2.1. Cotton Phenology

In Texas typical cotton yields are 0.75 bales per *acre* in the blacklands of Central Texas to 2.5 bales per acre in the Brazos River and Rio Grande Valleys of eastern and southern Texas. At typical planting densities, the yields are 3–10 mature bolls per plant. The standard commercial varieties in the Brazos River Valley (Stoneville 213 and TAMCOT SP37) will initiate squares (floral buds) over an approximately 40-day period reaching a peak number of 25–30 per plant. As the squares increase in size until bloom occurs (28 days), the number and size of the buds cause a decline in the rate of bud initiation due to an increasing photosynthate demand. From bloom, the bolls (fruit) rapidly increase in size causing a further decrease in square initiation. The number of bolls per plant peaks at about two-thirds the maximum number of squares, about 50 days after first square initiation. Essentially all squares and about one-third of the bolls are abscised by the plant, mainly due to photosynthate demand. Only fruit (squares and bolls) within a restricted age interval (12–40 days) are dropped in an attempt to balance the photosynthate supply–demand relationship. Once bolls reach the age beyond which abscission does not occur, they generally mature into open bolls or lint, although they possibly may be reduced in size due to photosynthate availability. This process takes another 40 days under typical eastern Texas conditions.

10.2.2. The Weevil Life Cycle

The female boll weevil attacks squares and bolls within the 7–24 mm diameter class by making a feeding puncture, depositing an egg into the fruit via the puncture, and then sealing it with frass. The fruit are abscised by the plant 4–8 days later due to the damage and/or chemicals produced by the immature larva feeding within the fruit. Weevils selectively avoid small squares and large bolls. Small-diameter squares have excessive drying rates and thus inadequate food supplies for proper development; older bolls exhibit a hardening of the epidermal layer and have significantly lower average nitrogen contents. The female weevil lays 150–250 eggs during her adult life, while the adult male is much more sedentary, feeding and damaging only a small number of fruit, mainly squares. The other major insect pests in Texas are the cotton bollworm (*Heliothis zea*) and the tobacco budworm (*H. virescens*) (Figure 10.1).

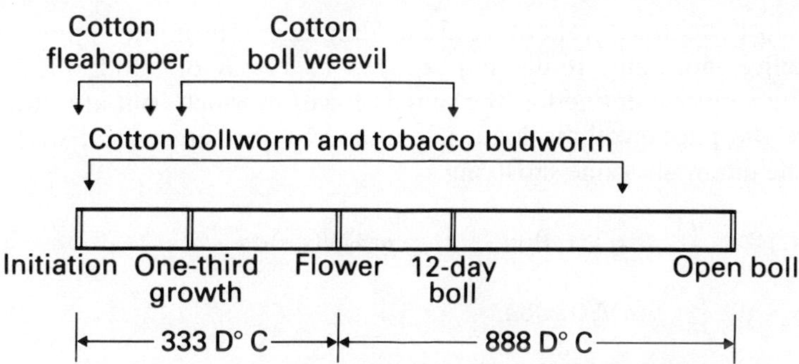

FIGURE 10.1 Phenology and pests of the cotton fruit in Texas.

The boll weevil overwinters in the adult reproductive–diapause state in leaf and plant litter near cotton fields. As air temperatures rise in the spring, the weevil migrates into the cotton fields. When the population of squares of acceptable size reaches 10–15 000 per acre, the weevils aggregate and reproduce within the cotton fields. Migration between fields apparently decreases and the populations develop within each field according to their initial size and the crop status. As fall nears, the crops begin to mature and a migration between fields is again observed. During this phase almost all the surrounding area becomes infested and a significantly increasing proportion of the weevils diapause and seek overwintering sites. During the midsummer period, the developmental time of the weevil varies from 2–3 weeks with a median of about 18 days. This rate is highly dependent on the nitrogen level of the food source, with development in low-nitrogen bolls taking approximately 15% longer than in squares that are high in nitrogen. The nitrogen level also apparently affects the reproductive status and levels. Weevils fed on squares and pollen grains from flowers have high reproductive rates and almost no incidence of diapause, while those fed on bolls have low reproductive levels and produce a high proportion of diapausing adults.

10.3. THE MODEL

10.3.1. Crop Fruiting

The crop fruiting model uses von Foerster's (1959) time–age partial differential equation approach to characterize an average plant. Since fruit ages from different initiation times it is necessary to use both physiological time t and physiological age a as independent coordinates. Thus the number of fruits at time between ages a_1 and a_2 is given by

$$\int_{a_1}^{a_2} N(t,a)\mathrm{d}a.$$

Fruit is lost due to natural or forced shedding caused by limited photosynthate availability, normally 16-day-old squares to 12-day-old bolls. The forced shedding window, defined as the range of ages in which fruit are dropped to balance the photosynthate load, is denoted by the interval $[a, s(t)]$, where $s(t)$ is the minimum value satisfying:

$$A(t) \geqslant \int_0^{a} p(a)N(t,a)\mathrm{d}a + \int_{s(t)}^{\bar{a}} p(a)N(t,a)\mathrm{d}a + \int_{\bar{a}}^{\infty} p(a)N(t,a)\mathrm{d}a, \tag{10.1}$$

where $A(t)$ is the available photosynthate at time t, $p(a)$ is the photosynthate demand for fruit of age a. If eqn. (10.1) has no solution with $s(t)$ less than $\bar{a}$, then, by definition, $s(t)$ is set equal to $\bar{a}$; the interval $[\underline{a}, \bar{a}]$ is the maximum possible shedding window. The equations for $N(t,a)$, for $t \geqslant 0$ and $a > 0$, can now be given:

$$\frac{\partial N}{\partial t} + \frac{\partial N}{\partial a}\frac{\mathrm{d}a}{\mathrm{d}t} = \mu(a)\, N(t,a) \qquad \text{for } \underline{a}\varepsilon(a,s(t)] \tag{10.2}$$

$$N(t,a) = 0 \qquad \text{for } a\varepsilon[\underline{a}, s(t)] \tag{10.3}$$

Equation (10.2) arises due to natural shedding where $\mu(a)$ is the natural shedding rate and eqn. (10.3) represents instantaneous forced abscission for all fruit in the shedding window.

The initiation of new fruiting sites is given as a function of photosynthate availability as one of the initial conditions for eqns. (10.2) and (10.3). This specifies the potential yield.

The change in fruit numbers with respect to time and age is then approximated by a numerical solution in which eqns. (10.2), (10.1), and (10.3) are iteratively evaluated in sequence. The crop is then modeled as eleven average plants based on the timing of boll weevil damage. One of the plants is always the average unattacked plant and, thus, gives a dynamic basis for yield comparisons. The number of plants in each category is maintained so that the average field result can be computed. Figure 10.2 shows the single-plant model against experimental results for Stoneville 213 in the Brazos River Valley.

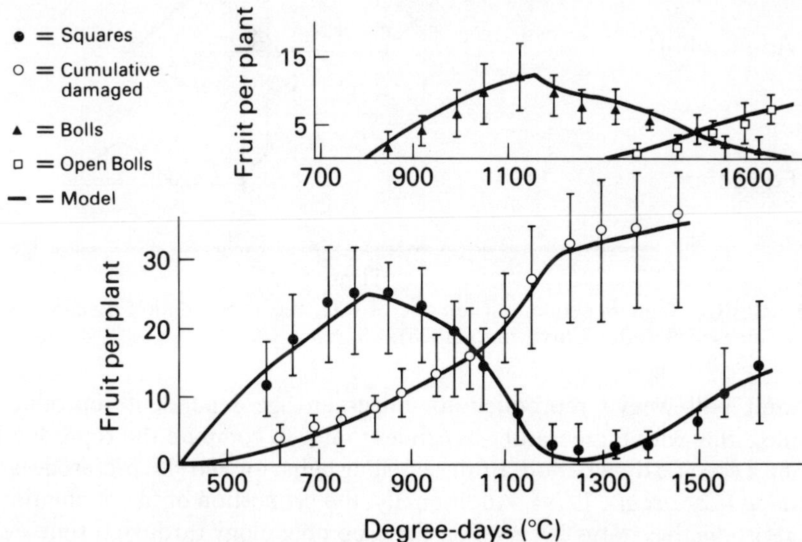

FIGURE 10.2 Model comparisons with Keener's 1974 experimental data for Stoneville 213 (after Curry *et al.*, 1980).

10.3.2. The Boll Weevil

The boll weevil population is modeled on a physiological time scale as a collection of cohorts, each cohort consisting of all those insects starting a specific stadium during the same time interval (Figure 10.3). The discrete cohort concept utilized here is analagous to the numerical solution procedures necessary to solve the von Foerster-type model (Wang *et al.*, 1977).

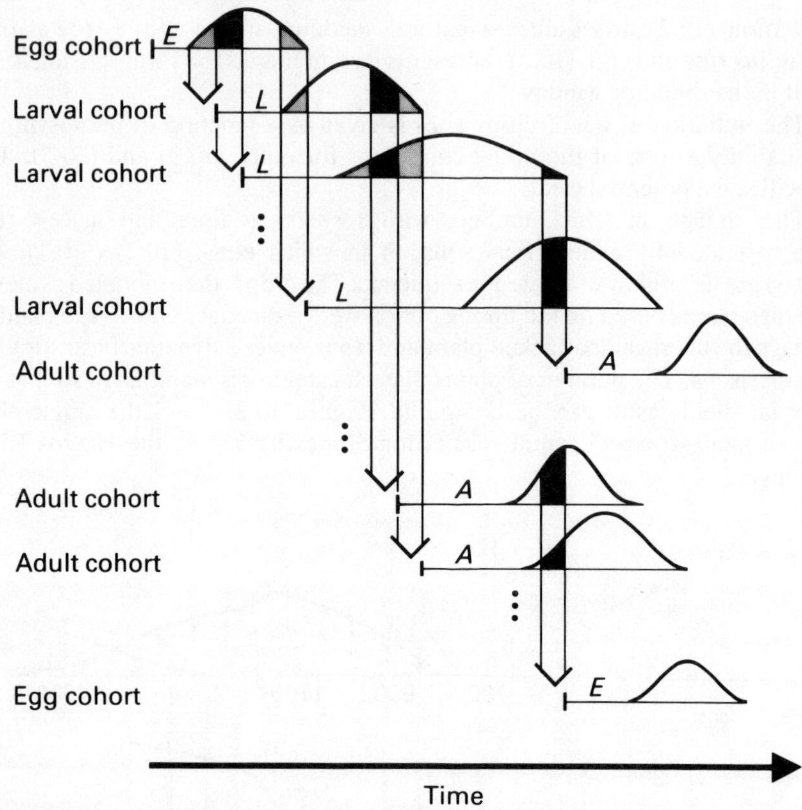

FIGURE 10.3 Time-dependent schematic of boll weevil population dynamics with stochastic aspects (after Curry *et al.*, 1978).

Adult boll weevil reproduction utilizes an age-dependent reproductive profile along with temperature-dependent rates to compute the reproduction during a specific time period. Fruit searching behavior and site preferences are included (Cate *et al.*, 1979). Additionally, the proportion of the population in the reproductive status is based on the crop phenology through a time delay functional relationship between the proportion of squares and bolls attacked. Adult natural survival is modeled by an age-independent survival fraction obtained from insect data.

Immature weevil development requires several distinct developmental scales based on the proportions of the population in each possible temperature environment. Additionally, the developmental rates are different in squares and bolls. The different microclimates are necessitated since abscized fruit can fall into a fully shaded region, partial shade, or mid-row, where they receive considerable direct sunlight. The proportions of the fruit in the different environments is a dynamic property of the crop (Mann *et al.*, 1980).

Immature survival is a function of the environment and fruit type via the associated developmental rates and the diameter of the fruit. Fruit size is important due to differing drying rates and resulting mortality of immatures if drying outruns development. This characteristic of the system necessitates the age–number model for fruit described above.

A comparison of adult counts with model predictions for four years (1965–68) of data is given in Figure 10.4. A 1978 data set, including counts of plant fruit and egg, immature, and adult boll weevils, is compared with the model results in Figure 10.5. From these and extensive validation studies on individual process submodels, the cotton-boll weevil model appears to provide a comprehensive and biologically correct description of the system. The model is certainly realistic enough to permit its use in preliminary pest management control strategy evaluations.

10.4. CONTROL STRATEGY EVALUATIONS

The effects of control strategies such as biological control, host–plant resistance, cultural modifications and dynamic spraying thresholds have so far not been assessed in detail in terms of population response, either in theory or practice. However, short-season varieties in conjunction with overwintering weevil spraying control has been shown to yield excellent results in the field in certain years (Walker *et al.*, 1978). Here we analyze several of these forms of control strategies using the cotton-boll weevil model.

10.4.1. Insecticides

Insecticides applied to control the boll weevil kill only the adults, since larvae are enclosed in the flower bud or fruit. The statewide standard for indicating that chemical control of the boll weevil is needed is when 10% of the squares on the plants have ovipositional punctures. This threshold approach results in a varying number of applications of insecticide, depending on the crop yield potential and the number and timing of overwintering weevil infestations. The use of a fixed threshold also frequently indicates that control is needed when the crop has already set the fruit that ultimately produce the yield. For example, when plants slow their bud initiations due to growing fruit photosynthate demand, the late-season weevil population increase produces a high relative abundance of weevils to squares and high damage percentages. Insecticides are then applied but they have little or no effect on the ultimate crop yield. Figure 10.6 shows the results of a spraying policy based on a threshold of 10% damaged squares, together with the number of squares (buds) per acre for the unattacked (weevil-free) field and an equivalent field receiving the prescribed treatment. The shaded region highlights the difference in the fruiting patterns. The numbers associated with the arrows showing

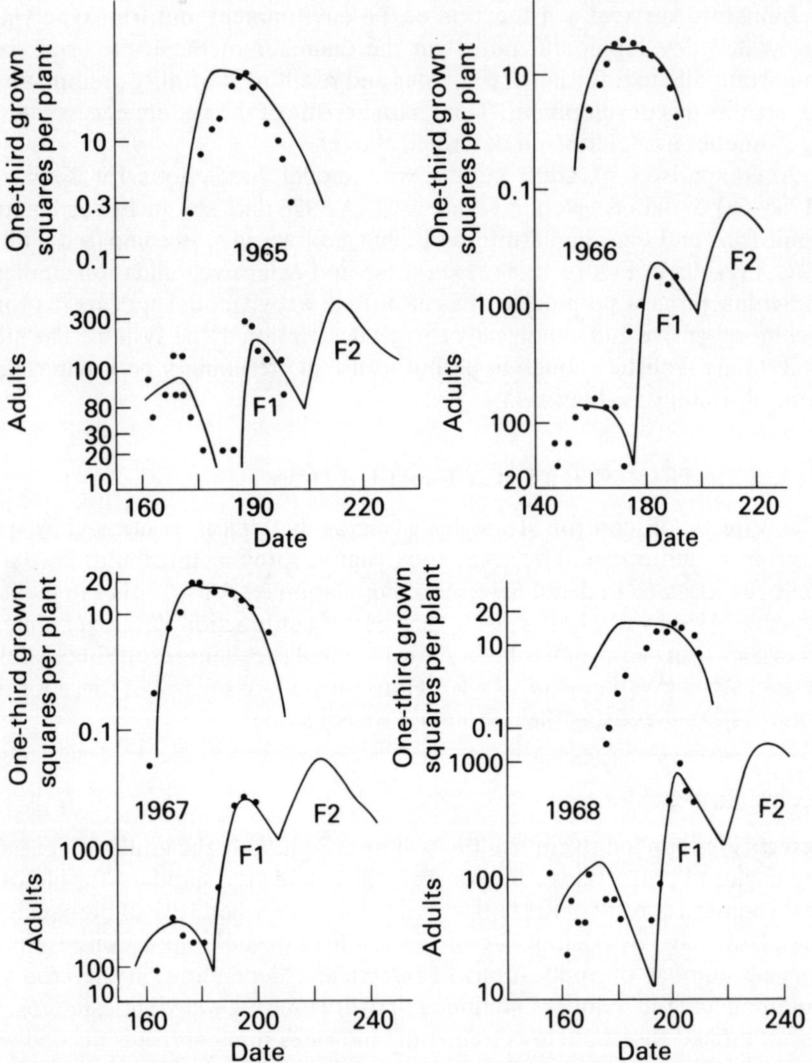

FIGURE 10.4 Model comparisons (full curves) with field data (•) of Walker for adult boll weevils per acre and one third grown squares per plant at College Station, Texas, 1965–68 (after Curry *et al.*, 1980).

insecticide timing are the percentage yields, as compared with those of the unattacked plant if each treatment were the last in the series. The last four treatments were made at low square density levels and contributed virtually nothing to crop yield. The timing of these sprays was such that a secondary outbreak of *Heliothis* spp. would almost certainly occur, necessitating a

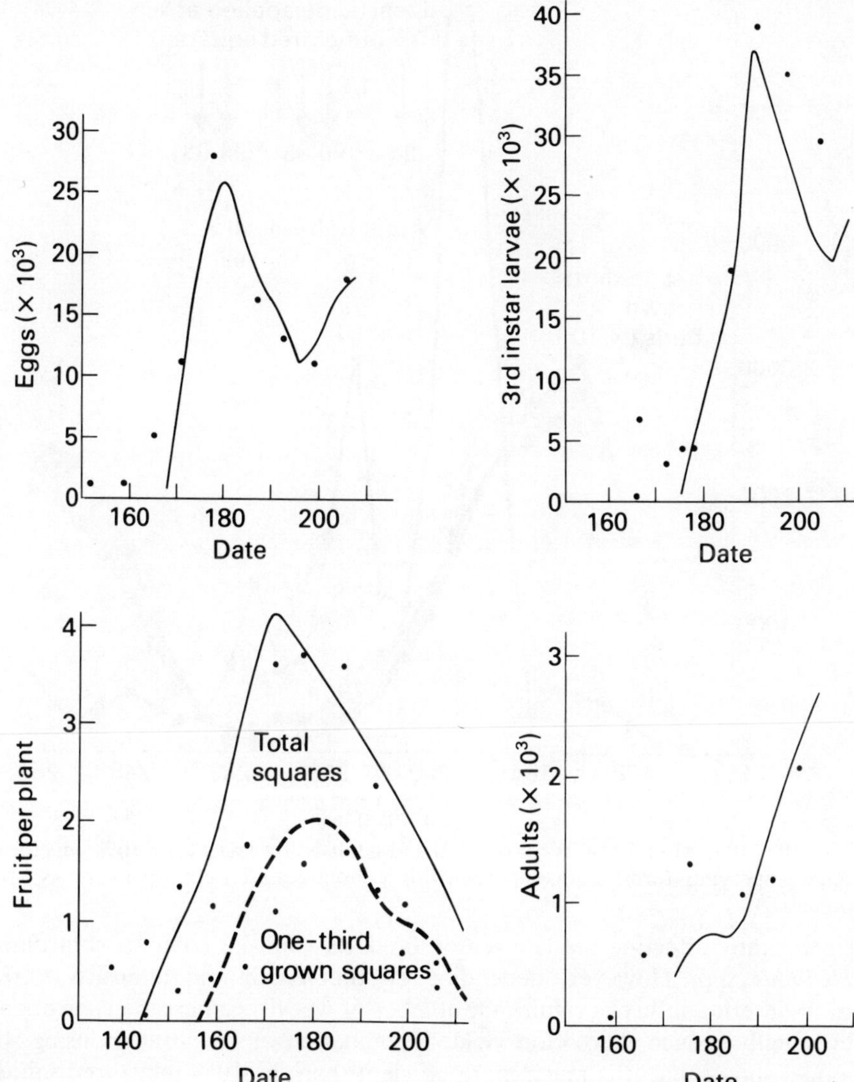

FIGURE 10.5 Model comparisons (full lines) with Cate experimental data (•) for a cotton field at Rosebud, Texas, 1978.

continuing treatment schedule beyond the six needed to protect against weevil damage.

An alternative control strategy is that of treating for overwintering boll weevils early in the season. This approach involves the use of three insecticide applications four days apart, starting with the first one-third grown square (12-day-old flower bud). This does result in improved control, without

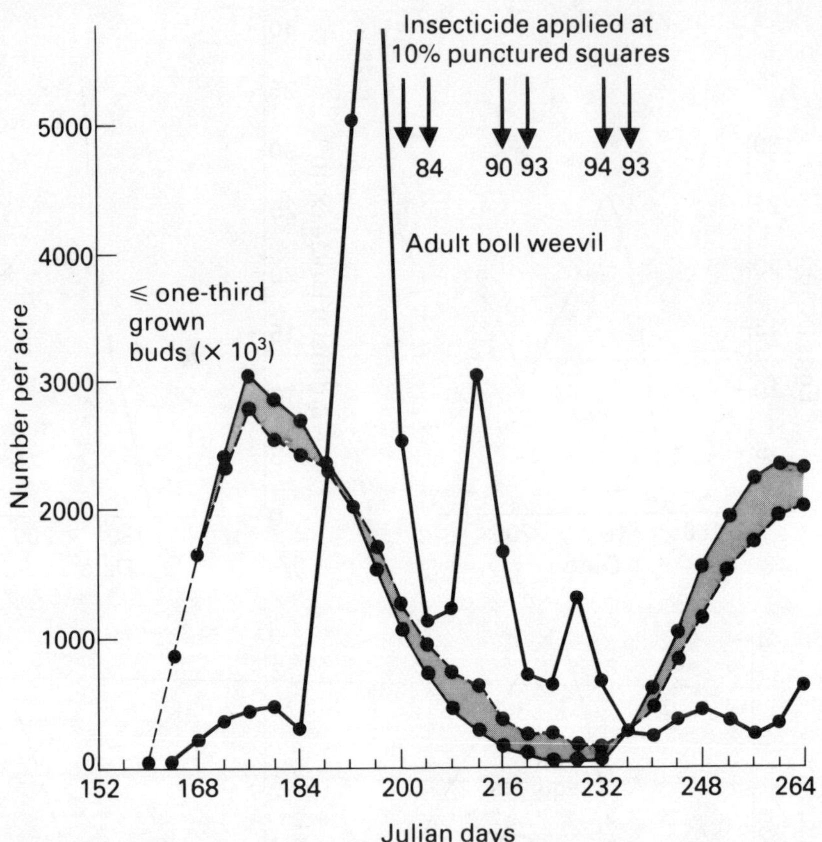

FIGURE 10.6 Model results for squares and adult boll weevils for a 10% punctured square spraying threshold compared with a noninfested field (yield = 93% of potential).

significantly affecting the late-season predator–parasite complex controlling *Heliothis* spp. However, depending on the length and intensity of the overwintering influx, a significant number of weevils can escape control and eventually reduce the cotton yield. Examination of this strategy using the model shows that it is much more efficient than the 10% punctured square threshold policy. The comparative efficiency in terms of actual versus potential revenue (minus insecticide cost) is illustrated in Figure 10.7. Different numbers of applications were obtained by varying the damage thresholds (full curve). Different thresholds generally cause the same number of applications to be made, but they produce differences in the timing of applications resulting in different levels of control. *Six or seven applications were needed to gain acceptable returns using the linear threshold of 10% oviposition punctures, whereas the treatments for overwintered weevils gave high rates of return with only three or four applications.*

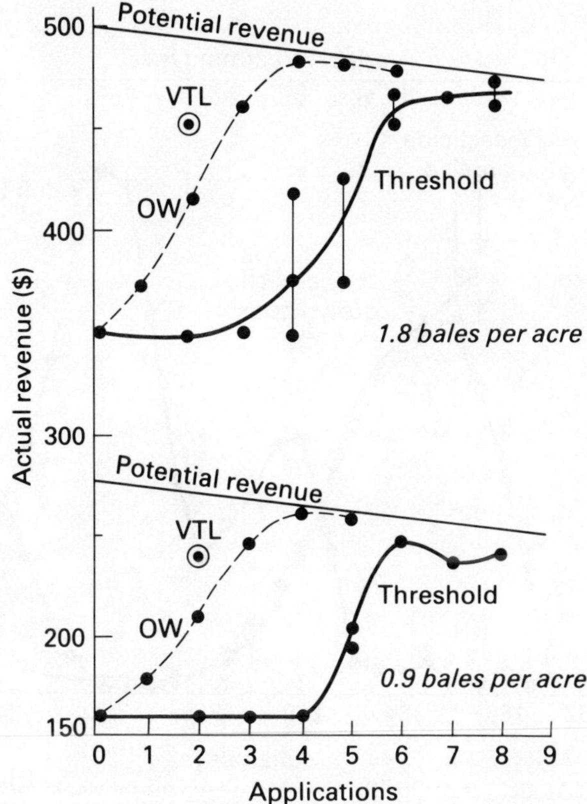

FIGURE 10.7 Economic yield comparisons for two production levels (0.9 and 1.8 bales per acre) for the variable treatment threshold (•), overwintering sprays – – –, and the fixed threshold policy —.

These results generally agree with observations from field situations throughout Texas. However, because of the dynamic nature of the fruit and weevil populations and their interactions, a much more appropriate threshold would be one that is itself dynamic. Such a threshold has been developed based on giving priority to the early fruit, with a decreasing protection for late-season buds. The treatment levels are 1% damaged buds until first bloom, then 25% damaged buds until bolls first reach the upper age region of weevil damage susceptibility (12-day-old boll), and a 75% damage threshold for the remainder of the season. Figure 10.8 presents the variable-threshold approach for the same conditions that was used for the 10% threshold approach in Figure 10.5. The variable treatment level threshold method results in better early-season control and essentially ignores late-season bud regrowth control, in line with short-season production practice; the results are shown in Figure 10.7 as circled data points (VTL). *Just two applications are as effective as three against overwintered weevils, and six treatments using the 10% linear threshold.*

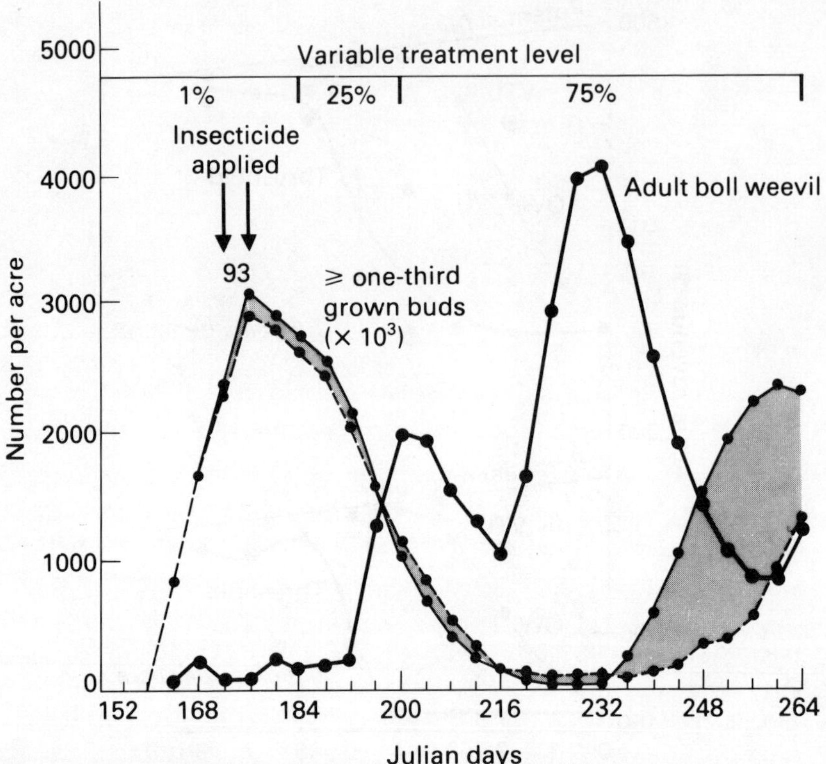

FIGURE 10.8 Model results for the variable treatment threshold method based on crop phenology compared with a noninfested field (yield = 93% of potential).

Additionally, the treatments for overwintered weevil populations are largely prophylactic in their initiation. A variable threshold permits proper evaluation of the potential damage and determination of the need for treatment based on insect population estimates. It is likely that many "automatic" treatments for overwintered weevils could be eliminated by using such a method. *It is evident from these comparisons that a variable treatment level threshold should be considered in boll weevil control. The reduction in the number of insecticide treatments required, together with increases in cotton yields and the avoidance of secondary pest outbreaks make this strategy particularly attractive*. Based on this analysis, a field evaluation of the variable-threshold method is planned.

10.4.2. Natural Enemies

At the present time, the braconid wasp (*Bracon mellitor* Say) is the only natural enemy of the boll weevil in Texas that appears to have any major impact on population numbers, although its full impact has not been adequately evaluated and has received little attention as a dependable biotic control agent. Several researchers are attempting to discover other natural enemies of the weevil and related species, particularly in Central America. However, we need to know what response to density these predators or parasites must have to produce successful weevil control, and a theoretical analysis of this has been made utilizing the cotton-boll weevil model.

A rectangular hyperbola was used to describe the increase in searching efficiency occurring with increasing host density. All known parasites of the boll weevil attack the third instar larvae (L), such that

$$p(L) = \frac{\beta L}{\alpha + L}, \tag{10.4}$$

where p is percentage parasitism, β is the maximum searching efficiency and α is the number of larvae at which one half of the maximum host-finding efficiency is reached. The lower the value of α, the more effective the parasite is at finding hosts at low densities. As a basis of comparison, data published by Bottrell (1976) indicate that *Bracon mellitor* has approximate parameter values of $\alpha = 0.24$ per 1000 larvae per acre, and $\beta = 0.85$. Density-dependent response curves are evaluated for potential weevil control in Table 10.1. For comparison, no control, the overwinter spraying policy, and the 10% damage threshold methods are also shown. *The control achieved by introducing natural enemies is considerably less than that using insecticides, but clearly it could make a significant contribution as part of an integrated pest management strategy.*

TABLE 10.1 Economic analysis of the effect of various density dependent response curves for immature weevil parasites.[a]

Density-dependent mortality			
% Parasitized at 1000 weevils/acre	Asymptote	Potential yield (%)	Crop value ($)[b]
No control		54	138
3 Insecticide applications to overwintered adults		87	206
10% damaged squares		91	208
0.07	0.25	56	141
0.07	0.60	59	147
0.15	0.60	62	154
0.15	0.85	67	169
0.24	0.85	72	182
—	—	—	—
0.35	0.95	82	205
0.50	0.90	83	208

[a]Production level at 0.8 bales/acre.
[b]Cotton price at 50¢/lb: insecticide application at $3.00/acre.

10.4.3. Host Plant Resistance

The boll weevil has essentially one host—cotton—in the agricultural regions of Texas. With such a close relationship between the host and the pest, variations in plant characteristics offer the possibility of adversely affecting the weevil to a significant degree. Okra leaf cottons, for example, produce slight microclimate changes, as does the low-input–nonirrigation production practice associated with short-season varieties. More significant, perhaps, are host plant resistance characteristics that affect reproduction, development, and immature weevil survival.

The cotton-boll weevil model is detailed enough in the mechanisms affecting immature weevil survival that changes in fruit density can be assessed at the crop population level. A change in the rate of square drying can thus be seen to alter the proportion of surviving immatures. Other plant-related changes, such as the nitrogen available in the fruit, affect the length of the weevil's developmental period and its reproductive potential. Reproduction is also adversely affected by a physical characteristic of the flower buds called the frago-bract. Cottons with this have bracts that are twisted away from the bud, thus greatly increasing the exposure of the insect while it attacks the bud. Thus there is an opportunity to reduce the weevil's reproductive rate by 50% if small-plot responses can be extended to the crop population level. However, none of the host plant resistance characteristics alone appear to facilitate weevil control, although they may be effective when applied in concert. Table 10.2 shows the effects of a 20% change in reproduction, development rate, and fruit drying rate, individually and in combination. When all three

TABLE 10.2 Economic analysis of the effects of various host–plant resistance characteristics on boll weevil control.[a]

Parameters evaluated				
Fecundity decreased	Developmental rate decreased	Fruit drying rate increased	Potential yield (%)	Crop value ($)[b]
No control			54	138
3 Insecticide applications to overwintered adults			87	206
Threshold of 10% damaged squares (%)			91	208
20	0	0	62	155
0	0	20%	60	149
0	20	0	73	184
20	20	0	82	205
0	20	20%	84	211
20	20	20%	89	224

[a]Production level at 0.8 bales/acre.
[b]Cotton at 50¢/lb: insecticide application at $3.00/acre.

characteristics are used, crop values are increased over traditional insecticide control. Even better results are obtained when a parasite is added to the system. Table 10.3 shows that *parasitism, together with a 20% reduction in the insect development rate produced by host resistance, provides excellent control and high yields, without the use of insecticides*.

TABLE 10.3 Economic analysis of a combined biotic and host–plant resistance characteristics.[a]

Developmental rate decreased	% Parasitism at 1000 weevils/acre	Asymptote	Potential yield(%)	Crop value ($)[b]
No control		0	65	318
3 Insecticide applications to overwintered adults			90	429
Threshold 10% punctured squares (%)			93	436
20	0.07	0.60	93	451
20	0.24	0.85	97	472

[a]Yield level at 1.6 bales/acre
[b]Cotton at 50¢/lb: insecticide application at $3.00/acre.

REFERENCES

Bottrell, D. G. (1976) Biological control agents of the boll weevil; and The boll weevil is a key pest, in *Boll Weevil Suppression, Management and Elimination Technology*, Proc. ARS-USDA Conf., Memphis, Tenn. pp22–5.

Cate, J. R., G. L. Curry, and R. M. Feldman (1979) A model of boll weevil ovipositional site selection. *Environ. Entomol.* 8(5):917–922.

Curry, G. L., R. M. Feldman, and K. C. Smith (1978) A stochastic model of a temperature dependent population. *Theor. Pop. Biol.* 13(2):197–213.

Curry, G. L., P. J. H. Sharpe, D. W. DeMichele, and J. R. Cate (1980) Towards a management model of the cotton–boll weevil ecosystem. *J. Environ. Management* 11:187–223.

von Foerster, H. (1959) Some remarks on changing populations, in F. Stohlman (ed) *The Kinetics of Cellular Proliferation* (New York: Grune and Stratton) pp382–407.

Mann, J. E., G. L. Curry, D. W. DeMichele, and D. N. Baker (1980) A model of light penetration in a cotton crop. *Agronomy* 72:131–42.

Walker, J. K., R. E. Frisbie, and G. A. Nies (1978) A changing perspective: *Heliothis* in short-season cottons in Texas. *ESA Bull.* 24(3):385–91.

Wang, Y., A. P. Gutierrez, G. Oster, and R. Daxl (1977) A population model for plant growth and development: coupling cotton–herbivore interaction. *Can. Entomol.* 109(10):1359–74.

11 Economic Thresholds for Cotton Pests in Nicaragua: Ecological and Evolutionary Perspectives

A. P. Gutierrez and R. Daxl

11.1. INTRODUCTION

Cotton (*Gossypium* spp.) worldwide receives massive amounts of pesticides for the control of its insect pests. In the southern cotton belt of the USA and in much of the Americas, boll weevil (*Anthonomus grandis* Boh.), cotton bollworm (*Heliothis zea* Boddie), and the tobacco budworm (*H. virescens* F.) are the principal pests, because all have high potentials for damaging cultivated cotton (*G. hirsutum* L.). The pink bollworm (*Pectinophora gossyiella* Saunders) is a serious pest only in desert areas. Generally, the boll weevil is considered a primary evolved pest, while the bollworm is induced by insecticide use (Phillips *et al.*, 1980). Many of our cultivated cotton varieties have recently been bred from wild cottons of North America (e.g., *G. hirsutum*), and this is significant for the economic analysis that follows.

Efforts to develop sound economic thresholds for these pests in cotton (see Stern *et al.*, 1959) have not been very successful. Moreover, the costs of collecting the extensive data required to estimate pest damage and various crop growth parameters are very high, and the data are often quite variable. The reasons for the highly variable nature of the data are poorly understood.

Falcon (1972) stressed that the cotton plant itself should be used to assess development, while several recent studies indicate that comparing the cotton plant's fruiting rate to the fruit loss rate from insect damage and other causes might prove instructive (Gutierrez *et al.*, 1975; 1977; 1979a Wang *et al.*, 1977). But specific conclusions could not be drawn from those studies because a necessary series of extensive data collected from the same fields in a similar manner over several years was unavailable. The cotton data from Nicaragua used here overcome those deficiencies and enable us to examine the economic (i.e., action) thresholds for bollworm and boll weevil in cotton. A preliminary report of those data has been published by Gutierrez *et al.* (1982b).

The concept of the *economic threshold* is a mix of a variety of notions, and those relevant to our work are briefly outlined here. The damage level where the crop's ability to compensate for damage is exhausted and yield reduction occurs is the *damage threshold* of Klemm (1940), the *critical figure* of Drachovska (1959), or the *physical threshold* of Carlson and Villagran (1977). A control measure, however, is profitable only if the loss prevented is greater than the control costs. Hence, crop loss greater than this level is preventable economic loss. This is the *economic injury level* of Stern *et al.* (1959), and Stern (1973), the *economic damage threshold* of Steiner and Baggiolini (1968), and the *critical injury threshold* of Sylvan (1968). Above the *physical threshold* (e.g., Carlson and Villagran, 1977) and below but near the *economic injury level* (Stern *et al.*, 1959) lies the *economic threshold. This threshold is the minimum pest level at which pest control measures should be instituted to prevent pest populations from reaching damage levels greater than the control costs.* Because pest resurgence and secondary pest outbreaks frequently result after pesticide use (see van den Bosch, 1978), additional treatments are often required to control these, and hence control costs often must include the additional costs of follow-up applications (Steiner and Baggiolini, 1968; van den Bosch *et al.*, 1971); Regev *et al.* (1976) included these costs in formulating an economic optimization model for the Egyptian alfalfa weevil in California.

Smith (1969) has also suggested that in each agroecosystem using a set of agrotechnical measures, a crop has a maximum potential yield (i.e., the *economic crop potential*). This is more likely to be true in areas where the influences of weather are minimal (possibly field corn in the midwestern US), or where the growth form of the crop enables it to have a high ability to compensate (e.g., Nicaraguan cotton, see below).

In this chapter we first present an analysis of detailed field data of the pattern of cotton growth and of the damage caused by its principal pests, the bollworm and boll weevil. Regression models are developed to estimate fruit point production and the rate of late-season damage. These estimates are then used as the basis of a decision rule to determine profitable pesticide application. In Section 11.5 we present a general discussion of the evolutionary context of cotton and its response to pest damage and draw conclusions on human management of this system.

11.2. METHODS

Extensive data sets on cotton growth and development (see Gutierrez *et al.*, 1975; Corrales, 1975), and on bollworm and boll weevil damage to fruits (Leon, unpublished data) were collected in commercial cotton fields near Managua, Prosoltega and Leon, Nicaragua, as part of an continuing informal FAO/University of California cooperative project. Also counted were the numbers of fruit abscission scars, bollworm larvae, and boll weevil adults, and fruits damaged by each pest that still remained on the plant. Twenty random samples consisting of all plants within a 1.44 m row of cotton (1/5000 manzana*) were mapped daily, and the per mz average computed. Because each pest species causes a distinctive type of feeding injury, each damaged fruit could be categorized. The pest density samples are probably the least accurate, since different samplers vary in their ability to find the pests by visual inspection, and different pest species (stages) have different within-plant distributions and patterns of behavior (see Byerly *et al.*, 1978 for a recent review of the sampling literature in cotton). On the other hand, cotton grows in a very definable way, hence the plant part counts are more accurate. This fact should be kept in mind in the subsequent analysis, when the rates of fruit production and insect damage are used to assess the impact of pest damage on yields.

11.3. PLANT–PEST INTERACTIONS

11.3.1. Phenology

Figure 11.1 shows a representative part of the extensive data sets used in these analyses, and illustrates the dynamics of healthy and damaged cotton fruits and of the bollworm and boll weevil. The *patterns* of crop formation were similar in all fields, but the *magnitude* of fruit production differed widely (Table 11.1). Boll yields in treated fields were uniform and close to 550×10^3 bolls per mz, but total early-(β_1) and lage-(β_2) season immature fruit insect damage rates were reasonably uniform in all treated fields, while they varied considerably in the untreated ones. Damage was highest in the Ojoche and La Calera fields near Managua; yields in the untreated Escoto fields near Leon ranked $1971 < 1974 = 1976 < 1972 = 1973 < 1975$. The rates β_1, and β_2 are derived and discussed in greater detail below.

The principal pest in most of the fields was bollworm (Table 11.1). Boll weevil populations are known to thrive in warm moist weather, hence the frequent dry periods during the years under study, especially in 1972 and 1974, suppressed them. The severe drought in 1972 suppressed boll weevil so strongly that the insect appeared only late and in low numbers in 1973, a wet year. Rainfall is also summarized per year in Table 11.1.

11.3.2. Pest Density–Damage Relationships

The populations of boll weevil and bollworm larvae fluctuate with the currently available fruit numbers, which in turn are greatly influenced by pest numbers (see below). Figure 11.2(B,C,D,F,H) shows the linear relationships between

*The unit of land in Nicaragua is the manzana (mz). 1 mz = 0.7 ha.

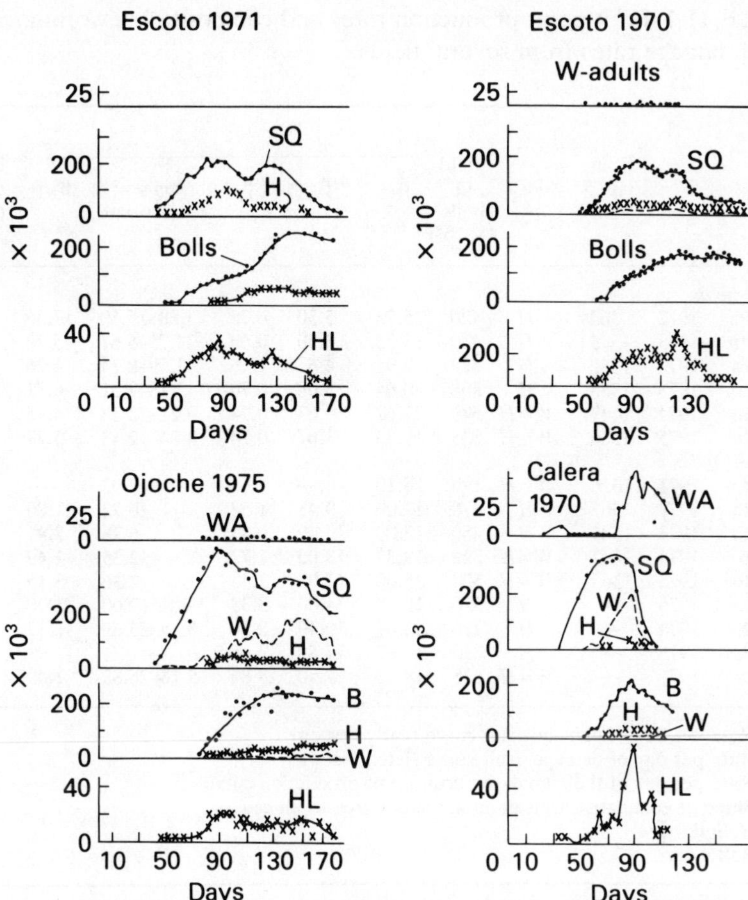

FIGURE 11.1 The dynamics of cotton undamaged fruiting parts (SQ, squares; B, bolls), boll weevil adults (WA), and bollworm larvae (HL) in four cotton fields plotted against days since planting. Fruit damaged by boll weevils and bollworms are indicated by W and H, respectively. The data are plotted per 0.7 ha i.e., $\times 10^3$).

bollworm damage and bollworm density found in all of the Escoto fields (Table 11.2). The relationships between boll weevil damage and weevil numbers (Figure 11.2A,E,G) are highly nonlinear and are clearly strongly density dependent. The form of the boll weevil curves may be explained by the fact that adults will not attack damaged squares and hence the diminished supply of healthy squares decreases the searching efficiency of females. By contrast, bollworm larvae are less selective in their food preferences and actively search the plant for food (squares, bolls, and terminal buds). Plots of total *damaged* fruits per pest numbers on total fruits per pest numbers suggest type I functional responses for *Heliothis* and type II for the weevil (Table 11.2C) (Holling, 1966; see Hassell, 1979 for a review). These latter data are not illustrated.

TABLE 11.1 Fruit point production rates and composite bollworm and boll weevil damage rates from several fields.

	Year	Rain-fall (mm)[a]	Pest	Yield in bolls ($\times 10^3$)	β_T ($\times 10^2$)	β_2 ($\times 10^2$)	β_1 ($\times 10^2$)	β_S Early–late ($\times 10^2$)	β_T/β_2	Plant density ($\times 10^3$/ 0.7 ha)
Treated fields										
Huerta	1972	362	*H*	420	25.75	3.50	0.75	3.68–15.50	7.36	53
Huerta	1973	1593	*H*	430	18.22	3.50	0.60	4.75–5.67	5.20	24
Escoto	1972	362	*H*	524	19.91	4.67	1.16	2.59–8.14	4.26	23
Escoto	1973	1585	*H*	508	20.64	3.08	0.76	2.18–9.31	6.71	24
Escoto	1974	1447	*W*=*H*	540	22.67	3.63	0.74	3.21–12.33	6.18	29
Escoto	1975	1366	*W*=*H*	535	19.33	3.67	0.74	3.04–12.43	5.27	28
Untreated fields										
Escoto	1971	1326	*W*=*H*	170	15.10	—	—	7.93	—	23
Escoto	1972	362	*H*	370	17.00	9.45	1.58	10.72	1.80	23
Escoto	1973	1585	*H*	450	13.00	5.40	0.87	6.78	2.41	24
Escoto	1974	1447	*W*=*H*	320	19.37	13.00	1.72	12.35	1.49	29
Escoto	1975	1366	*W*=*H*	520	25.00	7.93	1.43	12.36	3.15	28
Escoto	1976	—	*H*	320	18.18	14.00	3.33	12.00	1.30	—
Ojoche	1974	—	*H*	220	22.67	10.40	2.30	6.45–33.69	2.17	26
Ojoche	1975	—	*H*	220	17.67	17.88	3.08	—	.99	26
Calera	1971	—	*W*=*H*	125	18.13	27.50	3.84	5.16–26.83	1.49	21

β_1 = Rate per day of damage–fruit curve (early season)
β_2 = Rate per day of damage–fruit curve (late season)
β_T = Rate per day of the cumulative fruit point production curve
β_S = Slope of cumulative fruit point abscission due to all causes
H = *Heliothis zea*
W = Boll weevil

[a]Rainfall between planting and harvest.

On closer examination of the weevil data, the scatter of points in Figure 11.2A, E and G near the origin indicate that on average more than 10 fruits are destroyed per weevil per day. This estimate agrees reasonably well with that of Mitchell and Cross (1969), who showed that each female attacks 13.5 fruits per day (i.e., 0.97 oviposition punctures per hour).

Estimating the Cumulative Number of Damaged Fruit

Both pests damage squares and small bolls causing them to abscise. Squares and small bolls damaged by both species remain on the plant for approximately 10 days (Daxl and Hernandez, unpublished report). Older bolls are damaged directly via feeding and indirectly via introduced secondary fungal infections and rots, but they are rarely shed. The latter data are not discussed in this

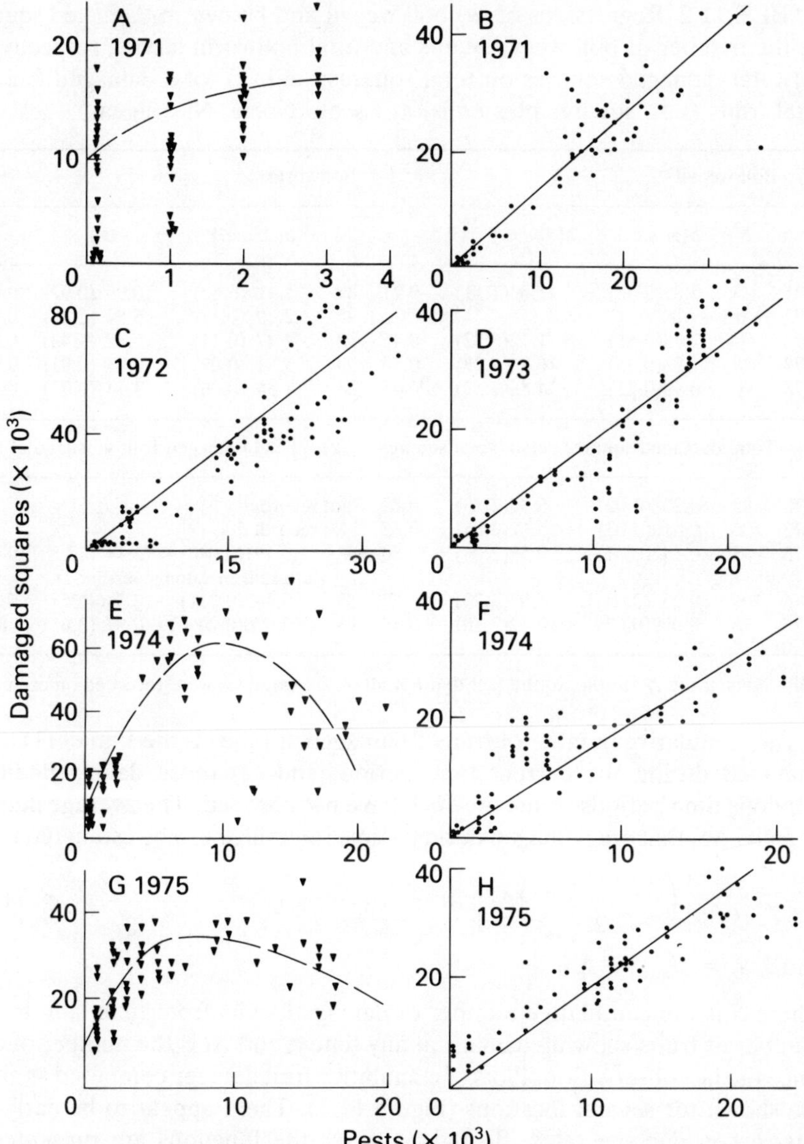

FIGURE 11.2 Relationship of boll weevil- and bollworm-damaged squares to boll pest numbers. The linear relationships (B,C,D,F,H) are for bollworm, while the nonlinear ones (A,E,G) are for boll weevil. The data are plotted per 0.7 ha (i.e., × 10^3).

chapter. While it is relatively simple to count the numbers of fruit scars, we cannot determine what caused the fruit to shed. Furthermore, counts of damaged fruit at any point in time are only snapshots of a dynamic process.

TABLE 11.2 Regressions of (*a*) boll weevil and bollworm-damaged squares on the number of boll weevil adults and total bollworm larvae, respectively; (*b*) total damaged squares on total squares; and (*c*) total damaged fruit on total fruits (i.e., squares plus bolls) at Escoto Norte, Nicaragua.[a]

(*a*) Boll weevil					Bollworm			
Year	*N*	*b*(± s.e.)	*a*(± s.e.)	*r*	*N*	*b*(± s.e.)	*a*(± s.e.)	*r*
1971	85	0.81(0.42)	11.63(1.11)	0.21	89	1.41 (0.62)	0.99 (0.97)	0.93
1972	No weevils				95	2.79 (0.1)	−3.57 (2.38)	0.90
1973	99	2.48(0.51)	1.72(0.32)	0.45	98	2.47 (0.11)	5.42 (1.43)	0.91
1974	98	1.73(0.35)	16.26(3.28)	0.46	99	1.81 (0.097)	0.73 (0.91)	0.89
1975	90	0.99(0.21)	14.58(1.53)	0.45	86	1.54 (0.06)	3.14 (0.77)	0.93
(*b*) Total damaged squares versus total squares					(*c*) Total damaged fruit versus total fruits			
1971	82	0.231(0.02)	8.35(2.23)	0.82	Not recorded			
1972	98	0.410(0.03)	−15.62(3.85)	0.85	Not recorded			
1973	95	0.142(0.01)	0.58(2.04)	0.80	98	0.191(0.040)	72.14 (17.59)	0.43
					(late-season damage lessened)			
1974	100	0.23 (0.02)	6.97(3.70)	0.77	85	0.192(0.011)	−18.50(4.25)	0.894
1975	98	0.099(0.02)	19.12(3.50)	0.55	93	0.070(0.0050)	– 0.37 (3.65)	0.82

[a]Discrepancies in *N* are due to the fact that not all of the variables were recorded on each date.

The cumulative number of fruits *S* damaged at time *t* is the sum of (1) those damaged during the current time period; and (2) those damaged during previous time periods, some of which have not abscised. The average number of fruits ΔS_t that are damaged during a time interval Δt can be estimated by

$$\Delta S_t = \left[\left(\frac{N_t + N_{t-\Delta t}}{2}\right)\Delta t\right]/10 \tag{11.1}$$

$$\text{and} \quad S_t = S_{t-\Delta t} + \Delta S_t, \tag{11.2}$$

where *S* is the cumulative number of damaged fruits from time t_0 to t; N_t is number of fruits showing damage at any time t; and Δt is the number of days since the last observation. Plots of cumulative fruit damage computed as above are shown for several locations (Figure 11.3). There appear to be early-and late-season damage rates, β_1 and β_2 (n.b., the functions are presented as piecewise linear for convenience.) The slopes of straight lines β_1 and β_2 fitted through the two sets of data for each location are given in Table 11.1. The shift from one phase to another in a field may occur during the interval 75–90 days.

Fruit Bud Production

The rate of fruit point production β_T was estimated from field data by

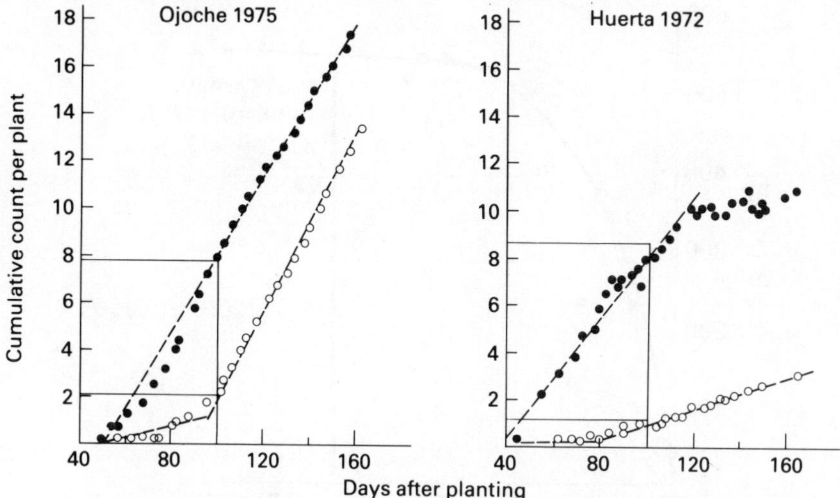

FIGURE 11.3 Cumulative fruit production • and cumulative insect-damaged fruit ○ plotted for untreated (Ojoche) and treated (Huerta) cotton fields in Nicaragua. The data are plotted per 0.7 ha, i.e., $\times\ 10^3$.

regressing counts of the total number of damaged and undamaged fruit, plus the number of scars of fruit shed due to all causes against time. Cumulative fruit bud production is known to be a linear function of plant density and temperature at least until the plant's carrying capacity has been reached (see Gutierrez *et al.*, 1975). Julian time was used in these plots simply because the daily variations in temperatures in this area are minimal and the error encountered using days rather than degree days is small relative to the error obtained by using weather data from local sources. Note that the rates of fruit point production β_T tend to slow at time 100–120 days (i.e., the plant's carrying capacity is reached) in those data sets where the fruit damage rates are relatively low (Figure 11.3), while fruit bud production continued unabated in those fields with high late-season damage rates β_2. These results are consistent with the carbohydrate stress model proposed by Gutierrez *et al.* (1975) and Wang *et al.* (1977) (see also McKinion *et al.*, 1974). The differences in the time of first fruit and in β_T are due to differences in planting density, some varietal differences and, in a minor way, weather. The rectangles in each graph were included to illustrate the differences in the slopes of β_T and β_2. Small differences in the slopes indicate low fruit survival.

Predicting Cotton Yields

The plants produce fruit at a rate β_T and the pests deplete them at rates β_1, β_2. Hence the final yield should be some function of the interplay between these rates. Figure 11.4 shows a plot of final yield Y on the ratio of β_T/β_2 for several

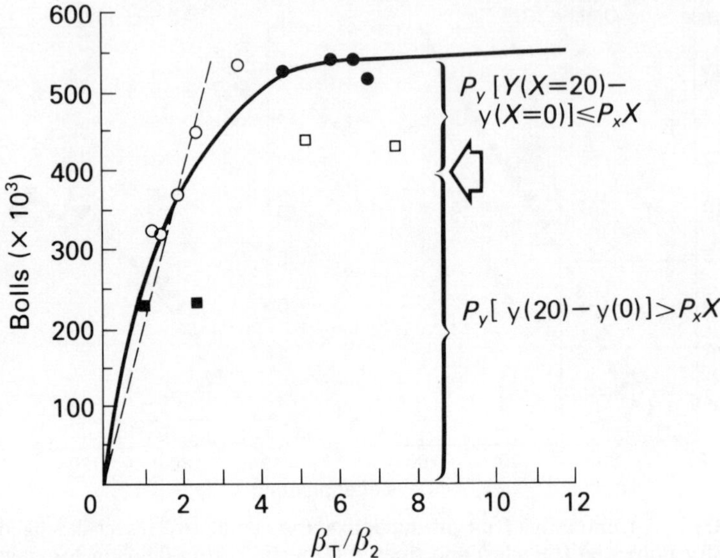

FIGURE 11.4 Relationship of final boll yields per 0.7 ha on the ratio of the rate of fruit point production (β_T) to the rate of late-season square damage (β_2). Also shown is a hypothetical economic pest control decision rule. ○ untreated Escoto fields; ● treated fields; □ and ■ are data from two other fields.

untreated and insecticide-treated Escoto crops (see Table 11.1). Fields in this area are normally treated with insecticides X, 20 or more times per season. The line fitted through the points (circles only) must pass through the origin because $\beta_T/\beta_2 = 0$ must result in zero yield. The broken line was fitted by eye, while the hypothetical dose response function (11.3) was estimated via an iterative method. Equation (11.3) is intuitively more satisfying because it is a continuous function through all of the Escoto data:

$$Y = Z[1 - \exp(-\alpha\beta_T/\beta_2)], \qquad (11.3)$$

where Y is the number of bolls ($\times 10^3$) per 0.7 hectares, in both treated and untreated fields (i.e., Escoto); Z is the maximum previously observed yield in the field, and is equivalent to Smith's (1969) economic crop potential; and $\alpha = 0.63$ for the Escoto data.

The model is a reasonable fit to the data, and suggests that some damage does not reduce yields because the plants compensate for the damage by producing far more fruit than they can mature. Premature abscission of fruit caused by pests results in reallocation of dry matter to fruit that will mature (see Gutierrez *et al*., 1975). Increased yields often result, as shown by Falcon *et al*. (1971) and by Gutierrez *et al*. (1977) for *Lygus hesperus* damage in cotton. The data clearly show that yield losses will accrue only if $\beta_T/\beta_2 < 3$ (i.e., if more than 30% of the fruit are attacked). The weakness of the model is that estimates of β_2 can be made only after considerable damage has occurred, hence, we require an earlier assessment of damage potential. This problem is addressed below.

A plot of yields on β_T/β_2 from Huerta (1972, 1973) and Ojoche (1974, 1975) on the same graph suggests similar relationships (Figure 11.4), and illustrates that the yield capacities Z of fields may differ.

Forecasting Insect Damage Potential

Estimating the survivorship of boll weevil and bollworm populations between seasons is as yet an intractable problem, further complicated by pest migration. For this reason, we must obtain the earliest indication of damage and use it to forecast the late-season damage potential of the pest. Figure 11.5 shows that the rate of early-season damage β_1 is related to the late-season rate β_2, and is described by a linear regression:

$$\begin{aligned} \beta_2 &= a(\pm \text{ s.e.}) + b(\pm \text{ s.e.})\beta_1 \\ \beta_2 &= -0.86(\pm 1.26) + 6.09(\pm 0.65)\beta_1 \end{aligned} \tag{11.4}$$

$$r^2 = 0.88, \qquad N = 16 \qquad \beta_2 = 5.72\,\beta_1 \text{ (i.e., forced through zero).}$$

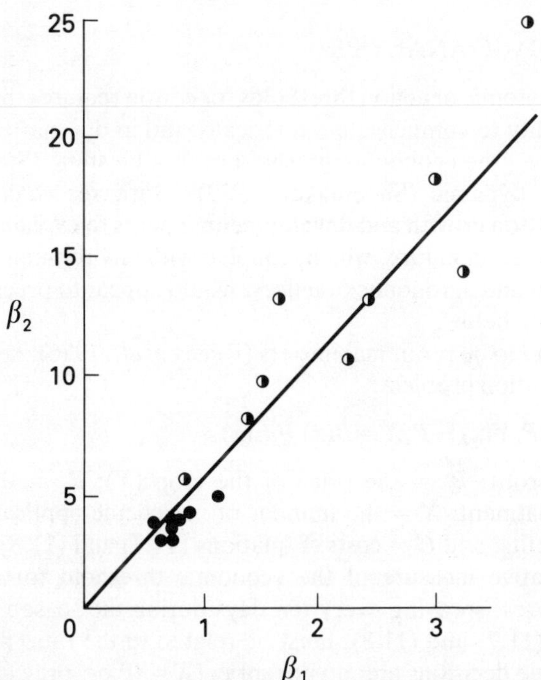

FIGURE 11.5 Relationship between late-season (β_2) and early-season (β_1) insect damage rates. • treated fields; ○ untreated fields (see Table 11.1).

The results are consistent across both unsprayed, $Y(X = 0)$, and sprayed fields, $Y(X = 20)$, only because insecticide applications X appear to be spaced throughout the season, so as to cause the same background mortality to both early and late populations (i.e., the relationship between β_2 and β_1 is unaffected). The relationship further implies that individuals which damage squares early give rise to populations that cause 5.72 times more damage later in the season (i.e., a numerical response).

Because the early-season damage rates are low, serious estimation errors can result, hence a great deal of care must be taken to minimize the error. But with this proviso, early-season estimates of β_1 can be used to predict yields under insecticide-free conditions $[Y(0)]$. Of course, if β_1 is too high, pesticide applications can be used to increase yields, $Y(X = 20) \Rightarrow Z$. The model for assessing potential impact using β_1 then becomes

$$Y = Z[1 - \exp(-0.63\beta_T/5.72\,\beta_1)]. \tag{11.5}$$

But estimating β_1 and β_2 still involves some cumbersome computations. We therefore tried using the rate of total shed β_s, which can be easily obtained by counting, but unfortunately no satisfactory relationship was found. The slopes β_s were computed using data from days 60–120.

11.4. ECONOMIC ANALYSIS

Estimating economic or action thresholds for cotton requires that we can assess the crop's ability to compensate numerically and in dry-matter allocation for lost fruits. Thus, the economic threshold cannot be static (Stern *et al.*, 1959) but, rather is dynamic (Shoemaker, 1979); Gutierrez *et al.*, 1979b). The plasticity of cotton growth and development appears to explain the wide range of yields observed in nature, which, coupled with environmental factors, field characteristics, and agronomic practices, would appear to preclude any simple analysis, but see below.

Farmers tend to be profit maximizers (Regev *et al.*, 1976); hence eqn. (11.6) is the maximization problem:

$$\text{Max}\,\Pi = P_y Y(x) - P_x X = B(x) - C(x) \tag{11.6}$$

where Π = profit; P_y = the price of the crop (Y); P_x = the cost of each insecticide treatment; X = the number of insecticide applications ($x = 0$ or 20); B = benefits; and C = costs. Equations (11.7) and (11.8) can be used to provide a relative measure of the economic threshold for instituting pest control measures, spraying every ten days during the season ($X = 20$). The decision rule (11.7) and (11.8), must be related to the ratio β_T/β_2 via (11.3), and the possible decisions are: do not spray ($X = 0$) or spray ($X = 20$). In this case the yield is approximately Z if insecticides are used, while $Y(0)$ results if $X = 0$; i.e., $Y(0) = f(\beta_T/\beta_2)$, as in eqn. (11.3). The decision rule is:

$$\text{do not spray if} \quad P_y[Y(X) - Y(0)] < P_x X; \tag{11.7}$$

$$\text{spray if} \quad P_y[Y(X) - Y(0)] \geqslant P_x(X), \tag{11.8}$$

where X is determined such that $P_y \partial Y/\partial X = P_X$. The decision rule thus determines the level of pesticides at which the last unit would cost no more than the damage prevented, and an additional one will cost more than that; that is when the marginal cost equals the marginal return. The gains following this rule are to be compared with the gains when no pesticide is used, $P_y[y(0)]$, as indicated by eqns. (11.7) and (11.8).

For a sufficiently low level of pests, eqn. (11.7) will hold. But as the pest W levels grow, $Y(X) - Y(0)$ increases until the threshold level is reached for which spraying becomes economical. Thus the threshold level of pests W could be mathematically written as:

$$W \text{ such that for } X \text{ satisfying } P_y[Y(X) - Y(0)] = P_x X \text{ will hold.} \tag{11.9}$$

The number may be reduced by better timed applications. Indeed, it is the task of the agricultural entomologist to minimize $P_x X$ using less expensive methods and/or reducing the frequency of control actions, in order to maximize Π and environmental quality. It is this minimization of $P_x X$ that economic entomology, and especially integrated pest control is all about.

11.4.1. An Assessment

Assuming a price of $17.14 per 100 lb of seed cotton and a cost of $10 for each pesticide application, only three of the five untreated Escoto fields sustained economic damage (Table 11.3 and Figure 11.4), suggesting that the current 20 insecticide applications are excessive. Remarkably, *the highest net gain and the two highest investment returns (profit/production costs) occurred in untreated fields, but the net returns fluctuated widely in the untreated (high-risk) fields*. Experience has shown that cotton in the Managua (Calera) and Ojoche area cannot be produced without insecticides given current agronomic practices, and this is borne out by our data (Table 11.3). In the Leon area, however, even untreated fields yielded profits, except during the severe 1972 drought. *While these data suggest that spraying is necessary in the Calera and Ojoche areas, experience has also shown that economic loss would have been avoided with fewer insecticide applications*.

Yields and gains in the treated fields were quite consistent, even during the dry 1972 season: pesticides lowered the short-term monetary risk. However, the current pest control system of extensive insecticide usage is expensive and environmentally unsound, and it does not optimize net returns (Villagran, 1978).

TABLE 11.3 An economic analysis of 12 cotton fields in Nicaragua, assuming: 10^4 bolls = 100 lb of seed cotton, each worth \$17.14; each of the 20 pesticide applications costs \$10; and that the treated fields and production costs without insect control are \$443/mz.

	Year	Boll yield (10^3/mz)	Value (\$)	Production costs	Net gains	Yearly investment returns (%)
Treated fields[a]						
Escoto	1972	535	917	643	274	42.6
Escoto	1973	500	857	643	214	33.3
Escoto	1974	525	900	643	257	40.0
Escoto	1975	550	943	643	300	46.7
Calera	1971	560	960	643	317	49.3
Untreated fields[b]						
Escoto	1972	170	291	443	−152	—
Escoto	1973	400	686	443	243	54.9
Escoto	1974	320	549	443	106	23.9
Escoto	1975	500	857	443	414	93.5
Ojoche	1974	150	257	443	−186	—
Ojoche	1975	240	411	443	− 32	—
Calera	1971	160	274	443	−169	—

[a]Break-even point = 370 × 10^3 bolls/mz for Escoto fields.
[b]Break-even point = 258 × 10^3 bolls/mz for Escoto fields.

11.4.2. Other Implications for Cotton Pest Management

The data clearly show that separating the damage caused by each species is irrelevant to predicting final yields. What really mattered was the rate of damage compared with the rate of fruit point production (i.e., a measure of the plant's ability to compensate for fruit loss). The early-season pest damage rate β_1 was much lower than and related to the late-season rate β_2. The obvious outcome discussed above is that with some effort we can determine our pest control needs early in the season. A less obvious outcome is that if the damage rate β_1 can be reduced by a factor γ, then the reduction in late-season damage is 5.72γ. This would suggest that early (i.e., for this season) and late-season (i.e., for next season) trap cropping for boll weevil on an area-wide basis would be highly desirable, since it would influence β_2 in this and the next year respectively. Furthermore, the resultant lower pesticide use would also decrease the potential for secondary pest outbreaks of *Heliothis zea* and other lepidopterous and mite pests (see Phillips *et al.*, 1980). Bollworms in our study were more important than boll weevils during the generally dry period 1972–76 (see Table 11.1), and were greatly aggravated by pesticides. Thus, trap cropping for boll weevil could be used to suppress their populations, and the early-season monitoring for cumulative pest damage would serve to determine whether season-long treatment will be required. These integrated pest management (IPM) tactics could result in considerable economic and environmental gains, as some fields currently treated on a routine basis would not be. These observations are similar to the results obtained by Gutierrez *et al.* (1977) for *Lygus hesperus* damage on cotton and it would appear that these results are applicable to cotton cultivation worldwide. In general, this analysis shows how a detailed understanding of a complex problem yields a simple solution.

11.5. CONCLUSIONS

11.5.1. Man's Point of View

Three of the five untreated Escoto fields sustained economic damage, but none of the treated fields did; hence, it is questionable whether the current pest control practice requiring 20 insecticide applications per season makes economic sense. If the addition of severe pollution and environmental disruption costs further mitigates against this policy, why does it continue? As we shall see, IPM is complicated and, more important, farmers are risk averse (see Chapter 27).

For example, the yield capacity of fields and the prices of both Y and X are variable, and hence the economic action level will also vary. However, if IPM techniques are to be used, some cost K for the service must be incurred, and the profits must be the same or better:

$$\Pi_{STD} = P_y Y(X) - P_x X = P_y Y(X^*) - P_x X^* - K = \Pi_{IPM} \quad (11.9)$$

$$B(\hat{X}) - C(\hat{X}) = B(X^*) - C(X^*) - K - D(X^*), \quad (11.10)$$

where Π_{STD} and Π_{IPM} are the profits from standard and integrated pest management approaches. The maximum number of sprays $X^{*'}$ which can be applied to satisfy (eqn. (11.9) is $X^* = X - K/P_x$. However, if some punitive cost D is levied for each unit of pesticide used, then $X^* = \hat{X} - K/(P_x + D)$, where $\hat{X}$ is the new value for X sufficient to satisfy eqn. (11.10). The cost $D = g(X)$ could be some environmental damage cost assessed to the grower, and if $g(N)$ is increasing, then the number of pesticide applications would be further reduced.

The various cost–benefit functions are illustrated in Figure 11.6. Increasing concave benefit and linear cost functions are assumed (see Headly, 1972). The private optimal solution (i.e., for a single farmer) to Figure 11.6a is shown in Figure 11.6b as the intersection of the derivatives of the two functions. This solution is the same as eqn. (11.6), and is shown as solid lines in b, c, and d. Increased costs via environmental degradation penalties D decreases the

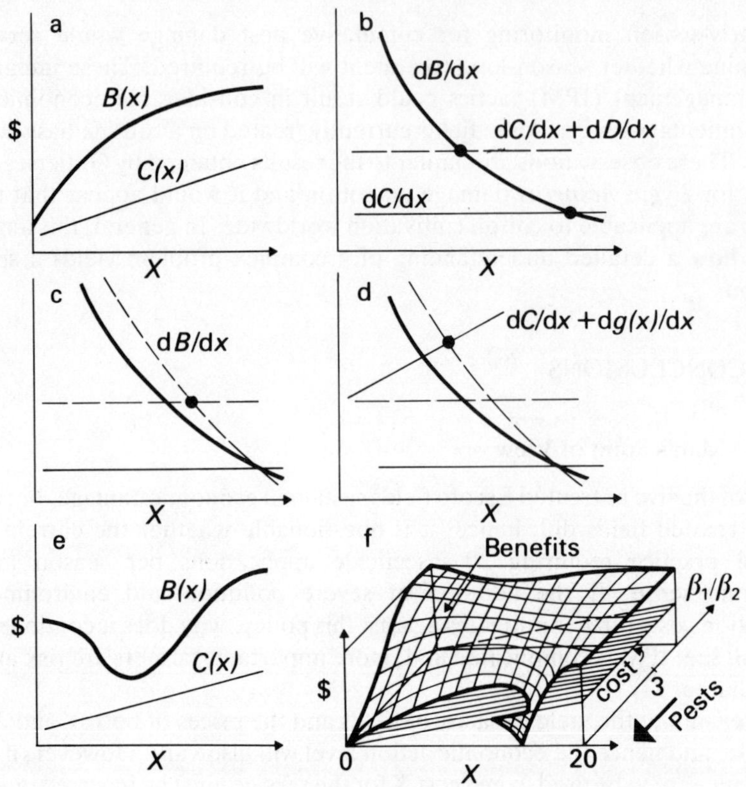

FIGURE 11.6 Various hypothetical decision rules for pesticide (X) use: (a) the economic notion of benefit B and cost C functions (see Headly, 1972); (b) the optimal decision with and without considering environmental damage costs D; (c) the optimum given a better benefit function; (d) a modified benefit function that includes disruptive effects of pesticides (see Gutierrez *et al.*, 1979a); and (f) the benefit and cost functions combining the notions in Figures 11.4 and in (c) above.

number of applications that can be made (Figure 11.6b, the broken line is the public solution; Regev *et al.*, 1976). The effect of a better benefit response (i.e., B^* for some technical advancement such as a new variety) serves to increase pesticide use (Figure 11..6c, broken line, dB^*/dx), while increasing penalties with increasing insecticide use decreases its use (Figure 11.6d, $dc/dx + dg(x)/dx$. But in reality the benefit function may be more complex, and not always increasing in X as insufficient or untimely insecticide application may cause additional yield reductions due to other induced pest outbreaks (Figure 11.6e, and see Gutierrez *et al.*, 1979b). The phenomenon of pest resurgence and secondary pest outbreaks have been well documented in cotton (van den Bosch, 1978).

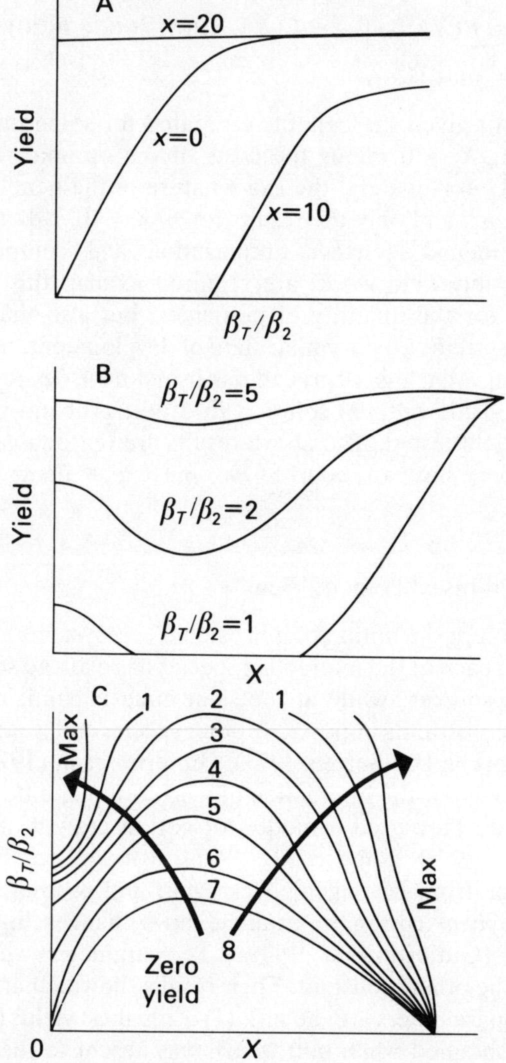

FIGURE 11.7 Decomposition of the benefit function in Figure 11.6(f): A, cross sections through the Y on β_T/β_2 plane; B, cross sections through the Y on X plane; and C, cross sections through the β_T/β_2 on X plane. The numbers in C merely depict the highest yield (1) from lower ones (> 1).

If we combine the observed results from Nicaragua (Figure 11.4) with this latter notion (Figure 11.6e), we obtain the qualitative results in Figure 11.6f. Figure 11.7A–C decomposes Figure 11.6f, and is included as an aid. In this case, the optimal solution is

$$\max \Pi_{IPM} = B(X^*, \beta_T/\beta_2) - C(X^*, \beta_T/\beta_2) - K > B(20) - C(20) \\ = \max \Pi_{STD} \tag{11.11}$$

or maximize profit given the benefits estimated for some initial measure of yield potential at $X = 0$ minus the costs of the optimal rate of pesticide application X^*. Unfortunately, the exact nature of the benefit surface is not well known *a priori*, and only the results for $B(X = 0, \beta_T/\beta_2)$ and $B(X = 20, \beta_T/\beta_2)$ can be judged. Further optimization and computer simulation studies, and possibly field work, are required to map this surface (Figure 11.7C) not only for the quantity of pesticides, but also their timing. Most analyses of crop systems are in similar state of development, and only after we solve this problem will a high degree of risk aversion be desirable. Regev *et al.* (1976) formulated the optimal solution to a pest problem in alfalfa that is currently being field tested. The above results are reasonable solutions from man's perspective, but, of course, we must put them into ecological perspective.

11.5.2. The Plant–Insect Point of View*

In ecology, we accept the notion that in "stable" ecosystems (*sensu* Wang and Gutierrez, 1980), each of the interacting species has evolved so that it does not overexploit its resources, while at the same time maximizing its ecological fitness given the constraints imposed by other species and weather (see Gilbert *et al.*, 1976; Gilbert and Gutierrez, 1973). Gutierrez *et al.* (1979a) showed that this generality holds for the interaction of wild cotton and boll weevil; in that optimization study they used a model for cotton growth and development (Gutierrez *et al.*, 1975; Wang *et al.*, 1977), a model for boll weevil, and observed weather from Managua, Nicaragua, and estimates of six growth parameters of *Sylvan* cotton to examine how changes in the parameters affected "fitness" (Gutierrez *et al.*, 1979a). The parameters were altered one at a time, keeping the others constant. Their results showed that without the boll weevil (only two parameters are shown): (1) the highest yields (Figure 11.8, full lines) could be obtained when boll weevil was absent (a fact well known to farmers); but (2) higher yields could be obtained if the observed parameters had evolved to some other value (fitness not maximized under pest-free conditions at the observed parameter values). With boll weevil (Figure 11.8, broken lines): (1) the model predicted greatly reduced yields; but (2) given boll weevil damage, the plant maximizes ecological fitness. (These, it should be noted, were the initial results, not those achieved after much adjusting of the model.)

*This section owes much of its development to discussions with Dr Uri Regev, Department of Economics, Ben Gurion University, Israel.

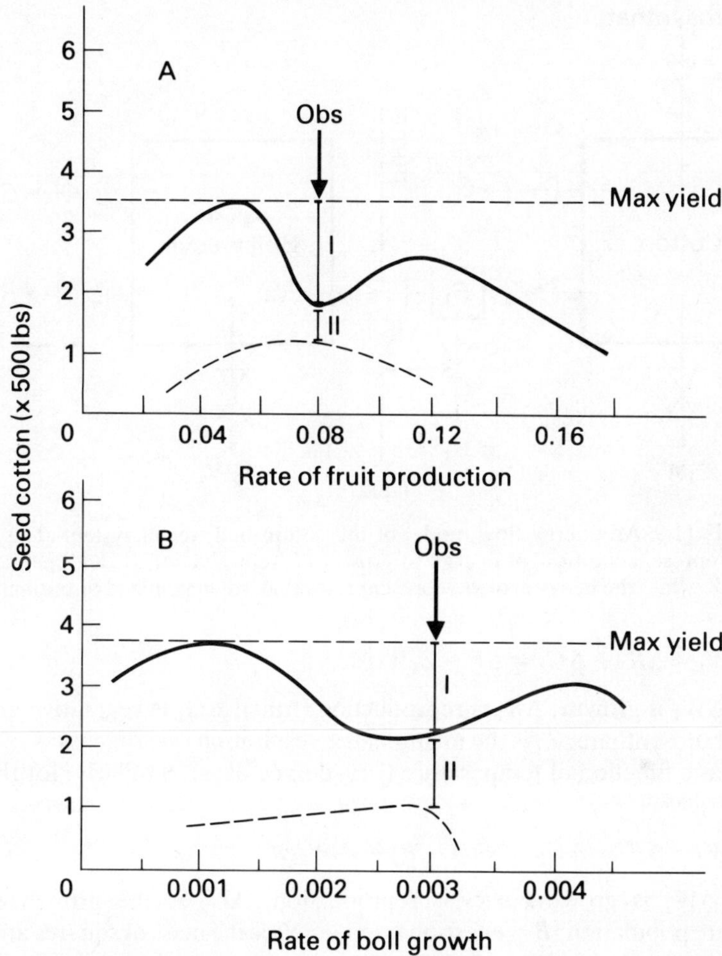

FIGURE 11.8 Cotton yields as a function of A, the rate of fruit point production, and B, the rate of boll growth without (full curve) and with broken curve) boll weevil predation. Line segments I and II represent the amount of energy (seed cotton equivalents) that cotton will expend to grow in its present configuration expecting boll weevil attack, and that additional energy expended to achieve optimal accommodation with boll weevil. The observed parameter is indicated (Obs→).

Figure 11.9 is a model for the plant's allocation of energy to maintenance respiration, vegetative growth (leaves, stem, and root) and reproduction (fruit). Also shown in Figure 11.9 is the allocation of energy consumed (quares) to growth, reproduction, and respiration in the weevil. For complete details of the mathematical form and plant physiology of the model see Gutierrez *et al.* (1975) and Wang *et al.* (1977). Mathematically, these relationships for each time step can be described as (c.f. Gutierrez *et al.*, 1982b):

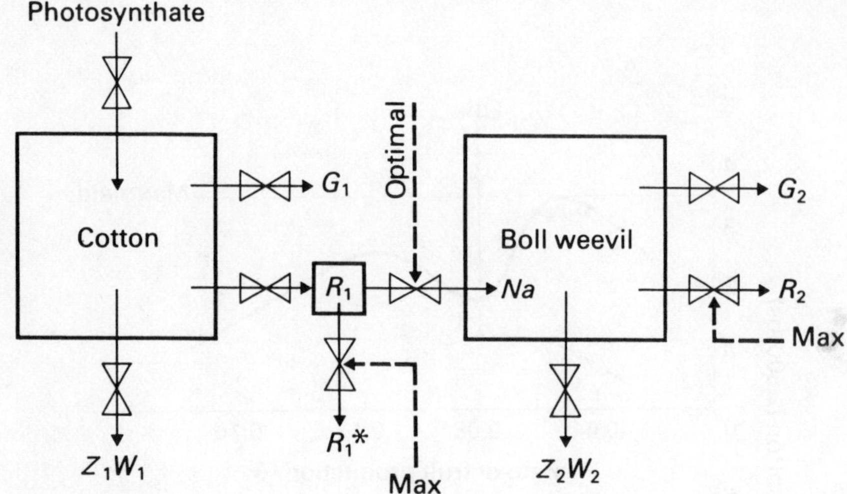

FIGURE 11.9 An energy flow model of the cotton boll weevil system. The valved arrows represent the flow of energy to growth G, reproduction R, and maintenance costs ZW, while the heavy arrows represent rates that are maximized or optimal in the system.

$$\Delta W_1 = \Delta R_1 + \Delta G_1 = \Delta P - Z_1 W_1 \Delta t, \tag{11.12}$$

where ΔW_1 is growth; ΔR_1 is reproduction (fruit); ΔG_1 is vegetative growth; ΔP is photosynthate; Z_1 is the maintenance respiration rate of mass W_1; and Δt is time as a function of temperature (i.e., degree days). Similarly, for the boll weevil:

$$\Delta W_2 = \Delta R_2 + \Delta G_2 = Na(1 - B) - Z_2 W_2 \Delta t \tag{11.13}$$

where ΔW_2 is growth; ΔR_2 is reproduction; ΔG_2 is the growth of the immature population; B = egestion fraction; Na is the mass of squares attacked via feeding and oviposition; and Z_2 is the maintenance respiration rate of mass W_2.

Over a season of length T the plant attempts to maximize fitness R_1 given its evolved growth parameters $V = V_1, V_2 \ldots V_n$ (i.e., evolutionary decision variables, Gutierrez *et al.*, 1979a):

$$\max V_1,\ldots,V_n \int_0^T R_1(V)\,\mathrm{d}t = \max \int_0^T R_1(V)\,\mathrm{d}t = \int_0^T [P(V) - Z_1 W_1(V) - G_1(V)]\mathrm{d}t. \tag{11.14}$$

Given that boll weevil predation eqn. (11.15) is the optimal solution for the plant:

$$\max V_1,\ldots,V_n \int_0^T R_1^*(V)\,dt = \int_0^T [P(V) - Z_1W_1(V) - G_1(V) - Na(\hat{V})]dt \quad (11.15)$$

$$\int_0^T \text{(Photosynthesis} - \text{maintenance costs} - \text{vegetative costs} - \text{fruit predation)}\,dt,$$

where R_1^* is the energy in surviving fruit. In our example, eqn. (11.14), R_1 was not maximized for the observed parameters, but R_1^* was in eqn. (11.15). A similar maximization expression could be written for R_2^* (i.e., the weevil's "point of view" with its set of $\hat{V}_i$).

But how can both species maximize their respective fitness? Returning to Figure 11.8, we see that the line segment I is the amount of energy that the plant is "willing" to lose via excess fruiting capacity given that its Darwinian memory expects boll weevil damage, while II is the additional energy the plant is "willing" to lose to fruit predation simply to reach optimal accommodation with boll weevil (the optimal flow of energy between the two systems), and represents the optimal long-term solution from both the plant's and pest's "points of view" (i.e., a constrained, maximization solution). Of course, this solution is not static, but represents the average equilibrium position, as both the pest and plant populations are in the process of evolving.

As boll weevil has a high intrinsic potential for damage, it is not surprising that cotton evolved excess fruiting capacity to cope with it (Gutierrez *et al.*, 1979a), while at the same time minimizing energy losses in the shed fruits (Gutierrez and Wang, 1979). Furthermore, the plant "cares little" which pest is depleting its future stock, it merely "seeks" to replace the losses and maximize fitness. The growth form of domestic cottons is very similar to that of sylvan cotton, man has merely selected for various high-yield and other agronomic characteristics (e.g., Figure 11.8, solid lines). So long as man protects these crops with insecticides and fertilizers, high yields accrue. But if resistance develops or climatic uncertainties occur, crop failures are likely to result because the crop has lost much of its capacity to compensate (broken lines in Figure 11.8).

This analysis has enabled us to take a detailed understanding of a complex plant–pest–climate interaction and distill out not only practical results, but also what we consider to be important evolutionary ones. The commonalities between the economies of man and nature is the subject of another paper currently in progress (Gutierrez and Regev, 1983).

REFERENCES

Byerly, K. F., A. P. Gutierrez, R. E. Jones, and R. F. Luck (1978) A comparison of sampling methods for some arthropod populations in cotton. *Hilgardia* 46(8): 257–82.

Carlson, G. and E. Villagran (1977) *CAITI Report* (Turialba, Costa Rica).

van den Bosch, R. (1978) *The Pesticide Conspiracy* (New York: Doubleday).

van den Bosch, R., T. F. Leigh, L. A. Falcon, V. M. Stern, D. Gonzalez, and K. S. Hagen (1977) The developing program of integrated control of cotton pests in California, in C. B. Huffaker (ed) *Biological Control* (New York: Plenum) pp377–94.

Corrales, D. (1975) *5th Seminerio Tecnica Algonodero Managua, Nicaragua*. (Banco Nacional de Nicaragua).

Drachovska, M. (1959) Die kritische Zahl als Mass zur Beurteilung von Befallsstufen. *Prognosa a diagnosa v ochraně rostlin* (Praha) p216.

Falcon, L. A. (1972) *Integrated Control of Cotton Pests in the Far West*, Proc. Beltwide Cotton Production Research, Memphis.

Falcon, L. A., R. van den Bosch, J. Gallagher, and A. Davidson (1971) Investigations on the pest status of *Lygus hesperus* in cotton in central California. *J. Econ. Entomol.* 64:56–61.

Gilbert, N. E. and A. P. Gutierrez (1973) An aphid–parasite plant relationship. *J. Animal Ecol.* 42:323–40.

Gilbert, N. E., A. P. Gutierrez, B. D. Fraser, and R. E. Jones (1976) *Ecological Relationships* (London: Freeman).

Gutierrez, A. P., J. U. Baumgaertner, and K. S. Hagen. (1982a) Predator growth, development and reproduction: *Hippodamia convergens* Guerin (Coccinellidae: Coleoptera) — An example. *Can. Entomol.* 113:21–33.

Gutierrez, A. P., R. Daxl, G. Leon Quant and L. A. Falcon (1982b) Estimating economic thresholds for bollworm (*Heliothis zea* Boddie) and boll weevil (*Anthonomus grandis* Boh.) damage in Nicaraguan cotton (*Gossypium hirsutum* L.). *Environ. Entomol.* 10:873–79.

Gutierrez, A. P., L. A. Falcon, W. Loew, P. A. Leipzig and R. van den Bosch (1975) An analysis of cotton production in California: A model for Acala cotton and the effects of defoliators on its yield. *Environ. Entomol.* 4(1):125–36.

Gutierrez, A. P., T. F. Leigh, Y. Wang, and R. Cave (1977) An analysis of cotton production in California: *Lygus hesperus* (Heteroptera: Miridae) injury — An evaluation. *Can. Entomol.* 109:1375–86.

Gutierrez, A. P. and U. Reger (1983). The economics of fitness and adaptedness: The interaction of sylvan cotton (*Gossypium hirsutum* L.) and boll weevil (*Anthonomus grandis* Boh.): An example. *Oecol. Geuer*. 4(3):271–287.

Gutierrez, A. P. and Y. Wang (1979) Applied population ecology: Models for crop production and pest management, in G. A. Norton and C. S. Holling (eds) *Pest Management*, Proc. Int. Conf., October 1976. IIASA Proceedings Series Vol. 4 (Oxford: Pergamon) pp255–80.

Gutierrez, A. P., Y. Wang, and R. Daxl (1979d) The interaction of cotton and boll weevil (Coleoptera: Curculionidae) — A study of co-adaptation. *Can. Entomol.* 111:357–66.

Gutierrez, A. P., Y. Wang, and U. Regev (1979b) An optimization model for *Lygus hesperus* (Heteroptera: Miridae) damage in cotton: The economic threshold revisited. *Can. Entomol.* 111:41–54.

Hassell, M. P. (1979) *The Dynamics of Arthropod Predator–Prey Systems*, Monogr. in Population Ecology (Princeton: Princeton University Press).

Headley, J. C. (1972) Defining the economic threshold, in *Pest Control Strategies for the Future* (Washington, DC: National Academy of Sciences) pp100–8.

Hidalgo, O., A. Sequeira, R. Bodan, and R. Daxl (1975) *5th Seminario Tecinco Algodonero, Managua, Nicaragua* (Banco Nacional de Nicaragua).

Holling, C. S. (1966) The functional response of invertebrate predators to prey density. *Mem. Entomol. Soc. Can.* 48:1–86.

Klemm, M. (1940) Ernteverluste, Schadensschatzung und Pflanzenschutzstatistik. *Der Forschungsdienst* 10:265–75.

Leon, G. (1977) *6th Seminario Tecnico Algodonero, Managua, Nicaragua* (Banco Nacional de Nicaragua).

McKinion, J. M., J. W. Jones, and J. D. Hesketh (1974) Analysis of SIMCOT: Photosynthesis and Growth, *Res. Conf. Proc. Beltwide Cotton Prod.*, Memphis, pp117–24.

Mitchell, H. C. and W. H. Cross (1969) Oviposition by the boll weevil in the field. *J. Econ. Entomol.* 62:604–5.

Phillips, J. R., A. P. Gutierrez, and P. L. Adkisson (1980) General accomplishments toward better insect control in cotton, in C. B. Huffaker (ed) *New Technology of Pest Control*. (Chichester: Wiley) pp123–53.

Regev, U., A. P. Gutierrez and G. Feder (1976) Pests as a common property resource: a case study of alfalfa weevil control. *Am. J. Agric. Econ.* 58:187–97.

Shoemaker, C. A. (1979) Optimal management of an alfalfa ecosystem, in G. A. Norton and C. S. Holling (eds) *Pest Management*, Proc. Int. Conf. October 1976. IIASA Proceedings Series Vol. 4 (Oxford: Pergamon) pp301–15.

Smith, R. F. (1969) The importance of economic injury levels in the development of integrated pest control programs. *Qualitas Plantarum et Materiae Vegetabiles* 17(2):81–92.

Steiner, H. and M. Baggiolini (1968) *Anleitung zum integrierten Pflanzenschutz im Apfelbäume* (Stuttgart).

Stern, V. M. (1973) Economic thresholds. *Ann. Rev. Entomol.* 18:259–80.

Stern, V. M., R. F. Smith, R. van den Bosch, and K. S. Hagen (1959) The integrated control concept. *Hilgardia* 29: 81–101.

Stinner, R. E., R. L. Rabb, and J. R. Bradley (1974) Population dynamics of *Heliothis zea* (Boddie) and *H. virescens* (F) in North Carolina: A simulation model. *Environ. Entomol.* 3:163–8.

Sylvan, E. (1968) Threshold values in the economics of insect pest control in agriculture. *Medd. Vaxtskyddsanst. Stockholm* 14:118,69–79.

Villagran, E. (1978) *Evaluacion de las prácticas de control de plagas del algodonero en Nicaragua, 1977–1978*, FAO-UNDP Integrated Pest Control Project, Managua, Nicaragua.

Wang, Y. H. and A. P. Gutierrez (1980) An assessment of the use of stability analyses in population ecology. *J. Animal Ecol.* 49:435–52.

Wang, Y., A. P. Gutierrez, G. Oster and R. Daxl (1977) A population model for plant growth and development: Coupling cotton–herbivore interactions. *Can. Entomol.* 109:1359–74.

Wilson, L. T. and A. P. Gutierrez (1980). Fruit predation submodel: *Heliothis* larvae feeding upon cotton fruiting structures. *Hilgardia*. 48 (2):24–36.

PART TWO

TACTICS

12 Tactical Models

Gordon R. Conway

12.1. INTRODUCTION

Tactical models have a much more difficult task than strategic models. They aim to provide a particular farmer or veterinary or medical practitioner (or, in the case of human illness, the patient himself) with highly reliable advice, at the very least on an annual or seasonal basis, but ideally for day-to-day decision making. They may do this for the farm or the sub population under the practitioner's general care or even for the field or the individual patient affected. Farmers, for example, may want to know whether or not to spray a particular field in a given year, and how much and on what dates. Farmers may also require advice on the precise mix of control strategies appropriate for their particular farm; similarly, medical practitioners need to know how general disease control strategies have to be modified for their sub-population of patients. This, of course, requires not only precise ecological information but also precise knowledge of the relevant socioeconomic factors. For example, is the farmer a profit maximizer or is he risk averse? What are the disease-relevant socioeconomic conditions of the patients being treated?

Strategic models, precisely because they aim at a level of generality, hopefully capturing and evaluating a limited set of critical elements and processes, can rarely be expected to answer such tactical questions. They may tell us what kind of variables are likely to be most important, but they may be vague as to how they should be measured or how they relate functionally to the pest or pathogen population parameters so as to give precise predictions. The only real exceptions are complex detailed simulation models, such as those described in Chapter 11 which try to bridge the gap, serving both strategic and tactical objectives.

The problems of tactical models are twofold: First, in order to establish and verify their predictive relationships they necessarily have to be based on laborious, detailed, and often lengthy (spanning several seasons) laboratory and field experiments. Second, in order to be implemented they require data on the critical variables relating either to the specific location or environment or, more particularly, to the specific time for which decisions have to be made. In Chapter 22, for example, Croft and Welch discuss the problem of the suitability of standard meteorological data and the extent to which tailor-made information is required for pest and disease modeling. However, as they and Zadoks and his colleagues (Chapter 21) show, it is possible through skillful organization to get farmers closely involved in the tactical modeling process and in this way make the relevant location-specific data readily available and provide the feedback necessary to correct and improve the model.

12.2. FORECASTING AND MONITORING

Inevitably, a great deal of tactical modeling in pest and pathogen control has centered on the problems of forecasting and monitoring, and this is the subject of Chapters 12–16. Forecasting is the process of predicting future pest or pathogen population sizes $\hat{N}_{t+1}$ and the resulting damage or disease, i.e.,

$$\hat{N}_{t+1} = N_t + \Delta N, \tag{12.1}$$

where ΔN is some function of present numbers, the host dynamics (crop phenology or human population change) H, and a range of environmental variables E:

$$\Delta N = f(N_t, H, E). \tag{12.2}$$

Monitoring is the process of measuring such variables as N_t, H, and E needed for forecasting. Much of forecasting in the past has been concerned with obtaining order of magnitude estimates of $\hat{N}_{t+1}$ over the medium or long term. They answer such questions as whether there will be an outbreak in the coming year, or in the following month. Such forecasts are usually functions of limited monitoring of environmental variables alone, made once or, at most, a few times per year, although there has recently been a growing demand for more precise and up to date estimates of $\hat{N}_{t+1}$, and the imminence of damage or disease. These rely on repeated or continuous monitoring of N_t and other variables (H and E). They are typically used to estimate the progress of the current infestation and how it will develop over the next few weeks.

12.2.1. Forecasting

Plant pathology has a long history of successful forecasting procedures (see recent reviews by Bourke, 1970; Krause and Massie, 1975; Waggoner, 1960, and Chapter 13). The classic examples were developed some 50 years ago for potato blight (*Phytophthora infestans*) and apple scab (*Venturia inaequalis*) and were the product of long experience and careful field observations. These demonstrated the close correlation between disease outbreaks and climatic factors such as humidity and temperature, responsible for the production and survival of spores and the success of infection. The forecasts were formulated as a set of simple, practical rules to warn farmers of an outbreak of disease should a well defined "period" (a certain combination of weather conditions over a set number of days) occur (van Everdingen, 1935; Beaumont, 1947; Mills, 1944). These rules have subsequently been modified for different countries and made more sophisticated. A particularly successful version developed for the FRG retains the simplicity of the early rules, providing farmers with a clear indication of when not to spray, although it is based on a fairly complex multiple regression analysis (Ullrich and Schrödter, 1966).

Most modern forecasts are based on explicit regression models. For example, the number of lesions may be forecast as a function of winter temperature or the numbers of spores in the vicinity of the field at the beginning of the season.

$$\hat{N} = b_0 + b_1 X, \tag{12.3}$$

where $\hat{N}$ is an estimate of the mean number of lesions, X the spore dose or winter temperature. More commonly, multiple regression relationships are used (Butt and Royle, 1974).

$$\hat{N} = b_0 + b_1 X_1 + b_2 X_2, \tag{12.4}$$

where, for example X_1 may be the spore count and X_2 atmospheric humidity. One such model, which performed satisfactorily for a number of years, forecast the incidence of the virus beet yellows in eastern England, which is carried by the green peach aphid (*Myzus persicae*). The regression analysis predicted the percentage of plants infected at the end of August as a function of temperature, because this influences the survival of the aphid over winter and its breeding in spring (Watson *et al.*, 1975).

It became clear, however, that regression models have a number of potentially serious defects. A fundamental problem is that they are only likely to be reliable within the range of data sets from which they were originally derived. They may work well for many years but then a combination of circumstances outside the original data sets may arise, causing the models to fail. For example, the beet yellows model greatly underestimated a major virus outbreak in 1974 (Watson *et al.*, 1975). If the important causal relationships are essentially linear or the predictions short-term, regression techniques may prove highly reliable. But where significant nonlinearities are present and the timescale of prediction is insufficient to allow their expression, the models may fail.

In chapter 13, Waggoner discusses more fully the problem of predicting major invasions and epidemics. He questions whether models could have forecast the potato blight epidemic in Ireland of the nineteenth century, or the southern corn leaf blight epidemic of 1970. He suggests that a possible method is to start looking for minimum values of the critical variables, replacing the regression equation by a predictor of the form

$$\hat{N} = \min (a_1X_1, a_2X_2, \ldots). \tag{12.5}$$

This method, of course, is similar to those used in several chapters in Part I of this volume. Presumably most rare epidemics are the product of threshold phenomena and if regression models are derived from data characteristic of the more common endemic state, they will inevitably fail to register an epidemic when a threshold value of a critical variable is breached.

One answer to this problem is to include more details of the essential causal relationships in the model and to incorporate explicitly all apparently important nonlinearities, although even this may be insufficient. Recent work has shown, for example, that deterministic and often quite simple population models will generate apparently chaotic behavior if the nonlinearities, such as those produced by density-dependent relationships, are sufficiently severe. The chaotic fluctuations so produced may be indistinguishable from those provided by random environmental fluctuations or sampling errors, and this may render long-term forecasting very difficult or even impossible for certain systems (May, 1981 and personal communication). This effect and its implications for weather forecasting have been investigated by Lorenz (1963); its occurrence in biological population models is discussed by Guckenheimer, *et al.* (1976), May (1974), and May and Oster (1976).

12.2.2. Monitoring

The desire to incorporate more causal detail in forecasting models has stimulated a number of detailed analyses of population changes in pests and diseases, over weekly, daily, or even shorter timescales. These are related to weather and other environmental variables and, particularly in insect pest models, related to the phenology and physiology of the crop plant. As a result of this painstaking work we now have a number of detailed simulation models for various crops, pests, and pathogens (see Aust *et al.*, 1980; Curry *et al.*, 1980; Haynes *et al.*, 1973; Kranz *et al.*, 1973; Rabbinge *et al.*, 1979; Tummala and Haynes, 1977, Tummala *et al.*, 1976; Waggoner and Horsfall, 1969; Waggoner *et al.*, 1972; Wang *et al.*, 1977).

Such models inevitably have heavy monitoring requirements. In Chapter 14, Kranz and his colleagues discuss these in relation to their simulation model (EPIGRAM) of barley powdery mildew, and in Chapter 15, Rabbinge and Carter describe the form and requirements of their model for wheat aphids. In both cases population growth is assumed to be exponential and the key monitoring requirements are measurements of the immigrants (using spore traps for fungi and suction traps for insects), temperature, and crop phenology, which determine the rate of development of the pest and pathogen populations. Kranz and his colleagues make the important observation that, since monitoring is so critical to success, the measurability and reliability of the variables to be monitored must be assessed very early in the modeling exercise. In fact, they go so far as to suggest that this should be the first step, the eventual form of the model being determined by what *can* monitored, rather than the model determining what *should* be monitored.

12.2.3. Estimation Procedures

Forecasting and monitoring necessarily imply a sampling and estimation procedure, the design and planning of which is an essential statistical exercise that tends to be neglected. In Chapter 16, Kuno looks at a statistical design problem where the challenge is to predict the damage that will be caused at some time in the future by a pest population (in this case the rice brown planthopper) which is several generations removed from the generation currently being sampled. For example,

$$N_4 = N_1 + \Delta N_1 + \Delta N_2 + \Delta N_3, \tag{12.6}$$

where N_1 and N_4 are the first and fourth generations. Using quadrat sample data gathered over eight years, Kuno shows how fairly robust estimates can be obtained by a sequential sampling procedure and suggests how this can be used in a decision-making scheme for control, once a precise yield–density relationship is obtained.

In the case of vector-borne disease, where monitoring is aimed at determining actual or potential rates of disease progress, the estimation procedure tends to focus on the rate of change rather than the absolute size of the pest (vector) population. In malaria, for example, the classic model developed by Ross (1911) and MacDonald (1957) gives (in simplified form):

$$R = \frac{ma^2p^n}{-r \ln p}, \tag{12.7}$$

where R is the basic reproductive rate (see Chapter 2), m is the density of mosquitoes per human, a is the man-biting frequency, p is the survival rate of the mosquitoes, n is the length of the parasite cycle in the mosquito, and $1/r$ is the length of infectiousness in man. R is most sensitive to changes in the survival rate of the vector, and this parameter must be monitored from place to place and time to time.

$$\hat{N}_s = N - N_R = N_{t+1} - N_t - N_R \tag{12.8}$$

where $\hat{N}_s$ and N_R are numbers of survivors and recruits.

The challenge, as in all forecasting and monitoring schemes, is to find a standardized sampling procedure that is simple but robust. In Chapter 17,

Birley examines one estimate of survival rate, known as the mean parous rate, and shows how the estimation procedure can be improved to take account of fluctuations in recruitment to the mosquito population, and in the biting cycle, which affects the ability of the parous rate to estimate survival.

12.3. OPTIMIZING PESTICIDE APPLICATIONS

It may become costly and laborious if, on every occasion a pest or pathogen population is monitored, a simulation model has to be rerun with the updated information before a decision can be made. An alternative approach is to use a model to explore all the possible situations that may arise (for example, in a crop season) and hence to prepare a set of decision rules in advance to cover every eventuality. But the problem here, too, is that the number of potential conditions of the pest or pathogen population and its crop or host rapidly becomes very large, particularly when decisions are to be made on a daily or weekly basis. Even with modern computers the computations rapidly become physically, if not financially, infeasible.

12.3.1. Dynamic Programming

One approach, however, is to use a technique of dynamic programming (Bellman, 1957), which essentially employs the trick of beginning the computation at the final stage of the decision process, for example, with the numbers of pests or level of damage at the end of the crop season, and then working backwards along an optimal pathway to the beginning. This considerably reduces the number of combinations that have to be assessed. Figure 12.1 shows the results of a dynamic programming exercise where the state variables are the population size N, divided into four grades, and the stage in the phenology of the plant P^m. The optimization problem is one of minimizing the sums of the costs C of control actions V and the losses from damage D. The function

$$F_k(N^k P^k) = \underset{m=k,\ldots,M}{\overset{\min}{V^m}} \left[\sum_{m=k}^{M} D^m(N^m P^m) + \sum_{m=k}^{M} (CV^m) \right] \tag{12.9}$$

gives this minimum from period k to the end of the season M. The solution is found from the recursive relation

$$F_k(N^k P^k) = \underset{m=k,\ldots,M}{\overset{\min}{V^k}} [D^k(N^k P^k) + (C^k) + F_{k+1}(N^k P^k)]. \tag{12.10}$$

One of the advantages of this technique is that if for any reason (adverse weather conditions, poor spray applications, etc.) the optimal path is not followed between two decision periods, a new optimal path is provided. For example, in Figure 12.1 if no spraying between periods 2 and 3 results in 200

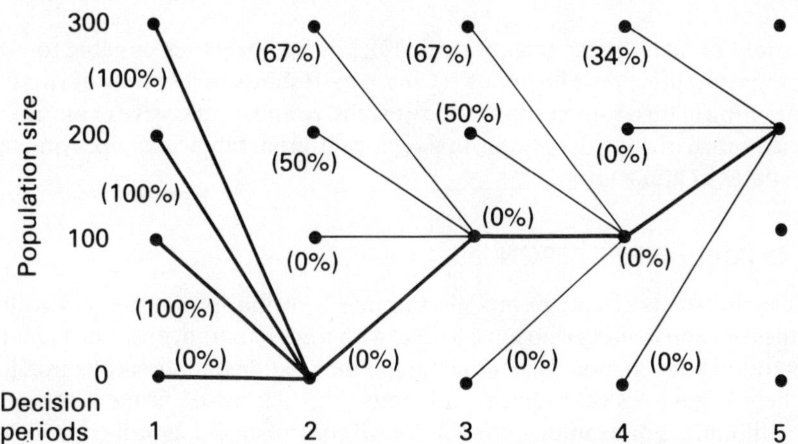

FIGURE 12.1 Optimal pathways generated by dynamic programming for a simple pest control decision problem. Damage is a function of pest numbers and the phenological state of the plant. The pests cause relatively less damage later in the season. The pathway gives the minimum total summed losses from damage and cost of control over the season and the paths show the percentage level of kill (V) to be aimed at with any given population size.

rather than 100 pests, an optimal path can be regained by a 50% kill, which minimizes the future total losses and costs.

Dynamic programming has been applied to a number of pest problems (Conway *et al.*, 1975; Shoemaker, 1979; Winkler, 1975) but, because it is unable to handle more than four state variables at the very most, it has not been used for situations in which, for example, age classes of the pest population differ in their pesticide susceptibility or propensity to cause damage, and hence have to be accounted for separately. In Chapter 18, Shoemaker shows that a good way around this problem is to take the date of last spraying as the state vector and, given relations for the pesticide mortality of different age classes, the damage caused, and the decay in residual toxicity, then to determine whether or not to spray at each decision period. She shows how the results, as illustrated by the Egyptian alfalfa weevil, can be displayed in readily understood tables for use by the farmer.

12.3.2. Other Optimization Techniques

In the simpler case of a choice between one or two pesticide applications in a season, El-Shishiny (Chapter 19) shows that solutions can be obtained by combining gradient and quadrat techniques. He illustrates this for the greenhouse whitefly, and in the same chapter Rodolphe explains the result more fully, using simulation in order to get a better feel for the robustness of the solution. Where the change in the state variable can be represented by relatively simple logistic or exponential equations, as is true of epidemic progress of many pathogens, for example, it even becomes possible to use an analog computer. As Matsumoto (Chapter 20) shows in his study of rice blast, computing in this fashion enables very extensive and inexpensive exploration of a wide range of control options, in this case different fungicides, spraying rates, and dates of application.

12.4. IMPLEMENTATION

The main effect of strategic models is primarily on the approaches and attitudes of the decision makers who have to deal with pests or pathogens. The guidelines generated by such models aim to improve the recommendations of agricultural or health advisors and hence, indirectly, the actions of farmers or general practitioners. For example, the boll weevil control model described in Chapter 11 may be deemed successful if its general recommendation of early spraying is adopted by advisors and cotton growers, and if this in turn results in more effective and profitable control of the cotton–pest complex as a whole. The implementation of strategic models is thus largely effected by publicizing the results of modeling exercises through books, leaflets, training courses, etc.

12.4.1. On-line Models

In contrast, the implementation of the results of tactical models, because they aim at more precise recommendations specific to particular locations and times of the year, necessarily involves close liaison between the modelers and decision makers (farmers) who have ultimate responsibility for control programs. Implementation is consequently considerably more complex and difficult, and requires network of permanent or semipermanent contacts between farmers and the modeling team, and an administrative machinery to oversee the process. Not surprisingly, few such systems exist at present, but, as Zadoks and his colleagues show in Chapter 21, the task can be approached successfully by starting on a small scale with well defined objectives, and providing limited but reliable advice. Their EPIPRE program with a well researched model of yellow rust and its damage to wheat is aimed at helping a small group of farmers directly on a field-by-field basis, with the timing and appropriateness of fungicide application. The benefits can be clearly seen in terms of savings in the costs of pesticide. Participation by the farmers is made easy by a simple, time-saving monitoring scheme and a straightforward method of communicating data to the modeling team. The success of this kind of approach has now encouraged an expansion of the advisory system to many more farmers and includes other wheat diseases and cereal aphids (see Chapter 15).

The on-line pest management system described by Croft and Welch (Chapter 22), although it focuses on a different set of problems concerning the control of apple pests, is based on a similar approach. The aim is initially limited to reducing costs and avoiding unwanted side effects by advising farmers when pesticide applications are not needed. As with EPIPRE, the advice consists essentially of either "treat" or "don't treat", with an in-between response necessitating further simulation studies and/or further monitoring. With apple mites the advice essentially depends on the population size and dynamics of the mites and their predators as determined in the field. The recommendation to treat or not to treat can be arrived at relatively inexpensively using simple decision rules previously derived from the simulation model. Costs start to rise rapidly, however, when individual farmers require detailed simulator runs and repeated intensive monitoring.

The management/advisory system described by Croft and Welch has now reached a considerable degree of sophistication, and the question of relative costs and benefits has become increasingly critical. They therefore carried their systems studies one logical step further by applying systems analysis to the management system itself. First they looked at the costs and benefits of the integral biological and environmental monitoring components of the system (reflecting the concern of Kranz and his colleagues in Chapter 14). One of their results is that the pursuit of too high a degree of model accuracy may easily increase costs to a point where the benefits are outweighed: a rather finely balanced optimum has to be sought. They then examine the overall organization of the management/advisory system. Here their analysis and experience suggest that a very centralized system is likely to incur very high communication costs and low reliability. They are consequently experimenting with a more hierarchical mode of system organization using small and cheap data processors, shifting the processing and modeling to regional, county, or even farm level.

12.4.2. Strategies and Tactics

The issues of cost and effort involved in implementing detailed, site-specific tactical models raise the question of whether such an approach is universally applicable or appropriate. As Norton and his colleagues argue in Chapter 23, it may be possible to use strategic models to search for "robust" strategies that perform adequately over a large range of parameter values. In the case of the cattle tick in Australia, tactical models would have to be tailored to a large number of individual cattle-raising properties for which detailed monitoring information is not available at present. Their aim has been to see if strategies can be developed that hold, not only for average values or for a narrow set of assumptions, but also result in reasonably satisfactory control at more extreme values or under more relaxed assumptions. The emphasis is then less on finding optimal strategies and more on identifying the boundaries within which the robust strategies give a good performance. Implementation then consists of simply examining individual properties with respect to these performance boundaries, and providing advice accordingly.

There is an underlying assumption that, although the biological, environmental, and economic dimensions of pest and disease systems are complex, their essential behavior, as far as management is concerned, is dominated by a few key processes and, accordingly, there is limited and constrained set of satisfactory management responses. The challenge of this kind of strategic modeling is thus to identify these processes and through analysis and simulation to discover the key management responses (see also Chapter 1).

In earlier work on the cattle tick, Sutherst *et al.* (1979) demonstrated the robustness of a strategy of dipping European cattle with acaricides on five occasions in the year, separated by three-week intervals. This was shown to be remarkably robust to ostensibly significant changes in such variables as initial population size, survival rate, density dependence, and the damage relationship, resulting at most in the addition or subtraction of a single dip. In particular, the three-week interval was robust to management errors such as, incomplete mustering of the cattle, a lower than recommended concentration of acaricide, and mistiming the dipping. Moreover, the strategy was also reasonably robust to changes in the cost of dipping and the price of beef.

In Chapter 23, Norton *et al.* turn their attention to the problem of controlling ticks on cross-bred cattle and again demonstrated a robust acaricide dipping strategy, this time requiring three dips per year. Because of the increasingly critical problem of acaricide resistance they have also searched for and found a robust strategy of pasture spelling that relies on moving the cattle from one paddock to another in order to break the life cycle of the tick. They show that very high levels of control can be obtained by spelling periods of only four weeks when carried out in spring or summer. The aim now is to implement these findings essentially by means of an extension "handbook", which allows each farmer to identify his own appropriate management strategy in terms of where it fits in the overall space of certain critical parameters identified by modeling analysis.

REFERENCES

Aust, H. J., B. Hau, and J. Kranz (1983) EPIGRAM—a simulator for barley powdery mildew. *Z. Pflanzenkrankh. Pflanzenschutz* 90: 244–50.

Beaumont, A. (1947) The dependence on the weather of the dates of outbreak of potato blight epidemics. *Br. Mycol. Soc. Trans.*, 31:45–53.

Bellman, R. B. (1957) *Dynamic Programming* (Princeton: Princeton University Press).

Bourke, P. M. A. (1970) Use of weather information in the prediction of plant disease epiphytotics. *Ann. Rev. Phytopathol.* 8:345–76.

Butt and Royle D. J. (1974) Multiple regression analysis in the epidemiology of plant diseases, in J. Kranz (ed) *Epidemics of Plant Diseases: Mathematical Analysis and Modeling* (New York: Springer-Verlag) pp78–114.

Conway, G. R., G. A. Norton, A. B. S. King, and N. J. Small (1975) A systems approach to the control of the sugar froghopper, in G. R. Dalton (ed) *Study of Agricultural Systems*. (London: Applied Science Publishers) pp193–229.

Curry, G. L., P. J. H., Sharpe, S. W. De Michele, and J. R. Cale (1980) Towards the management of the cotton-boll weevil ecosystem. *J. Environ. Management* 11:187–223.

van Everdingen, E. (1935) Net verbrand tusschen de weer gesteldheidgen de aardappelziekte (Tweede mededeling). *Tijds. Plziekte.* 41:129–33.

Guckenheimer, J., G. F. Oster, and A. Ipaktchi (1976) The dynamics of density-dependent population models. *J. Math. Biol.* 4:101–47.

Haynes, D. L., R. K. Brandenburg, and P. D. Fisher (1973) Environmental monitoring for pest management systems. *Environ. Entomol.* 2:889–99.

Kranz, J., M. Mock, and A. Stumpt (1973) EPIVEN — ein Simulator für Apferschorf. *Z. Pflanzenkrankh.* 80:181–7.

Krause, R. A. and L. B. Massie (1975) Predictive systems: Modern approaches to disease control. *Ann. Rev. Phytopathol.* 13:31–47.

Lorenz, E. N. (1963) Deterministic non-period flow. *J. Atmos.* 20:130–41.

Macdonald, G. (1957) *The Epidemiology and Control of Malaria* (London: Oxford University Press).

May, R. M. (1974) Biological populations with nonoverlapping generations: stable points, stable cycles and chaos. *Science* 186: 645–7.

May, R. M. (1981) Models for single populations, in R. M. May (ed) *Theoretical Ecology: Principles and Applications* 2nd edn (Oxford: Blackwell) pp5–29.

May, R. M., and G. F. Oster (1976) Bifurcations and dynamic complexity in simple ecological models. *Am. Nat.* 110:573–99.

Mills, W. D. (1974) Efficient use of sulfur dusts and sprays during rain to control apple scab. *Cornell Univ. Agric. Exp. Station Bull.* 630.

Rabbinge, R., G. W. Ankersmit, and G. A. Pak (1979) Cereal aphids, epidemiology and simulation of population development in winter wheat. *Neth. J. Plant Pathol.* 85:197–220.

Ross, R. (1911) *The Prevention of Malaria* 2nd edn (London: Murray).

Shoemaker, C. A. (1979) Optimal management of an alfalfa ecosystem, in G. A. Norton and C. S. Holling (eds) *Pest Management*, Proc. Int. Conf., October 1976. IIASA Proceedings Series Vol. 4 (Oxford: Pergamon) pp301–15.

Shoemaker, C. A. (1981) Applications of dynamic programming to pest management. *IEEE J. Automatic Control.*

Sutherst, R. W., G. A. Norton, N. D. Barlow, G. R. Conway, M. Birley, and H. N. Comins (1979) An analysis of management strategies for cattle tick (*Boophilus microplus*) control in Australia. *J. Appl. Ecol.* 16:359–82.

Tummala, R. L. and D. L. Haynes (1977) On-line pest management systems. *Environ. Entomol.* 6:339–44.

Tummala, R. L., D. L. Haynes, and B. A. Croft (eds) (1976) *Modeling for Pest Management: Concepts, Applications and Techniques* (East Lansing: Michigan State University).

Ullrich, J. and Schrödter (1966) Das Problem der Vorhersage des Auftretens der Kartoffelkrautfäule (*Phytophthora infestans*) und die Möglichkeit seiner Lösung durch eine "Negativprognose". *Nachrichtenbl. Dtsch. Pflanzenschutzd.* 18:33–40 (Braunschweig).

Waggoner, P. E. (1960) Forecasting epidemics, in J. G. Horsfall and A. E. Dimond (eds) *Plant Pathology* Vol. 3 (New York: Academic Press) pp291–313.

Waggoner, P. E. and J. G. Horsfall (1969) EPIDEM, a simulator of plant disease written for a computer. *Connecticut Agric. Exp. Station Bull.*, 698.

Waggoner, P. E., J. G. Horsfall, and R. J. Lukens (1972) EPIMAY a simulator of southern corn leaf blight. *Connecticut Agric. Exp. Station Bull.*, 729.

Wang, Y., A. P. Gutierrez, G. Oster, and R. Daxl (1977) A population model for plant growth and development: coupling cotton–herbivore interaction *Can. Entomol.*, 109:1359–74.

Watson, M. A., G. D. Heathcote, F. B. Lauckner, and P. A. P. Sowray (1975) The use of weather data and counts of aphids in the field to predict the incidence of yellowing viruses of sugar-beet crops in England in relation to the use of insecticides. *Ann. Appl. Biol.*, 81:181–98.

Winkler, C. (1975) *An Optimization Technique for the Budworm Forest–Pest Model* RM-75-11 (Laxenburg, Austria: International Institute for Applied Systems Analysis).

13 Models for Forecasting Disease Control

Paul E. Waggoner

13.1. INTRODUCTION

13.1.1. A Blight of Potatoes

On 23 August 1845 a newspaper for gardeners and farmers spread the alarm:

> A fatal malady has broken out amongst the potato crop. On all sides we hear of the destruction. In Belgium the fields are said to have been completely desolated. There is hardly a sound sample in Covent Garden Market.

More reports came, telling of the spread of the disease across northern Europe and relating that every kind of potato in England was succumbing. Only three weeks later, on 13 September, the newspaper announced that

> the potato murrain has unequivocally declared itself in Ireland. The crops about Dublin are suddenly perishing....where will Ireland be, in the event of a universal potato rot?

The population of Ireland, sustained by potatoes, had grown from 4.5 million in 1900 to over 8 million in 1841, but by 1851, after the famine, it had dropped to 6.5 million, far less than the 9 million extrapolated from the 1841 census. Thus there was a loss of 2.5 million people, about a million to emigration and about 1.5 million to famine and diseases brought on by hunger. What model could have forecast this disaster?

The weather had been unusual; up to the beginning of July the weather had been almost ideal—hot and dry, beautiful hay-making weather.

> It then suddenly changed to the most extraordinary contrast that I ever witnessed in this climate, the atmosphere being for upwards of three weeks one of continued gloom, the sun scarcely ever visible during the time, with a succession of most chilling rains and some fog, and for six weeks the temperature was from 1½ degrees to 7 degrees below the average for the past 19 years.

However, the weather of the preceding 3–6 years had been even more favorable to the malady, but it had not struck then.

With hindsight, one can see clues to the disaster. An outbreak of a similar disease had affected potato crops in North America the year before, and a traveler had informed the French Academy that it had long been known in Colombia, where the Indians, like the Irish, lived chiefly on potatoes. But even with these clues added to the experience of 1845, could a model have forecast the rapidity of a second tragedy only two years later?

> On July 27 I passed from Cork to Dublin, and this doomed plant bloomed in all the luxuriance of an abundant harvest. Returning on August 3, I beheld with sorrow one wide waste of putrefying vegetation. In many places the wretched people were seated on the fences of their decaying gardens, wringing their hands and wailing bitterly at the destruction which had left them foodless.

Did they only lack a computer that could have foretold their troubles? (This description of the epidemics of 1845 and 1847 is taken from Bourke, 1964; Large, 1940; and Woodham-Smith, 1962).

13.1.2. A Blight of Corn

On 17 August 1970 in the USA the headline in a business newspaper stated: "Much of Nation's Crop of Corn Called Perilled by Southern Leaf Blight". The next morning the same journal said, "Grains, Soybeans, Rise Daily Limit on Corn Blight". The next day a headline announced, "Blight Kills Off 25% of Corn in Illinois"—bad news that was scarcely lessened by another headline in the same paper: "Grains, Soybeans, Rise on Monday Reversed by Profit Takers". The bad news spread as the headlines went on: "Corn-Product Prices Are Raised As Blight Imperils Crop and Pushes Grain Quotes Up", and "Blight in Cornfields May Bring Sharp Rises in Prices of Meat, Eggs".

The pathogen *Helminthosporium maydis* devastated the American corn crop in its first year on the scene. In the southern US many corn fields were totally lost and some were plowed under and planted to other crops. In the corn belt yields were often reduced by half, ears and stalks of corn continued to rot until harvest, and stalks fell, making harvesting difficult. Desperate applications of fungicides came too late. Corn futures rose from $1.35 to $1.68 per bushel, and, as the Irish had turned to corn instead of potatoes over a century earlier, British farmers in 1971 shifted from corn to barley, thus driving up its price. What model could have forecast this loss in America's major crop, formerly so unmolested by pests?

With hindsight we now see that the Texas strain of cytoplasmic male sterility, by reducing the labor of producing hybrid seed had created a subcontinent of homogeneous host. We now know that *H. maydis* was present in the corn belt in 1969, and over winter, Hooker *et al.* (1970) had actually named the new strain race T for its ability to destroy the Texas strain of corn. We also know that in the southern US the spring of 1970 was wet, and that in the corn belt the summer was warm and wet. Reinforcing the similarity to the Irish epidemic, a national conference was called by President Nixon in 1970, just as Prime Minister Sir Robert Peel set up a scientific commission 125 years earlier. (This description of the southern corn leaf blight of 1970 is taken from Ullstrup, 1972.)

When the southern corn leaf blight struck plant pathologists had been devising forecast rules for more than a generation, and they had even taken to using computers. But what model could have forecast the events of 1970?

13.1.3. A Mold of Tobacco

On 1 August 1979 The Connecticut Agricultural Experiment Station opened its farm to the public, and a reporter photographed the director tossing dusty soil in the air to illustrate a story of drought. On the same day, a farmer brought to the station laboratory 30 miles away a tobacco leaf with spots. Held in the correct light, the leaf showed a faint blue mold. G. S. Taylor, who diagnosed the disease, had studied tobacco blue mold during previous epidemics, but these had been 25 years ago (Waggoner and Taylor, 1958).

By 1 September the expensive and precious leaves of shade-grown tobacco were being abandoned, and those growers who tried to salvage the top half of the crop wondered if they were wasting their time. A young man brought a plant from his first tobacco crop, asking what the spots were. When told they were blue mold, he said "I hope the bank understands".

Again, with hindsight, we could see the source of this epidemic. It had appeared in Cuba during the winter of 1979 and in Pennsylvania and even Canada before it reached Connecticut. Rain showers that had passed other localities had fallen where the blue mold first appeared in Connecticut, and it was realized that the dewpoint was high everywhere. A scheme for warning of the onset of blue mold had been employed in Connecticut 25 years previously (Waggoner and Taylor, 1955), but what model could have been kept going for such a long time, and what observer would have fed into the model climatic data of the farm where the disease first appeared?

13.2. MODELS AND EXPERIENCE

The answer of a practical man is clear: although prediction models take much of the guesswork out of pest management and make it a science rather than an art, a pest manager wrote that he continued to use experience and knowledge acquired over the years and concluded, "I have no knowledge of commercial pest managers utilizing predictive models at the present time" (Barnett, 1976).

A representative of an agricultural chemical association wrote, "The pesticide chemicals industry does not maintain nor coordinate predictive modeling systems and does not have access to data banks or other repositories concerned with pest management, and it probably never will." He also wrote, "My company conducted a trial program several years ago in an attempt to define some of the parameters of pest population prediction systems. This exploratory endeavor was directed at a single pest on a single crop in a limited, well-defined area of one state... The information gained from this experiment was very elementary. We did learn that we could not afford the cost of collecting and programming the data, especially since we could not place complete confidence and reliability in the predictive system" (Conner, 1976).

Plant pathologists with a vested interest in forecasting have naturally behaved differently. In reviewing disease forecasting in 1960, I introduced the subject of forecasting but then turned to epidemiology, using the equation for eddy diffusivity. Finally, I returned to forecasting and described a method for predicting tobacco blue mold, but this had fallen into disuse before the epidemic of 1979 (Waggoner, 1960).

A decade later, a very readable review (Bourke, 1970) lamented my statement that farmers have made disappointingly little practical use of forecasts, but concluded with a hopeful anecdote: "In Italy, where local warnings to spray against vine mildew are transmitted dramatically by church bells, growers have been known to abandon a political meeting en masse on hearing the signal, leaving the speaker in mid-sentence." Even optimistic Bourke, however, wondered how many reservations the average maker of forecast models calls to mind after his simulator, warts and all, has disappeared from sight into the belly of a computer and when the machine is turning out its oracular conclusions with all the bland assurance of a confidence trickster. He wondered if the imposing computer may rob us of that small residue of commonsense that we all have and asked, "What hope remains of scrutiny by the lacklustre eye when the model is buried in the bowels of the machine?"

Two reviews (Krause and Massie, 1975; Shrum, 1978), however, take cheer from the very computers that Bourke fears. They believe that the late blight forecasts of Hyre and Wallin would be used by farmers if they had been provided by a computer installed in their own fields and called "Blitecast". There are reasons for hope—Blitecast has been successful at 41 locations in four states (Krause and Massie, 1975), and is now being manufactured at a cost of $1200. A pessimist, however, would remember that Hirst and Stedham (1956) found no advantage in forecasting late blight from observations among rather than above the potatoes. Croxall *et al.* (1961) found that the weather at a group of farms, rather than at an individual farm was a more accurate basis for blight forecasts. Finally, mechanization alone will not make forecasting successful, as Grainger (1953) showed.

13.2.1. Some Classic Forecasting Models

I now take a more conventional view and discuss some classic forecasting models. These embody decades of experience, and, as Krause and Massie (1975) have emphasized, old methods can be useful in the new computerized approach. The best forecasting results in plant diseases have been obtained with *Phytophthora infestans*, where the Dutch rules (van Everdingen, 1935) were modified for the UK (Beaumont, 1947). In the US, Hyre (1955) and Wallin and Schuster (1960) invented and perfected a system of weighting favorable periods; this is the heart of computerized Blitecast. The "Negativprognos" was developed in the FRG to tell farmers when they need not spray with fungicides (Ullrich and Schrödter, 1966). Synoptic weather maps have been exploited in forecasting late blight, since they show at a glance air flow and the extent and persistence of foul weather conditions that encourage disease (Bourke, 1957; Wallin and Riley, 1960). With late blight one also finds thorough tests of forecasts, showing the consequences in terms of disease controlled or fungicide saved (Frost, 1976).

It has been possible for a long time to forecast outbreaks of apple scab. In fact, it is common jargon to say that a "Mills period", which favors apple scab, has occurred (Mills, 1944). Scab prediction has now been mechanized (Jones *et al.*, 1979). In addition to weather conditions, the scab fungus requires open buds and mature ascospores so that forecasting therefore uses the phenology of the tree (Lake, 1956) and of the fungus (Massie and Szkolnik, 1974). Systems for forecasting many other diseases have been developed. Schemes in the US, encouraged by a disease survey and forecasting bureau within the Department of Agriculture, have been reviewed by Miller and O'Brien (1952, 1957), and those in the UK have been surveyed by Large (1955). In 1960 I surveyed the stock of forecasts (Waggoner, 1960), and this was later done by Bourke (1970) and by Krause and Massie (1975).

The advent of the computer evoked two new types of forecasting models. Regression analysis had been used since Sir Francis Galton (born 1822) analyzed the heights of fathers and sons, but its application to plant disease awaited computers; the effect of environmental factors on wheat leaf rust (Eversmeyer and Burleigh, 1970) and hop downy mildews (Royle, 1973) exemplify the use of linear regression equations from observations of disease, spores, and weather.

Another forecasting model also awaited the advent of the computer. Bourke (1970) thought that the computer merely facilitated arithmetic in statistical analyses, but it was actually an integral part of "the full-blooded models of plant disease, so detailed as to merit being ranked as a simulation of the real thing." Again, *Phytophthora infestans* was the first example—an early simulator mimicked its control by weather and fungicides (Waggoner, 1968). The other classic subject of forecasting, apple scab, was also the subject of an early simulation (Kranz *et al.*, 1973). The mathematical analysis and modeling of plant disease epidemics by both regression analysis and simulation are described in Kranz (1974).

13.3. FORECASTING THE UNEXPECTED

I now turn to observations on the original far-reaching topics. How can we forecast epidemics of previously rare pathogens? How can we forecast extreme events rather than routine recurrences? What types of forecasts, and thus what types of models, are of practical use? Why concentrate on forecasting epidemics of rare pathogens? One could argue that it would be more profitable to concentrate on forecasting the regular recurrence of common diseases on which we expend tons of pesticides, hoping to reduce costs and consumption. It is precisely these common diseases that have produced our forecast methods.

Let us consider the alternative—forecasting the unexpected; after all, the opposite takes little skill and presents little challenge. Berger (1977) showed that in Florida, *H. maydis* follows the same course year after year, so the best forecast is one based on the previous year's experience. A far more important reason for forecasting unexpected pathogen epidemics is, however, to predict unexpected changes in crops and thus in human food supply. Our channels of trade, and even society, are adjusted to the usual. Chemicals and varieties are developed to control common diseases. Markets are attuned to delivering the usual chemicals and varieties to the farmer and taking away the usual crop, and the consumer receives the usual produce. An unusual epidemic throws all into confusion.

Unexpected outbreaks of plant diseases have been catalogued by Klinkowski (1970), who divided catastrophes into two classes. In one class, a crop is moved to a different environment and its pathogens are left behind; an epidemic occurs when the pathogen finally catches up. This can be exemplified by the arrival of *P. infestans* and *Peronospora tabacina* in the potato and tobacco fields of Europe long after they had occurred in North America. The other class of catastrophe is caused by the emergence of a wholly new pathogen, as exemplified by the outbreaks of races 38 and 56 of stem rust and of *H. maydis* in the US.

Forecasting these unusual events is much more demanding than forecasting the annual recurrence of late blight in a humid climate, for instance, but a useful first approach to forecasting epidemics is to consider the use of simulators. EPIMAY, a simulator used for the southern corn leaf blight, was thoroughly tested in 1971 by comparing blight strains in Texas and Indiana.

The greatest difficulty was in establishing the level of the initial inoculum so that the simulator could integrate the effect of weather on the increase of that initial inoculum. The greatest challenge was not the main computer program but the subroutine that established the initial condition (Shaner *et al.*, 1972). Thus there is a resemblance between mathematical simulation of familiar diseases such as southern corn leaf blight in 1971, forecasting a pandemic caused by a pathogen catching up with an old host, and a new pathogen finding a new host. When has the inoculum arrived, and how much is there?

13.3.1. Trapping Spores or Disease Monitoring?

It follows that logical disease forecasting models should take into account the concentration of propagules in the air, water, or soil. For example, spores in the air have been trapped since 1917 in North America, but despite its attraction spore trapping does not provide an accurate initial condition for models. For example, it failed to forecast downy mildew because identification of the pathogen was difficult, its viability was unknown, and its variability, even within one field, was great (Hyre, 1950). Even in the flyway of wheat rust in the Great Plains, spore trapping has been unsuccessful because the traps primarily detect the spores rising from the infections nearby that are easily detected by strolling through a field (Rowell and Romig, 1966). The disappointing results of trapping compared with strolling was confirmed by Eversmeyer and Burleigh (1970). In two years, for two varieties of wheat scattered over seven locations in three states, the best eight-day forecast of wheat leaf rust was obtained by combining data on prior disease severity, temperature, and rainfall, rather than the number of spores trapped, temperature, and rainfall.

Thus, the first step in forecasting the success of a new pathogen is knowing when and how much has landed. Economy and experience indicate that the best way to watch for the invader is by surveying crops rather than sampling air. Disease monitoring is an old science, which probably began in Denmark in 1884 and was certainly carried out in the US in 1917. Although it requires vigilance and sharp eyes, and goes against the tendencies of pathologists to work in the laboratory rather than in the field (Horsfall, 1969), it is nonetheless the only reasonable way of obtaining the initial condition for disease forecasting models, and it is still used today (Young *et al.*, 1978).

The initial conditions for an epidemic should be very carefully established, in order to build a model that can integrate them into a forecast. If a disease is new to an area, historical data are obviously not relevant to a model. We therefore have to simulate the development of a pathogen in its natural state from its indoor behavior. This was almost accomplished for the southern corn leaf blight epidemic. Within months of the initial outbreak, a simulation of *H. maydis* was built up from indoor observations and during the following season it forecast the course of the blight successfully from week to week (Felch and

Barger, 1971). The simulator also accurately predicted the spread of the disease into New England (Waggoner *et al.*, 1972).

The specter of climatic change increases the need for epidemic forecasting. For example, if increases in atmospheric carbon dioxide cause temperatures in the northern temperate region to rise and global precipitation patterns to shift (Manabe and Weatherald, 1975), we can expect epidemics in regions where they have not previously occurred. The history of disease in those regions is then irrelevant or even misleading. Again, the answer is disease monitoring to establish the initial condition and then integration by a simulator based on the physiology of the pathogen.

Forecasting an extreme epidemic, like forecasting an unusual one, merits special attention. Certainly there is a different type of interest, for example, in forecasting and avoiding an extreme diet of 100% less food for one month than a diet of 8% less for one year. Our models, I believe, have not been designed to forecast extremes with particular accuracy.

Considering historical models by regression, we see first the assumption that the level of a disease, or its logarithm, is normally distributed around an average, and is determined by such independent variables as climate and the numbers of initial inoculum. The method then estimates the average by minimizing the sum of squares of the disease or its logarithm around the average. It is not obvious that this is the best method for discriminating between levels at which the disease is tolerable and those at which it is disastrous.

13.3.2. Regression or the Law of the Minimum?

We could possibly forecast extreme outbreaks if we assumed that each epidemic was limited by a single factor such as temperature or level of inoculum, rather than that each epidemic was determined by a linear function of the effects of all factors. Therefore, instead of the regression equation

$$y = a_0 + a_1x_1 + a_2x_2 \ \ldots \ , \tag{13.1}$$

we could use

$$y = \min(a_1x_1, a_2x_2, \ldots), \tag{13.2}$$

where disease y is related to factors x by the parameter a. In the case of the new equation, 'min' is the FORTRAN command to take the smallest value. The new equation is, of course, merely the law of the minimum or law of limiting factors. Neither equation is wholly logical. The regression equation adds the effects of the factors, saying, for example, that a disease can occur without inoculum x_1 if rainfall x_2 is sufficient. The law of the minimum, on the other hand, says that if there is no inoculum there will be no disease, which is logical. The law also says, however, that if there is little inoculum, the amount of moisture will be unimportant, which is illogical.

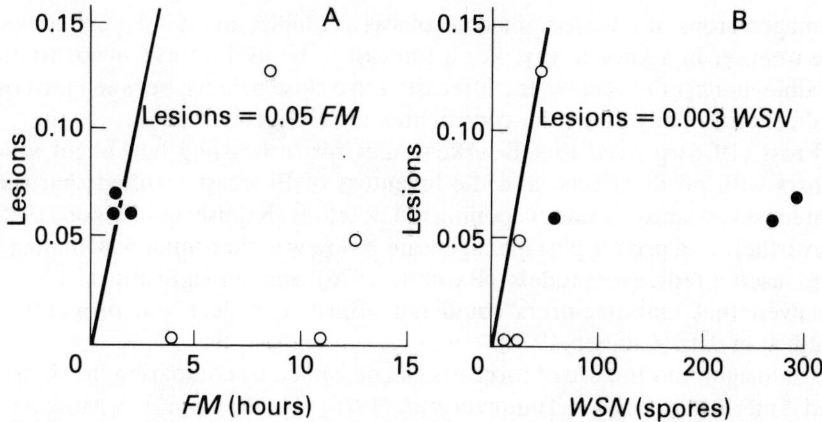

FIGURE 13.1 Fit of the law of the minimum to lesions of leaf rust per 100 wheat leaves at Colby, Kansas, in 1968. A, three observations • limited by hours of free moisture FM are fitted by $y = 0.05FM$. B, four observations ○ limited by the number of spores *WSN* are fitted by $y = 0.003WSN$. The parameters of a multiple linear regression fitted to these data indicate little increase in lesions with increasing moisture, and a decrease in lesions with more spores. The observations are from Eversmeyer and Burleigh (1970) and the analysis from Waggoner *et al.* (1980).

Regression and the law of the minimum have been compared by Waggoner *et al.* (1979) using observations of hop downy mildew (Royle, 1973) and wheat leaf rust (Eversmeyer and Burleigh, 1970). The 12 analyses conducted showed some advantages of the law (Figure 13.1). *The fit to observations, as shown by the coefficient of determination, was closer for the law than for regression in eight out of 12 cases. Compared with regression, fitting the law estimated larger values for the effects of the factors (parameters a in the equations), making them closer to the effects observed in controlled experiments on single factors. Further, the effects a of the factors estimated for the law of the minimum did not vary erratically as in the regression equations because of correlations between inoculum, temperature, and moisture.*

Forecasting extremes rather than the average is the problem here. When I examined the deviations of the predictions from observation, of the 10 most severe of 51 infections with hop downy mildew, I found that both the law and regression generally predicted too little disease. Happily, in these few examples of extremes, predictions by the law came closer than those of regression.

We have only begun, however, to think about forecasting severe rather than average epidemics. We shall have to make large changes in our models and not merely fiddle with transformations of data for existing models.

13.4. CONCLUSIONS

We come now to the question of what sort of models can be used in practice. Bourke (1970) thought that forecasts would be worthwhile if the disease damaged crops, if it varied, if a control was available, and if it depended upon the weather in a known way. For a forecast to be useful there needs to be a flexible manager to receive the forecast, and a close balance between the costs and benefits of any particular control measure (Waggoner, 1960).

Frost (1976) proved that Bourke's rules for forecasting late blight saved sprays with no ill effects, and the inventors of Blitecast testified that their system saved sprays without spoiling the potatoes (Krause and Massie, 1975). Nevertheless, a private pest manager said he knew of no commercial managers who used predictive models (Barnett, 1976) and an agricultural chemist believed that manufacturers could not afford to collect and program the prediction data (Conner, 1976).

An insight into the use of forecasts can be gained by comparing the Russian and American reviews in Tummala *et al.* (1976). The Russian, I. I. Minkevich, took a broad view and anticipated development of a disease over a wide area as he made his forecasts before the beginning of vegetation. As an American, I took a much narrower view, forecasting the success of a specific inoculum from the physiology of *Phytophthora infestans* and prevailing climatic conditions. At first it seemed that Minkevich was "planning" rather than "forecasting", but then I reread a paper by a pesticide manufacturer (Conner, 1976). After discarding forecasting in the sense used in the *Annual Review of Phytopathology*, the manager went on to demonstrate that he had to forecast because "the role of industry is to ensure that adequate stocks of pesticide chemicals be in the channels of trade, at the site of need and at the time of need." No doubt a seedsman would have written the same about resistant varieties; he would like forecasts to guide research, development, and marketing. Clearly the American manufacturer was seeking what the Russian called a long-term forecast to help him pick the right research and researchers, and then to manufacture the most useful or profitable products (Polyakov, 1976).

When evaluating the practical use of forecasts, one comes up against economics. Because the demand for food crops is often inelastic, the market can amplify the harm of plant diseases. Thus in 1970, the year of *H. maydis*, farmers in the US produced 14% less corn, but the cost to buyers was only 4% less (in constant dollars). A model of economic loss is likely to give special weight to the unusual and extreme epidemics discussed above.

Although there is ample use for models that will forecast when pesticide spraying is needed, using regression analysis with historical data or a laboratory simulator, there seems to be an even greater role for models that forecast unusual or extreme epidemics, created by new circumstances of husbandry or even climate. While refining, and even mechanizing, the forecasts of next week's disease encouraged by this week's weather, we must explore how we could have forecast the Irish blight of 1845, the American blight of 1970, and the Connecticut mold of 1979.

REFERENCES

Barnett, W. W. (1976) Status and prospects for use of modeling in integrated pest management—a private agricultural advisor's view, in, R. L. Tummala, *et al.* (eds) *Modeling for Pest Management* (East Lansing: Michigan State University Press) pp230–33.

Beaumont, A. (1947) The dependence on the weather of the dates of outbreak of potato late blight epidemics. *Br. Mycol. Soc. Trans.* 31:45–53.

Berger, R. D. (1977) Application of epidemiological principles to achieve plant disease control. *Ann. Rev. Phytopathol.*, 15:165–83.

Bourke, P. M. A. (1957) The use of synoptic weather maps in potato blight epidemiology. *Irish Meteorol. Serv. Tech.* Note 23.

Bourke, P. M. A. (1964) Emergence of potato blight, 1843–46. *Nature* 203:805–8.

Bourke, P. M. A. (1970) Use of weather information in the prediction of plant disease epiphytotics. *Ann. Rev. Phytopathol.*, 8:345–70.

Conner, J. T. (1976) Status and prospects for use of modeling in integrated pest management, in, R. L. Tummala, *et al.* (eds), *Modeling for Pest Management.* (East Lansing: Michigan State University Press) pp227–9.

Croxall, H. E., W. A. Davey, D. C. Gwynne, and W. Johnson (1961) Potato blight forecasting in Cumberland, 1953–57. *Plant Pathol.*, 10:127–32.

van Everdingen, E. (1935) Het verbrand tusschen de weer gesteldheidgen de aardappelziekte (tweede mededeling) *Tijds. Plziekte* 41:125–33.

Eversmeyer, M. G. and J. R. Burleigh (1970) A method of predicting epidemic development of wheat leaf rust. *Phytopathology* 60:805–11.

Felch, R. E. and G. L. Barger (1971) EPIMAY and southern corn leaf blight. *Weekly Weather Crop Bull.*, 58(43):13–17.

Frost, M. C. (1976) Potatoes, *Phytophthora infestans* and weather. *Ann. Appl. Biol.* 84:271–2.

Grainger, J. (1953) Potato blight forecasting and its mechanization. *Nature* 171:1012–14.

Hirst, J. M. and O. J. Stedham (1956) The effect of height of observation in forecasting potato blight by Beaumont's method. *Plant Pathol.* 5:135–40.

Hooker, A. L., D. R. Smith, S. M. Lim, and J. B. Beckett (1970) Reaction of corn seedlings with male-sterile cytoplasm to *Helminthosporium maydis. Plant Disease Rep.* 54:708–12.

Horsfall, J. G. (1969) Relevance: Are we smart outside? *Phytopathol. News*, 3(12): 5–9.

Hyre, R. A. (1950) Spore traps as an aid in forecasting several downy mildew type diseases. *Plant Disease Rep. Suppl.*, 190: 14–18.

Hyre, R. A. (1955) Three methods of forecasting late blight of potato and tomato in northeastern United States. *Am. Potato J.* 32:362–71.

Jones, A. L., S. L. Lillevik, P. D. Fisher, and T. C. Stebbins (1979) A microcomputer-based instrument to predict apple scab infection periods. *Plant Disease* 64:69–72.

Klinkowski, M. (1970) Catastrophic plant diseases. *Ann. Rev. Phytopathol.* 8:37–60.

Kranz, J. (1974) *Epidemics of plant diseases: Mathematical analysis and modeling.* (Berlin: Springer-Verlag).

Kranz, J., M. Mogk, and A. Stumpf (1973) EPIVEN-ein Simulator for Apfelschorf. *Z. Pflanzenkrankh.* 80:181–7.

Krause, R. A. and L. B. Massie (1975) Predictive systems: Modern approaches to disease control. *Ann Rev. Phytopathol.*, 13:31–47.

Lake, J. V. (1956) The effect of microclimate temperature upon the date of flowering of the apple. *J. Hort. Sci.* 31:244–257.

Large, E. C. (1940) *The Advance of the Fungi.* (London: Cape).

Large, E. C. (1955) Methods of plant-disease measurement and forecasting in Great Britain. *Ann. Appl. Biol.* 42:344–54.

Manabe, S. and R. T. Weatherald (1975) The effect of doubling the CO_2 concentration on the climate of a general circulation model. *J. Atmos. Sci.* 32:3–15.

Massie, L. B. and M. Szkolnik (1974) Prediction of ascospore maturity of *Venturia inaequalis* utilizing cumulative degree-days. *Proc. Am. Phytopathol. Soc.* 1:140.

Miller, P. R. and M. J. O'Brien (1952) Plant disease forecasting. *Bot. Rev.* 18:547–601.

Miller, P. R. and M. J. O'Brien (1957) Prediction of plant disease epidemics. *Ann. Rev. Microbiol.* 11:77–110.

Mills, W. D. (1944) Efficient use of sulfur dusts and sprays during rain to control apple scab. *Cornell Univ. Agric. Exp. Station Bull.*, 630.

Polyakov, I. Y. (1976) Development and role of forecast models for pest control programs, in R. L. Tummala, *et al.* (eds) *Modeling for Pest Management.* (East Lansing: Michigan State University Press) pp40–9.

Rowell, J. B. and R. W. Romig (1966) Detection of urediospores of wheat rusts in spring rains. *Phytopathology 56:807–11.*

Royle, D. J. (1973) Quantitative relationships between infection by the hop downy mildew pathogen, *Pseudoperonspora humuli*, and weather and inoculum factors. *Ann. Appl. Biol.* 73:19–30.

Shaner, G. E., R. M. Peart, J. E. Newman, W. L. Stirm, and O. L. Loewer (1972) EPIMAY an evaluation of a plant disease display model. *Purdue Univ. Agric. Exp. Station*, RB-890.

Shrum, R. D. (1978) Forecasting of epidemics, in J. G. Horsfall and E. C. Cowling (eds) *Plant Disease* Vol. 2. (New York: Academic Press) pp223–38.

Tummala, R. L., D. L. Haynes, and B. A. Croft (1976) *Modeling for Pest Management* (East Lansing: Michigan State University Press).

Ullrich, J. and H. Schrödter (1966) Das Problem der Vorhersage des Auftretens der Kartoffelkrautfäule (*Phytophthora infestans*) und die Möglichkeit seiner Lösung durch eine "Negativprognose". *Nachrichtenbl. Dtsch. Pflanzenschutzd.* (*Braunschweig*) 18:33–40.

Ullstrup, A. J. (1972) Impacts of the southern corn leaf blight epidemics of 1970–71. *Ann Rev. Phytopathol.* 10:37–50.

Waggoner, P. E. (1960) Forecasting epidemics, in J. G. Horsfall and A. E. Dimond (eds) *Plant Pathology* Vol. 3 (New York: Academic Press) pp291–313.

Waggoner, P. E. (1968) Weather and the rise and fall of fungi, in W. P. Lowry (ed) *Biometeorology* (Corvallis: Oregan State University Press) pp45–66.

Waggoner, P. E., J. G. Horsfall and R. J. Lukens (1972) EPIMAY, a simulator of southern corn leaf blight. *Connecticut Agric. Exp. Station Bull.*, 729.

Waggoner, P. E., W. A. Norvell, and D. J. Royle (1979) The law of the minimum and the relation between pathogen, weather and disease. *Phytopathology* 70:59–64.

Waggoner, P. E. and G. S. Taylor (1955) Tobacco blue mold epiphytotics in the field. *Plant Disease Rep.* 39:79–85.

Waggoner, P. E. and G. S. Taylor (1958) Dissemination by atmospheric turbulence: spores of *Peronospora tabacina. Phytopathology* 48:46–51.

Wallin, J. R. and J. A. Riley (1960) Weather map analysis—an aid in forecasting potato late blight. *Plant Disease Rep.* 44:227–34.

Wallin, J. R. and M. L. Schuster (1960) Forecasting potato late blight in western Nebraska. *Plant Disease Rept.*44:896–900.

Woodham-Smith, C. (1962) *The Great Hunger*. (New York: Harper and Row).

Young, H. C. Jr., J. M. Prescott and E. E. Saari (1978) Role of disease monitoring in preventing epidemics. *Ann. Rev. Phytopathol.* 16:263–85.

14 Monitoring in Crop Protection Modeling

J. Kranz, B. Hau, and H. J. Aust

14.1. INTRODUCTION

Immanuel Kant, the philosopher, once observed that there is nothing more practical than a good theory. Systems modeling in general has already made considerable progress towards this goal, and this applies equally to plant disease epidemiology. The development of good theories ensures more mature concepts, better understanding of systems, and improved methodology in modeling. And all this is imminently of practical value as we have now reached a stage in modeling where mathematics and computer facilities, as well as appropriate experience and abilities, are no longer limited.

On the other hand, many contemporary mathematical models suffer from a lack of realism and a tendency to be somewhat arbitrary—in the choice of topics, the underlying assumptions, and in the selection of relevant information. As Wang *et al.* (1977) state. "... most analytical models tend to be highly simplistic, having neither realism nor predictive value." This is clearly unacceptable in a subject such as epidemiology, which has to serve practical ends. Our need is for decision models in which both the simulation and optimization components are realistic (Loucks, 1977). Unfortunately, few models fit these requirements.

The most imminent risk resulting from noncompliance with these requirements is that modeling—with all its potentials—is discredited not only amongst prospective users, but also amongst fellow scientists. Credibility is essential for further funding of research projects, as well as for the acceptance of our results by extension workers and farmers. Several authors have recently voiced these misgivings about the present state of modeling for pest management, but they are not new. The continuing challenge is how to produce practical, useful models and so avoid the risk of nonacceptance. There appear to be three possible solutions to the problem:

- Modelers need to clearly state the objectives, boundaries, and the status of their exercises.
- Modelers need to borrow less from literature and rely more on consistent experimentation.
- Modelers need to give the monitoring aspect high priority in their experiments, bearing future implementation in mind.

In this chapter our main emphasis is on the third aspect—monitoring in relation to modeling. We first describe how we developed our model and then show how we are trying to ensure proper monitoring techniques for the practical employment of the model.

14.2. DEVELOPMENT OF A MODEL

Our example is a model of barley powdery mildew (*Erysiphe graminis* DC. f. sp. *hordei* Marchal) that we have developed and tentatively called "EPIGRAM" (Aust *et al.*, 1983). We define the objective for this modeling exercise as the development of a computer model to predict the further progress of the disease. The model has emerged from a process of systems analysis following Hall and Day's (1977) procedure of building the final model following a sequence of conceptual, diagrammatic, mathematical, and finally, computer models. These four steps are repeated several times, feeding in new results from experiments to replace assumptions or leaving out elements that prove to be unimportant.

Our conceptual model embraces both the essential relationships of the pathosystem and relevant information on the biology of the epidemic interactions from our own extensive data from growth chamber and field experiments, as well as from the literature. The infection chain (Gäumann, 1951), comprising the main components of the pathogen's life cycle, constituted the underlying diagrammatic model. Using this model as our starting point, together with the inputs and outputs then known, we began a process of interplay between experimental verification and progressive adaptation of the model, employing sensitivity tests along the way.

14.2.1. Components of the Model

In the course of this interplay we found we were able to omit a number of components of the infection chain from the model, which eventually consisted of the following submodels (Aust *et al.*, 1983):

(1) A function to determine the infection efficiency of the pathogen in relation to temperature, rainfall and adult plant resistance reflected in the leaf position.
(2) A function describing the relation between temperature and incubation period corrected for the leaf position.

(3) A function describing the colony development in relation to temperature, rainfall and leaf position.

Thus the final model consists only of a few core (Patten, 1971) or key variables of which some functions are complexes as they include variables from host, pathogen, and environment.

The resulting computer model is deterministic and dynamic, written in FORTAN 77. For its validation we compared its output and measurements of the disease progress curves in the field.

14.2.2. Inputs and Outputs

In its present state the model needs the following inputs: (1) daily spore catches per cm^2 leaf surface with trap plants; (2) daily maximum and minimum temperatures (°C) and rainfall (mm); and (3) weekly leaf area development in the leaf "storeys", 1 to 8 of the host (cm^2).

The output for the different leaf "storeys" and the whole plant are: (i) diseased leaf area (cm^2); (ii) disease severity (%); and (iii) number of colonies per leaf.

During the analysis we initially measured any variable that appeared to be biologically relevant, irrespective of whether it would be of use in the implementation of the eventual model. However, acceptance of the model, provided it is sufficiently reliable, will eventually depend on the simplicity of input acquisition and output verification. Both of these are monitoring problems.

An adaptation phase then followed the modeling phase, aimed at ensuring proper monitoring and hence the eventual practical value of the model as a possible component of a barley pest management system.

14.3. DEVELOPING APPROPRIATE MONITORING

It should be made clear we are not referring here to Apple and Smith's (1976) "biological monitoring phase", which consists essentially of disease or pest surveillance or supervised crop protection, using levels of infestation to determine the level of pesticide application. This is often unrelated to any model in the sense used here, although modelers could well learn from the endeavors and experiences of those engaged in surveillance. By contrast, we are dealing here with biological and environmental monitoring as input and output values for mathematical models.

This monitoring can be done either by private field consultants or pest management scouts employed by farmers on a regional basis, or by the use of mobile monitoring systems (Croft *et al.*, 1979). In most cases, however, the farmer himself will be forced to do the actual monitoring, due to time and cost constraints. But this is not a disadvantage—by carrying out his own monitoring the farmer achieves a better understanding of the pest management process and, more important, receives more appropriate outputs for his own situation. It is thus necessary that the farmers understand and accept the systems models, overcoming such reservations that are due to lack of understanding or familiarity with the approach. These are essential prerequisites.

The actual use of a system will depend heavily on whether the monitoring is practical for the user, requiring simple but substantially correct techniques. The parameters that are eventually monitored under practical field conditions may thus differ from those measured during the modeling phase. The model will not need to be changed, however, since for most users it will remain a "black box".

Monitoring is concerned with both sensing and sampling techniques, the former often determining the choice of the latter.

14.3.1. Present Status of Monitoring in Modeling

Some extensive pest management systems have been developed in the USA that use computers, models, and a well developed communication network (see Chapter 22). But, as Haynes *et al.* (1973) state, one of the principal factors limiting the usefulness of prediction models in such systems is the acquisition of real-time information about the pest–crop system. For the implementation of pest management such information from monitoring is indispensable, preferably as a direct input into the models. Deficient monitoring increases errors and may seriously affect efficiency and reliability of a model, and can render even impressive pest management networks ineffective.

Despite these obvious needs, surprisingly little effort has been made in research to establish adequate sensing and/or sampling techniques for the validation of models, as well as for their implementation (Bottrell and Adkisson, 1977).

14.3.2. Monitoring Input Variables

Prospective users will accept even the most useful models only if they are not too laborious or time consuming. Hence automatic and, where possible, remote sensing of physical (environmental) or biological parameters is likely to be preferred. Thus, for field-oriented objectives in modeling the ease as well as substantial reliability and accuracy by which a certain variable can be monitored should determine its eventual inclusion in a model.

When adapting model inputs for practical applications, adequate monitoring facilities may be available already or devices may have to be designed specifically to provide the inputs that a model needs. Whatever may apply, the proposed monitoring should be nonambiguous and readily accessible to other scientists or users.

14.3.3. Environmental Monitoring

Waggoner and Horsfall (1969) took the right direction when they based the first simulator in plant pathology entirely on weather parameters that were available from synoptic stations throughout the world. However, the required information may not always be so readily available or appropriately placed, and programmed weather monitoring stations may have to be built. In many cases, however, apparently inadequate weather information may be easily converted by interfaces or computer subroutines into the parameters actually needed for the model. The model underlying PHYTPROG (Schrödter and Ullrich, 1965), a nationwide negative prognosis for potato late blight (*Phytophthora infestans*), has proved the value of such standard weather information.

Modern developments in dataloggers and sensors may also allow a wider use of more sophisticated appliances such as thermocouples, thus replacing thermometers etc., in weather huts placed directly in the field. Weather information lends itself easily to use in microcomputers that may be placed within the plant canopy, like the apple scab warner of Richter *et al.* (1973), and subsequent developments.

14.3.4. Biological Monitoring

Host Growth and Development

Host growth and development data are easily and reliably obtainable, but they are less amenable to automation. In some cases, however, they may be linked with a measure of time, for example with a "zero-time" as in the PHYTPROG model, or with degree days. Nevertheless, the relevance of host stages—in terms of the ontogenetic changes in susceptibility, and of trapping efficiency, as well as their influence on microclimate—have to be established experimentally. For our model, we measured weekly the leaf surface of randomly selected barley plants (or stems) over the entire vegetation period and obtained curves for the development of green leaf mass and for the leaf area already removed. Such information is valid for a given cultivar in most circumstances, provided the function is scaled over growth stages. So far growth stages can be monitored only visually, but there is some hope that crop models may someday replace these assessments.

Disease Monitoring

Inputs of disease or pest status can be obtained from the intensity of attack or number of spores or insects caught with trapping devices. Disease intensity of barley powdery mildew, for example, is best reflected by the disease severity, i.e., the proportion of the diseased host surface. During the modeling phase we computed this proportion by means of a programmed pocket calculator from numbers and diameters of the colonies, as well as from the length and width of leaves. From a practical point of view such a procedure is too time-consuming and the high degree of accuracy produced is not needed. It is sufficent to assess the proportion of the diseased area directly. In our laboratory, Amanat (1977) showed by means of leaf analogues that, with a bit of training, visual assessment yields substantially the same results as the above-mentioned exact measurement. Koch (1978), another member of our team, proved that certain values of the decimal system are preferred to others by test persons, which permits the design of more helpful standard diagrams to assist growers in making their assessments. To count the number of diseased leaves in a sample (disease frequency) and relate it to disease severity by a certain factor (James and Shih, 1973; Koch, 1978) is adequate early in an epidemic.

14.3.5. Sampling

Sampling is a widely neglected aspect of pest and disease modeling. Authors such as Haynes *et al.* (1973), for instance, only refer in passing to the need for appropriate sampling techniques in connection with model implementation. Croft (1975), Rabbinge and Carter (Chapter 15), Zadoks *et al.* (Chapter 16), and Kuno (Chapter 21) are to our knowledge the only workers who have developed a biological monitoring plan for an integrated control program. Croft's plan comprised nested sampling for orchard, tree and leaf units, and minimal sample sizes (for leaves and trees), evaluated sequentially on 25 leaves (five trees and five leaves per tree). He also proposed the need for accurate-life distribution estimates as starting conditions for management model simulations.

For practical purposes, only those sampling schemes that maximize the accuracy within the cost and time constraints are useful (Analytis and Kranz, 1972; Croft *et al.*, 1976). This often rules out random sampling because of the expense involved. Koch (1978) compared several sampling schemes for barley powdery mildew and proposed sampling along a transect, taking into consideration the horizontal distribution of the disease in the stand, i.e., stratified sampling. He found the main culm (with all its leaves) to be representative as a sampling unit. For the sampling size, Koch (1978) found 20–30 plants per plot or field to be sufficient; these were equally distributed in the usual two or three strata in a field reflecting external influences (e.g., neighboring winter barley, shade-throwing woodland, etc.).

Spore trapping is perhaps a more appealing method of monitoring since it can be more readily automated, provided a realistic relation with actual disease progress has been established. If so, the widespread misgivings concerning the microscopic size and huge numbers of fungal spores as important constraints on monitoring are no longer justified.

During the modeling phase of our work we established the influence of the spore catch input on the model output by measuring inoculum landing in three different leaf layers. For practical monitoring this method would be far too clumsy, but we did find a high correlation coefficient between the different levels in the crop canopy, so that for practical purposes the catch of one leaf layer is sufficient to calculate the others. Studies are now under way to replace trap plants by a mechanical device such as a spore trap. Although trap plants are most appropriate to determine the amount of spores landing on field plants, slide traps, or similar devices are more easy to handle. Since there is an obvious relation between trap plants and slide traps, we are also experimenting with still rod traps (Jenkyn, 1974) for spore monitoring. (Volumetric spore traps do not appear to be of use for this purpose.)

Finally, there is a critical issue in biological monitoring that relates both to the variability of spore catches and disease assessment and to the errors involved in monitoring them. Instead of a single mean disease severity value, a range (confidence intervals, etc.) of likely disease severity should be monitored. Such a range is mainly determined by the calculated sampling error (Yates, 1971), which in turn depends on sample size and sampling technique, and this sampling error has to be included in the instructions to the user in the sampling plan. All of this is particularly relevant in the comparison of model output with what appears to be the real situation in the field.

14.3.6. Monitoring the Model Output

Model inputs lead to a prediction that has an error attached and it is therefore desirable to check to what extent the real world has reacted in the way the model predicts. The monitoring or testing of the model output in comparison to reality is called the management or feedback loop (Croft *et al.*, 1976).

Model input usually differs in form from model output and, in any case, a model output intended for practical use should be straightforward. Tactical models should provide information in terms of spraying advice, predicted increases in disease intensity, or generate other such nonambiguous information. Curves and figures should be in real values, not in misleading transformations. A reference base to which a model's output actually refers should be defined and established by appropriate research, and it should culminate (as mentioned already) in detailed instructions as to how this should be sampled and assessed.

The field will be the reference base for the barley powdery mildew model. With the earlier model on apple scab epidemics, EPIVEN (Kranz *et al.*, 1973), we modeled disease progress on 100 leaves that constituted our reference base. If a model is based on climatic parameters, its dimension equals the region for which the climatic situation monitored is valid. Again, sampling questions are involved here and an appropriate placement of measuring devices has to be ensured. Only with appropriate and convenient sampling techniques for the output can a model succeed in practical pest management.

14.4. CONCLUSION

In this chapter we have stressed the point that models are simply tools for the implementation of pest management. If we wish to effectively forge such tools we may have to reverse our traditional approach. *When planning research relevant to pest management modeling we should perhaps ask ourselves first what is there in the real situation we are trying to manage that can be measured. Biological elements of the resulting model may in this way be influenced or even determined by the monitoring practicalities. Hence, in turn, so will be the mathematical techniques employed in the modeling.* If, however, a model is available or in progress already, as in our case, a two-phase research program emerges: a modeling phase, and an adaptation phase, during which proper monitoring is established for the implementation of the model in practical pest management.

REFERENCES

Amanat, P. (1977) *Modellversuche zur Ermittlung individueller und objektabhängiger Schätzfehler bei Pflanzenkrankheiten* PhD Thesis, Universität Giessen.

Analytis, S. and J. Kranz (1972) Bestimmung des optimalen Stichprobenumfanges für phytopathologische Untersuchungen. *Phytopathol. Z.* 74:344–57.

Apple, J. L. and R. F. Smith (1976) Progress, problems, and prospects for integrated pest management, in J. L. Apple and R. F. Smith (eds) *Integrated Pest Management* (New York: Plenum) pp179–96.

Aust, H. -J., B. Hau, and J. Kranz (1983) EPIGRAM—a simulator for barley powdery mildew. *Z. Pflanzenkrankh. Pflanzanschutz* 90: 244–50.

Bottrell, D. G. and P. L. Adkisson (1977) Cotton insect pest management. *Ann. Rev. Entomol.* 22:451–81.

Croft, B. A. (1975) Tree fruit pest management, in R. L. Metcalf and W. H. Luckmann (eds) *Introduction to Insect Pest Management* (New York: Wiley) pp471–507.

Croft, B. A., R. L. Tummala, H. Riedl, and S. M. Welch (1976) Modelling and management of two prototype apple pest sub-systems, in R. L. Tummala, D. L. Haynes, and B. A. Croft (eds) *Modeling for Pest Management* (East Lansing: Michigan State University Press) pp97–119.

Croft, B. A., S. M. Welch, D. J. Miller, and M. L. Marino (1979) Developments in computer-based IPM extension delivery and biological monitoring system design, in D. J. Boethel and R. D. Eikenbury (eds) *Pest Management Programs for Deciduous Tree Fruits and Nuts* (New York: Plenum Press) pp 223–50.

Gäumann, E. (1951) *Pflanzliche Infektionslehre* 2nd edn (Basel: Birkhäuser).

Hall, C. A. S. and J. W. Day, Jr. (1977) Systems and models: Terms and basic principles, in C. A. S. Hall and J. W. Day, Jr. (eds) *Ecosystem Modeling in Theory and Practice* (New York: Wiley) pp5–36.

Haynes, D. L., R. K. Brandenburg, and D. P. Fisher (1973) Environmental monitoring network for pest management systems. *Environ. Entomol.* 2:889–99.

James, W. C., and C. S. Shih (1973) Relationship between incidence and severity of powdery mildew and leaf rust on winter wheat. *Phytopathology* 63:183–7.

Jenkyn, J. F. (1974) A comparison of seasonal changes in deposition of spores of *Erysiphe graminis* on different trapping surfaces. *Ann. Appl. Biol.* 76:257–67.

Koch, H. (1978) *Die Verteilung von Erysiphe graminis DC. f. sp. hordei Marchal in Gerstenbeständen und ihr Einfluß auf die Stichprobenauswahl für Befallserhebungen* PhD Thesis, Universität Giessen.

Kranz, J., M. Mogk, and A. Stumpf (1973) EPIVEN—ein Simulator für Apfelschorf. *Z. Pflanzenkrankh. Pflanzenschutz* 80:181–7.

Loucks, O. L. (1977) Emergence of research on agro-ecosystems. *Ann. Rev. Ecol. Syst.* 8:173–92.

Patten, B. C. (1971) Introduction to modeling, in B. C. Patten (ed) *Systems Analysis and Simulation in Ecology* Vol. I (New York: Academic Press) pp1–2.

Richter, J., H. Steiner and W. Schipke (1973) Ein elektronisches Schorfwarngerät im Obstbau-Arbeitsweise und Aussichten für die Zukunft. *Mitt. Biol. Bundesant. Land Forstwirtnh. Berlin – Dahlem* 151:281–2.

Schrödter, H. and F. Ullrich (1965) Untersuchungen zur Biometeorologie und Epidemiologie von *Phytophthora infestans* (Mont.) de By. auf mathematisch-statistischer Grundlage. *Phytopathol. Z.* 54:87–103.

Waggoner, P. E. and J. G. Horsfall (1969) EPIDEM, a simulator of plant disease written for a computer. *Connecticut Agric. Exp. Station Bull.* 698.

Wang, Y., A. P. Gutierrez, G. Oster and R. Daxl (1977) A population model for plant growth and development: coupling cotton–herbivore interaction. *Can. Entomol.* 109:1359–74.

Yates, F. (1971) *Sampling methods for censuses and surveys* (London: Griffin).

15 Monitoring and Forecasting of Cereal Aphids in the Netherlands: A Subsystem of EPIPRE

R. Rabbinge and N. Carter

15.1. INTRODUCTION

Cereal aphids have become increasingly important pests in Western Europe; during the last few years losses have often exceeded 1000 kg/ha of wheat. Much of this damage is caused by secondary effects, since the actual loss of sap caused by aphid feeding is only of minor importance in plants growing under optimal conditions. Model calculations, confirmed by experimental measurements, show that the major reason for damage is honeydew produced by the aphids, which covers the epidermis, limits CO_2 diffusion, and probably promotes leaf senescence (Rabbinge and Vereijken, 1979; Carter and Dewar, 1981; Vereijken, 1979). This type of damage also explains why there is such a low correlation between aphid load, whether expressed as maximum number or as aphid days (integrated aphid numbers throughout the infestation period) and yield loss ($r = 0.69$, $n = 21$; Figure 15.1). This low correlation makes the definition of fixed damage thresholds impossible, and the prediction of expected yield losses possible only when more quantitative knowledge of the dynamics of aphid–host plant interrelations are included. Detailed studies of these damage effects have shown the complex nature of the yield losses, and have also demonstrated that the role of ants as mutualists of aphids is irrelevant, but that saprophytic fungi on the honeydew may have considerable effects.

Three species of cereal aphid are usually involved: the English grain aphid (*Sitobion avenae*), the rose grain aphid (*Metopolophium dirhodum*), and the bird cherry-oat aphid (*Rhosalosiphum padi*). *S. avenae* is generally considered the most important; it prefers to feed on wheat ears, where it interferes with the translocation of assimilates to the kernels. In 1978 and 1979, however, cereal losses exceeded 1000 kg/ha, caused mainly by large numbers of *M. dirhodum*, while *S. avenae* was uncommon.

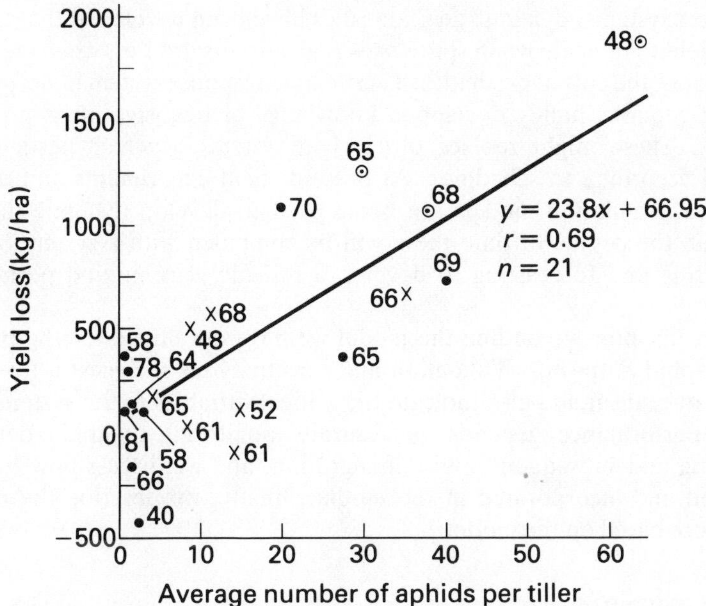

FIGURE 15.1 Yield loss as a function of the peak number of aphids per year. • more than 80% of *M. dirhodum*; × more than 80% of *S. avenae*; and ◉ both species present; numbers represent the yield in the control (× 10^2 kg ha^{-1})

Yield losses of this size caused by aphids frequently occur in Europe, probably due to the considerable changes in wheat cultivation during the last decade (Potts, 1977). High sowing densities, early sowing times, large applications of split nitrogen, and the use of growth regulators and control measures for leaf and ear diseases have resulted in wheat yield levels approaching about 10 000 kg/ha. These high production levels depend on a healthy wheat crop with an extended maturation period of three or four weeks.

Each day lost in kernel filling and ripening will decrease the yield by 200 kg/ha. The management strategy of many Dutch farmers is to keep crops free of diseases during spring and early summer, so that after flowering a sound crop will start kernel filling. Accordingly, pesticide use in wheat crops has had to be greatly increased to protect the crop up to flowering; spray applications using fixed schedule schemes have, in some places, resulted in five or more treatments of fungicides, insecticides, and herbicides being applied.

The EPIPRE system was initiated to prevent further increases in pesticide usage on wheat, thus reducing the chance of problems associated with insecticide resistance encountered in other crops, such as cotton and apples. The EPIPRE system, which is discussed further in Chapter 21, aims at flexible crop protection, based on detailed knowledge of crop growth and the prevailing pests and diseases. By integration of this knowledge in large computer systems, dynamic decision rules have been developed that are used in the field to indicate when spraying is really necessary, i.e., when yield gains will balance the cost of pesticides. This flexible response system limits pesticide use but requires highly developed knowledge of intensive plant protection systems. These might replace other rigid systems in which pesticides are applied according to schedules. At present, field experiments and dynamic crop–pest simulation models are being used to develop dynamic threshold levels; in the course of time these will be combined with a system of aphid monitoring and forecasting to develop a reliable warning and pest control system.

In this chapter we outline the model we have developed for the principal wheat aphid *S. avenae*. Validation and sensitivity tests suggest a reasonable degree of realism and also indicate the critical variables in the system. Good model performance depends on accurate monitoring of aphid density at flowering and subsequent aphid immigration, and we discuss how these are obtained and incorporated in the model. Finally, we describe the decision procedure based on the model.

15.2. APHID BIOLOGY

In 1978 and 1979 *M. dirhodum* was the most common aphid species on cereals. It overwinters in the egg stage on roses. The fundatrigenae (founder adults) are either apterous (wingless) or alate (winged), but the third generation is completely alate (Hille Ris Lambers, 1947), and can also overwinter viviparously on grasses and cereals (Dean, 1978; George, 1974). Once it has migrated to cereals it feeds on the leaves, moving up the plant as the crop develops. As the crop flowers this species stays on the leaves, mainly on the flag leaves, especially in the short straw varieties. Its absence on the ear is the major reason why many researchers consider it less important than *S. avenae* in causing yield losses.

R. padi is the most important vector of barley yellow dwarf virus, a disease that up to now has not been important in Western Europe and is therefore not considered in this study. *R. padi* is one of the commonest aphids caught in suction traps in Europe, and is a potentially serious pest, but it is only important in Scandinavia (Markkula, 1978). It can overwinter either on bird cherry in the egg stage, or viviparously on cereals and grasses. Fundatrices hatch from eggs in April, around the time of bud-burst (Dixon, 1971). The number of aphids increases rapidly and many alates are produced that colonize grasses. *R. padi*, like *M. dirhodum*, settles on the leaves but it is also encountered on the ears and stems.

S. avenae is monoecious on Gramineae, on which it may overwinter as viviparae or as eggs, but very little information is available on the relative importance of the methods (Dean, 1974). It is usually very difficult to find either *S. avenae* aphids or eggs on grasses in winter, so that it is impossible to monitor spring populations for use in an early warning scheme. Alates usually colonize winter wheat in preference to most other cereals, from the end of May until the end of June (Carter, 1978), and are caught in suction traps before they are encountered in the field (George, 1974). At the start of immigration the wheat has not headed and the alates settle on the leaves. As the ears emerge they are colonized by the alates and their nymphs. Most of these nymphs develop into apterous adults whose reproductive rate is higher than that of the alate adults and hence population growth is maximized (Wratten, 1977). The reproductive rate of *S. avenae* is higher on the young ears (up to the end of the watery ripe stage) of cereals than on the leaves or the older ears (Vereijken, 1979; Watt, 1979), and its survival rate is also highest on these young ears (Watt, 1979). As the aphid density rises and the crop ripens an increasing proportion of the nymphs born to these apterae develop into alate adults. Also, as the aphid density increases the reproductive rate decreases, and this contributes to the population decline. From the milky ripe stage onwards, alates leave the crop, resulting in a rapid decline in field populations and also by the large catches in suction traps at this time. This emigration is probably induced by a combination of the high aphid densities, by the ripening of the crop, and possibly by the large numbers of natural enemies in the crops, which destroy any remaining aphids.

15.2.1. Natural Enemies

These can broadly be divided into four groups: aphid-specific predators, polyphagous predators, parasitoids, and fungal pathogens. The aphid-specific predators (Coccinellidae, Coleoptera; Syrphidae, Diptera; Chrysopidae, Neuroptera) are usually rare before flowering but their numbers increase after this stage. Only occasionally are their numbers very high. The contribution of these predators to population control of cereal aphids is limited although the few detailed studies of natural enemies using cereal aphids as the prey indicate that their predation capacity is high. The fourth and most voracious instar of *Coccinella 7-punctata* consumes 40 third instar *S. avenae* nymphs per day at 20 °C (McLean, 1980). The predation capacity of *Syrphus corollae* is even higher and amounts to 200 L1–L3 larvae of *S. avenae* or 100 fourth instar larvae (Bombosch, 1962). The searching capacity of this predator is low, however, and it immigrates late. The predation activity of *Chrysopa carnea* is of the same order of magnitude but it is even less common than the Syrphidae and Coccinellidae, the latter group being common at some places at the very end of kernel filling.

Potts and Vickerman (1974) stress the importance of polyphagous predators in controlling aphid populations; if they are already present in fields when aphids arrive they could reduce aphid numbers considerably and thus prevent outbreaks. Potts (1977) and Vickerman and Sunderland (1977) have noted that these predators are declining in cereal fields in the UK, perhaps due to the increasing use of pesticides. There is very limited information available on the searching behavior or consumption rates of these predators. To evaluate their quantitative effect in aphid control more detailed studies on their predation characteristics should be carried out.

Parasitoids belonging to two families of Hymenoptera: Aphelinidae and Aphidiidae, are always present in cereal fields. The latter group is the more important one in Europe. The numbers and dominance of each species is different each year. For example, in the Netherlands the parasitoids found in 1976 were, in order of abundance, *Aphidius uzbekistanicus* (type), *Praon volucre*, and *A. picipes*, while in 1977 the order was *A. ervi*, *A. picipes*, and *A. uzbekistanicus* (type). Their effect is usually limited due to their low numbers early in the season and the high proportion of hyper-parasitism.

Late in the season fungal pathogens of the aphids may occur, belonging to the genus *Entomophthora*. Especially when weather conditions are suitable the fungal disease may become epizootic and contribute to the collapse of the aphid population.

15.3. MODELING

15.3.1. Description

Two models to simulate the population development of *S. avenae* have been developed independently in the UK (in FORTRAN IV; Carter, 1978) and in the Netherlands (in CSMP III; Rabbinge *et al*. 1979). A simple diagram for the development of *S. avenae* is given in Figure 15.2. Juvenile aphids with three larval stages develop into alatiform or apteriform L4. The maturing apterous females may produce a new generation of aphids, whereas the alates emigrate and disappear from the system. Wing formation is a result of the combined effects of crowding and the stage of the plant, and is introduced as such in the population model. Further studies of these effects have been made in the laboratory, although these are still of a preliminary character and the results are not very consistent. The influence of temperature on development reproduction rates is shown in the relational diagrams by the arrows from the driving variable, temperature, to the rates. These relations are derived from Dean's (1974) data and our own experiments using young wheat plants as a food source instead of barley leaf disks on wet cotton wool. Dispersion during development and its dependence on temperature are introduced in the Dutch model. For example, larvae of the same age are reproductive at different times. A special "boxcar" routine is developed with which growth through different stages is mimicked. This subroutine mimics the dispersion in time during development and ageing and adapts it to external conditions. Basically, this is

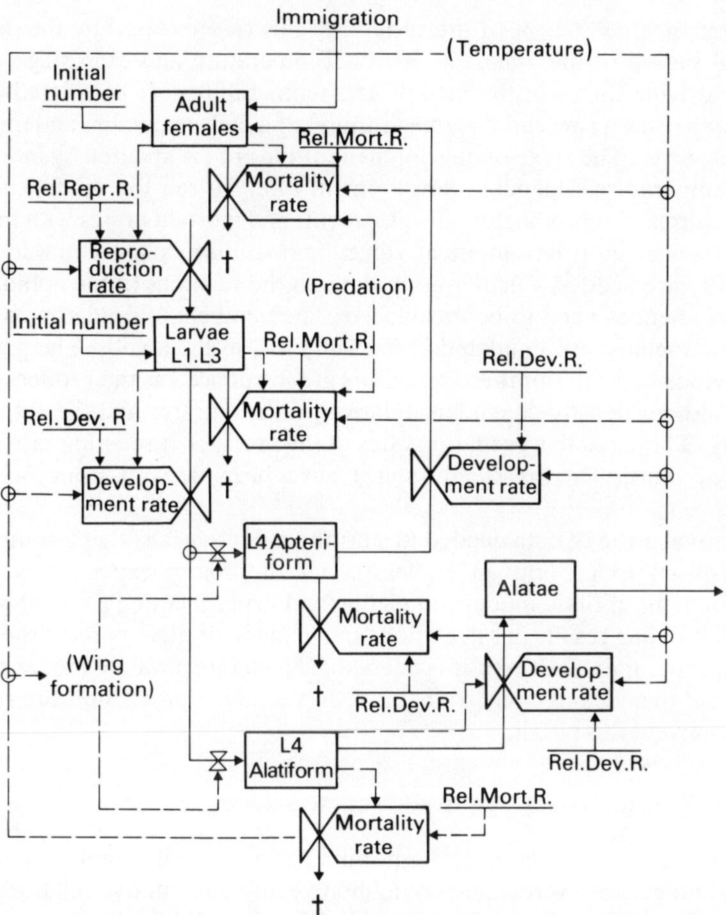

FIGURE 15.2 Relational diagram of population growth and development of *S. avenae*.

done by distinguishing artificial age classes within the morphological age classes, which are reached at different rates. Dispersion, however, seems to be of minor importance, as indicated by sensitivity analysis. The relatively small size of the standard deviation explains this.

Both models simulate aphid population growth up to the time of its collapse. The period before the aphids migrate into the winter wheat fields and the winter period are not taken into account. The models are started with the first numbers of aphids in the crop or suction trap catches, respectively. Initial inputs are maximum and minimum temperatures, latitude of the site, initial stage of crop development, initial numbers of aphids and natural enemies, and migration data. From then on, population growth is simulated numerically using time steps of one hour or 15 minutes; this latter is determined by the smallest time coefficient of the system, in this case dictated by the developmental period of the aphids at 30 °C. Temperature and crop stage are the major driving forces of the model. The temperatures are calculated at each time step using a sine curve passing through the daily maximum and minimum temperatures. The stage of development of the crop is updated by integration of a temperature-dependent development rate. To run the models, data on the relations of reproduction, development, and mortality rates with morphological stage, crop development stage, temperature, plant conditions, and humidity are needed. Furthermore, data on the relations of the aphid and its natural enemies need to be introduced. The numbers of predators, syrphids and coccinellids, are simulated in the same way as the aphids. The predation and oviposition rates of these predators are introduced as the product of prey and predator density-dependent relative predation rates and the actual prey density. These relative predation rates of the predator (predation rate divided by prey density in the steady state) have been derived from Bombosch (1962).

The avalanche of data needed to start the models means that a compromise is necessary to prevent an endless series of process experiments. In the development of these models, modeling and experimenting go hand in hand. Model building results in calculations showing the decisive parts of the system where experimental emphasis is needed. This iterative way of model building may lead to a vicious circle so that careful validation procedures are required to circumvent this pitfall.

15.3.2. Validation of the Model

During model construction and experimentation several hypotheses, assumptions, and guesses were made on qualitative and quantitative relations within the system and on effects of forcing variables. To validate the structure of the model and the incorporated relations a comparison has to be made with the results of independent experiments. Moreover, a final analysis of the relative impact of different relations used has to be determined (sensitivity analysis).

Models have to be validated at different levels of integration. Data at the field level are available while data to verify the implicit hypotheses and model outcomes at lower levels of integration are still being collected. Population density curves for aphids of different instars and morphs are calculated by the models. Some results from the Dutch model are given in Figure 15.3, compared with actual field counts for different seasons at three locations in the Netherlands. The experimental data are given with 95% confidence intervals. For each year and location the simulated and observed total population density curves are in good agreement during the rapid growth phase. For nearly all simulated situations the age composition of the aphid population is also reasonably well simulated. The last part of the population

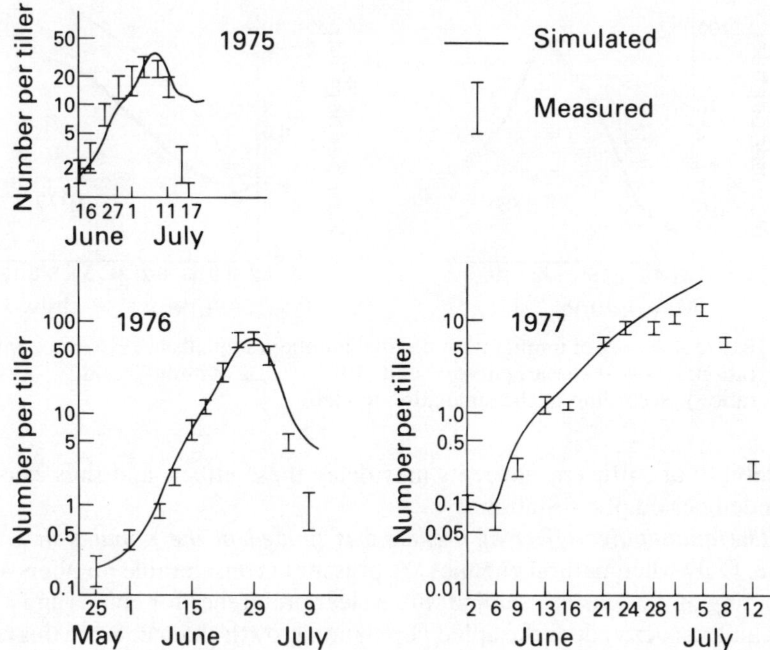

FIGURE 15.3 Simulated and actual numbers of *S. avenae* in 1975, 1976, and 1977. The actual numbers are given in terms of 95% confidence intervals.

curve is only reasonably simulated when there is a rapid decrease in reproduction and an increase in mortality rates (Watt, 1979; Vereijken, 1979).

The preliminary results of the simulated predator density in the Dutch model correspond reasonably well with field observations. This is encouraging, but it should be emphasized that the introduced relations on functional and numerical response of the predators are still based on estimates and incomplete experimental data.

The good overall agreement of model output and experimental results justifies a sensitivity analysis of the model. This serves both to improve our insight into the system and guide management and to pinpoint further laboratory or field studies that need to be done. The results of this sensitivity analysis show the following:

(1) *The role of immigration in the population upsurge is different from year to year*. For example, the omission of immigration after flowering in 1976 had only a slight effect on the population density curve, whereas similar changes in 1977 caused a major effect on the population growth of the aphids (Figure 15.4).
(2) *Emigration in combination with decreased reproduction and development and increased mortality are the major reasons for the flattening of the population*. An extended maturing period of the crop due to relatively cold weather and

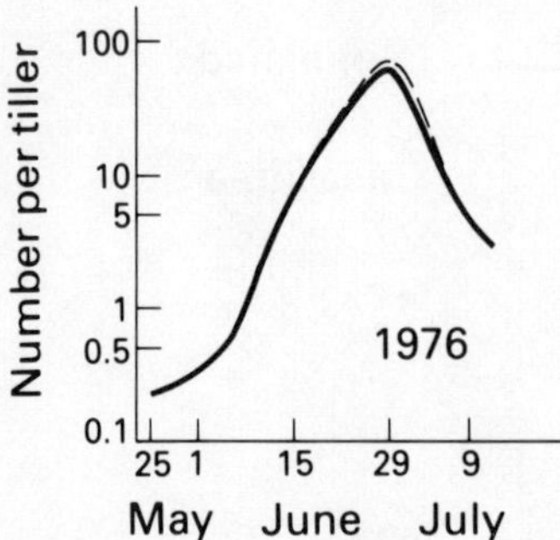

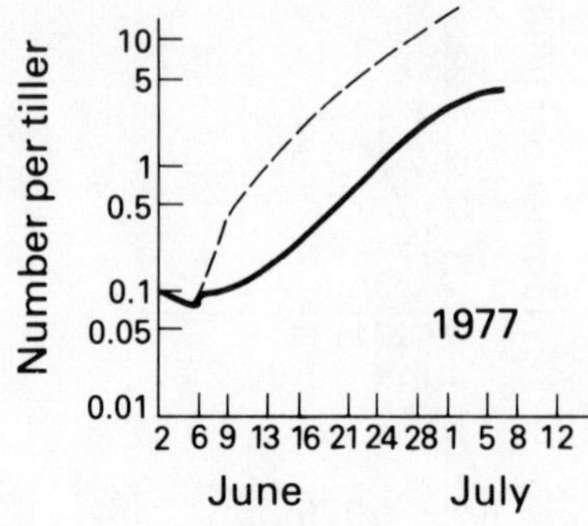

FIGURE 15.4 Effect of immigration on total number of aphids in 1976 (—— without immigration; ----- basic curve); and 1977 (—— without, and ---- with immigration), according to the simulation model.

availability of sufficient nutrients may delay these effects and thus cause an extended period of infestation.

(3) *The quantitative effect of predators is limited in the population growth phase*. Only when natural enemies are present in considerable numbers at the beginning of the season, coupled with a clear preference for aphids and a high searching capacity, does the aphid population growth change. Since this rarely occurs the role of natural enemies is only important during the flattening and collapse of the aphid population. The contribution of predators becomes clearer in Table 15.1, where the simulated numbers of aphids lost per tiller due to predation in 1976 are given.

TABLE 15.1 Disappearance of aphids (numbers per tiller) in 1976 caused by predation, abiotic mortality, and emigration based on model calculations.

Date	Time	Growth stage of plant[a]	Mortality		Emigrants (winged females)	Actual number of aphids (total)
			Predation and parasitism	Abiotic		
24 May	5	10	0	0.001	0	0.3
29 May	10	10.1	0	0.004	0	0.4
	15	10.2	0	0.03	0	1.5
	20	10.3	0	0.05	0	5.6
	25	10.5.1	0.2	0.3	0.02	13.6
	30	10.5.2	4.0	1.0	0.27	40.6
	35	10.5.3	25.0	5.9	5.4	68.4
	40	11.1	99.8	12.2	20.0	22.3
9 July	45	11.1	119.6	13.3	24.1	5.0
	50	11.2	121.3	13.4	26.0	4.1

[a]Feekes scale.

(4) *Evaluation of the effect of initial conditions and the relations of the different rates with environmental conditions show that initial population densities around flowering may vary by at least 20% without having a considerable effect on population growth during the rest of the season.*

15.4. MONITORING AND SAMPLING

Model calculations have shown that knowledge of the population densities at the time of flowering suffices to start the simulation model and to predict the population upsurge when additional information on immigration is supplied. To assess the initial aphid population densities and the size of immigration Carter and Dewar (1981) describe how suction trap catches may be used to determine the amount and timing of immigration. These findings are now being compared with the number of aphids collected in the field using insect suction samplers. After immigration has started, farmers are advised to inspect their fields for aphids; these assessments may then be used to update the decision models of EPIPRE and enable prediction on the course of the aphid population in time.

Advice as to whether spraying is needed is based on the expected population peak and the still unreliable corresponding damage assessments (see Figure 15.1). Estimates of population densities should be found with simple but reliable methods that are not labor intensive. To derive such methods, the aphid distribution in the field was considered. In 200 out of 225 cases the aphid distribution fits a negative binomial distribution, with k-values ranging from 0.5 to 2. When average numbers are lower than 0.3 per tiller, determination of the distribution in the field requires more than 1000 tillers to be searched since the colonies are then scattered. Very rarely does a Poisson distribution give a better fit (20 out of 225 cases). Tests were made of the relation of the probit value of the infestation level and the logarithm of the average number of aphids per tiller of *S. avenae*, *M. dirhodum*, and *R. padi*, and combinations of these species. In all cases a linear relationship exists; see Figure 15.5 (correlation coefficients in all cases > 0.92; number of cases $>$ 225). These linear relationships enable a simple, less labor-intensive sampling method to be used. The infestation level is determined, giving the average number of aphids per tiller that is used to start the decision model.

15.5. THE DECISION PROCEDURE

The procedure used in EPIPRE is as follows. At flowering, farmers are asked to determine the aphid infestation level by inspecting 100 tillers taken at random over a diagonal of a field. When infestation levels are lower than 70%, farmers may delay any action for two to three weeks. At infestation levels higher than 70%, the economic damage level will be exceeded (350 kg of wheat/ha), and so farmers are advised to spray. The timing of the second observation by the

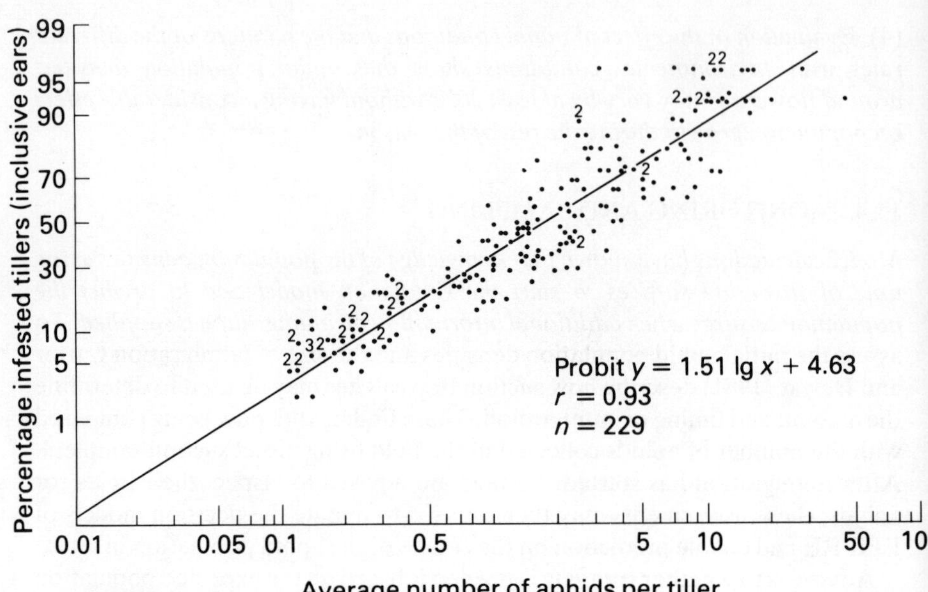

FIGURE 15.5 The percentage of infected tillers as a function of the average number of *M. dirhodum*, *R. padi*, and *S. avenae* per tiller.

farmer depends on computer calculations with the simplified simulation models. This period may vary from 10 to 20 days after flowering, at which time farmers are asked to determine the proportion of tillers with over ten aphids. These proportions, again after transformation, are linearly related to the average density per tiller. They provide supplementary information on the number of colonies, and the potential for emigration since population density is one factor that induces wing formation. All field data are sent to the forecasting research team on preprinted cards, and these are then used to make the decision on whether or not to spray.

The weakest point in the scheme is the determination of the damage threshold, since the timing of nitrogen top dressing and many other environmental factors may affect the actual damage. At present, therefore, threshold adjustment is more or less guesswork and additional research of the type discussed by Carter *et al.* (1982) is needed to solve these problems.

REFERENCES

Bombosch, S. (1962) Über den Einfluss der Nahrungsmenge auf die Entwicklung von *Syrphus corollae* Fabr. (dipt. Syrphidae) *Z. angew. Entomol.* 50:40–5.

Carter, N. (1978) *A Simulation Study of English Grain Aphid Populations*, PhD Thesis, University of East Anglia.

Carter, N. and Dewar, A. (1981) The development of forecasting systems of cereal aphid outbreaks in Europe, *Proc. 9th Int. Congr. Plant Protection, Washington, 1979.*

Carter, N., A. F. G. Dixon and R. Rabbinge, (1982) *Cereal Aphid Populations, Biology, Simulation and Prediction*, Simulation Monographs, (Wageningen: Pudoc).

Dean, G. J. (1974) The overwintering and abundance of cereal aphids. *Ann. Appl. Biol.* 76:1–7.

Dean, G. J. (1978) Observations on the morphs of *Macrosiphum avenae* and *Metopolophium dirhodum* on cereals during the summer and autumn. *Ann. Appl. Biol.* 89:1–7.

Dixon, A. F. G. (1971) The life-cycle and host preference of the bird cherry-oat aphid *Rhopalosiphum padi* L. and their bearing on the theories of host alternation in aphids. *Ann. Appl. Biol.* 68:135–47.

George, K. S. (1974) Damage assessment aspects of cereal aphid attack in autumn- and spring-sown cereals, Proc. Assoc. Appl. Biologists. *Ann. Appl. Biol.* 77:67–74.

Hille Ris Lambers, D. (1947) Contribution to a monograph of the Aphididae of Europe. *Temminckia* 7:179–319.

McLean, I. F. G. (1980) *Ecology of the Natural Enemies of Cereal Aphids*, PhD Thesis, University of East Anglia.

Markkula, M. (1978) Pests of cultivated plants in Finland in 1977. *Ann. Agric. Fenn.* 17:32–5.

Potts, G. R. (1977) Some effects of increasing the monoculture of cereals, in J. M. Cherrett and G. R. Sagar (eds) *Origins of Pest, Parasite, Disease and Weed Problems*, Proc. 18th Symp. Br. Ecol. Soc. (Oxford: Blackwell) pp183–202.

Potts, G. R. and G. P. Vickerman (1974) Studies on the cereal ecosystem. *Adv. Ecol. Res.* 8:107–97.

Rabbinge, R., G. W. Ankersmit, and G. A. Pak (1979) Epidemiology and simulations of population development V of *Sitobion avenae* in winter wheat. *Neth. J. Plant Pathol.* 85:197–220.

Rabbinge, R. and P. H. Vereijken (1979) The effect of diseases upon the host. Spalte 128, Proc. 3rd Int. Congr. Plant Pathology, Munich, 1978. *Z. Pflanzenkrankh.* 87(7): 409–22.

Vereijken, P. H. (1979) Feeding and multiplication of three cereal aphid species and their effect on yield of winter wheat. *Agr. Res. Rep.* 888, Pudoc, Wageningen.

Vickerman, G. P. and K. D. Sunderland (1977) Some effects of dimethoate on arthropods in winter wheat. *J. Appl. Ecol.* 14:767–77.

Watt, A. D. (1979) The effect of cereal growth stages on the reproductive activity of *Sitobion avenae* and *Metopolophium dirhodum. Ann. Appl. Biol.* 91:147–57.

Wratten, S. D. (1977) Reproductive strategy of winged and wingless morphs of the aphids *Sitobion avenae* and *Metopolophium dirhodum. Ann. Appl. Biol.* 85:319–31.

16 The Design of Sampling for Pest Population Forecasting and Control

Eizi Kuno

16.1. INTRODUCTION

The regular monitoring of pest population densities is an essential routine in any system of insect pest management. The rational planning of monitoring, however, is not simple, and has so far received little attention, although currently there are various population estimation techniques that are applicable to insects (e.g., see Southwood, 1978). The difficulty arises because the real objective of monitoring is not to assess the current population itself, but to forecast the possible future damage to crops by the resulting pest population, which is a highly stochastic variable.

In this chapter I attempt to solve this problem using some simple assumptions, and to develop fundamental monitoring schemes for population forecasting and control. As an illustration, planning for the rice brown planthopper control will be used, although it is still largely hypothetical due to lack of data.

16.2. POPULATION FORECASTING

16.2.1. The Basic Scheme

Let the pest population density at a given time (stage or generation) when damage will occur be N_0. This is to be predicted from the density some time before, N_{-t}. Then

$$X_0 = X_{-t} + R_t \tag{16.1}$$

where X_0 and X_{-t} are the densities in natural logarithms ($X_0 = \ln N_0$ and $X_{-t} = ln\ N_{-t}$) at the time injury occurs (0), and at the time of prediction ($-t$), respectively, and R_t is the rate of population change during that period. R_t is a stochastic variable with normal distribution, and we further assume that it is independent of the density X_{-t}, i.e., that the regression of X_0 on X_{-t}, $b = 1$ (see May *et al.*, 1974; Chapter 2). Then, if both the mean $\bar{R}_t$ and the variance $V(R_t)$ of R_t are known, together with X_{-t}, the mean or expected value and the predicted range for, say, $P = 0.90$ of X_0 are given by:

$$\hat{X}_0(-t) = X_{-t} + \bar{R}_t \tag{16.2}$$

and

$$\hat{X}_0(-t) - 1.64[V(R_t)]^{1/2} \leqslant X_0 \leqslant \hat{X}_0(-t) + 1.64[V(R_t)]^{1/2}, \tag{16.3}$$

since $\quad V[\hat{X}_0(-t)] = V(R_t)$.

The width of the range of eqn. (16.3) represents the precision of prediction, which becomes narrower as the time lag between the prediction and the occurrence of injury approaches zero.

16.2.2. The Brown Planthopper

Now we consider the case of the rice brown planthopper, *Nilaparvata lugens* Stål, as an example. This insect is one of the most important rice pests in Japan, and its ecology has been studied in detail (e.g., Kisimoto, 1977; Kuno, 1979). The population in rice fields is established by long-range migrants that appear within a restricted period in early summer, and it grows steadily through three generations until the autumn. The annual fluctuations in density are large and in outbreak years the population often causes serious damage due to "hopperburn", which mainly occurs in the last generation. Since the population peaks of different generations are easily distinguishable, we can conveniently define N_{-t} as the peak density t generations prior to the last or fourth generation (e.g., N_0 is the density of the last generation, while N_{-3} that of the first generation). *N. lugens* is highly tolerant to crowding so that population growth can be regarded as virtually density-independent. In fact the b-values for the regression of X_0 on X_{-1}, X_{-2}, and X_{-3} from our eight-year data (Kuno and Hokyo, 1970) are 0.710 ± 0.305, 0.969 ± 0.569, and 0.946 ± 0.805, respectively (differences from 1 are all insignificant). *N. lugens* populations thus provide a model case that satisfies the simple assumptions described above. The annual fluctations in the mean and variance of X_0 and R_t (based on the data by Kuno and Hokyo, 1970) are given in Table 16.1.

On the basis of Table 16.1 and eqns. (16.2) and (16.3), we can now obtain a basic scheme of forecasting for this insect pest (Figure 16.1).

TABLE16.1 Means and variances of population parameters in *N. lugens*.

	X_0	R_1	R_2	R_3
Mean	2.97	1.32	3.51	7.53
Variance	1.71	0.23	0.40	0.68

Some corrections are made to the original values according to the assumption of $b = 1$.

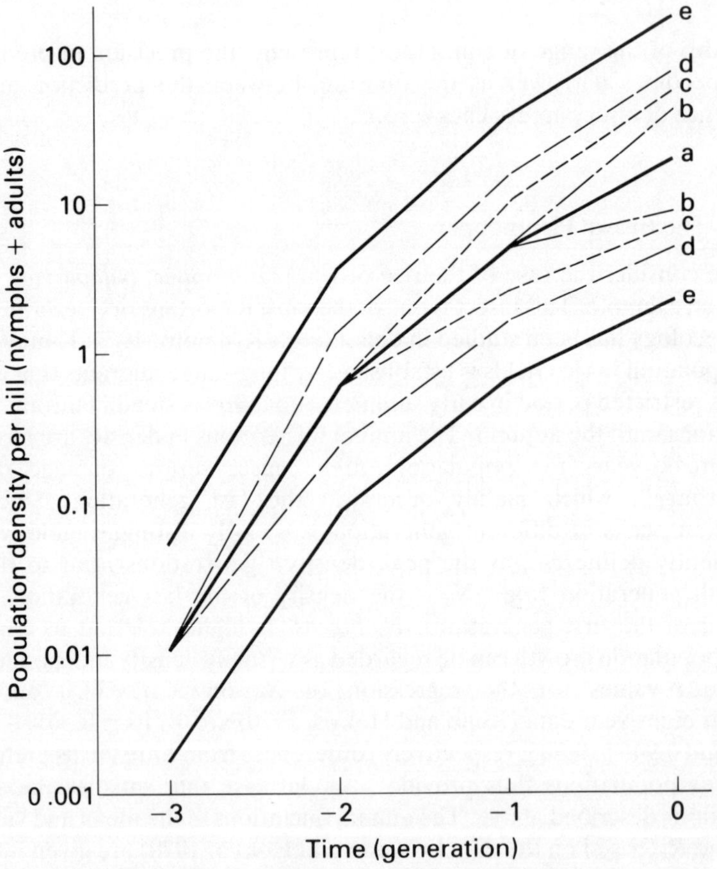

FIGURE 16.1 Forecasts of population size of *N. lugens* (as probable ranges for P = 0.90) at different time stages. a, t = 0; b, t = 1; c, t = 2; d, t = 3; e, t > 3.

Before the appearance of the initial invading population ($t > 3$), the only possible way to predict X_0 (the density of the last generation) is to use the observed variance of X_0 to calculate the probable range (*e* in Figure 16.1). But, as time elapses and as the population develops, the predicted range of X_0 gradually becomes narrower because of the reduction in $V(R_t)$ with decreasing t, and at the third generation ($t = 1$), the range b is reduced to nearly one-third of the initial one (e).

16.2.3. Effect of Sampling Error

In the above discussion we assumed no error for X_{-t} for use in forecasting X_0. But since we can only obtain X_{-t} by sampling, we must also consider the effect of sampling error in estimating X_{-t} on the prediction of X_0.

If we use the sample estimate X'_{-t}, instead of X_{-t}, for predicting X_0, the variance of $\hat{X}_0(-t)$ ($= X'_{-t} + \bar{R}_t$) is now increased to $V(X'_{-t}) + V(R_t)$, where $V(X'_{-t})$ is the error variance of the sample estimate X'_{-t}. Since the density X_{-t} here is expressed in natural logarithms (i.e., $X'_{-t} = \ln \bar{N}$, where $\bar{N}$ is the mean density), the relation $V(X'_{-t}) \approx D^2$ holds, where D is a conventional index of precision as standard error/mean ($S_{\bar{N}}/\bar{N}$) for the original or pre-transformation density estimate (Kuno, 1971). Thus we have:

$$V[\hat{X}_0(-t)] = V(R_t) + D^2. \tag{16.4}$$

This indicates that the predicted range of X_0 will more or less increase with the value of D, corresponding to eqns. (16.2) and (16.3):

$$\hat{X}_0(-t) = X'_{-t} + \bar{R}_t \tag{16.5}$$

and

$$\hat{X}_0(-t) - 1.64[V(R_t) + D^2]^{1/2} \leqslant X_0 \leqslant \hat{X}_0(-t) + 1.64[V(R_t) + D^2]^{1/2}. \tag{16.6}$$

These relations can now be used for forecasting X_0 based on the sample estimate, X'_{-t} provided that the precision of estimate D, is known in terms of the standard error for the untransformed density.

16.2.4. The Sampling Scheme

Equation (16.4) above indicates that if $V(R_t)$, the variability of population change during the corresponding period, is large, the effect of D on the precision of forecasting X_0 in terms of $V[X'_0(-t)]$ will be small, in which case it may be of little use to estimate N_{-t} precisely. In the planning of the sampling scheme it is thus prudent to determine the required precision in terms of $D(D_a)$ corresponding to the value of $V(R_t)$ at that time. The condition is thus quite different from that of planning sampling for the usual analyses of population dynamics where it is recommended that the precision levels for all the stages or generations involved are equalized (Kuno, 1971). An appropriate criterion in determining D_a at a given t is to keep the increment of $[V\{\hat{X}_0(-t)\}]^{1/2}$ due to a sampling error, $\delta = [V(R_t) + D^2]^{1/2} - [V(R_t)]^{1/2}$, constant for any t, which gives:

$$D_a = \{\delta(\delta + 2)\,[V(R_t)]^{1/2}\}^{1/2}, \tag{16.7}$$

in which case the resulting predicted range of X_0 is

$$\hat{X}_0(-t) - 1.64\{[V(R_t)]^{1/2} + \delta\} \leqslant X_0 \leqslant \hat{X}_0(-t) + 1.64\{[V(R_t)]^{1/2} + \delta\}. \tag{16.8}$$

δ can be determined according to the condition concerned, but usually 0.1 or 0.2 may be taken as a tentative criterion, corresponding to the same value of D when $V(R_t) = 0$. Table 16.2 gives values of D_a calculated from eqn. (16.7) for each generation of *N. lugens*, for $\delta = 0.1$ and 0.2.

TABLE 16.2 Values of D_a.

t	0	1	2	3
$\delta = 0.1$	0.10	0.33	0.37	0.42
$\delta = 0.2$	0.20	0.48	0.54	0.61

The next problem is how to attain the level of precision assigned in actual sampling procedures. Since the precision of the estimate depends on the sample size n and on the density, in theory, the best way is to adopt a plan using the principle of sequential estimation (Anscombe, 1953), in which n is determined through continual feedback of the interim census results. Such sampling schemes have been developed for various situations, e.g., simple quadrat sampling (Kuno, 1969; 1972), multistage quadrat sampling (Kuno, 1976), and capture–recapture censuses (Kuno, 1977b), in which the total number of individuals so far observed, T_n is successively plotted against n. The census is stopped when T_n intersects the specific boundary corresponding to the assigned level of precision.

In the case of the brown planthopper, population censuses are usually done by simple quadrat sampling, in which a hill of rice plants is taken as the "quadrat". With detailed information on the distribution of individual

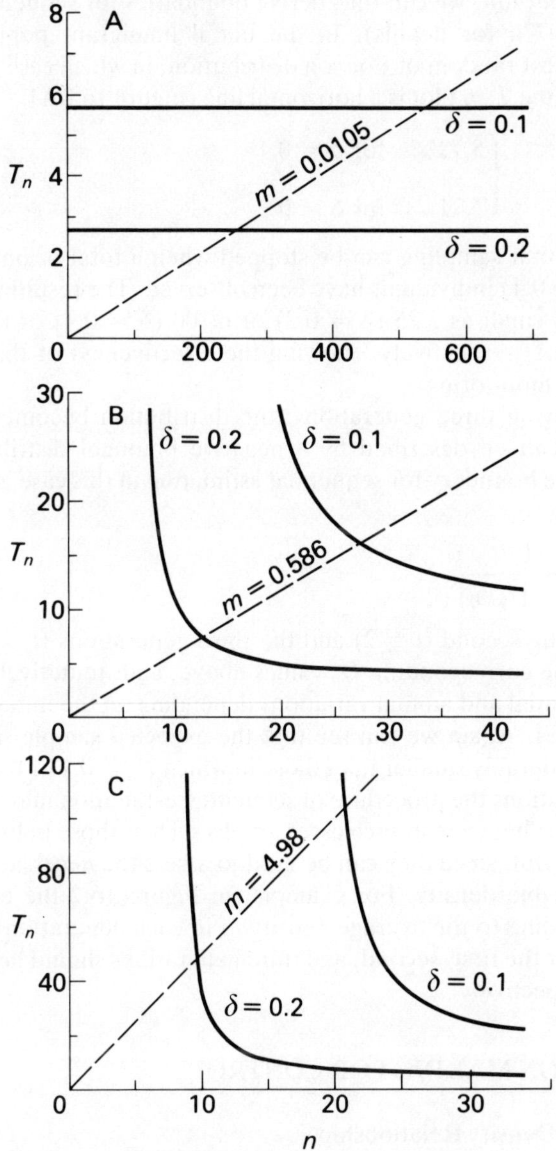

FIGURE 16.2 Sequential estimation charts for forecasting the peak generation ($t = 0$) using different preceding generations. n is sample size and T_n cumulative total individuals observed. The broken line in each graph shows the expected T_n/n for the average density m at that generation. A, first generation (Poisson distribution); B, second generation (negative binomial, $k = 0.5$); C, third generation (negative binomial, $k = 0.5$).

planthoppers per hill, we can thus derive boundaries for sequential estimates (see Kuno, 1977a for details). In the initial immigrant population, N_{-3}, there is an almost random or Poisson distribution, in which case the boundary to be drawn in the T_n/n plot is a horizontal line (Figure 16.2A):

$$T_n = \frac{1}{D_a^2} = \begin{cases} 5.72 & \text{for } \delta = 0.1 \\ 2.71 & \text{for } \delta = 0.2 \end{cases} \tag{16.9}$$

This indicates that sampling can be stopped when a total of only three (δ = 0.2) or six (δ = 0.1) individuals have been observed. The resulting sample size will thus be as small as 3/25 (δ = 0.2) or 6/100 (δ = 0.1) of that for D_a = 0.2 or D_a = 0.1, respectively, showing the effectiveness of this scheme for labor saving in monitoring.

In the following three generations, the distribution becomes remarkably clumped and can be described by a negative binomial distribution with a common k. The boundary for sequential estimation in this case is a hyperbolic curve given by

$$T_n = \frac{1}{D_a^2 - 1/(kn)} . \tag{16.10}$$

The lines for the second (t = 2) and the third generations (t = 1) were thus drawn using the corresponding D_a values above, and, tentatively, k = 0.5 (k shows some spatial and annual variations depending on the initial density; see Figure 16.2B,C). Again we can see that the expected sample sizes for these plans are considerably smaller than those in which D_a = 0.1 or 0.2 is used.

In some situations the procedure of sequential estimation may be difficult to apply in practice but even in such cases graphs such as those in Figure 16.2 will be still meaningful, since they can be used to assess the necessary sample size for some probable density. For example, in Figure 16.2 the expected T_n/n plot corresponding to the average density m in each generation suggests that sample sizes for the first, second, and third generations should be roughly 600, 30, and 20, respectively.

16.3. DECISION MAKING FOR CONTROL

16.3.1. Yield–Density Relationship

Next we consider a more specialized system of monitoring, the purpose of which is to judge whether or not insecticide spraying is necessary. It is first necessary to elucidate the yield–pest density relationship, together with the economic threshold in terms of pest density at the time of injury occurrence. The "economic threshold" here represents the density level at which the cost of spraying equals the increase in revenue so produced.

Unfortunately, we have not yet obtained a precise yield–density relationship for the brown planthopper, so for the present we use a hypothetical relationship based on an arbitrary assumption that the insects completely kill the rice plant hill, resulting in no yield, if 100 or more individuals attack it in the last generation, but no damage is done otherwise. The rice yield (as the percentage of the rice untreated) will then be given as the cumulative frequencies for $N \geq 100$ of the negative binomial distribution, with a common $k = 0.5$. The resulting yield–density relationship is shown in Figure 16.3 (curve a). The line RC represents the revenue when insecticide spray is used, on the assumption that the cost of spray is 10% of the maximum yield, and that the insecticide kills all the insects present. The crossing point of these two lines gives ET, the economic threshold.

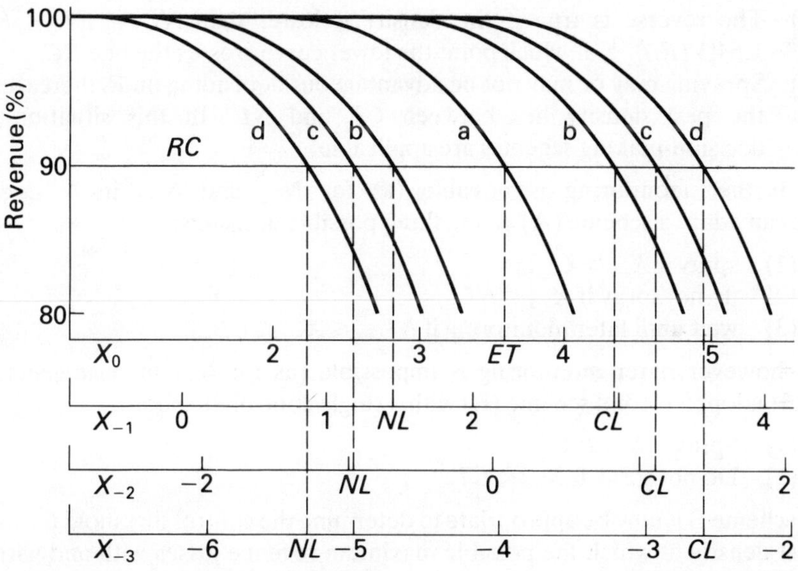

FIGURE 16.3 Hypothetical yield–density relationship for *N. lugens* at different generations. a, $t = 0$; b, $t = 1$; c, $t = 2$; d, $t = 3$. The two curves for b, c, and d correspond to the upper and lower limits ($P = 0.90$) of R_t at that time. ET, economic threshold; CL, control level; NL, noncontrol level; RC, expected revenue when insecticide is sprayed.

If, however, the pest density at some time prior to the time of injury occurrence (i.e., X_{-1}, X_{-2} or X_{-3} in this case) is used for the assessment, the yield–density relation can no longer be represented by a single curve, since the rate of population change, R_t, is a stochastic variable. Instead it may be shown in terms of an assigned probability range (say, $P = 0.90$), the upper and lower limits of which are the yield values corresponding to the limits of X_0 in inequality (16.6), as shown in Figure 16.3 (curves b, c, and d for $t = 1$, 2, and 3).

16.3.2. Basic Schemes

In this monitoring system the decision to introduce pest control will almost always be made in a situation in which the yield–density relation is stochastic (Figure 16.3; curves b, c, and d), since the purpose is to prevent crop damage before it occurs. A single threshold such as *ET* for X_0 no longer exists, but the following conclusions can be drawn.

(1) Spraying is generally preferred to not spraying if the pest density is higher than $CL = ET - \bar{R}_t + 1.64[V(R_t)]^{1/2}$, at which point the upper curve crosses the line *RC*.

(2) The reverse is true if the density is lower than $NL = ET - \bar{R}_t - 1.64[V(R_t)]^{1/2}$, at which point the lower curve crosses the line *RC*.

(3) Spraying may or may not be advantageous depending on R_t thereafter, if the pest density lies between *CL* and *NL*. In this situation two decision-making schemes are applicable:

If further monitoring is possible (as for N_{-2} and N_{-3} in *N. lugens*) we can adopt a scheme (A) giving three possible decisions:

(1) spray if $X_{-t} > CL_{-t}$
(2) do not spray if $X_{-t} < NL_{-t}$
(3) wait until later monitoring if $NL_{-t} \leqslant X_{-t} \leqslant CL_{-t}$.

If, however, later monitoring is impossible (as for N_{-1} in *N. lugens*), we must adopt a second scheme (B) with a single control threshold:

(1) Spray if $X_{-t} \geqslant CT$
(2) Do not spray if $X_{-t} < CT$.

In scheme B it may be appropriate to determine the control threshold *CT* as the pest density at which the possible maximum revenue losses with and without spraying balance (i.e., when both the upper and lower yield limits are the same distance from the line *RC*; see Figure 16.5B). But there does remain the possibility of loss of revenue due to subsequent unpredictable fluctuations in R_t. The only way to reduce this is to decrease $V(R_t)$ by delaying the time of last monitoring.

The values of various threshold densities determined hypothetically for *N. lugens* are listed below (antilog values are given in parentheses).

ET	NL_{-3}	CL_{-3}	NL_{-2}	CL_{-2}	CT_{-1}
3.58 (36)	−5.30 (0.0050)	−2.60 (0.074)	−0.97 (0.38)	1.11 (3.0)	1.90 (6.7)

16.3.3. Decision Making by Scheme A

The problem here is how to test the difference between X_{-t} and CL_{-t} or NL_{-t} in actual population censuses. Since both L and NL are known, we can of course directly apply a conventional test method or Iwao's (1975) convenient sequential method based on the same principle. A more efficient method, however, can be derived bearing in mind that our true aim is not to determine X_{-t}, but to predict X_0. If X_{-t} is assumed to be CL_{-t}, then the predicted lower limit of X_0 (for $\alpha = 0.05$) will be $X'_{-t} + \bar{R}_t - 1.64[V(R_t) + D^2(n,NL_{-t})]^{1/2}$, where $D(n,NL_{-t})$ is the D-value for sample size n and mean density $e^{NL_{-t}}$. Similarly, the predicted upper limit of X_0 on the assumption that $X_{-t} = CL_{-t}$ will be $X'_{-t} + \bar{R}_{-t} + 1.64[V(R_t) + D^2(n, CL_{-t})]^{1/2}$.
The resulting decision-making scheme is therefore

(1) *Spray insecticide if*

$$X'_{-t} + \bar{R}_t - 1.64[V(R_t) + D^2(n,CL_{-t})]^{1/2} > ET \tag{16.11}$$

(2) *Do not spray if*

$$X'_{-t} + \bar{R}_t + 1.64[V(R_t) + D^2(n,NL_{-t})]^{1/2} < ET \tag{16.12}$$

(3) *Reserve decision for later monitoring if otherwise.*

For the brown planthopper this scheme is applicable to the first two generations ($t = 3, 2$). Here, since the census is made by simple quadrat sampling, $X'_{-t} = \ln(T_n/n)$, where T_n is the total number of individuals in a sample size n, and $D^2(n,CL)$ or $D^2(n,NL)$ for the first and second generations are obtained from:

$$D^2(n,X) = 1/(ne^X), \tag{16.13}$$

and

$$D^2(n, X) = \frac{1}{n}\left(\frac{1}{e^X} + \frac{1}{k}\right) = \frac{1}{n}\left(\frac{1}{e^X} + 2\right), \tag{16.14}$$

corresponding to the respective distribution patterns, the Poisson and the negative binomial with $k = 0.5$. Thus, using known values of ET, $\bar{R}_t$, $V(R_t)$, CL_{-t}, and NL_{-t}, we can put the scheme into practice.

This decision-making scheme is also applicable to the census by the frequency-of-zeros method. For *N. lugens* this method may be useful in practice in the second and later generations when the pest density rises and precise counting of individuals in each hill becomes difficult. In this case, if the distribution is a negative binomial, we have:

$$X'_{-t} = \ln\left\{k\left[\left(\frac{T_n}{n}\right)^{-1/k} - 1\right]\right\}, \tag{16.15}$$

where T_n here is the number of empty or unoccupied quadrats (hills) in a sample size n, and:

$$D^2(n,X) = \frac{1}{ne^{2X}}\left(1+\frac{1}{k}e^X\right)^{k+2}\left[1-\left(1+\frac{1}{k}e^X\right)\right]^{-k} \tag{16.16}$$

(Kuno, 1977a). This type of estimation, however, is vulnerable to bias in parameter k, so that in its application care must be taken to use as precise a k-value as possible.

These procedures can be easily converted to a form suitable for sequential sampling. The two boundaries to be drawn on the T_n/n plot for sequential test corresponding to eqns. (16.11) and (16.12) are:

$$T_n = n\exp\ \{ET - \bar{R}_t + 1.64[V(R_t) + D^2(n,CL_{-t})]^{1/2}\} \tag{16.17}$$

$$T_n = n\exp\ \{ET - \bar{R}_t - 1.64[V(R_t) + D^2(n,NL_{-t})]^{1/2}\}. \tag{16.18}$$

Combining these with eqns. (16.13) and (16.14) we get sequential test graphs for the first two generations of *N. lugens* (Figure 16.4A,B). The zone above the line a, eqn. (16.17), corresponds to "spray", that below the line b, eqn. (16.18), to "non-spray", and the zone between the lines to "reserve". Similarly, the boundary lines for the census by frequency-of-zeros method are derived from eqns. (16.11), (16.12), and (16.15) as:

$$T_n = n\left(\frac{1}{k}\ \exp\{ET - \bar{R}_t + 1.64[V(R_t) + D^2(n,CL_{-t})]^{1/2}\} + 1\right)^{-k} \tag{16.19}$$

$$T_n = n\left(\frac{1}{k}\exp\{ET - \bar{R}_t - 1.64[(V(R_t) + D^2(n,NL_{-t})]^{1/2}\} + 1\right)^{-k}, \tag{16.20}$$

where T_n is the number of zeros among n quadrats sampled, and $D^2(n,CL_{-t})$ and $D^2(n,NL_{-t})$ are given by eqn. (16.16) if the distribution is a negative binomial. Figure 16.4C shows the results obtained for the second generation (t = 2) of *N. lugens*. In this case eqn. (16.19) corresponds to the lower boundary (b) below which is the "spray" zone, whereas eqn. (16.20) corresponds to the upper boundary (a) above which is the "non-spray" zone.*

Generally speaking, it is not advisable to take a large sample at this stage of monitoring because it may be possible to delay the decision, especially if $V(R_t)$ is large. In such a situation the sequential plans as discussed above may often not be

*Given the non-normality of the distribution in small samples, in practice it may be safer to add one to the upper (16.17 or 16.20) and subtract one from the lower (16.18 or 16.19) boundaries in the right-hand side.

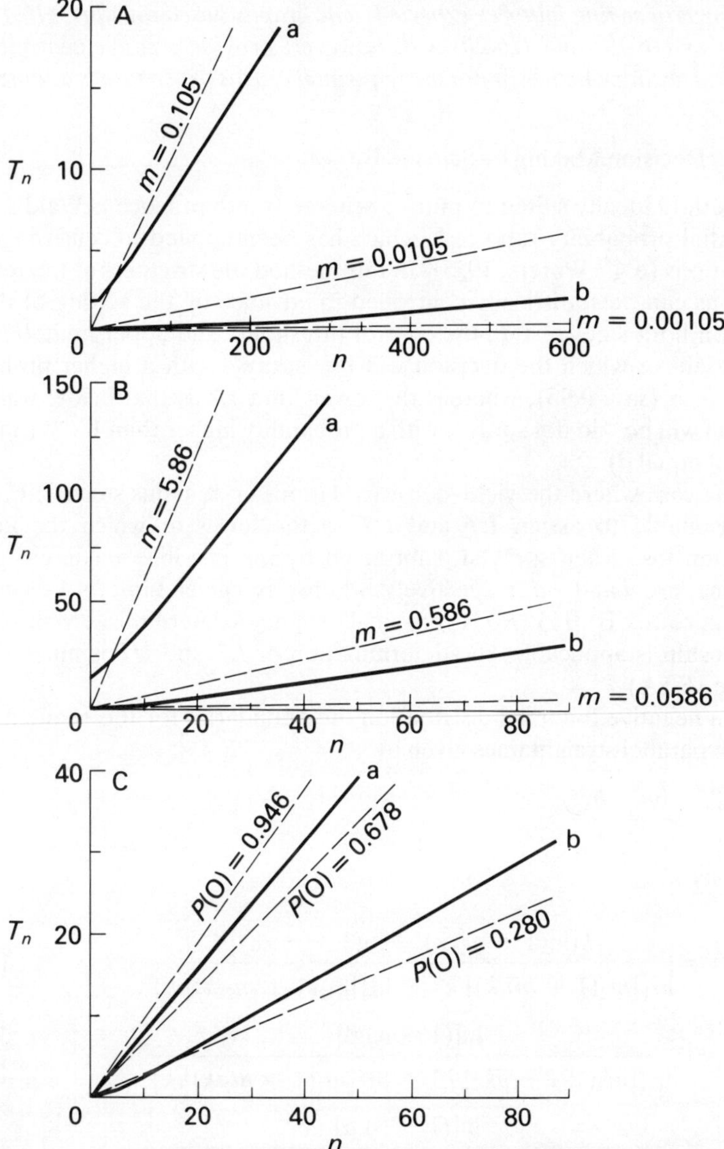

FIGURE 16.4 Test charts for control decisions using scheme A (for first and second generations of *N. lugens*). A, first generation (Poisson distribution) Above a, "spray" zone; below b, "nonspray" zone; between a and b, "reserve" zone. B, second generation (negative binomial, $k = 0.5$). Above a, "spray" zone; below b, "non-spray" zone; between a and b, "reserve" zone. C, second generation (frequency-of-zeros method). Above a, "non-spray" zone; below b, "spray" zone; between a and b, "reserve" zone. Broken lines show expected trajectories of T_n/n for three different densities.

as efficient in saving labor as expected. The graphs based on eqns. (16.17) and (16.18) or (16.19) and (16.20) will, however, provide a convenient means of graphical testing whether or not the sequential census procedure is adopted.

16.3.4. Decision Making by Scheme B

The method ideally suited to putting scheme B into practice is Wald's (1948) sequential probability ratio test, which has been applied to censusing insect populations (e.g., Waters, 1955). In this method the strictness of the resulting decisions can be statistically controlled in advance by the setting of the two limits on both sides of CT, the control threshold. The upper limit UT is the density above which the decision will be "spray", with a higher probability than $1 - \alpha$ (say, 0.95); whereas the lower limit LT is that below which the decision will be "do not spray", with a probability higher than $1 - \beta$ (α may or may not equal β).

In the case where the yield–density relationship remains stochastic, it may be reasonable to assign LT and UT as the levels at which the possible maximum loss when sprayed, subtracted by the possible maximum gain at that time, are d and $-d$, respectively, where d is a given limit for loss increase (see Figure 16.5B). If $V(R_t)$ is very small, so that a deterministic yield–density relationship is applicable, the determination of LT and UT is much simpler (Figure 16.5A).

For a negative binomial distribution the boundaries for this sequential test are two parallel straight lines given by:

$$T_n = bn - h_1 \tag{16.21}$$

$$T_n = bn + h_2, \tag{16.22}$$

where

$$b = \frac{k[\ln(1 + m_2/\mathrm{k}) - \ln(1 + m_1/\mathrm{k})]}{\ln\{[m_2(1 + m_1/k)]/k\} - \ln\{[m_1(1 + m_2/k)]/k\}} \tag{16.23}$$

$$h_1 = \frac{\ln[(1 - \alpha)/\beta]}{\ln\{[m_2(1 + m_1/k)]/k\} - \ln\{[m_1(1 + m_2/k)]/k\}} \tag{16.24}$$

$$h_2 = \frac{\ln[(1 - \beta)/\alpha]}{\ln\{[m_2(1 + m_1/k)]/k\} - \ln\{[m_1(1 + m_2/k)]/k\}}. \tag{16.25}$$

$$(m_1 = \mathrm{e}^{LT}, m_2 = \mathrm{e}^{UT})$$

Figure 16.6A shows the sequential test graph thus obtained for the third generation of *N. lugens* using $k = 0.5$, $m_1 = 5.1$, and $m_2 = 9.2$. Figure 16.6B shows the corresponding graph for a test based on the frequency-of-zeros method (binomial distribution). T_n here is the number of planthopper-

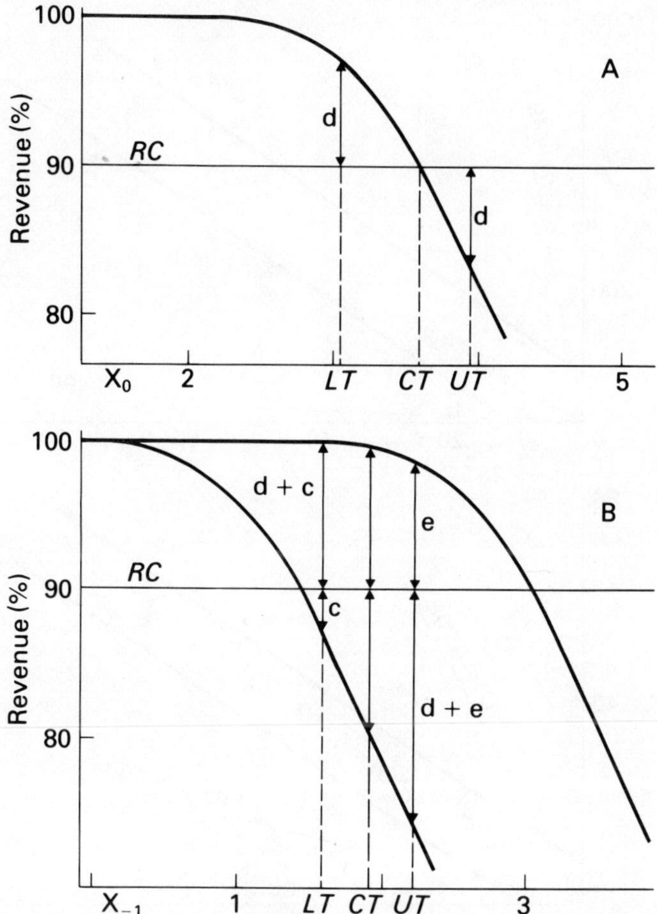

FIGURE 16.5 Determination of control threshold (CT) and practical limits of allowance for nonspray (UT) and spray (LT) for decision making by scheme B. d, assigned limits of loss increases. A, $V(R_t) = 0$; B, $V(R_t) > 0$.

free hills for sample size n; b, h_1, and h_2 substituted in eqns. (16.21) and (16.22) are:

$$b = \frac{\ln q_1 - \ln q_2}{\ln p_2 q_1 - \ln p_1 q_2} \tag{16.26}$$

$$h_1 = \frac{\ln[(1 - \alpha)/\beta]}{\ln p_2 q_1 - \ln p_1 q_2} \tag{16.27}$$

$$h_2 = \frac{\ln[(1 - \beta)/\alpha]}{\ln p_2 q_1 - \ln p_1 q_2}, \tag{16.28}$$

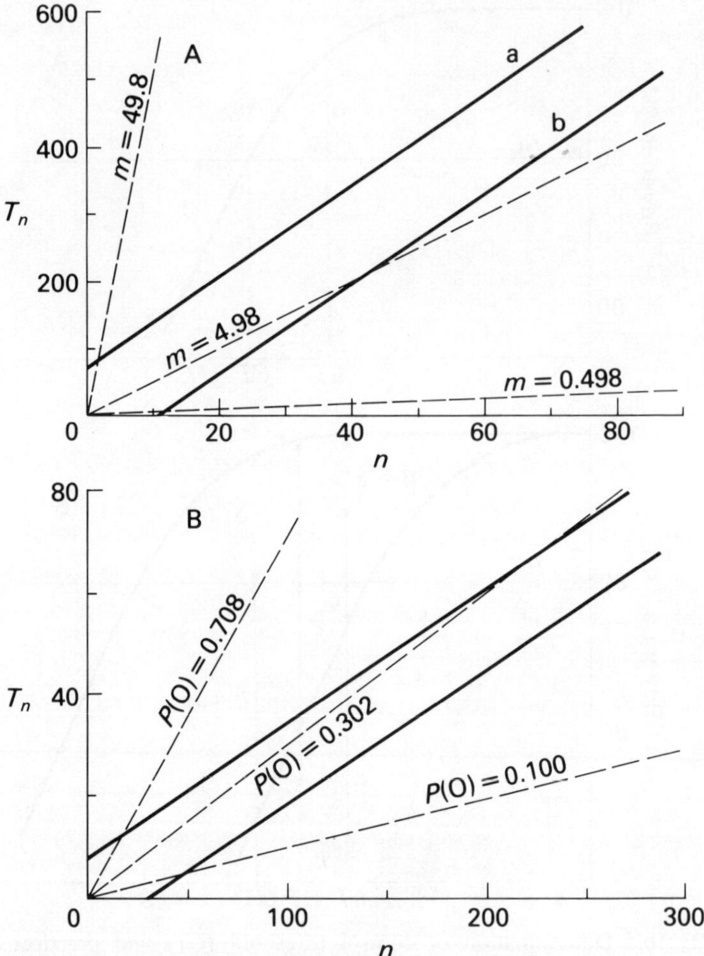

FIGURE 16.6 Test charts for control decisions using scheme B; sequential probability ratio test (for third generation of *N. lugens*). A, Sample mean method (negative binomial, $k = 0.5$; $m_1 = 5.1$, $m_2 = 9.2$). Above a, "spray" zone; below b, "non-spray" zone; between a and b, "continue sampling" zone. B, Frequency-of-zeros method ($P_1 = 0.23$, $P_2 = 0.30$). Above a, "non-spray" zone; below b, "spray" zone; between a and b, "continue sampling" zone. Broken lines show expected trajectories of T_n/n for three different densities.

where $p_1 = (1 + m_2/k)^{-k}$, $p_2 = (1 + m_1/k)^{-k}$, $q_1 = 1 - p_1$, and $q_2 = 1 - p_2$.

In the former test, the area below (16.21) is the "non-spray" zone and that above (16.22) the "spray" zone, whereas the reverse is the case in the latter. In a situation where one cannot adopt sequential census procedure but can utilize the graph for decision making, it may sometimes be necessary to make decisions *while T_n remains between both boundaries. In such a case it will be generally safer to spray.*

16.4. MODIFICATION FOR DENSITY-DEPENDENT POPULATIONS

All the methods discussed above are for density-independent populations in which the slope b in the regression of X_0 on X_{-t} is 1. We should thus make some modification to the basic equations if this condition is not satisfied. Here we assume the linear regression relationship (May *et al.*, 1974):

$$X_0 = bX_{-t} + a \tag{16.29}$$

where b is a constant and a, a stochastic variable with mean $\bar{a}$ and stable variance $V(a)$. Then, the predicted value of X_0 and its variance based on a sample estimate of X_{-t} are given by:

$$X_0(-t) = bX'_{-t} + \bar{a} \tag{16.30}$$

and

$$V[\hat{X}_0(-t)] = V(a) + b^2V(X'_{-t}) = V(a) + b^2D^2, \tag{16.31}$$

so that the inequality representing the probable range of X_0 (for $P = 0.90$), to be used in place of (16.6), is

$$bX'_{-t} + \bar{a} - 1.64[V(a) + b^2D^2]^{1/2} \leq X_0 \leq bX'_{-t} + \bar{a} + 1.64[V(a) + b^2D^2]^{1/2}. \tag{16.32}$$

It also follows that (16.7), which represents the necessary value of D for keeping the increment δ of $[V\{\hat{X}_0(-t)\}]^{1/2}$ due to sampling error constant, now becomes:

$$D_a = (1/b)[\delta\{\delta + 2[V(a)]^{1/2}\}]^{1/2}. \tag{16.33}$$

This equation leads to the important conclusion that the more intensely the population becomes density-dependent (i.e., $b \to 0$), the less precision is required in estimating X_{-t} for fixed δ and $V(a)$, i.e., the less effective the information on X_{-t} becomes for the prediction of X_0.

The criteria for decision making on insecticide spraying should also be modified by using eqns. (16.30) and (16.31). The resulting boundaries for use in the graphical tests for decision making using scheme A are:

$$T_n = n \exp\left(\frac{1}{b}\{ET - \bar{a} + 1.64[V(a) + b^2D^2(n,CL_{-t})]^{1/2}\}\right) \tag{16.34}$$

$$T_n = n \exp\left(\frac{1}{b}\{ET - \bar{a} - 1.64[V(a) + b^2D^2(n,NL_{-t})]^{1/2}\}\right) \tag{16.35}$$

for the usual sample mean method, corresponding to eqns. (16.17) and (16.18), and:

$$T_n = n\left[\frac{1}{k} \exp\left(\frac{1}{b}\{ET - \bar{a} + 1.64[V(a) + b^2D^2(n,CL_{-t})]^{1/2}\}\right) + 1\right]^{-k} \quad (16.36)$$

$$T_n = n\left[\frac{1}{k} \exp\left(\frac{1}{b}\{ET - \bar{a} - 1.64[V(a) + b^2D^2(n,NL_{-t})]^{1/2}\}\right) + 1\right]^{-k} \quad (16.37)$$

for the frequency-of-zeros method, corresponding to eqns. (16.19) and (16.20), where $CL = (1/b)\{ET - \bar{a} + 1.64[V(a)]^{1/2}\}$ and $NL = 1/b\{ET - \bar{a} - 1.64[V(a)]^{1/2}\}$. The boundaries for decision making by scheme B (Wald's probability ratio test) are also readily obtainable from eqns. (16.21) and (16.22) after the critical density levels, m_1 and m_2, are determined graphically (as in Figure 16.6B) from the yield–density relationship based on eqn. (16.32).

16.5. CONCLUSIONS

Population monitoring in pest management systems almost always involves forecasting of an uncertain future situation because the aim is to avoid crop damage before it occurs. In such a situation, as we have seen, efficient planning for monitoring is possible only when the information on this uncertainty is incorporated into the scheme. A practical obstacle to the use of such a scheme is that it requires detailed empirical data on pest population dynamics, and this usually requires intensive field studies covering at least several years. Moreover, such data should ideally be obtained for different locations and conditions. At earlier stages of the study, one can of course use some tentative plans based on provisional estimates of pest population parameters such as $\bar{R}_t$ and $V(R_t)$. However, such data are a prerequisite for any objective forecasting systems in which the range of pest density is to be predicted, and hence are worth obtaining, despite the work involved, at least for some important pest species.

REFERENCES

Anscombe, F. J. (1953) Sequential estimation. *J. R. Stat. Soc.* B15:1–21.

Iwao, S. (1975) A new method of sequential sampling to classify populations relative to a critical density. *Res. Pop. Ecol.*, 16:281–288.

Kisimoto, R. (1977) Bionomics, forecasting of outbreaks and injury caused by the rice brown planthopper, in *The Rice Brown Planthopper* (Taipei: Food and Fertilizer Technology Center for the Asian and Pacific Region) pp27–34.

Kuno, E. (1969) A new method of sequential sampling to obtain the population estimates with a fixed level of precision. *Res. Pop. Ecol.* 11:127–36.

Kuno, E. (1971) Sampling error as a misleading artifact in key factor analysis. *Res. Pop. Ecol.* 13:28–45.

Kuno, E. (1972) Some notes on population estimation by sequential sampling. *Res. Pop. Ecol.* 14:58–73.

Kuno, E. (1976) Multi-stage sampling for population estimation. *Res. Pop. Ecol.* 18:39–56.

Kuno, E. (1977a) Distribution pattern of the rice brown planthopper and field sampling techniques, in *The Rice Brown Planthopper* (Taipei: Food and Fertilizer Technology Centre for the Asian and Pacific Region), pp135–46.

Kuno, E. (1977b) A sequential estimation technique for capture–recapture censuses. *Res. Pop. Ecol.* 18:187–94.

Kuno, E. (1979) Ecology of the brown planthopper in temperate regions, in *Brown Planthopper: Threat to Rice Production in Asia* (Laguna, Philippines: International Rice Research Institute) pp45–60.

Kuno, E. and N. Hokyo (1970) Comparative analysis of the population dynamics of rice leafhoppers, *Nephotettix cincticeps* Uhler and *Nilaparvata lugens* Stål, with special reference to natural regulation of their numbers. *Res. Pop. Ecol.* 12:154–84.

May, R. M., G. R. Conway, M. P. Hassell, and T. R. E. Southwood (1974) Time delays, density dependence and single species oscillations. *J. Animal Ecol.* 43:747–70.

Southwood, T. R. E. (1978) *Ecological Methods* 2nd edn. (London: Chapman and Hall).

Wald, A. (1948) *Sequential Analysis* (New York: John Wiley).

Waters, W. E. (1955) Sequential sampling in forest insect surveys. *Forest Sci.* 1:68–79.

17 Estimation, Tactics, and Disease Transmission

M. H. Birley

17.1. INTRODUCTION

In this chapter I am concerned with three processes that are central to the control of vector-borne diseases: population dynamics, sampling, and estimation. A study of vector population dynamics produces a number of deductive inferences that purport to predict whether diseases may be transmitted. But these inferences are based on unknown parameters and are consequently of limited value without the additional processes of sampling and estimation. The three processes, moreover, are intimately linked in both theory and practice. For example, the practical limitations of sampling restrict the choice of population model, while deductions from the model indicate what should be sampled. The estimation process, in turn, is restricted by model configurations and by the nature of sampling error distributions.

The principal objective of this chapter is to analyze and discuss the properties of an ideal estimation process for the key parameters of disease vector populations.

17.1.1. Parameters of Vector Populations

The principal parameter of a vector population is the average survival rate, which determines life expectation and places an upper limit on potential disease transmission rate. It has achieved almost universal importance in vector population analysis, even though estimation methods often seem crude or cumbersome. The estimate of transmission rate is often extremely sensitive to small variations in survival rate and consequently a high level of precision is required.

The average survival rate characterizes the distribution of survival rates within a population. It does not presuppose that survival is homogeneous with respect to age or genetic composition, although it may be so interpreted if no other option is available. It is a slow variable, remaining relatively constant for a matter of weeks, in contrast to the population density, which may vary rapidly. The average survival rate may be interpreted as the success of the population in withstanding the exigencies of the average environment. Catastrophic events such as desiccation, flooding, or spraying must be treated separately.

However, survival rate is only one component of a complex pattern of vector behavior. Another component is the frequency with which insects or other vectors seek a blood-meal from man or another host. This affects the rate of disease transmission, and the proportion of the population that is available to be sampled.

In this chapter I begin by discussing the traditional, practical estimator of survival rate, known as the mean parous rate. In a series of models I show how this may be improved in robustness to take account of fluctuations in recruitment and biting rate. I also outline methods for testing density dependence and sampling bias. I finally discuss the relationship between the parous rate and the key concept of vectorial capacity in disease transmission and suggest that surveys based on sequential decision rules can now be used to map the potential for disease transmission.

17.2. THE SAMPLING PROCESS

Methods of sampling insect populations in the field have been reviewed by Southwood (1978). Gilbert (1973) and Gilbert *et al.* (1976) have referred to the particular problem of unbiased sampling. Either every insect in the population must be assumed equally likely to be sampled or significant departures from that assumption need to be made explicit. By contrast with plant feeders, insects that feed on blood, such as mosquitoes, may only be found on their host for short periods of time, separated by several days.

There are three main methods in general use for sampling adult female anthropophilic mosquitoes and other biting flies; none is free from bias (Gillies, 1974; Service, 1976, 1977). First, the total number of mosquitoes resting in a house or shelter may be removed using aspirators, nets, or sprays. Second, a light trap or carbon dioxide trap may be operated for a fixed period during the diel and attracted mosquitoes may be caught by suction. Third, human bait may be exposed for fixed periods during the diel and landing/biting mosquitoes caught. A typical sampling program of the third kind may be envisaged as follows. A portion of the body is exposed for a fixed period at a particular time in the evening and the mosquitoes that land are trapped. The regime is followed daily for a period of about one month. The relative numbers of individuals caught per unit effort are assumed to reflect changes in the population density or size. There are sampling errors, of course, so that the daily samples are drawn from a sampling distribution. For example, the probability of any mosquito arriving during a small time period is low and the number caught could be Poisson-distributed. There are also random disturbances due to changing environmental conditions.

The sample collected each evening is then examined under a low-power microscope and is sorted into species. The females of the key species are dissected and ovaries and tracheoles are examined to determine parity and the total number of parous individuals (i.e., those that have laid at least one egg batch) is recorded, together with the total sample size. Because of the nature of this sorting process the number of parous individuals must follow a binomial distribution. The probability that any individual is parous is assumed equal to the proportion of parous individuals in the population in the behavioral phase where they can be sampled.

17.3. PROPERTIES OF AN IDEAL ESTIMATION PROCESS

Before considering practical methods of estimating survival rates from sample data it is useful to list the properties of an ideal estimator. First, it is assumed that the estimation process is to form part of a tactical or strategic management procedure. Consequently it must be standardized and simple to use by those with diverse skills, such as technicians, medical auxiliaries, as well as entomologists. Second, it must be robust so that the estimate is insensitive to changes in the underlying assumptions. Further, the precision must be specified so that the range or interval estimate can be interpreted with confidence. It should then be possible to make comparisons between populations in different locations and times. The assumptions should be biologically realistic and allow for processes such as density-dependent mortality. Finally, and most important, it should be interpreted in real time so that survival estimates can be derived on the same day that each new datum is collected. *In summary, the search is for a simple, sequential, standardized, robust procedure for determining the importance of a vector population from sample surveys.*

One of the older practical estimators of mosquito survival is the mean parous rate. This has some of the properties of the ideal estimator as well as a number of problems; it also forms the basis for a discussion of the population process and is therefore presented here in some detail.

17.3.1. The Mean Parous Rate

The central problem of estimating insect survival rates is that the age of insects cannot be accurately determined. Instead individuals are grouped into life stages that may each last a considerable and variable number of days. In order to estimate the survival rate within a life stage it is essential to distinguish, as further subgroups, between new and old individuals in the stage (Birley, 1977). Estimates of survival rate may then be obtained by comparing the proportions of new and old individuals on different occasions. The binary nature of this sampling process is critical for the analysis of the results.

It is a general evolutionary requirement that adult females should mate and reproduce as soon as practicable after reaching maturity. Nulliparous adults are females that have not reproduced and are therefore generally young, while parous or multiparous females, which have reproduced, are old individuals. Methods of distinguishing nulliparous and parous individuals are well developed in the Culicidae (Detinova, 1962), Ceratopogonidae (Dyce, 1969), and Muscidae of the genus *Glossina* (Saunders, 1960), among other insect groups.

If the recruitment rate of nulliparous flies to a population is constant then the population should reach a steady state with a constant proportion of parous individuals. The proportion parous is then equal to the probability that flies survive the average length of time between parous grade transitions, i.e., the time interval between blood-meals assuming one gonotrophic cycle (e.g., the laying of one egg batch) per blood-meal. The proportion parous may be estimated by sampling flies at random and determining the mean parous rate Q_0 as:

$$Q_0 = \frac{\text{total parous}}{\text{total caught}} . \tag{17.1}$$

Since the population density is assumed constant, the samples may be obtained over a number of days and the totals pooled. This estimate of survival rate has a number of advantages: it is simple, sequential and requires only a small amount of sampling to obtain a high level of precision. The standard error may be estimated by the binomial formula:

$$\text{s.e. } Q_0 = \left[\frac{Q_0(1 - Q_0)}{\text{total caught}}\right]^{1/2} . \tag{17.2}$$

However, there are a number of disadvantages. Recruitment rates are seldom constant in nature and the consequent changes in the proportion parous may bias the estimate of survival rate. No account of density-dependent mortality may be obtained since the density is assumed constant. Differences in behavior between nulliparous and parous individuals may ensure that they cannot be sampled at random. Finally, no information may be obtained about the average length of time between parous grade transitions. As a preliminary step to overcoming some of these difficulties we can view the underlying population process as follows.

17.3.2. The Population Process

The mosquito population consists of highly mobile adult females with a behavioral rhythm composed of daily and multiple-day elements (Figure 17.1).

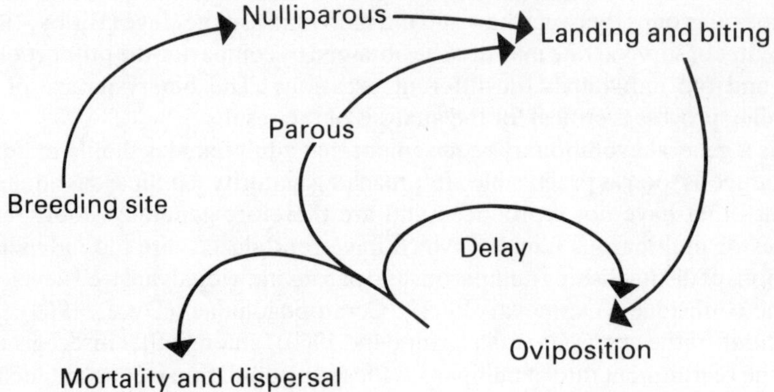

FIGURE 17.1 Essential elements in the population process.

After emergence, the females mate over the course of a few days, and as nullipars they join their parous sisters in seeking a blood-meal at a particular part of the diel. Together they then retire to rest and mature an egg batch. During subsequent days they fly in search of oviposition sites and lay their eggs, after which they may again rest and, now all parous, the survivors seek another blood-meal. This rhythm of behavior is referred to as the gonotrophic cycle and represents the time period between parous grade transitions. It should not be confused with the daily behavioral rhythm that is sometimes referred to as the biting cycle. The inverse of the cycle is called the biting rate or frequency. It may have a mean of several days duration although many factors may modify the distribution of cycle length.

To complete the picture it is assumed that the mosquitoes are only sampled as they seek a blood-meal and that this only occurs once per cycle. It is also assumed that variations of survival rate with age may be ignored. Four models can now be derived from this simple scheme, which try to capture the essentials of the process.

17.4. MODEL 1

It should be clear from Figure 17.1 that if the cycle is of fixed duration then the number of parous individuals biting on any day is equal to the total number one cycle earlier, multiplied by the proportion surviving:

$$M_t = P_u T_{t-u} \tag{17.3}$$

$$T_t = N_t + M_t \tag{17.4}$$

where T_t is the total biting population on day t, composed on N_t nulliparous and M_t parous individuals. The proportion P_u is the survival rate per cycle, which has a mean duration of u days. Although this model is simplistic it provides a number of useful predictions about the population process. First, when recruitment N_t is constant there will be a steady state value for M_t and T_t and the best estimate of P_u is then the mean parous rate.

Second, when the recruitment rate varies, a better—and possibly the best—estimate of P_u is:

$$Q_u = \frac{\sum M_{t_i}}{\sum T_{t_i - u}} \tag{17.5}$$

where M_{t_i} is the parous sample in the total sample T_{t_i} collected on day t_i, and there are m pairs of sampling occasions separated by u days. Garrett-Jones (1973) suggested a similar formula but does not appear to have described its properties in detail. By analogy, a possible standard error for Q_u may be obtained from the binomial formula when the sample size is small.

Equation (17.3) is essentially a linear regression through the origin and eqn. (17.5) estimates the slope of the regression by the ratio of means. Since M_{t_i} is a binomial variable the residual variance is a function of total sample size and in this case the ratio of means is probably the best estimate of the regression (Armitage, 1971). When the sample size is large the binomial distribution approximates a normal distribution and an alternative estimator may be derived from least-squares theory. A third estimator is the geometric mean, which may be useful when the sampling distribution is more complex and a logarithmic transformation should be considered. This is also a useful precursor to tests for density dependence that are discussed in relation to model 4.

The introduction of the time lag u associated with the cycle conveniently solves the problem of varying recruitment rate, which biases the mean parous rate formula. It therefore extends the use of the simple, sequential parous rate procedure to the analysis of short runs of data and short-term fluctuations in population size. However, the value of the integer u, the mean length of the cycle, is unknown and it, too, must be estimated from the data.

17.4.1. Mean Length of the Gonotrophic Cycle

Model 1 predicts that the time series $M(t)$ and $T(t - j)$, lagged by j days, are correlated. A cross-correlation coefficient may therefore be derived to compare the degree of correlation. A suitable coefficient would conveniently distinguish between recruitment rates that are roughly constant, in which case the ordinary parous rate formula applies and biting rate cannot be estimated, and recruitment rates that are varying, in which case the coefficient should peak at the correct time lag. The lag at which this peak occurs is then the best estimate of the mean length of the cycle.

If the coefficient is very small for all time lags then the sampling method may be dominated by sampling errors or the population may be incoherent. In an incoherent population, density and age structure are dominated by migration rates or environmental disturbances so that the age composition on any day is not correlated with previous days. In this case it is not possible to estimate survival rate. There are a number of other special cases that could cause problems. For example, if the population has been sampled from a time when recruitment was zero to a later time when it is again zero, then one or more discrete generations have been observed and, intriguingly, the ordinary parous rate formula is again applicable. If the population is increasing exponentially then the time series may exhibit the property of nonstationarity and trend removal techniques must be applied.

17.4.2. Simulating the Population and Sampling Process

The internal validity of handling the estimation properties of the model in this way may be investigated very simply. First all a pair of time series are generated using eqns (17.3) and (17.4) and driven by the kind of recruitment process under consideration. For example, recruits may be chosen from a uniform random distribution of a convenient size. Second, the population is sampled so as to remove a constant mean proportion but drawing actual numbers from a suitable sampling distribution. For example, the Poisson distribution may mimic the sampling process and the proportion chosen may be varied to correspond to either large or small sample sizes. Third, the sample is sorted into two categories corresponding to nulliparous and parous individuals. This is a binomial sampling process where the probability of choosing a parous individual is equal to the true proportion of parous individuals in the population.

These three stages of randomization yield time series with binomial errors when the sample size is small and normal errors when the sample size is large. An appropriate length of time series may then be chosen and the original parameters extracted by the estimation procedure. The properties of the estimation procedure are then determined. For example, when the sample size is small the least-squares estimate is indeed biased and a time series of length at least 20 is required before a stable estimate of the parameter is obtained from eqn. (17.5). This is illustrated in Figure 17.2, which indicates that in field surveys a minimum of 20 consecutive days of sampling are required.

17.5. MODEL 2

In model 1 the mean or expected length of the cycle was determined from an examination of the cross-correlation index, which is an integer value for which no measure of dispersion is available. The question therefore arises as to its robustness: how variable is it and how does its variability affect both the estimation of survival and the potential for disease transmission?

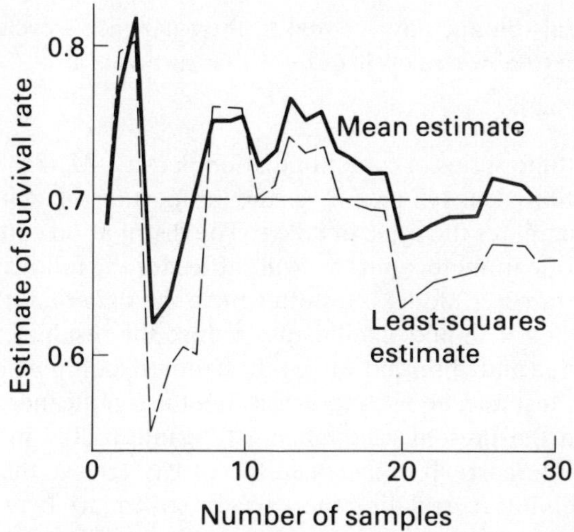

FIGURE 17.2 Sequential estimates of survival rate obtained from simulated data with small sample size. The true value of survival rate was 0.7.

The behavioral cycle of adult female biting flies is a complex of responses to host, habitat, and environment, movements between resting, feeding and breeding sites, and physiological states of hunger, engorgement, and gravidity (Muirhead-Thomson, 1951; Mattingly, 1969; Gillett, 1962, 1965; Hocking, 1971; WHO, 1972; Ungureanu, 1974). Many components of behavior are associated with a particular time of day or night: maturation of ovaries is a function of temperature; host-seeking and flight may depend on climatic conditions. The cycle is therefore a stochastic process based on a discrete, daily time unit. The length of this cycle may be expected to vary between individuals and have a frequency distribution throughout the population. Evidence for a frequency distribution of biting period in mosquito populations may be deduced from field data presented by Gillies and Wilkes (1965), Pant and Yasuno (1970) and Conway *et al.* (1974). Similar effects have been observed in mark–recapture experiments with *Phlebotomus ariasi* (Killick-Kendrick, personal communication) and *Glossina fuscipes fuscipes* (Rogers, 1977).

An alternative hypothesis, proposed by Conway *et al.* (1974), was that the interval between primary blood-meals is constant but that there are also some secondary blood-meals in each cycle and a number of distinct subpopulations feeding out of phase. The secondary blood-meal component is important, although it will not be discussed further at present. However, the subpopulation hypothesis does not fully take account of the stochastic nature of population behavior, which would induce subpopulation mixing. The possibility of a distribution in cycle period is indicated in Figure 17.1 by the addition of a loop marked "delay" and in this case a proportion of survivors feed at different times. If the delay only amounts to one extra day then the number of parous individuals on any day is equal to the total with a cycle of i days that survive, plus the total with a cycle of $i + 1$ that survive:

$$M_t = P_i T_{t-i} + P_{i+1} T_{t-i-1}, \tag{17.6}$$

where T_t is the time series of total population density, M_t is the time series of parous population density, and P_i is the proportion of a population that survives and completes the cycle in i days. The distribution could be extended to any number of parameters but two will suffice for the following discussion.

If the two-parameter model is a better fit to the data than the single-parameter model, then it should significantly reduce the residual sum of squares between predicted and observed values. In standard multiple linear regression a variance ratio test may be used to determine the significance. The same test may be used in the present case but must be interpreted more empirically because of the intercorrelation between successive terms in the time series T_t. As a result of this intercorrelation there will be covariance between parameter estimates, and the point estimates of P_i and P_{i+1} will lie within an ellipsoid confidence region in parameter space. Figure 17.3 illustrates a hypothetical

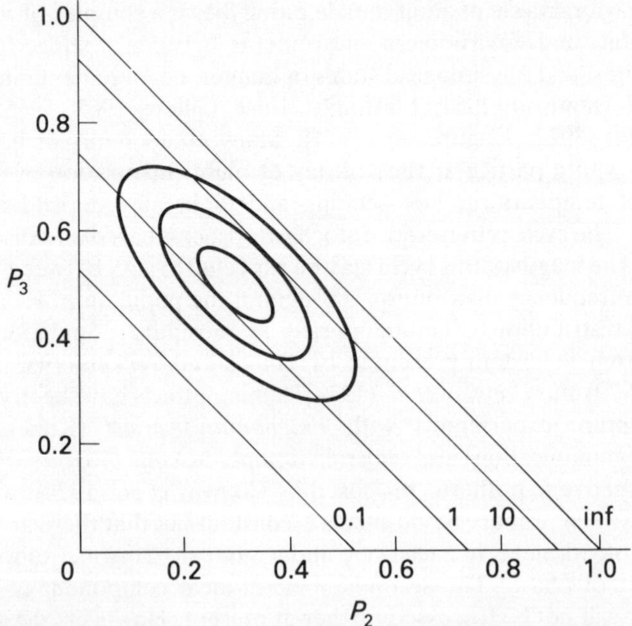

FIGURE 17.3 An example of distributed survival rate. A proportion P_2 of the population survives and completes the cycle in two days, and a proportion P_3 in three days. The longevity factor or vectorial capacity per fly V is calculated assuming an incubation period of 10 days. The joint confidence region for the survival rate estimates is ellipsoid and oriented as illustrated. The three multiples of the standard ellipse correspond to increasing levels of confidence.

example of such a region (the straight line contours of constant vectorial capacity may be ignored at present). The orientation of the ellipse is determined by the strong intercorrelations between successive terms in T_t and the slope of the major axis is approximately parallel to lines of constant $P = P_i + P_{i+1}$. P is the total survival rate per cycle and the standard error of the estimate of P is determined approximately by the minor semi-axis of the ellipse.

The basic dimensions of the ellipse are a function of the standard deviation of the residuals and confidence regions are specified by multiples of these dimensions. The first three multiples are indicated in Figure 17.3. If the sampling errors are independently and normally distributed the second and third multiples would approximate 95% and 99.8% joint confidence regions (large sample, two-tailed test). However, M_t is sampled as a binomial variable and consecutive terms in T_t are correlated so that the normality assumption may not apply unless the sample size is very large. More consecutive estimates must be employed and the worst case is determined by the Chebyshev (or Bienyamé-Chebyshev) inequality (Bard, 1974). The second and third multiples would then approximate 75% and 89% confidence regions. The true confidence regions probably lie between these two extremes and simulation studies are in progress to determine them more accurately.

From the viewpoint of pest management theory the total survival rate per cycle P is more important than its component parts. Figure 17.3 suggests that the estimate of P is insensitive to the choice of a single-parameter model with either P_i or P_{i+1}. However, the multiple parameter model reduces or removes the bias previously mentioned in relation to the least squares estimation procedure.

Examination of the two-parameter population process also illuminates certain features observed in mark–recapture and multiple age-grading experiments. Specifically, when the cycle period is distributed in time then the number of flies completing the kth cycle, or in the kth parous grade, at a specific age is even more widely distributed in time. The effect may be calculated using the multinomial theorem and this reduces to the binomial theorem when there are two parameters—P_i and P_{i+1}. Let proportion $B(k,j)$ of the population reach the kth parous grade at age j from the first blood-meal.

$$B(k,j) = \binom{k}{d} (P_i^{k-d})(P_{i+1}^{d}), \tag{17.7}$$

where $d = j - i.k$ and lies in the range (o,k). Table 17.1 illustrates how flies distribute with age and grade when $P_2 = 0.3$ and $P_3 = 0.4$. The rows and columns of matrix $B(k,j)$ may be summed in order to extract two other aspects of the population process. The sum of columns is $S(j)$, the age-specific survival and biting rate. This function has several distinct peaks that merge and decay

TABLE 17.1 Proportion of each parous grade k, biting at each age j, when $P_2 = 0.3$ and $P_3 = 0.4$. The results are rounded to three decimal places.

Parous grade (k)	Age j																
	0	1	2	3	4	5	6	7	8	9	10	11	12	13	14	15	16
0	1.0																
1	0	0	0.30	0.40	0	0											
2			0	0	0.09	0.24	0.16	0	0								
3					0	0	0.027	0.108	0.144	0.064	0	0					
4							0	0	0.008	0.043	0.086	0.077	0.026	0	0		
5									0	0	0.002	0.016	0.043	0.058	0.038	0.010	0
.											0	0	.	.	.	.	.
.													0	0	.	.	.
Proportion of any grade feeding at age j, $S(j)$	1	0	0.30	0.40	0.09	0.24	0.187	0.108	0.152	0.107	0.088	0.093	.	.	.	.	.

until they form a smooth exponential. The sum of rows is $G(k)$, the parous grade specific survival and biting rate, which may be formulated as:

$$G(k) = \sum_{j=0}^{\infty} B(k,j) = (P_i + P_{i+1})^k. \tag{17.8}$$

These results are of value when longevity is calculated later in the discussion.

Models 1 and 2 uncover certain features of the population process while leaving others obscured and two models that highlight additional features may now be mentioned. The first concerns differences between nulliparous and parous behavior.

17.6. MODEL 3

There is often doubt in mosquito sampling surveys as to whether the relative numbers of nulliparous and parous individuals caught truly reflects their prevalence in the population (Gillies, personal communication). Behavior may differ with age and lead to differences in sampling efficiency, survival, or biting rate. These differences may be taken into account by the following modification of models 1 and 2.

$$M_t = P1_i N_{t-1} + P2_j M_{t-j} \tag{17.9}$$

where $P1_i$ represents the proportion of nullipars surviving a cycle of mean length i days, multiplied by the relative sampling efficiency; and similarly for $P2_j$. In principle, both parameters may be estimated in the same manner as in the previous model. In practice, the covariance appears to be excessive and the binomial features of the sampling process compound the errors. An alternative would be to test the null hypothesis that $P1 = P2$ and $i = j$, but such a test has not yet been devised. Birley and Mutero (unpubl.) have formulated a modification for the case where a proportion of the nullipars require 2 blood-meals.

Equation (17.9) can be expanded and generalized in the form:

$$M_t = \sum_{j=0}^{\infty} S_j N_{t-j}, \tag{17.10}$$

where S_j represents the proportion of individuals that may be sampled, in any parous grade, at age j from recruitment to the nulliparous grade. This general form was discussed by Birley (1977) and used to estimate the age-specific survivorship function of an agricultural pest. In that case the insect behavior was relatively simple and it could be assumed that S_j was a monotonically decreasing function. It was not necessary to assume that the survival rate was constant with age or that new recruits were sampled with equal efficiency as older individuals. In the present case insect behavior is relatively complex and

S_j need not be monotonically decreasing (see, for example, the specific case depicted in Table 17.1). When S_j is a complex, many-peaked function it does not appear to be feasible to obtain an estimate directly from the time series of M_t and N_t.

17.7. MODEL 4

The final model addresses the problem of regulation. It is generally accepted that insect populations must be regulated by density-dependent mortality, although this need not operate primarily in the adult stage. If adult density is regulated then the survival rate should be a function of density. There are many possible models of which the simplest, and perhaps oldest, is:

$$M_t = P_u T_{t-u}^{1-b} \tag{17.11}$$

This formula (see Chapter 2) was originally used, in a modified form, to describe population changes between generations (Morris, 1959; Varley and Gradwell, 1960; May, *et al.*, 1974), although in that case there were generally insufficient data points for a precise analysis. In the present case the data concern daily changes in density during a single generation and the same restrictions do not apply. The model may be expressed as a linear regression by logarithmic transformation of the variables. A test of the null hypothesis that $b = 0$ may then indicate whether either the population or sampling process is regulated by density. If the null hypothesis cannot be rejected and $b = 0$ the estimate of survival rate obtained is the geometric mean.

17.8. DISEASE TRANSMISSION

The immediate relationship between a pest or vector population and the injury that it causes is often relatively simple (Conway, 1976). The future consequences, however, may be more complex as the original injury is either compensated for or compounded by dynamic processes. Insect vectors of disease organisms, feeding on infectious hosts, may become infected. An incubation period then ensues after which the infectious insect may transmit the disease organism to hosts on which it subsequently feeds. The incubation period may vary from a few days, in the case of some arboviruses, to a few weeks, in the case of some protozoans. Transmission may be complicated by variations in vector susceptibility and survival as well as host immunity and degree of anthropophily. However, an upper limit for potential disease transmission may be determined by counting the average number of blood-meals that the potentially infectious insect consumes during its life.

Macdonald (1952, 1973) explored the epidemiological consequences of disease transmission by insect vectors and combined his results into a single formula that has had important implications for the broad strategies of malaria control (Conway, 1977). However, Macdonald's model was too aggregated and biologically unrealistic to cope with the tactical considerations of a particular control project (Najera, 1974). Garrett-Jones (1964, 1968) provided an important step of disaggregation when he defined vectorial capacity, separating the epidemiological and ecological aspects of parasite transmission.

17.8.1. Vectorial Capacity

The vectorial capacity was defined as the average number of inoculations with a specific parasite, originating from one case of disease in unit time, that a vector population would distribute to man if all the vectors biting the case became infective (Garrett-Jones, 1964; Garrett-Jones and Grab, 1964). It is a single parameter, derived from a vector population, that predicts the future consequences of introducing one infected host into an uninfected population for one day. It measures the capacity of an ecosystem to transmit disease. Vectorial capacity is calculated as the product of the man-biting rate, the man-biting habit, and the expectation of infective life.

Implicit in the formula for vectorial capacity there lies an assumption that the average survival rate and behavior of vectors only varies slowly with time, in contrast with the vector population density. The distinction between slow and fast variables is important because it enables a further disaggregation. The vectorial capacity per fly, which may also be referred to as the expectation of infectious life or longevity factor, is a function of the slow variables and independent of the actual number of vectors biting each category of person on each day. The longevity factor may be calculated from model 1, assuming a single host species, using the formula

$$V_n = P^h/(1 - P), \tag{17.12}$$

where n is the incubation period of the disease in the insect, P is the survival rate for u days; u is the mean cycle length; and h is the number of cycles in the incubation period, i.e., an integer larger than n/u. Equation (17.12) may be compared with Garrett-Jones' (1964) formula for vectorial capacity:

$$V_g = (ma)ap^n/{-\ln p}, \tag{17.13}$$

where ma is the man-biting habit; a is the biting frequency; and p is the daily survival rate. Set $ma = 1$, $a = 1/u$, $p = P^a$ and after substitution:

$$V_g = P^h/{-\ln P}. \tag{17.14}$$

The differences in the denominators of eqns. (17.12) and (17.14) are the result of assumptions made in the underlying models. Garrett-Jones, following Macdonald (1952), assumed that the probability of blood-feeding was a continuous variable. When feeding is a discrete process that occurs during a specific phase of the diurnal cycle a summation is required in place of an integration. In this case the survival function remains relatively unspecified throughout most of the diurnal and gonotrophic cycle. Only the proportion of vectors alive at the time of biting, or sampling, is defined. The absolute difference in expectation of biting life predicted by the two formulae is always less than or equal to unity. At normal parameter values the difference is less than half a bite. In the limit as $P \to 1$ the expressions become identical. But at the other extreme, when the survival rate is zero, the underlying assumptions are illustrated more clearly. Macdonald would assume that when survival is zero there must be zero bites. The new formula assumes that there is always one bite, since this is the point from which survival is measured.

The choice of formulae may depend on assumptions that are made about behavior of the vector during the cycle. Blood-feeding may be a high-risk activity for the vector so that most of the average mortality occurs after the commencement of feeding, when the host has potentially been inoculated. In this case the discrete formula should be more appropriate. Residual pesticide may be applied to the inside of houses and vectors may rest on sprayed surfaces before or after feeding. If the vector is prevented from feeding through contact with insecticide then the continuous formula could apply. However, if the vector generally survived long enough to commence feeding or rested after feeding then the discrete formula could be more appropriate. The single-parameter calculation of the expectation of infective life, or longevity factor, may be extended to the distributed biting cycle of model 2. It may be recalled that S_j was defined as the proportion of flies surviving and biting at age j from a prior blood-meal, which is the age-specific survival and biting function when each cycle is assumed to be the same. The longevity factor may be determined for any function S_j by summing terms for j greater than n, the incubation period. Fortunately there is also a simpler approach. If P, the total survival rate per cycle, is distributed between i and $i + 1$ days, so that $P = P_i + P_{i+1}$, then

$$\frac{P^h}{1 - P} \leqslant V_n \leqslant \frac{P^g}{1 - P}, \tag{17.15}$$

where h is the number of cycles of length i in an incubation period of length n, and g is the number of length $i + 1$. Contours of constant V_n on a graph of P_i versus P_{i+1} are approximately straight lines that are also approximately parallel to lines of constant P. This is illustrated in Figure 17.3 for the case $i = 2$, $n = 10$, and $V_n = 0.1, 1.0, 10.0$, and infinity.

In Figure 17.3 the contours are superimposed on the ellipsoid confidence region containing the estimates of P_i and P_{i+1}. As previously discussed, the major axis of the ellipse is also approximately parallel to the line of constant P Therefore, the range of estimated V_n varies only along the minor ellipse axis and may be read directly from the graph.

These results may be generalized to a cycle distributed between any number of discrete time durations and it will remain true that the estimate of potential disease transmission is sensitive to only the total estimated survival rate and the mean length of the cycle.

17.9. CONCLUSIONS

This chapter reports on a search for a simple, sequential, standardized, robust procedure for determining the importance of a vector population from sample surveys. It may be viewed within the context of the multidisciplinary ecosystem research procedure discussed by Conway (1977) and Walker *et al.* (1978); see Chapter 1. The procedure starts by formulating a set of key ecological and management issues based on available knowledge about the system under study. In the present case the ecosystem consisted of interactions between populations of hosts, vectors, and disease-causing organisms. A key ecological question concerned the capacity of the vector population to survive and transmit the disease-causing organism between members of the host population. A key management question was formulated in tactical and strategic terms. At the strategic level, how do vectorial capacities differ between locations, seasons, and species? At the tactical level, how can the daily fluctuations of vector population density be filtered so as to provide an immediate estimate of the longevity factor during a sample survey? These questions stimulated research into the three processes of vector population dynamics, sampling, and estimation.

The measurement of parous rate is already a well known, standard, and relatively simple procedure in vector population surveys. The modification proposed is robust with respect to fluctuations in both recruitment and gonotrophic period.

The method is now being used as a practical experimental design for vector surveys (Birley and Rajagopalan, 1981; Birley and Boorman, 1982; Birley, Walsh and Davies, 1983; Birley, Braverman and Frish, in press). A sequential decision rule has been incorporated in order to optimize the effort devoted to each survey. Surveys may be standardized and combined so as to map the potential of disease transmission by region and by season. The result may be fed back through the multidisciplinary procedure as an answer to a key ecological question and the management of an important pest problem may, hopefully, be advanced.

ACKNOWLEDGEMENTS

This work commenced while I was research fellow at Imperial College and benefited from discussions with Drs G. R. Conway, G. A. Norton, R. M. Anderson, and Professor T. R. E. Southwood. I received much useful comment on an earlier manuscript from Drs M. T. Gillies, J. Boorman, C. Garrett-Jones, L. Molineaux, and K. Dietz. I owe my present position to Dr

M. W. Service, who has encouraged me to continue my research and this has been made possible by a generous grant from the Wellcome Trust.

REFERENCES

Armitage, P. (1971) *Statistical Methods in Medical Research (Oxford: Blackwell).*

Bard, Y. (1974) *Nonlinear Parameter Estimation* (New York: Academic Press).

Birley, M. (1977) The estimation of insect density and instar survivorship functions from census data, *J. Animal Ecol.* 46:497–510.

Birley, M. H. and P. K. Rajagopalan (1981) Estimation of the survival and biting rate of *Culex quinquefasciatus* (Diptera: Culcidae). *J. Med. Entomol.* 18(3):181–186.

Birley, M. H. and J. P. T. Boorman (1982) Estimating the survival and biting rate of haematophagous insects, with particular reference to the *Culicoides obsoletus* group (Diptera, Ceratopogonidae) in southern England. *J. Animal Ecol.* 51:135–148.

Birley, M., H., J. F. Walsh and J. B. Davies (1983) Development of a model for *Simulium, damnosum* s.l. recolonisation dynamics at a breeding site in the Onchocerciasis Control Programme Area when control is interrupted. *J. Appl. Ecol.* 20:507–519.

Birley, M. H., Y. Braverman and K. Frish (in press) A survey of the survival and biting rates of some *Culicoides* in Israel. *Environ. Entomol.* (in press).

Conway, G. R. (1976) Man versus pests, in R. M. May (ed) *Theoretical Ecology: Principles and Applications* (Oxford: Blackwell) pp257–81.

Conway, G. R. (1977) Mathematical models in applied ecology. *Nature* 269:5626, 291–7.

Conway, G. R., M. Trpis, and G. A. H. McClelland (1974) Population parameters of the mosquito *Aedes aegypti* (L.) estimated by mark-recapture in a suburban habitat in Tanzania. *J. Animal Ecol.* 43:289–304.

Dctinova, T. S. (1962) Age grouping methods in Diptera of medical importance, with special reference to some vectors of malaria. *WHO Monograph* 47:1–216.

Dyce, A. L. (1969) The recognition of nulliparous and parous *Culicoides* (Diptera: Ceratopogonidae) without dissection. *J. Aust. Entomol. Soc.* 8:11–15.

Garrett-Jones, C. (1964) Prognosis for interruption of malaria transmission through assessment of the mosquito's vectorial capacity. *Nature* 204:4964, 1173–5.

Garrett-Jones, C. (1968) Epidemiological entomology and its application to malaria. *WHO/MAL*/68.672, 1–17.

Garrett-Jones, C. (1973) Prevalence of *Plasmodium falciparum* and *Wuchereria bancrofti* in *Anopheles*, in relation to short term female population dynamics. *9th Int. Congr. Trop. Med. and Malaria, Athens* Vol. 1, pp298–300.

Garrett-Jones, C. and B. Grab (1964) The assesment of insecticidal impact on the malaria mosquito's vectorial capacity, from data on the proportion of parous females. *WHO Bull.* 31:71–86.

Gilbert, N. (1973) *Biometrical Interpretation* (Oxford: Clarendon) pp1–125.

Gilbert, N., A. P. Gutierrez, B. D. Frazer, and R. E. Jones (1976) *Ecological Relationships* (Reading: Freeman) pp1–157.

Gillett, J. D. (1962) Contributions to the oviposition-cycle by the individual mosquitoes in a population. *J. Insect Physiol.* 8:665–81.

Gillett, J. D. (1965) Analysis of the overall laying-cycle by the individual mosquitoes in a population of insects. *Proc. 12th Int. Congr. Entomol., London* pp789–90.

Gillies, M. T. (1974) Methods for assessing the density and survival of blood-sucking Diptera. *Ann. Rev. Entomol.* 19:345–363.

Gillies, M. T. and T. J. Wilkes (1965) A study of the age-composition of populations of *Anopheles gambieae* Giles and *A. funestus* Giles in north-eastern Tanzania. *Bull. Entomol. Res.*, 56:237.

Hocking, B. (1971) Blood-sucking behaviour of terrestial arthropods. *Ann. Rev. Entomol.* 16:1–26.
Macdonald, G. (1952) The analysis of equilibrium in malaria. *Trop. Disease Bull.* pp813–28.
Macdonald, G. (1973) *Dynamics of Tropical Disease: The Late G. Macdonald*, eds C. J. Bruce-Chwatt and V. J. Glanville (London: Oxford University Press).
Mattingly, P. F. (1969) *The Biology of Mosquito-Borne Disease* (London: Allen and Unwin).
May, R. M., G. R. Conway, M. P. Hassell, and T. R. E. Southwood (1974) Time delays, density dependence and single species oscillations. *J. Animal Ecol.* 43:747–70.
Southwood, T. R. E. (1978) *Ecological Methods* 2nd edn. (London: Chapman and Hall).
Wald, A. (1948) *Sequential Analysis* (New York: Wiley)
Waters, W. E. (1955) Sequential sampling in forest insect surveys. *Forest Sci.* 1:68–79.

18 The Optimal Timing of Multiple Applications of Residual Pesticides: Deterministic and Stochastic Analyses

Christine A. Shoemaker

18.1. INTRODUCTION

In the field of agricultural pest management, the most widely applied population modeling techniques have been simulation methods. However, other types of mathematical models, including those using analytical or optimization techniques, can also provide very useful tools for the analysis of pest management programs. Table 18.1 summarizes the advantages and disadvantages corresponding to each of the three types of models.

TABLE 18.1 Advantages and disadvantages of different types of models used to analyze the behavior and management of dynamic ecosystems.

Type of model	Suitability for analysis		
	Response to changes in parameters	Large number of variables	Large number of management alternatives
Analytical	Excellent	Poor	Fair
Simulation	Fair	Excellent	Poor
Optimization	Fair	Fair	Excellent

Because of the large number of variables that can be incorporated into simulation models, they can describe an ecosystem in greater detail than is possible with analytical or optimization models. The major disadvantage of simulation models is the need to recompute them to evaluate the impact of each change in parameter values or in management strategies. Because of the expense of these computations, numerical results are usually calculated for relatively few sets of parameter values. Hence, one has a limited view of the response of the pest ecosystem to changes in management strategies, weather inputs, and other factors that are described by model parameters.

Analytical model is a term usually used to describe those models for which a closed-form solution can be obtained, i.e., the results of the model can be written as an algebraic expression involving parameter values. For example, the logistic model $dN/dt = rN(1 - N/K)$ has the closed-form solution:

$$N(t) = \frac{K}{1 + c\exp(-rt)}, \tag{18.1}$$

where $c = K/N(0) - 1$. Thus we can see that an $x\%$ increase in K would be expected to increase $N(t)$ by:

$$\frac{Kx}{[1 + c\exp(-rt)] \cdot 100}$$

Hence, we have an immediate understanding of the impact of changes in the parameter K on the population density. This is one of the greatest advantages of analytical models.

Unfortunately, analytical models are seriously limited in their application to many situations because it is usually not possible to find closed-form solutions for a series of nonlinear equations with a large number of variables. For agricultural pests there has been a greater emphasis on simulation modeling, but, as this volume indicates, most work on human disease epidemiology emphasizes analytical models. I believe that the difference in emphasis here is partly a result of the fact that more variables are measurable for agriculturalists. For example, agriculturalists can allow an insect pest or pathogen to destroy a crop in an experiment. Epidemiologists could not allow a human disease epidemic to go unchecked simply for the purpose of conducting a controlled experiment. The difficulty of predicting human behavior and its influence on disease transmission also complicates attempts to measure the factors governing the effectiveness of disease control procedures. There is little point in developing a detailed model if the data supporting such a complex model do not exist. For those cases where a model of few variables is appropriate, analytical methods can be quite effective.

Optimization methods are a body of techniques that are designed to choose the best management techniques from a range of alternatives. Models using such methods are much more efficient than simulation models in evaluating a large number of management options. Numerical optimization methods can analyze systems with many more variables than is typically possible with analytical models. However, for nonlinear systems, optimization models are not able to incorporate as many dynamic variables as are commonly used in an ecosystem simulation model.

18.1.1. Computational Difficulties Associated with Optimization Methods

A number of articles have been published that include a review of the application of optimization methods to pest management (Shoemaker, 1981; Varadarajan, 1979; Wickwire, 1977; Conway, 1977; Ruesink, 1976; Jaquette, 1972). It is clear from these reviews that the major challenge in applying optimization methods to pest management analysis is developing special procedures or problem formulations that can incorporate the level of detail necessary to utilize fully our knowledge of an agricultural ecosystem and the impact of its dynamics on crop yields. For example, Regev *et al.* (1976) utilized a reduced gradient method to calculate the optimal timing of insecticide applications for control of Egyptian alfalfa weevil (*Hypera brunneipennis* (Boheman)), which has seven distinct age classes. Regev *et al.* were only able to incorporate three age classes in their analysis because of computational difficulties (Regev, personal communication). These difficulties have also been experienced with other types of optimization methods. Dynamic programming, for example, has been used to analyze the management of spruce budworm (Winkler, 1975; Stedinger, 1977). In both these studies the authors encountered insurmountable computational difficulties in their attempts to incorporate spatial heterogeneity in pest densities and age structures of the host trees. In all of the studies mentioned above, various transformations of the problems were used to reduce the dimension of the optimization problem.

In this chapter I describe a procedure for incorporating a much more detailed description of the age structure of a pest population than has been possible with previous dynamic programming analyses of multiple applications of pesticide with residual toxicity. An earlier dynamic programming study by Shoemaker (1977) does incorporate a large number of pest age classes, as well as parasitism and cultural control methods. Unfortunately, the special algorithm developed to solve this problem is not designed to consider more than one pesticide application per generation. The computational details of this earlier algorithm are described in Shoemaker (1982).

18.2. OPTIMAL TIMING OF MULTIPLE APPLICATIONS OF PESTICIDE WITH RESIDUAL TOXICITY

The problem we wish to consider is the optimal timing of pesticide applications to control a pest population of mixed ages given that the pesticide has a residual toxicity. Incorporated in this analysis is the variation among age classes in their susceptibility to pesticide toxicity and in their ability to damage the crop.

In optimization models, the state vector is used to predict the dynamics of the

TABLE 18.2 Definitions of variables. Numbers in parentheses denote the equations in which the variable is used.

Variable	Definition
a	Age in unit period (18.2).
a_M	Maximum age of pests (18.3).
$A(k,t_1,t_2)$	Total number of attacks on crop in period k (18.3).
$C^k(v^k)$	Cost of pesticide treatment (18.12,15).
$D^k(t_1,t_2)$	Amount of damage occurring in period k (18.4,5,12).
$F_k(t_1,t_2)$	Minimum cost incurred in going from period k to period N given that pesticide applications have been made in periods t_1 and $t_2 (t_1 < t_2 < k)$ (18.15,17,21).
$G(x^N)$	Value of the yield given that the crop has incurred x^N units of damage (18.10)
j	A period of time, usually used to denote the period in which a cohort has been hatched or recruited into the population.
k	A period of time, usually used to denote the current period.
N	Total number of periods in the decision process (18.12,13,17).
r	Rate of decay of pesticide residuals (18.9).
$R(j)$	Number of pests hatched or recruited into the population during period j (18.2).
$S(j,k,t_1,t_2)$	Fraction of the cohort recruited in period j that will survive pesticide toxicity up to period k given that pesticide has been applied in periods t_1 and t_2 (18.3,8,9).
T^k	Residual toxicity in period k (18.22,23).
t_1^k,t_2^k	Times of the two most recent pesticide applications before period $k (t_1 < t_2 < k)$ (18.6,7,16,21).
v^k	Binary variable, which equals one if and only if pesticide is applied in period k (18.6,7,15,16,23).
X^k	Cumulative amount of damage that has occurred to the crop by the beginning of period k (18.4,5).
$\lambda(a)$	Fraction of the pest population that survives natural causes of mortality up to age a (18.3).
$\psi(a)$	Number of attacks made in one time period by one pest of age a (18.3).
$\emptyset(a)$	Fraction of pests of age a that survive pesticide toxicity for one stage given that a pesticide has been applied during that stage (18.9).
$\xi(k)$	Expected amount of feeding throughout its lifetime to be done by an egg laid in period k.
m	Residual toxicity.

system in response to management and other factors. In order to incorporate the age structure of the population it is typical in pest management optimization studies to have the state vector include components describing the size of each age group. The difficulty with this approach is that the dimension of the state vector is so large that numerical solution of the optimization model procedure is very difficult, if not impossible.

The approach we have developed differs from most earlier approaches in that the state vector is used to describe the times of previous pesticide applications (t_1,t_2) *rather than the sizes of population age groups*. In each period k, we use t_1 and t_2 to compute the age structure of the pest population in the current time period and the amount of damage that will occur. The number of individuals in age group a in time period k is:

$$R(k-a)S(k-a,k,t_1,t_2)\lambda(a), \tag{18.2}$$

where R,S and λ are defined in Table 18.2. Equation (18.2) is simply the original number of individuals in the cohort R multiplied by the pesticide survival factor S and a natural survival factor λ. Defining $\psi(a)$ as the rate of attack of pests of age a on the crop, we can determine A, the total rate of attack on the crop during period k, as:

$$A(k,t_1,t_2) = \sum_{j=L}^{k} R(j)S(j,k,t_1,t_2)\Psi(k-j)\lambda(k-j), \tag{18.3}$$

where $L = max(0,k - a_M)$ and a_M is the maximum age the pest can reach.

The amount of damage that can occur in a given period depends upon both the amount of attack and upon the previous damage. Given the definition of X^k in Table 18.2, the amount of damage occurring in period k is

$$D^k[A(k,t_1,t_2),X^k].$$

The cumulative amount of damage by the beginning of period $k + 1$ is

$$X^{k+1} = X^k + D^k[A(k)t_1,t_2),X^k].$$

For simplicity assume that D^k is independent of X^k and denote $D^k[A(k,t_1,t_2),X^k]$ as $D^k(t_1^k, t_2^k)$. Then

$$X^k = \sum_{m=1}^{k-1} D^m(t_1^m,t_2^m).$$

The optimization procedure for computing the deterministic optimal solution given for a more general form of D^k is discussed in Shoemaker (1979).

18.3. DETERMINISTIC DECISION MODEL

18.3.1. Decision Variables

The decision variables in the decision model are v^k, $k = 1, \ldots ,N$, which are binary variables describing whether or not pesticide has been applied. The function $C^k(v^k)$ is the cost of applying an insecticide. The state vector (t_1^k,t_2^k) changes as a function of the decision vector. Since t_2^{k+1} is defined as the most recent time of pesticide application on or before period k,

$$t_2^{k+1} = \begin{cases} t_2^k & \text{if } v^k = 0 \\ k & \text{if } v^k = 1 \end{cases} \tag{18.6}$$

Similarly,

$$t_1^{k+1} = \begin{cases} t_1^k & \text{if } v^k = 0 \\ t_2^k & \text{if } v^k = 1 \end{cases} \tag{18.7}$$

18.3.2. Pesticide Survival Function $S(j,k,t_1,t_2)$

The combination of eqns. (18.2) to (18.7) enable us to compute the impact of pesticide applications on the current and future crop damage caused by a pest population. However, the entire procedure depends upon being able to calculate the survival from pesticides as a function of the four variables: j, k, t_1, and t_2. This representation of the survival function is based on the following assumptions:

(1) The mortality rate from pesticides is a percentage that is independent of the pest density. The cumulative survival rate is the product of the survival rate during each time period.
(2) The toxicity of a pesticide decays exponentially after application, i.e., toxicity ratio $= \exp[-r(k - t_2)]$, where k is the current time period, t_2 is the time of application, and r can be constant or a function of time.
(3) The natural survival function $\lambda(a)$ is independent of pest density and of pesticide applications.
(4) None of the mortality factors is influenced by random factors such as weather.

Given (1), (2) and the definition of $\emptyset(k - j)$ in Table 18.1, the survival rate in period k from pesticide applied in period t_2 is

$$1 - \emptyset(k - j) \exp[-r(k - t_2)]$$

for a member of the cohort recruited in period j. If a single pesticide application has been made at time t_2, the cumulative survival from pesticide up to period k for a cohort entering the population in period j is

$$S(j,k,0,t_2) = \prod_{\tau=M_2}^{k} \{1 - \emptyset(\tau - j)\exp[-r(\tau - t_2)]\}, \tag{18.8}$$

where $M_2 = \max(t_2,j)$ and $t_2 < k$. The function M_2 is necessary because pesticide mortality to members of cohort j occurs only after their recruitment and after the pesticide application. Hence, $S(j,\tau,0,t_2) = 1$ if $\tau < \max(t_2,j) = M_2$.

If two pesticide applications have been made at times t_1 and t_2, then the survival rate is assumed to be the product of the survival rates from each of the applications. Hence, the survival rate within a single period k for a cohort j is:

$$\{1 - \emptyset(k - j)\exp[-r(k - t_1)\}\{1 - \emptyset(k - j)\exp[-r(k - t_2)]\}.$$

The cumulative survival is then

$$S(j,k,t_1,t_2) = \prod_{t=M_1}^{k} \{1 - \emptyset(t - j)\exp[-r(t - t_1)]\} \times \prod_{\tau=M_2}^{k} \{1 - \emptyset(\tau - j)\exp[-r(\tau - t_2)]\} \tag{18.9}$$

for $0 < t_1 < t_2 < k$, $M_1 = \max(t_1,j)$ and $M_2 = \max(t_2,j)$. $S(j,k,0,0) = 1$, since $t_1 = t_2 = 0$ indicates that no pesticide has been applied.

In theory, the survival from pesticides depends upon all of the pesticide applications preceding period k. However, pesticide survival rates are usually quite low and the probability of surviving two applications is typically very small. As a result, for most combinations of values of r and $\emptyset(a)$, the expected percentage survival resulting from exposure to pesticide can be approximated quite closely by considering only the two most recent applications. Numerical calculations discussed in Shoemaker (1979) indicate that determining S on the basis of only the last two applications is an adequate approximation in most situations.

18.3.3. Criteria for Determining Optimal Decisions

In order to calculate the optimal solution we must define a criterion by which we measure the success of a particular alternative. The most typical measure of success is some indication of net income, i.e., gross income minus costs. For example, if yield loss is linearly related to total pest attack and if price ρ is independent of the quantity produced, the value of the crop to the farmer is

$$G(x^N) = \rho(K - gx^N), \tag{18.10}$$

where K is the yield that would have been achieved in the absence of pest damage. In the more general case, G can be a nonlinear function of x. Such a representation can arise because x has a nonlinear impact on yield or because the farmer's criterion of success includes not only expected income, but also some measure of the risk associated with a management decision. In this paper we consider a criterion function G, which has the form of eqn. (18.10). The more general deterministic case is considered in Shoemaker (1979).

The optimal decision variables $\{v^k | k = 1, \dots, N\}$ are those that maximize $G(x^N)$ minus costs, or, equivalently,

$$\rho(K - gx^N) - \sum_{i=1}^{k} C(v^k) - f. \tag{18.11}$$

The second term in eqn. (18.11) incorporates only those costs that are related to pest management decisions. The constant f refers to the costs that are independent of pest management decisions and of the size of the yield. By substituting eqn. (18.5) for x^N in eqn. (18.11) and by omitting the constant terms ρ, K and f, which have no impact on the optimal decision, we obtain the following criterion function:

$$\max_{\substack{v^k \\ k=1,\dots,N}} \left[-p \sum_{i=1}^{N} D^k(t_1^k, t_2^k)\right] - \left[\sum_{i=1}^{k} C(v^k)\right], \tag{18.12}$$

where $p = \rho g$. Since the maximum of $(-z)$ equals minimum of z, eqn. (18.12) becomes

$$\min_{\substack{v^k \\ k=1,\dots,N}} \left[p \sum_{k=1}^{N} D^k(t_1^k, t_2^k)\right] + \left[\sum_{k=1}^{N} C(v^k)\right]. \tag{18.13}$$

Equation (18.13) is the minimum cost (crop damage plus pest management costs) that will occur over the N time periods.

There are 2^N possible combinations of values for $\{v^1, \ldots, v^N\}$, so it is not efficient to compute eqn. (18.13) for each possible value of $\{v^k\}$. For this reason we have used a dynamic programming procedure to compute the best choice of decision variables. The first step in the procedure is to define a function $F_k(t_1,t_2)$ as the minimum cost that will occur between period k and period N given that the last two pesticide applications were made at times t_1 and t_2. Mathematically, this can be expressed as

$$F_k(t_1^k,t_2^k) = \min_{\substack{v^m \\ m=k, \ldots, N}} \left[p \sum_{m=k}^{N} D^m(t_1^m,t_2^m)\right] + \left[\sum_{m=k}^{N} C(v^m)\right]. \qquad (18.14)$$

By using Bellman's principle of optimality (Bellman, 1957) and by substituting the definition of F_{k+1} for the last $N - k - 1$ terms of eqn. (18.14), we obtain the recursive relationship

$$F_k(t_1^k,t_2^k) = \min_{v_k}[pD^k(t_1^k,t_2^k) + C(v^k) + F_{k+1}(t_1^{k+1},t_2^{k+1})], \qquad (18.15)$$

where

$$t_1^{k+1} = (1 - v^k)t_1^k + v^k t_2^k \qquad t_2^{k+1} = (1 - v^k)t_2^k + v^k k. \qquad (18.16)$$

Equation (18.16) is equivalent to eqns. (18.5) and (18.6). Since the crop is harvested in period N and no further damage can occur after this time:

$$F_{N+1}(t_1^{N+1},t_2^{N+1}) = 0. \qquad (18.17)$$

Since the values of F_k are not given, there are unknown values on both sides of eqn. (18.15), and hence the equations cannot be solved directly. The procedure used in dynamic programming is to compute F_k recursively starting with the substitution of eqn. (18.17) in eqn. (18.15). This substitution yields

$$\begin{aligned} F_N(t_1^N,t_2^N) &= \min_{v_k}[pD^N(t_1^N,t_2^N) + C(v^N) + 0] \\ &= pD^N(t_1^N,t_2^N). \end{aligned} \qquad (18.18)$$

Now this relationship can be substituted into eqn. (18.15) to determine F_{N-1}. This procedure is repeated for each k by moving backward from $k = N - 1$ to $k = 1$.

The advantage of this procedure is that it can incorporate a large number of age classes into a very inexpensive optimization procedure that can be solved on a small computer or hand-held calculator. This latter compatibility is important since many extension agents and pest management consultants only have access to small microcomputers.

TABLE 18.3 Recruitment and oviposition rates for Egyptian alfalfa weevil (adapted from Gutierrez *et al.* 1976).

Time period j	Recruitment rate $R'(j)$	Age, a	Oviposition rate $\psi'(a)$
1	0.0125	1	164
2	0.0125		
3	0.0125	2	304
4	0.0115		
5	0.0160	3	350
6	0.0275		
7	0.0675	4	350
8	0.1600		
9	0.0600	5	347
10	0.0600		
11	0.0600	6	322
12	0.0850		
13	0.1900	7	301
14	0.0800		
15	0.0450	8	266
16	0.0400		
17	0.0300	9	231
18	0.0200		
		10	182

It should be noted that without approximation for pesticide survival (eqn. 18.9), it is not computationally feasible to calculate optimal pesticide timing over a large number of decision periods. Birley (1979) points out that the use of a state vector describing the preceding sequence of control decisions has at least 2^N possible values. The application discussed in the following section analyzes the control of Egyptian alfalfa weevil over 28 time periods. Without approximation for S, this problem would have required the solution of the dynamic programming problem for 268 million (2^{28}) values of the state vector. A problem of such a size is computationally infeasible.

18.3.4. Application

In order to illustrate the procedure discussed above, I applied the method to data for Egyptian alfalfa weevil published by Gutierrez *et al.* (1976). The rates at which female weevils enter the field are given in Table 18.3. In this application the action of the insecticide is directed against adult weevils, which are assumed to do no damage to the alfalfa. The purpose of the pesticide application is to prevent the deposition of eggs, which, upon maturation, can do considerable damage. Hence, the rate of attack $\psi(a)$ is the number of eggs laid per day by a female of age a. The values of $\psi(a)$ are given in Table 18.3. Let

TABLE 18.4 Optimal timing of pesticide applications[a].

m	Ratio $\gamma\beta/c$				
	0.001	0.002	0.003	0.006	0.012
0.1	none (1.68)[b]	14 (2.69)	9,14 (3.46)	9,13,17 (4.66)	8,11,14,18 (6.19)
0.3	none (1.68)	13 (2.60)	9,14 (3.26)	9,13,17 (4.38)	9,13,17 (5.75)

[a] $A'(j)$ and $\psi'(a)$ are given in Table 18.2.
[b] Figures in parentheses are minimum cost $F_1(0,0)$ for $\beta = 0.1$, $c = 1$, and $\emptyset(a) = 0.9$.

$\xi(k)$ be the expected amount of feeding throughout its lifetime to be done by an egg laid on day k. The function $\xi(k)$ is the sum of the amount eaten by each age class weighted by the probability of surviving that age class. Then the total damage is

$$\begin{aligned} D^k(t_1^k,t_2^k) &= \xi(k)A(t_1^k,t_2^k) \\ &= \xi(k) \sum_{j=L}^{k} R(j)\lambda(k-j)S(j,k,t_1^k,t_2^k)\psi(k-j), \end{aligned} \tag{18.19}$$

where $L = \max(0,k - a_M)$.

In order to evaluate the results under a range of parameter values, the input functions were parametrized in the following: $R(j) = \gamma R'(j), C(v^k) = cC'(v^k)$, and $p = \beta p'$ for fixed values of R', C', and p'. The optimal policy will not change with changes in the values of γ, c, and β as long as the ratio $\gamma c/\beta$ stays constant (Shoemaker, 1979). The optimal cost, however, will increase in proportion to c, i.e., for two sets of parameters, $\gamma^1\beta^1/c^1 = \gamma^2\beta^2/c^2$

$$F_1^2(0,0) = (c_2/c_1)F_1^1(0,0), \tag{18.20}$$

where the superscripts on F_1 correspond to the parameter set upon which the optimal cost is based. The optimal costs given in Table 18.4 are based on a cost of insecticide treatment $c = 1 = C'$. Hence the minimum cost is presented in multiples of the cost of an insecticide treatment.

In Table 18.5 the cost of the optimal management policy for $m = 1$ and $\gamma\beta/c = 0.003$ is compared with the costs for alternative strategies. The costs for all of the options, including the optimal one, were calculated by forward simulation of the equations for R, S, A, and D. The fact that the cost computed for the optimal policy (eqns. 18.9) and (18.14) given in Table 18.5 is the same as that computed by backward dynamic programming (Table 18.4) and that this cost is lower than all of the other alternatives examined support the accuracy of the computer program and the validity of the optimization procedure.

TABLE 18.5 Costs of alternative pesticide application schedules as calculated by simulation.[a]

Single application		Multiple applications	
Times	Cost	Times	Cost
None	5.03	5, 10	4.94 c
1	5.94	8, 14	3.33 c
5	5.65	9, 14[b]	3.26 c
9	4.29	10, 15	3.30 c
11	4.03	9, 13, 17	3.69 c
13	3.40	8, 11, 14, 18	4.45 c
15	3.59	2, 7, 13, 18	4.92 c
17	4.18	3, 6, 9, 12, 15, 18	6.27 c

[a] $A'(j)$ and $\psi'(a)$ given in Table 18.2; $m = 0.4$, $\gamma\beta/c = 0.003$, $\beta = 0.1$, $c = 1$, and $\emptyset(a) = 0.9$.
[b] Policy computed to be optimal for these parameter values.

Although the results presented in Table 18.4 indicate how optimal policies change with changes in parameter values, it is perhaps more important in practical applications to measure the actual performance of policies selected by the optimization procedure on the basis of incorrect parameter values. Table 18.6 examines the impact on costs and policies that will result from incorrect estimates of the parameter values. For example, assume that the total adult female recruitment, γ, has been underestimated with the result that it is assumed that $\gamma\beta/c = 0.002$, when the actual value of the ratio is 0.003. Assume as in example 2 in Table 18.6, that $m = 0.3$ and that $A'(j)$ and $\psi'(a)$ are given by the values in Table 18.3. Columns II and III for example 2 of Table 18.6 give the timings and costs that will be computed as optimal based on these parameter values. Column IV gives the actual cost (3.40 c) that will be incurred if the policy is to apply pesticide only in period 13. The values in column IV are taken from Table 18.5. The additional cost incurred because parameter values were not known exactly is $(3.40 - 3.26)c$, a 4% increase. Therefore, an original error of 33.3% in estimating γ resulted in an increase in cost of about 4%.

Examining the results for the other examples in Table 18.6 we see that generally an error of x% in value of $\gamma\beta/c$ or in m, results in a calculated optimal policy which has an actual cost that deviates by far less than x% from the true minimum cost. Hence, the policies generated by the optimization are quite good even when the parameter values upon which they are based are not known exactly.

18.4. DECISION MAKING IN A STOCHASTIC ENVIRONMENT

In many cases, the rate at which pesticides lose their toxicity depends upon

TABLE 18.6 Changes in optimal policies due to use of incorrect parameter values.†

Sample number	I (Incorrectly) Estimated parameter values	II Optimal solution for estimated parameter values: Application[b] timing	III Calculated[b] costs	IV Actual cost[c]	V Cost difference from true optimum (%)[d]
1[a]	$m = 0.3,\ \gamma\beta/c = 0.003$	9, 14	3.26 c	3.26 c	—
2	$m = 0.3,\ \gamma\beta/c = 0.002$	13	2.60 c	3.40 c	4
3	$m = 0.3,\ \gamma\beta/c = 0.006$	9, 13, 17	4.28 c	3.69 c	13
4	$m = 0.1,\ \gamma\beta/c = 0.012$	8, 11, 14, 18	6.19 c	4.45 c	37

†$A'(j)$ and $\psi'(a)$ are assumed to have the values given in Table 18.2 for all examples.
[a]Correct parameter values are assumed to be those given in Table 18.4.
[b]Values from Table 18.4.
[c]Values from Table 18.5.
[d]Percentage = (actual cost $-3.26\ c) \times 100/3.26\ c$.
[e]True optimum.

random factors such as weather. Rainfall, which washes pesticide off the leaves, is especially important. If such stochastic events are significant, the randomness of their occurrence should be incorporated into the decision analysis.

In order to solve the stochastic decision problem, it is necessary to incorporate into the state vector a variable T^k, which describes the level of pesticide toxicity still remaining on the foliage in period k. In the deterministic case T^k could be computed directly from t_1 and t_2; in the stochastic case T^k depends upon the occurrence of previous rainfall events as well as on t_1 and t_2. Thus, the optimal solution is a function not only of t_1^k and t_2^k but also of T^k.

The dynamic programming equation that needs to be solved to find the optimal policy is:

$$F_k(t_1^k,t_2^k,T^k) = \min_{v^k} \{pD[A(k,t_1,t_2,T)] + C(v^k) + \sum \sigma_i^{k+1} F_{k+1}(t_1^{k+1},t_2^{k+1},\omega_i)\}, \tag{18.21}$$

where A is computed as in eqn. (18.3) except that $S(j,k,t_1,t_2)$ is also a function of T. The sum inside the braces of eqn. (18.21) is the expected value of F_{k+1} given that T^{k+1} has the conditional probability distribution

$$\text{prob}\{T^{k+1} = \omega_i \big| t_1^k,t_2^k,v^k,T^k\} = \sigma_i^{k+1}, \qquad i=1, \ldots M. \tag{18.22}$$

Further details of the state vector equations and solution procedures are given in Shoemaker (1983).

TABLE 18.7 Optimal decision rules for periods 13, 15, and 17. The numbers in this table are the maximum residual toxicity T that would justify a pesticide application in period k. See text for explanation.

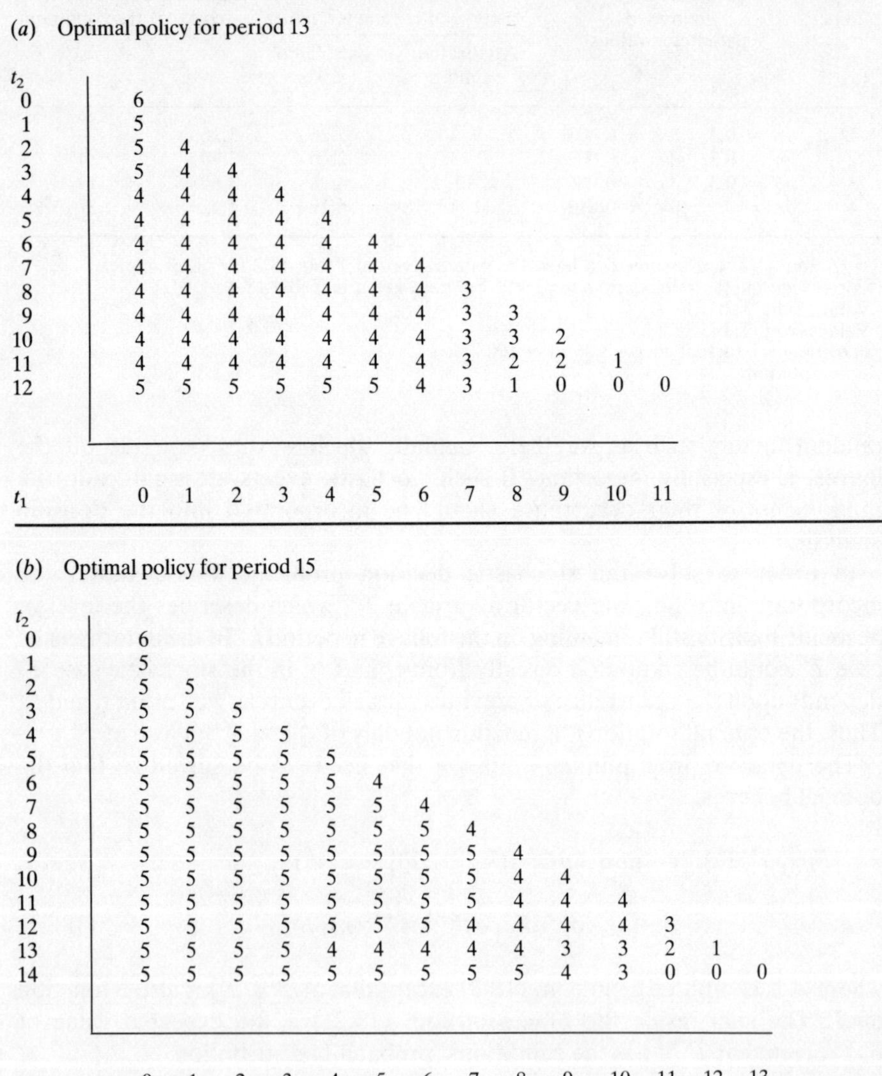

(*a*) Optimal policy for period 13

t_2 \ t_1	0	1	2	3	4	5	6	7	8	9	10	11
0	6											
1	5											
2	5	4										
3	5	4	4									
4	4	4	4	4								
5	4	4	4	4	4							
6	4	4	4	4	4	4						
7	4	4	4	4	4	4	4					
8	4	4	4	4	4	4	4	3				
9	4	4	4	4	4	4	4	3	3			
10	4	4	4	4	4	4	4	3	3	2		
11	4	4	4	4	4	4	4	3	2	2	1	
12	5	5	5	5	5	5	4	3	1	0	0	0

(*b*) Optimal policy for period 15

t_2 \ t_1	0	1	2	3	4	5	6	7	8	9	10	11	12	13
0	6													
1	5													
2	5	5												
3	5	5	5											
4	5	5	5	5										
5	5	5	5	5	5									
6	5	5	5	5	5	4								
7	5	5	5	5	5	5	4							
8	5	5	5	5	5	5	5	4						
9	5	5	5	5	5	5	5	5	4					
10	5	5	5	5	5	5	5	5	4	4				
11	5	5	5	5	5	5	5	5	4	4	4			
12	5	5	5	5	5	5	5	4	4	4	4	3		
13	5	5	5	5	4	4	4	4	4	3	3	2	1	
14	5	5	5	5	5	5	5	5	5	4	3	0	0	0

(*c*) Optimal policy for period 17

t_2 \ t_1	0	1	2	3	4	5	6	7	8	9	10	11	12	13	14	15
0	6															
1	5															
2	5	4														
3	5	4	4													
4	5	4	4	4												
5	5	4	4	4	4											
6	4	4	4	4	4	4										
7	4	4	4	4	4	4	4									
8	4	4	4	4	4	4	4	4								
9	4	4	4	4	4	4	4	4	4							
10	4	4	4	4	4	4	4	4	4	3						
11	4	4	4	4	4	4	4	4	4	4	3					
12	4	4	4	4	4	4	4	4	4	4	4	3				
13	4	4	4	4	4	4	4	4	4	4	4	3	3			
14	5	5	5	5	5	5	5	4	4	4	4	4	3	2		
15	5	5	5	5	5	5	5	5	5	5	4	4	3	2	1	
16	6	6	6	6	6	6	6	6	5	5	5	4	3	0	0	0

The optimal policy is computed for all values of k. Table 18.7 gives the optimal decision rules for periods $k = 13, 15$, and 17. The values given are the maximum levels of toxicity $T^{\max}$ that would justify an additional pesticide application in period k. For example, assume pesticide has been applied in periods 9 and 3, and we need to decide if pesticide should be applied again in period 13. Thus, $t_1^{13} = 3$, $t_2^{13} = 9$, and (from Table 18.7a) $T^{\max} = 4$. Hence, if the observed toxicity is $\leqslant 4$, pesticide should be applied. If the observed toxicity is > 4, no additional pesticide applications should be made in period 13. Therefore, we have a well defined decision rule based on the observations of the current level of toxicity in the field. In practice, the toxicity itself would not be measured; rather it would be estimated based upon observed rainfall events.

The decision rules given in Table 18.7 are based on a number of parameters. The values of $R(j)$ and $\psi(a)$ used for the stochastic decision analysis are those given in Table 18.3. To illustrate the impact of random occurrences on decisions, we replaced the linear cost function of eqn. (18.10) with a cost relationship that is assumed to be a quadratic function of the damage occurring in each period. The toxicity in period $k + 1$ is assumed to be

$$T^{k+1} = m_k \max[T^k, \emptyset(0)v^k], \tag{18.23}$$

where m_k is a random variable representing the fraction of the pesticide toxicity remaining after one period. The values of m_k depend upon rainfall events as given in Table 18.8. The last term $\emptyset(0)v^k$ is the toxicity present because of an application in period k. If $v^k = 0$ because no application has been made, then $T^{k+1} = m_k T^k$.

TABLE 18.8 Relationship between rainfall events and percentage residual m_k.

Event	Probability	Percentage residual m_k
Heavy rain	0.25	0.1
Light rain	0.50	0.5
No rain	0.25	0.9

Although the pesticide application policy has fixed operating rules, the implemented application schedule will depend upon the sequence of rainfall events that actually occur. Table 18.9 illustrates the variation in the timing of pesticide use that results from differences in rainfall patterns. Cases I, II, and III represent three different rainfall patterns and the resulting residual rates m_k. The column labeled *application schedule* indicates the policy that would be obtained by following the decision rules given in Table 18.7. "Spray" denotes an application and a blank indicates no application, i.e., $v(t_1,t_2,T) = 0$. Note that among the three cases, the number as well as the timing of the applications changes. These situtions were artifically generated so that in all cases the average of all $m_k = 0.5$, the predicted mean. During an growing season the average rainfall could vary considerably from the predicted mean. Therefore, the pattern of pesticide application would be expected to exhibit more variation than exhibited among cases I, II, and III.

In order to examine the impact of extreme events, in Table 18.10 we have given the optimal application schedules associated with no rain (case V), and with heavy rains every week (case IV), obtained by following the decision rules given in Table 18.7. We see that the application schedules range from two applications in case V to five applications in case IV. Hence, no sequence of m_k would result in more than five or less than two applications.

The optimal decision for each case is based on the prediction of future events and on the observations of events that have already occurred. For example, in period 3 in case I, the decision is based upon the fact that in week 1 there was no rain, in week 2 there was a light rain, and that the probability for rain in week 3 and later is given by the distribution in Table 18.8. *A major advantage of stochastic dynamic programming is that it can incorporate observations of the current value of the state vector, as well as the randomness associated with future events.*

The example in Table 18.11 shows the optimal policy when a light rain occurs every week. From the stochastic decision rule given in Table 18.7, the optimal policy is to make pesticide applications in periods 8, 12, and 15. Although the average event ($m_k = 0.5$) occurs in each period, optimal decisions are based

TABLE 18.9 Application schedules for three possible rainfall patterns over a 28-week season ($\emptyset \equiv 0.5$, $\beta = 0.01$, $\gamma = 0.12$, $c = 0.1$).

	Case I		Case II		Case III	
Period	Percentage residual m_k	Application schedule	Percentage residual m_k	Application schedule	Percentage residual m_k	Application schedule
1	0.90		0.50		0.90	
2	0.50		0.90		0.90	
3	0.50		0.50		0.50	
4	0.90		0.10		0.10	
5	0.50		0.10		0.50	
6	0.90		0.50		0.50	
7	0.50		0.50		0.10	
8	0.50	Spray	0.10	Spray	0.10	Spray
9	0.50		0.50		0.10	
10	0.50		0.90		0.50	Spray
11	0.10	Spray	0.90		0.90	
12	0.50		0.50		0.90	
13	0.10	Spray	0.10	Spray	0.50	
14	0.10		0.90		0.50	Spray
15	0.50	Spray	0.90		0.90	
16	0.10		0.50		0.10	Spray
17	0.50		0.10	Spray	0.10	
18	0.10		0.50		0.10	
19	0.10		0.90		0.50	
20	0.10		0.90		0.50	
21	0.50		0.50		0.90	
22	0.90		0.10		0.90	
23	0.50		0.10		0.50	
24	0.50		0.50		0.10	
25	0.90		0.10		0.90	
26	0.50		0.90		0.50	
27	0.90		0.50		0.50	
28	0.90		0.50		0.50	
AVG	0.50		0.50		0.50	

upon the assumption that in the future, rainfall events will be randomly distributed, as in Table 18.8.

It is informative to compare case VI to the deterministic optimal policy. With the assumption that a light rain will fall every week in the future as well as in the past, the policy computed by the deterministic optimization is to make applications in periods 8, 10, 13, and 15. Although the rainfall events in case VI are identical to those assumed in the deterministic case, the application schedules are distinctly different because the stochastic policy is based on the assumption that there is also a chance for either heavy rain or no rain in the future.

TABLE 18.10 Optimal application schedules for two extreme rainfall patterns over a 28-week season ($\emptyset \equiv 0.5$, $\beta = 0.01$, $\gamma = 0.12$, $c = 0.1$).

	Case IV		Case V	
Period	Percentage residual m_k	Application schedule	Percentage residual m_k	Application schedule
1	0.10		0.90	
2	0.10		0.90	
3	0.10		0.90	
4	0.10		0.90	
5	0.10		0.90	
6	0.10		0.90	
7	0.10		0.90	
8	0.10	Spray	0.90	Spray
9	0.10		0.90	
10	0.10	Spray	0.90	
11	0.10		0.90	
12	0.10		0.90	
13	0.10	Spray	0.90	
14	0.10		0.90	
15	0.10	Spray	0.90	
16	0.10		0.90	Spray
17	0.10	Spray	0.90	
18	0.10		0.90	
19	0.10		0.90	
20	0.10		0.90	
21	0.10		0.90	
22	0.10		0.90	
23	0.10		0.90	
24	0.10		0.90	
25	0.10		0.90	
26	0.10		0.90	
27	0.10		0.90	
28	0.10		0.90	
AVG	0.10		0.90	

18.5. DISCUSSION AND CONCLUSIONS

This chapter was introduced with a discussion of the integration of analytical, simulation, and optimization models in population management. Consider the relationship of analytical and simulation techniques to the optimization procedure developed above.

Mathematical analysis of equations was essential in the development of the optimization procedure. For example, the representation of $S(j,k,t_1,t_2)$ and $A(t_1,t_2)$ were developed by a mathematical analysis of the population equations. This analysis was based on the assumption that mortality rates are density independent. The recognition that the optimal policies would not

TABLE 18.11 Optimal application schedule for constant $m_k \equiv 0.5$ over a 28-week season ($\emptyset \equiv 0.5$, $\beta = 0.01$, $\gamma = 0.12$, $c = 0.1$).

Period	Case VI Percentage residual m_k	Stochastic optimization application schedule	Deterministic optimization application schedule
1	0.50		
2	0.50		
3	0.50		
4	0.50		
5	0.50		
6	0.50		
7	0.50		
8	0.50	Spray	Spray
9	0.50		
10	0.50		Spray
11	0.50		
12	0.50	Spray	
13	0.50		Spray
14	0.50		
15	0.50	Spray	Spray
16	0.50		
17	0.50		
18	0.50		
19	0.50		
20	0.50		
21	0.50		
22	0.50		
23	0.50		
24	0.50		
25	0.50		
26	0.50		
27	0.50		
28	0.50		
AVG	0.50		

change with perturbations in the values of γ, β, and c(as long as $\gamma\beta/c$ is constant) was based upon analyses of the mathematical equations describing cost and damage. Understanding the significance of $\gamma\beta/c$ is important because it greatly reduces the number of times the optimization model needs to be solved in a sensitivity analysis of the parameter values.

To most effectively examine the validity of an optimization procedure, we should test the "optimal" policies with a simulation model of the specific pest ecosystem. Table 18.5 illustrates the use of a simulation model to check the numerical acuracy of the optimization model. It is also possible to use a simulation model to evaluate more complicated relationships than those incorporated in the optimization model. For example, the optimization procedure developed earlier is based on several assumptions such as density independence. These assumptions may, in general, be reasonable for managed agricultural systems, but in some special cases they may not be satisfied. A simulation model of the population dynamics can be developed to incorporate the detail and nonlinearities not included in the optimization model and thereby to reduce the need for simplifying assumptions. As mentioned earlier, it is usually not computationally feasible to solve the simulation model for each of the 268 million ($=2^{28}$) possible pesticide application schedules. However, we can easily compute the simulation model for a few management alternatives, including the one that is computed to be best by the optimization model. Using both optimization and simulation models helps us screen for policies that can eventually be tested in the field.

ACKNOWLEDGMENTS

This research was supported in part by grants from the US Environmental Protection Agency (EPA CR-806277-020) and from the Forest Service of the US Department of Agriculture (E87-8397). The results presented in Tables 2–10 are based on computer programs written by King Au and Daniel Pei. I wish to express my gratitude to the agencies and programmers who made this research possible.

REFERENCES

Bellman, R. E. (1957) *Dynamic Programming* (Princeton, NJ: Princeton University Press).

Birley, M. H. (1979) The theoretical control of seasonal pests—a single species model. *Math. Biosci.* 43:141–57.

Conway, G. R. (1977) Mathematical models in applied ecology. *Nature* 269:291–7.

Gutierrez, A. P., J. B. Christensen, C. M. Merritt, W. B. Lowe, C. G. Summers, and W. R. Cothran (1976) Alfalfa and the Egyptian alfalfa weevil (*Coleoptera: Curculionidae*). *Can. Entomol.* 108:635–48.

Jaquette, D. L. (1972) Mathematical models for controlling growing biological populations: a survey. *Operations Res.* 20:1142–51.

Regev, U., A. P. Gutierrez, and G. Feder (1976) Pests as a common property resource: a case study of alfalfa weevil control. *Am. J. Agric. Econ.* 58:185–95.

Ruesink, W. G. (1976) Status of the systems approach to pest management. *Ann. Rev. Entomol.* 21:27–44.

Shoemaker, C. A. (1977) Pest management models of crop ecosystems, in C. A. Hall and J. W. Day (eds) *Ecosystem Modeling* (New York: Wiley). 545–574.

Shoemaker, C. A. (1979) Optimal timing of multiple applications of pesticides with residual toxicity. *Biometrics* 35:803–12.

Shoemaker, C. A. (1981) Applications of dynamic programming to pest management. *J. IEEE Automatic Control* 26:1125–1132.

Shoemaker, C. A. (1982) Optimal integrated control of pest populations with age structure. *Operations Res.* 30:40–61.

Shoemaker, C. A. (1983) Optimal timing of pesticide applications with stochastic rates of residual toxicity. School of Civil and Environmental Eng., Cornell University, technical report

Stedinger, J. R. (1977) *Spruce Budworm Management Models* PhD Thesis, Department of Applied Physics and Engineering, Harvard University, Cambridge, Massachusetts.

Varadarajan, R. V. (1979) *Applications of Modern Control Theory to the Management of Pest Ecosystems* PhD Dissertation, Department of Electrical Engineering and Entomology, Michigan State University, East Lansing.

Wickwire, (1977) Mathematical models for the control of pests and infectious diseases: A survey. *Theor. Pop. Biol.* 11:182–238.

Winkler, C. (1975) *An Optimization Technique for the Budworm Forest–Pest Model* Research Memorandum RM-75/11. (Laxenburg, Austria: International Institute for Applied Systems Analysis).

19 Optimal Chemical Control of the Greenhouse Whitefly

I The Optimization Procedure

Hisham El-Shishiny

19.1. INTRODUCTION

In recent years the cultivation of vegetables, fruits, and flowers in greenhouses has developed dramatically, taking advantage of the low price of energy. An important constraint, however, has been the damage caused by pests. Because of the fragile, artificial environment of greenhouses, minor initial contamination by insect pests exposes the crop to severe hazard, with accompanying potential yield losses.

Pest control practice is essentially empirical and commonly leads to a massive use of chemical insecticides, which in the long run greatly increases costs and may build up a pollutant contamination of the greenhouse. Additionally, the recent rise in the price of energy, resulting in increased costs of heating and insecticides, has produced a major change in the economics of greenhouse production. For these reasons optimization techniques are required to improve greenhouse management, particularly pest control operations.

A number of optimization techniques have been previously used for pest population models. Vincent (1972) and Leitman (1972) employed variational calculus while Watt (1963) and Shoemaker (1973) used dynamic programming methods. In the first part of this chapter I describe the use of a nonlinear programming technique to optimize a distributed parameter model for the greenhouse whitefly population. I first describe the population dynamics of the whitefly and the construction of the population model. Optimization of chemical control is determined by the short-term criterion of minimizing damage and by the longer-term criterion of reducing the population so that biological control may be effective. I use a gradient and quadratization technique and the optimum solutions are for very early insecticide applications, even though this finding is contrary to current control practice.

19.2. POPULATION DYNAMICS

Trialeurodes vaporariorum Westwood, the greenhouse whitefly, is a serious pest of both ornamental plants and the main vegetable crops. The adult females live for 30–40 days, and lay their eggs on the apical leaves of the host plant after a short preoviposition period. The life cycle comprises an egg stage and four larval stages. The duration is about 30 days under average greenhouse temperature and molting in each stage and egg hatching depends on age. There is no diapause (Di Pietro, 1977). Age is considered as physiological age based on energy accumulation (Rodolphe *et al.*, 1977).

The dynamics of this insect are relatively simple for the first generation because:

(1) The initial population is composed of adults only (there are some larvae in certain cases, but this is of minor importance in the first generation).
(2) Temperature is the only relevant exogenous parameter of the system and is considered as a daily sinusoidal curve.
(3) There is no interaction between the pest population and the host plant (population density is low during generation 1).
(4) Parasitism can be ignored during generation 1.

19.3. THE MODEL

Let $Z_i(t,\zeta)$ represent the population density with respect to age ζ at time t in stage $i(i = \omega, L1, L2, L3, L4, \alpha)$, where ω represents the egg sfage, L1–L4 the larval stages, and α the adult stage. The chosen model (El-Shishiny, 1978) is then composed of the following groups of equations:

The Ageing Equations

$$\frac{\partial Z_i(t,\zeta)}{\partial t} + h_i\{\theta(t)\} \frac{\partial Z_i(t,\zeta)}{\partial \zeta} = -Z_i(t,\zeta)\{m_i[\theta(t)] + r_i(\zeta)]h_i[\theta(t)], \quad (19.1)$$

where $i = \omega$, L1, L2, L3, L4, θ is temperature, h_i,r_i and m_i are the ageing, molting, and mortality rates.

$$\frac{\partial Z_\alpha(t,\zeta)}{\partial t} + h_\alpha[\theta(t)] \frac{\partial Z_\alpha(t,\zeta)}{\partial t} = -Z_\alpha(t,\zeta)m_\alpha(\zeta)h_\alpha[\theta(t)] \quad (19.2)$$

The Molting Equation

$$Z_{i+1}(t,0) = S_{i+1} \frac{h_i[\theta(t)]}{h_{i+1}[\theta(t)]} \int_{\zeta_0}^{\zeta_{max}} Z_i(t,\zeta) r_i(\zeta) d\zeta, \tag{19.3}$$

where $i = \omega$, L1, L2, L3, L4, and S_i is the probability of survival to molting.

The Laying Equation

$$Z_\omega(t,0) = \frac{h_\alpha[\theta(t)]}{h_\omega[\theta(t)]} \int_{\zeta_0}^{\zeta_{max}} Z_\alpha(t,\zeta) f(\zeta) d\zeta, \tag{19.4}$$

where f is the fecundity rate.

The Initial Conditions

$$Z_\alpha(0,\zeta) = B(\zeta), \tag{19.5}$$

where $B(\zeta)$ is the age distribution of females at $t =$ (taken to be normal).

$$Z_i(0,\zeta) = 0 \qquad i = \omega, L1, L2, L3, L4. \tag{19.6}$$

The model simulates the population dynamics over a 65-day period (generation 1 and the beginning of generation 2). We chose the model time $t = 0$ to coincide with crop planting. The model parameters were estimated using laboratory data on the greenhouse whitefly, and then adjusted to field survey data after a sensitivity analysis. Validation of the model against field data has shown that the model is a good prediction tool for real systems (Figures 19.1 and 19.2), and can be adapted to more complex situations for other insect pests.

19.4. THE CONTROL PROBLEM

The greenhouse whitefly indirectly seriously disfigures the host plant because of the black mold that grows on the honeydew excreted by the larvae (especially L4). If the relative humidity rises above 90% for 70 hours cumulatively (more than seven consecutive nights) black molds may develop where sufficient honeydew has accumulated. Fruits are disfigured and may have to be rejected or cleaned, thus adding to production costs.

The envisaged control action is an application of the insecticide Primiphos-M, which is widely used in greenhouses. The differential response of the larvae of the different stages to various doses of the insecticide is shown in Figure 19.3 after a laboratory experiment (Hennequin, personal communication). Only L1 larvae are affected by the residual effect of the insecticide due to their partial mobility. The other larval stages are immobile.

19.4.1. Residue Constraints

For health reasons, the latest legal date for insecticide application is at least 15 days before the estimated date of the first crop. The insecticide dose is preferably not to exceed a value of 75 g/hl(hectolitre).

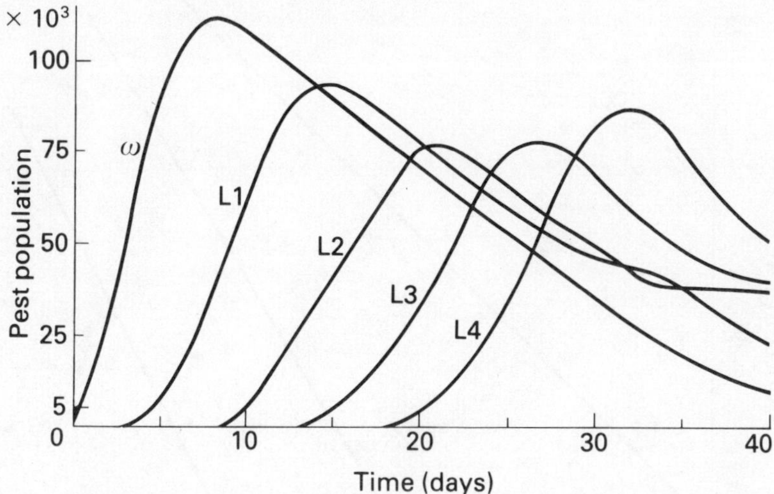

FIGURE 19.1 Simulation of the greenhouse whitefly population.

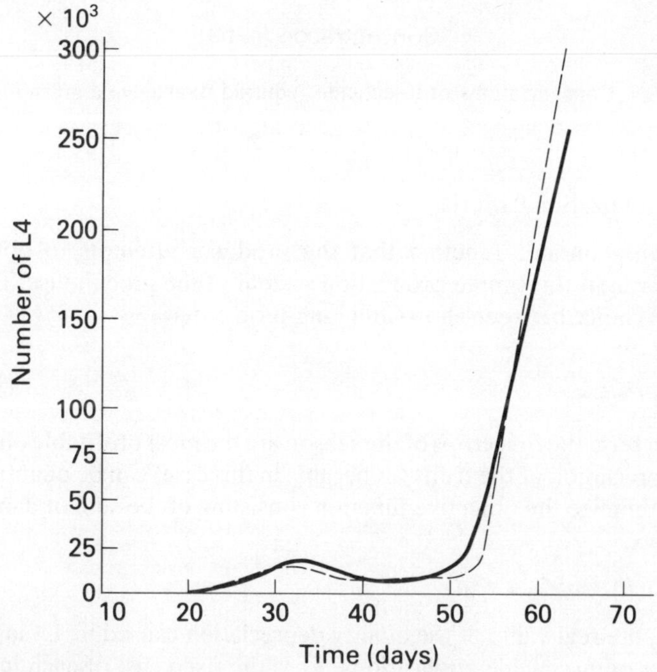

FIGURE 19.2 Final fit of the model to larval stage L4; ——simulated population, ---observed population.

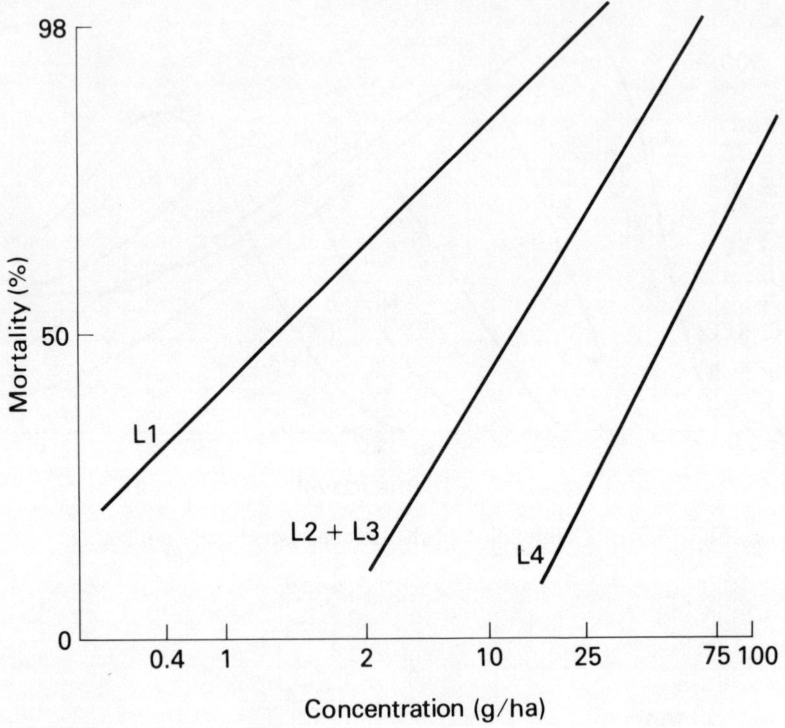

FIGURE 19.3 Concentrations of insecticide required to achieve a given mortality.

19.4.2. Optimization Criteria

Defining an economic function that the producer attempts to optimize is difficult. Even in the simple production system of the greenhouse, there is a problem of choice between short- and long-term criteria.

Short-term Criteria

In the short term the first crops of the season are the most profitable ones. If the quality depreciation of the fruits (aubergine in this case) can be quantified, our aim is to minimize the objective function consisting of the sum of damage and control costs:

$$\Omega = D(F) + n(C_0 + C_1 d), \tag{19.7}$$

where D is the real value of the quality depreciation caused by L4 larvae, n is the number of insecticide applications, C_0 is the fixed cost of each insecticide application, C_1 is the cost of application of a dose of 1 g/hl of the insecticide, and d is the dose used (g/hl). F is a function that helps determine the unsold proportion of the crop due to damage. At the date t_J:

$$F = \int_{t_{J-7}}^{t_J} h_{L4}(\theta) \int_{\zeta_0}^{\zeta_{max}} Z_{L4}(t,\zeta)\mathrm{d}\zeta\mathrm{d}t, \tag{19.8}$$

where $J-7$ is the seven-day period over which cumulative honeydew production is measured, and t_J is defined as the estimated date of appearance of the first crop.

Long-term Criteria

In the longer term, different considerations may apply. Even when later crops in the season are less profitable than the first ones, they constitute a marginal return that cannot be neglected. To protect them, biological control might be practiced by introducing a parasite into the greenhouse. The initial condition of the parasite–pest system depends upon the population of the pest's first generation. Due to the reduced potential multiplication rate of the parasite compared with that of the pest population, the initial aim is to reduce the pest population at the outset. Thus the total incidence of L4 larvae of generation 1 (Σ) is to be minimized for a specified insecticide cost. (The dosages in this case are predetermined empirically, and we look for the optimal application dates.)

$$\Sigma = \int_0^{t_f} \int_{\zeta_0}^{\zeta_{max}} Z_{L4}(t,\zeta)\mathrm{d}\zeta\mathrm{d}t, \tag{19.9}$$

where the time t_f corresponds to the end of the first generation.

19.4.3. Definition of the Controller

The problem now is to select a controller that will optimize the chosen criteria. Let us take, for instance, the case where two insecticide applications are required; we can define a control vector $\{U(t)\}$ with component $U_i(i =$ L1,L2,L3,L4).

$$U_{L1}(t) = \left| \begin{array}{ll} 0 & t < t_0 \\ M_{L1}(t - t_0,d) & t_0 \leqslant t < t_0' \\ M_{L1}(t - t_0',d) & t \geqslant t_0' \end{array} \right.$$

$$\begin{array}{l} U_i(t) = \\ i = \text{L2,L3,L4} \end{array} \left| \begin{array}{ll} 0 & t < t_0 \\ M_i(d) & t = t_0 \\ 0 & t_0' > t > t_0 \\ M_i(d) & t = t_0' \\ 0 & t > t_0' \end{array} \right. \tag{19.10}$$

where $0 < t_0, t_0' < t_s$, and $0 < d \leqslant d_s$. M_i is the mortality rate due to insecticide in stage i; t_0, t_0' are the dates of insecticide application; t_s is the last allowed date for insecticide application; and d_s is the maximum insecticide dose. The controller appears in the model in the ageing equation of the larval stages:

$$\frac{Z_i(t,\zeta)}{\partial t} + h_i\{\theta(t)\}\frac{\partial Z_i(t,\zeta)}{\partial \zeta} = -Z_i(t,\zeta)[h_i\{\theta(t)\}[m_i\{\theta(t)\} + r_i(\zeta)] + U_i(t)]$$
$$i = \text{L1, L2, L3, L4.} \qquad (19.11)$$

In fact, the control vector $\{U(t)\}$ depends only upon t_0, d_0, t_0', and d_0' (d_0 and d_0' are insecticide doses applied at times t_0 and t_0', respectively), so the problem is to determine application dates and corresponding insecticide doses that minimize the first criterion, or application dates that minimize the second criterion.

19.4.4. Optimization Techniques

The values of t_0, t_0', d_0, and d_0' were obtained using two different nonlinear programming techniques: a gradient and a quadratization technique (El-Shishiny, 1978). The gradient algorithm showed great improvements in the first few interations but had poor convergence characteristics as the optimal solution was approached. The gradient algorithm was used to improve the nominal solution to a point near the optimal one, and then the quadritization algorithm, which converges rapidly near this solution, was used to determine the optimal point precisely.

19.4.5. Results and Discussion

Figures 19.4 and 19.5 show the controlled population for the case of one insecticide treatment (compare with Figure 19.1). (Obviously these solutions depend on the market conditions, the differential response of the pest to the specified insecticide, and the initial level of contamination.) *The optimal control policy is to control the L1 larvae at a very early date. In the case of two insecticide treatments (Figure 19.6), the results show that the second application should come just after the effect of the first insecticide treatment has disappeared.* The optimal doses are relatively low since the L1 larvae are the most fragile and mobile.

This policy contrasts with current pesticide treatment where insecticide applications are reserved until the latest date allowed by residual precautions. Due to the relative tolerance of L4 larvae and their large numbers at this time, the doses utilized are usually high, based on the idea that the insecticide will be more effective the closer it is applied to the appearance of the damage.

19.5. CONCLUSIONS

Such discordance between standard practice and the "optimal" solution deduced from the model deserves comment. First, the optimal results are

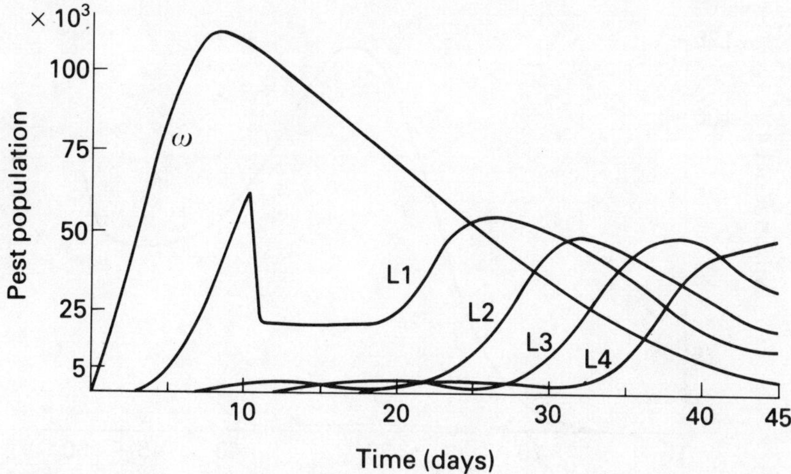

FIGURE 19.4 The controlled population after one insecticide treatment.

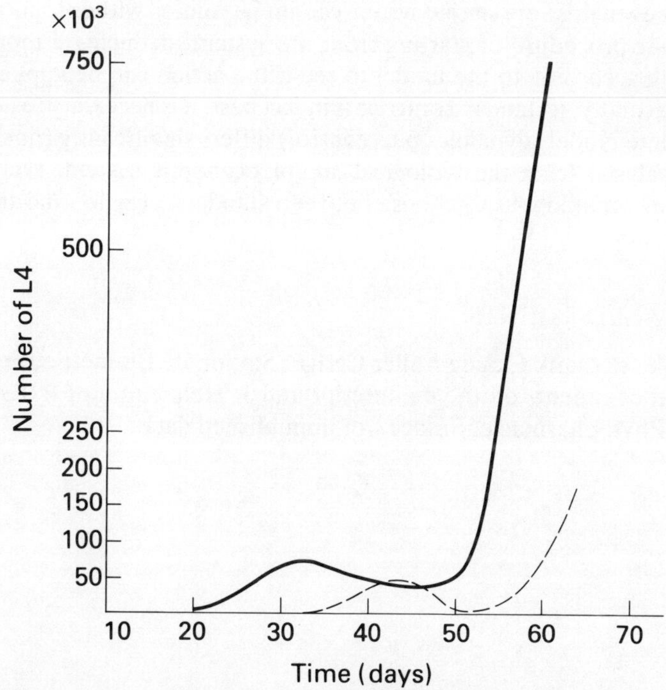

FIGURE 19.5 The controlled L4 population after one insecticide treatment: —uncontrolled population, ---controlled population.

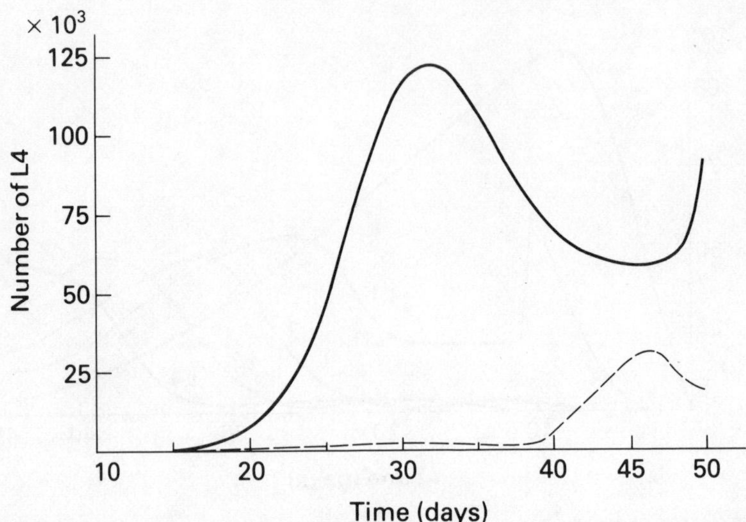

FIGURE 19.6 The controlled L4 population after two insecticide treatments: ——uncontrolled population, – – –controlled population.

validated by biological reasoning; it is necessary to act on the more fragile stage. But even this very simple system cannot be studied without a model. This justifies our procedure of starting from the system, deducing a model, then proceeding according to the model to see if the action can be applied on the system. Secondly, if standard practice is in fact best, it is necessary to admit that the complete model (dynamics plus control) differs significantly (not revealed by the analysis) from the biological and/or economic system. Only a field experiment can allow us to choose between standard practice and the model optimal policy.

ACKNOWLEDGMENTS

I would like to thank Claude Miller CNRF, Station de Biométrie, France, for his helpful comments on the manuscript and J. Hennequin of INRA-Laboratorie de Phytopharmacie, France, for unpublished data.

II Robustness of Optimal Chemical Control

F. Rodolphe

19.6. INTRODUCTION

The second part of this chapter presents numerical results on the robustness of the optimal control recommendations for control of the greenhouse whitefly in the face of changes in the economic environment. The basic population and control model is the same as in Part I (see also Rodolphe *et al.*, 1977; El-Shishiny, 1978), but here I am only concerned with the optimal application of single insecticide doses.

19.7. THE MODEL

A greenhouse constitutes a relatively well controlled environment, protected from major climatic fluctuations. Temperature was found to be the only important time-varying climatic parameter, and is the only driving function in the model of population dynamics. Moreover, because of the high value of greenhouse production, tolerance levels correspond to small pest population densities. In such conditions densities are not permitted where intraspecific competition for sites or food becomes important. Thus in the domain of validity, for practical purposes, the dynamics are linear with respect to state variables.

During host plant growth adults lay their eggs on the leaves of a well defined age class (one or two leaves per plant carry 95% of the total eggs laid). Larvae are fixed and located on leaves whose age is strongly correlated with their own age, so that interrelations between the host plant and the pest population are considered to be independent of plant growth.

For those pest population densities of practical interest, the physiological effect of the pest on the host plant is negligible, so that although the host plant influences the population dynamics through a number of variables, these can be regarded as constant. However, two population parameters have to be described precisely. First, since the host plant–pest population dynamic interaction can be ignored, population development is independent; but as the initial conditions are very far from the eventual stationary age composition a good prediction of population development requires consideration of the age distribution within each stage. Secondly, because insecticides have been shown to be efficient at specific short periods in the life cycle of the pest (during molting, for example), the transitions between stages have to be precisely represented. The insecticide modeled, in fact, is very active at the beginning of larval stage L1.

19.8. OPTIMALITY

The optimal strategy has been studied by El-Shishiny for both single and double insecticide applications. To get a clearer picture of the sensitivity of the optimal strategy to changes in the damage relationship, we went back to the process of damage. In practical conditions, the location of damage on the fruit are the flecks produced by a mold that develops on the honeydew excreted by the whitefly larvae. The risk of mold development, and hence the risk of crop damage, is related to the amount of honeydew present. The accumulated honeydew production over a seven-day period was thus chosen as the index of damage risk (or index of deterioration).

The optimization criterion is the sum of two terms: the cost of treatment, and the damage resulting from a particular pest density. But, irrespective of how the last term is defined, it is an increasing, positive functional on the set of all possible indexes. Thus if for the same dose (i.e., the same treatment cost) an insecticide application at time t_1 produces an index that is always less than that obtained by an application at time t_2, we can say that the first treatment is uniformly better than the second.

19.8.1. Simulations

As an initial condition, the pest population consists exclusively of adults (generation 0), most frequently produced either by a contamination of the greenhouse by an influx of wild adults, or by the introduction of plants already infested by a single generation of eggs.

The insecticide simulated, Pirimiphos-M, is very active on larval stage L1, active on adults (A), much less active on larval stages L2, L3, and L4), and inactive on the eggs (ϕ) (Hennequin, unpublished results). This is a quite common pattern of action for a contact insecticide. Eggs are protected by their chorion, as are larvae by their waxy secretions, except L1, which is mobile and poorly protected in the early stages.

All the simulations were made with a constant temperature of 22 °C. Temperature is the only factor of inhomogeneity, with respect to time, in the population dynamics; it affects the natural mortality only slightly, but strongly affects adult fecundity, and the speed of development of each stage. However, El-Shishiny used an observed temperature sequence for his optimization study, which is why the results presented here are different from his. But if a correction for temperature is made, namely, defining the time with respect to the speed of development, the optimal date is the same even though the dose is changed. This can be explained by the fact that a temperature variation has very little effect on the development of the population composition but influences, through fecundity, the number of individuals of the subsequent generation.

19.8.2. Results

A very high insecticide dose was first simulated. Figure 19.7 shows indexes for doses of 75 g/hl of insecticide applied on days 3, 8, and 13. Indexes corresponding to applications on days 8 and 3 have the same form, but applying on day 8 is uniformly better than on day 3, and on day 13, except during a small time interval (days 35 to 45) where the indexes are very low.

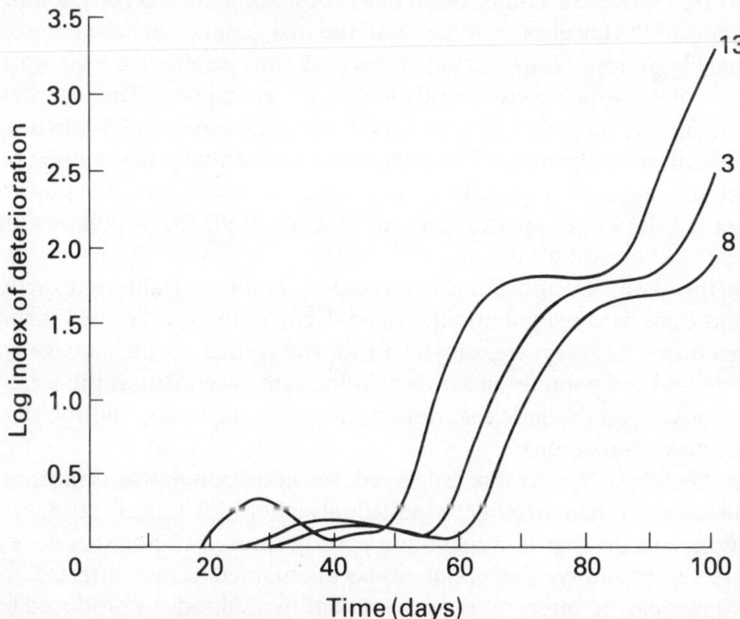

FIGURE 19.7 Indices of damage corresponding to the same dose (75 g/hl) and three different dates of application (days 3, 8, 13).

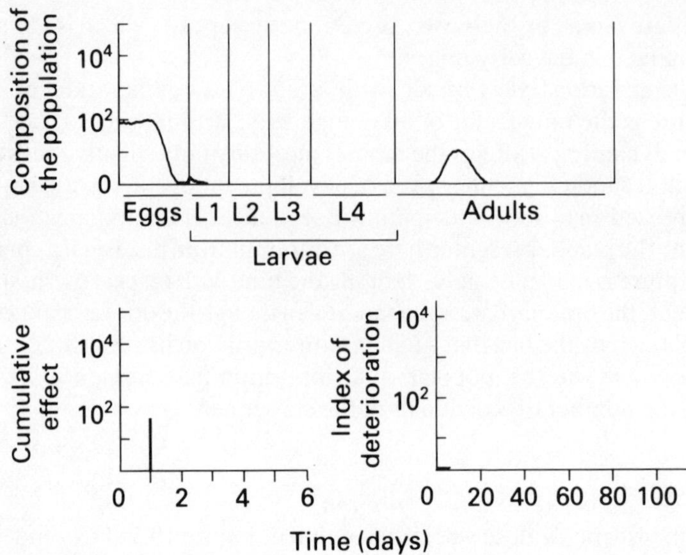

FIGURE 19.8 Pest population composition on day 8, before spraying. Vertical lines separate the different stages. Adults and eggs are present and the first-laid eggs have entered stage L1.

Figure 19.8 shows the composition of the pest population on day 8. Since on the first day adults lay eggs, individuals of the first generation have entered L1. Applying a high dose of insecticide at this time thus produces a high mortality of L1 and adults, which consequently lowers the laying rate. The insecticide is persistent and the mortality of adults and individuals entering L1 remains high for a while after application. The population is essentially maintained by the most recently laid eggs. Figure 19.9A shows the composition of the population on day 31 (22 days after application), and Figure 19.9B the population on day 31 if there has been no treatment.

When the date of application is advanced, adult mortality is exactly the same, and there is consequently the same decay in the laying rate. When the first eggs enter L1 they are subject to mortality due to the persistence of insecticide and the population survives in the same way. But as the treatment has been advanced the index is translated by five days too, and the result is evidently disadvantageous.

When the date of application is delayed, the adult population decay remains the same, as does the mortality of individuals entering L1 after the insecticide application. But the first-laid eggs have already entered L2 and therefore have a much lower mortality. The population is maintained as two different groups and the index can be interpreted as the sum of two subindices produced by the two groups of individuals: (1) the subindex produced by the last laid eggs and their descendants still has the same form but is translated to the right by five

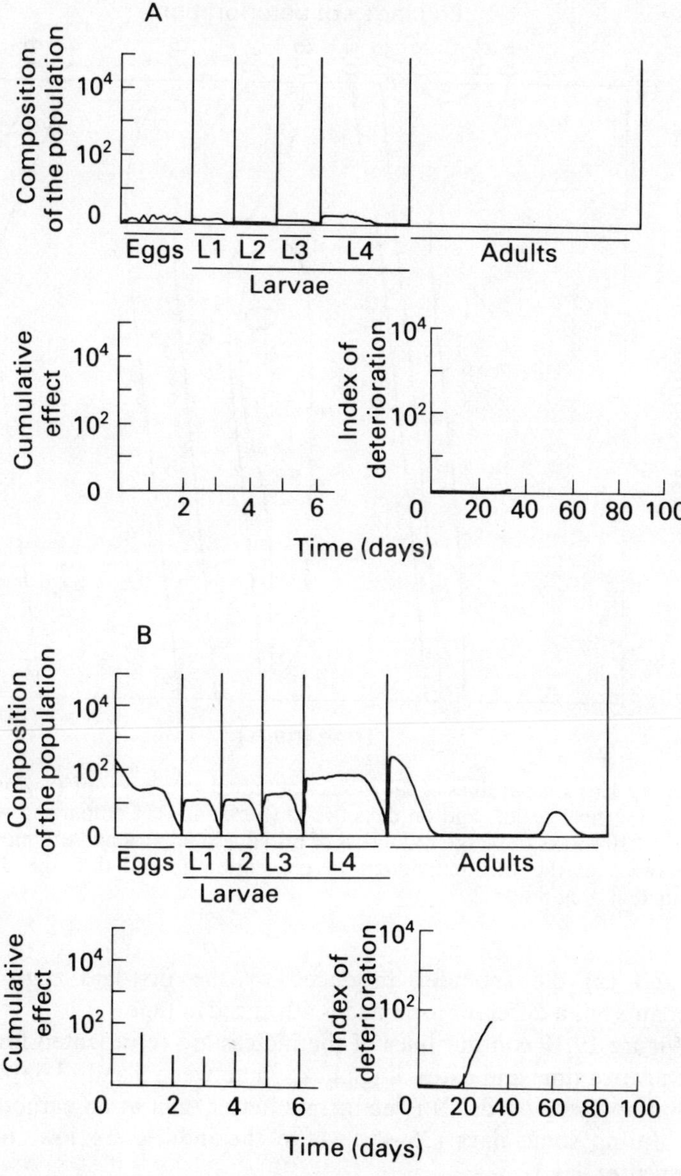

FIGURE 19.9 A: Pest population composition on day 31, after spraying 75 g/hl on day 8. Surviving adults are not apparent but are responsible for the continuing production of eggs and larvae. B: Pest population composition on day 31, with the same initial conditions and without treatment. The oldest adults are the migrants (generation 0); the first individuals from generation 1 have matured and are responsible for the increasing laying rate, representing the beginning of generation 2.

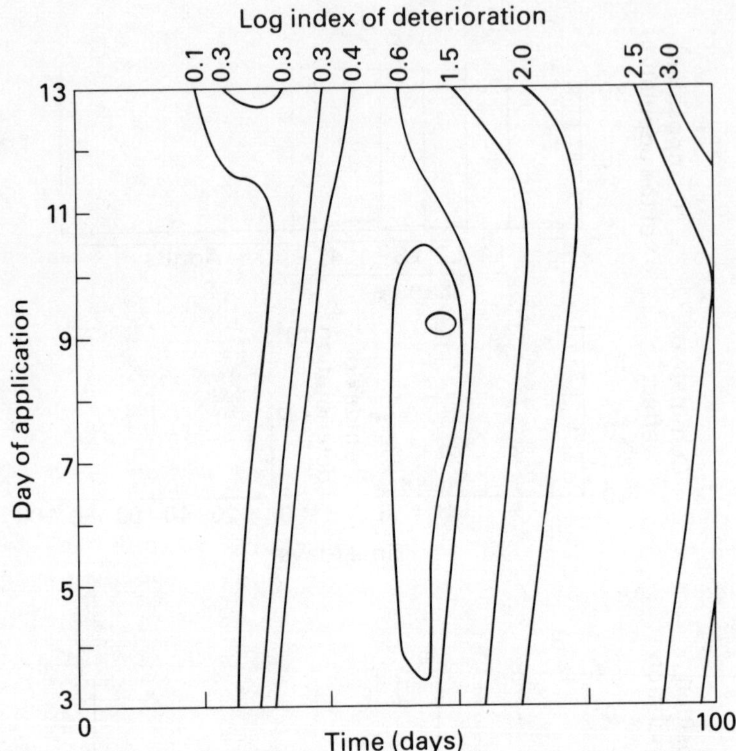

FIGURE 19.10 Contour lines of the indices obtained with doses of 75 g/hl applied on days 3–13 (vertical scale), and on days 0–100 (horizontal). Comparison with Figures 19.9 and 19.10 shows the large reduction of initial adults (around 98% mortality), the disappearance of the first individuals of generation 1, and thus the delay in the appearance of generation 2.

days; and (2) the subindex produced by the first-laid eggs and their descendants has a different form and is advanced in time.

On Figure 19.10 contour lines of the indices are represented for the same dosage against time and dates of application between 3 and 13 days. It can be seen that applying on day 9 is uniformly better than at an earlier date and, except during some days (35–45) where the indices are low, better than applying after day 12.

In Figure 19.11 the indexes obtained with a lower dose of 50 g/hl and application on days 3, 8, or 13 are plotted against time. If we examine the indices obtained with lower doses and application between days 3 and 13, the indices obtained from application on day 9 are higher than the corresponding indices obtained from application on day 12 (in the region of days 75–85), but the difference is never very great.

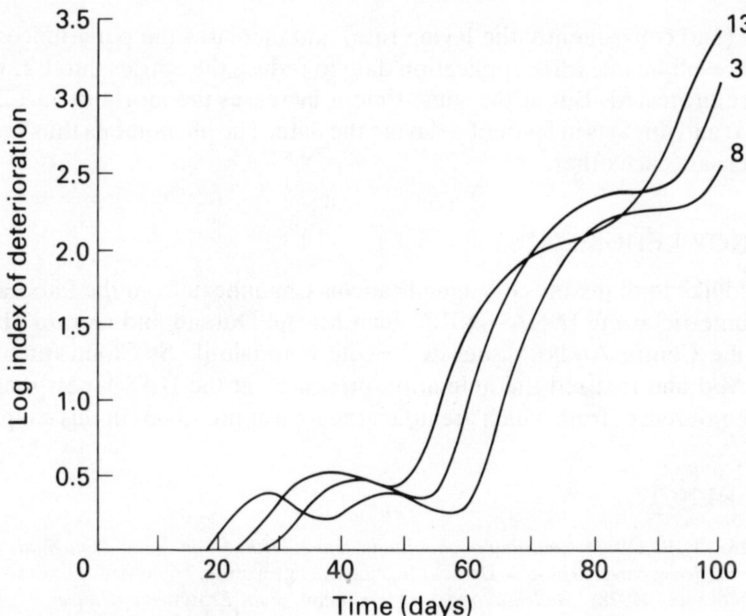

FIGURE 19.11 Indices of deterioration corresponding to a dose of 50 g/hl and three different application times (days, 3, 8, 13).

19.9. CONCLUSIONS

Comparison of the indices of deterioration corresponding to applications of different dosages on days 3–13 leads to the conclusion that, with some exceptions producing slightly better indexes between days 35 and 45, and 75 and 85, the date of application of the optimal one-application strategy will always lie within a limited range of days 8–12.

In practical circumstances, time cannot be defined with the precision of a single day. But the optimal date appears to be quite independent of the dose, and very robust to changes in the damage relationship. *The key to control is treatment of the entry of the first individuals to L1 before they enter L2 in significant numbers*.

The index is linearly dependent on the initial conditions (although the optimization criteria may not be), but the results are independent of the initial number of adults, and therefore *the optimal date will lie in the same time interval irrespective of the size of the founder population*, provided that the optimization criteria do not change too much the regions days 35–45 and 75–85. *By comparison, the dose is very sensitive to the damage relationship and to the initial number of adults*. Both date and dose are insensitive to changes in the age composition of the initial population (provided they are only adults).

The relative independence between dose and date can be intuitively understood in the following way. Increases in doses reduce the survival of adults (and consequently the laying rate) and increases the persistence. This would result in an earlier application date to reduce the entries into L2, which is more protected. But at the same time it increases the mortality in L2, L3, and L4, and this acts in favor of delaying the date. The phenomena thus seem to compensate each other.

ACKNOWLEDGMENTS

I would like to thank my colleague Francois Chahuneau from the Laboratoire de Biométrie at the INRA-CNRZ, Jean Michel Durand and Gerard Hachin from the Centre Audio Visuel de l'Ecole Normale de St Cloud for having conceived and realized the animation presented at the IIASA pest management conference, from which are taken the figures presented in this chapter.

REFERENCES

Di Pietro, J. P. (1977) *Contribution à l'Étude d'une Méthode de Lutte Biologique Contre l'Aleurode des Serres*. Thèse de Docteur Ingénieur, Université de Toulouse.

El-Shishiny, H. (1978) *Modelisation et Optimisation d'un Ecosystème Gouverné par des Equations aux Derivées Partielles*. Thèse de Docteur Ingénieur, Université de Nancy.

Leitmann, G. (1972) A minimum principle for a population equation. *J. Optimization Theory Appl.* 9:155.

Rodolphe, F., H. E. El Shishiny, and J. C. Onillon (1977) Modelisation de deux populations d'aleurodes ravageurs des cultures. *Congrès de l'AFCET*.

Shoemaker, C. (1973) Optimization of agricultural pest management. *Math. Biosci.* 16:143.

Vincent, T. L. (1972) Pest management program via optimal control theory. *Proc. 13th Joint Automatic Control Conf.* pp658–63.

Watt, K. E. F. (1963) Dynamic programming, "look ahead programming", and the strategy of insect pest control. *Can. Entomol.* 95:525.

20 Analog Simulation of Chemical Control of Rice Blast

Kazuo Matsumoto

20.1. THE MODEL

This chapter describes the use of an analog computer to analyze the efficiency of chemical control of rice leaf blast epidemics cause by *Pyricularia oryzae* Cavara (Matsumoto 1979). The progress of the disease is measured in terms of the proportion of the leaf area of rice seedlings showing lesions. Under nursery bed conditions the progress follows a typical sigmoidal curve that can be represented by van der Plank's formula:

$$\mathrm{d}x/\mathrm{d}t = rx(1-x), \tag{20.1}$$

where x is the proportion of the leaf area that is diseased, and r is the coefficient of disease progress. Figure 20.1 uses a block diagram to illustrate the analog computer used for this model in the rewritten form:

$$\mathrm{d}x/\mathrm{d}t = r(x-x^2). \tag{20.2}$$

Good fits to field data for both sprayed and unsprayed diseased rice seedlings could be obtained with constant r values (Figure 20.2), but in some cases where progress was variable, r was altered once or twice to provide a fit, using the weighted mean observed r values calculated for each data point (Figure 20.3). In the nursery, r ranged from 0.3 to 0.6.

I measures the area of the initial lesion at time t_0 and Figure 20.4(A,B,C) shows the curves produced by varying r and I. Progress of the disease is summarized by calculating the number of days to a given proportion of diseased leaf area D_x. Thus $D_{0\cdot 5}$ gives the time to 50% diseased leaf area. If r is constant between 0.05 and 1.0 the relation between D_x and I is given by

$$D_x = -b\lg[I/(1-I)] + c, \tag{20.3}$$

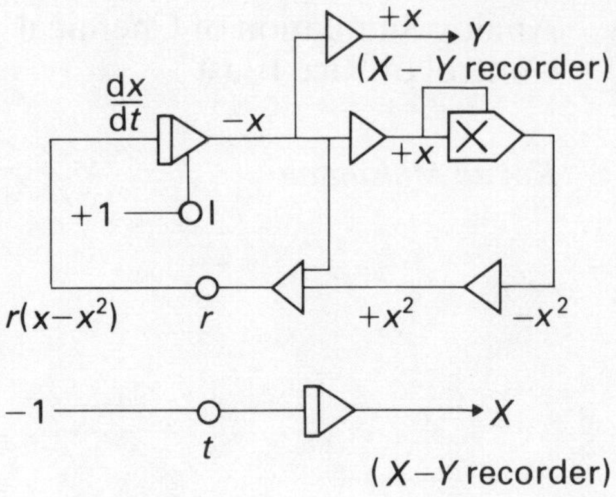

FIGURE 20.1 Block diagram of an analog computer for x (top) and for t (time, bottom).

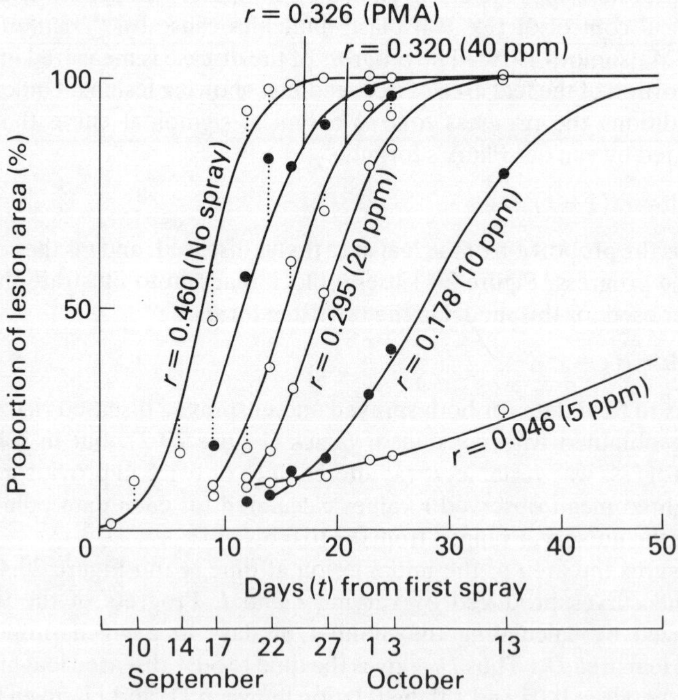

FIGURE 20.2 Effects of serial spraying of different dosages of Blasticidin-S and PMA on leaf blast progress. Disease progress curves were obtained by analog simulation.

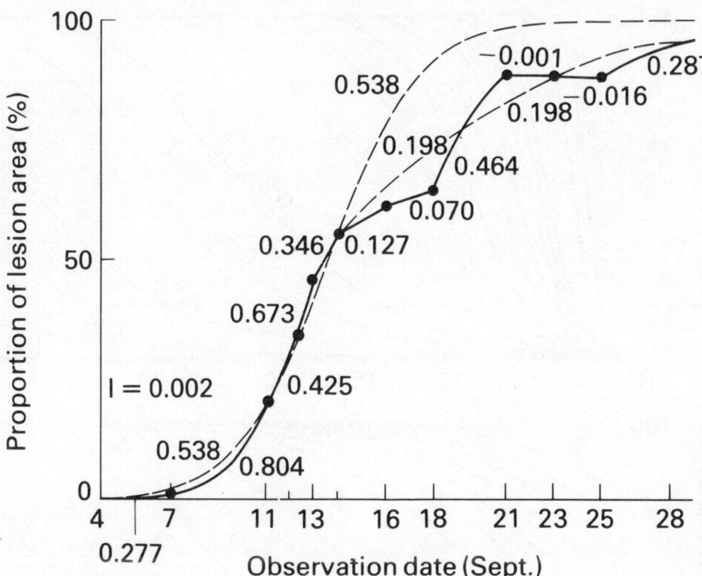

FIGURE 20.3 Changes in r of analog simulation curve in the course of disease progress (cultivar, Ginga. Check plot, high level of nitrogen fertilizer).

where b is a constant and $c = D_x$ at $I = 0.5$. For small values of r (< 0.5) and I (< 0.01) a lag phase is present before the logarithmic growth of the disease.

20.1. EVALUATION OF FUNGICIDAL ACTION

The effects of fungicides were evaluated in terms of five variables:

H	the decrease in x;	$1 - (x_f/x_c)$; (protective value)
H_i	the decrease in I;	$1 - (I_f/I_c)$;
H_r	the decrease in r;	$1 - (r_f/r_c)$;
H_{t_0}	the retardation in days of t_0; $1 - (t_{0-c}/t_{0-f})$; and	
$H_{D_{0\cdot5}}$	the retardation in days to 50% diseased leaf area.	

c and f refer to check and spray plots respectively. Fungicides may inhibit the germination of the initial invading spores and hence retard t_0. They may also inhibit the elongation of the lesions so reducing I, and finally may inhibit the formation of spores by the lesions. All of these actions will reduce r (Table 20.1).

Two fungicides, and their various combinations, were considered:

Blasticidin-S: B_1 at 40 ppm, B_2 at 20 ppm;

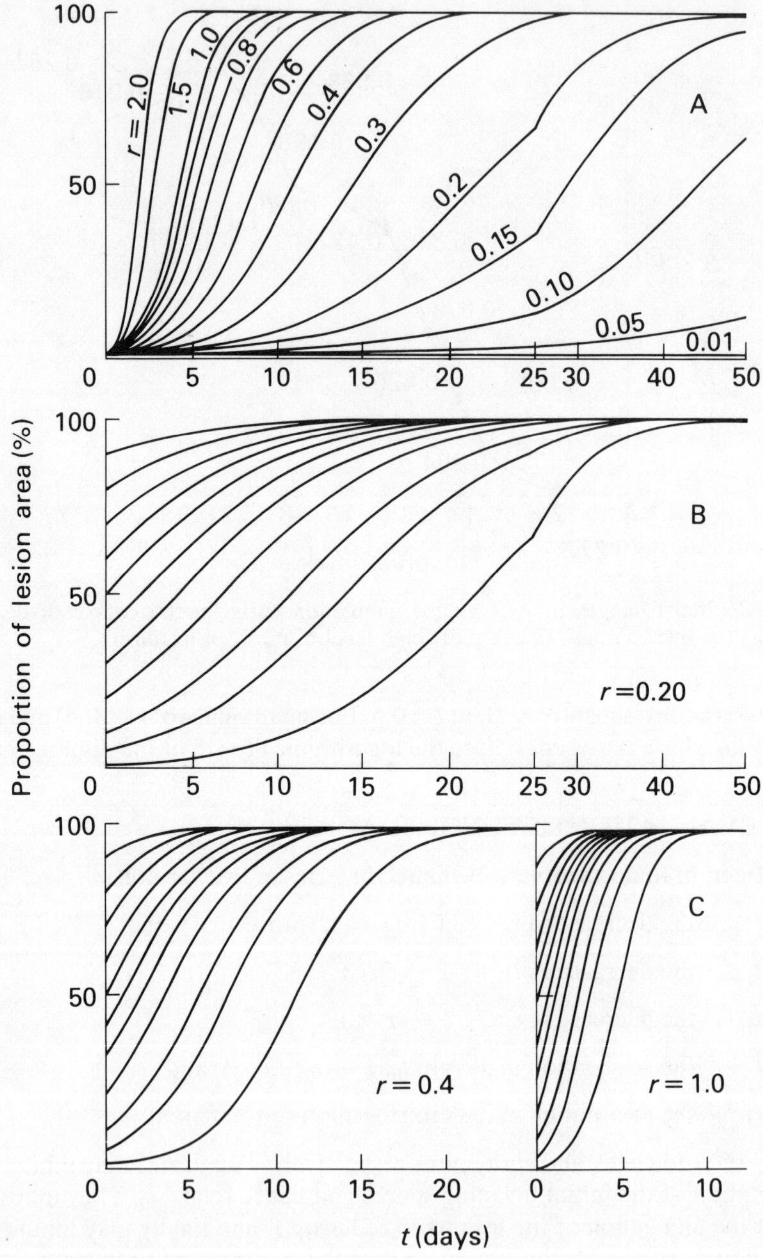

FIGURE 20.4 A: Disease progress curves of different r-values, starting from I = 0.01. B: Disease progress curves of r = 0.2, starting from different I values. C: Disease progress curves of r = 0.4 and 1.0, starting from different I values.

TABLE 20.1 Effects of fungicides on disease progress.

Characteristics fungicidal action	Spray at infection	Spray after infection
Inhibition of spore germination—infection	Retards t_0 (H_{t_0})	Decreases r (H_r)
Inhibition of lesion elongation	Decreases x (H_i)	Decreases r (H_r)
Inhibition of spore formation	—	Decreases r (H_r)

PMA: M_1 at 34.7 ppm, M_2 at 17.3 ppm, and M_3 at 8.7 ppm.

Figures 20.5 and 20.6 give the field data for two of these treatments compared with untreated disease progress (check) and show the various fitted curves using the analog computer. The effects are summarized in Tables 20.2 and 20.3.

Blasticidin-S is more effective in reducing the size of the initial lesion, whereas PMA reduces progress of the disease in the rate of increase in r. The overall progress of the disease in more effectively reduced by PMA. A combination of Blasticidin-S at 20 ppm and PMA at 17.3 ppm (Figure 20.7) shows the reduction in I from Blasticidin-S in the first 15 days, and thereafter the effect of PMA in reducing r.

20.1.1. Effect of Serial Spraying

A further experiment examined the effect of a single early spray of 40 ppm of Blasticidin-S as compared with up to eight sprays giving the same total concentration (Figures 20.2 and 20.8). The simulations clearly demonstrated the relatively greater impact of early spraying on r (Table 20.4). The results (Figure 20.9) suggest a relationship of the form:

$$H_r/(1-H_r)=aN^n \tag{20.4}$$

or

$$\lg[H_r/(1-H_r)]=n\lg N+\lg a, \tag{20.5}$$

where N is the number of sprays, and n is a constant measuring the intensity of the dosage–response relationship.

20.1.2. Effects of Antibiotic Fungicide

Kasugamycin, an antibiotic fungicide, was applied as a wettable powder and as a dust at four different dosages, G(0.6, 1.2, 2.4, and 4.8 g/1000m^2). The

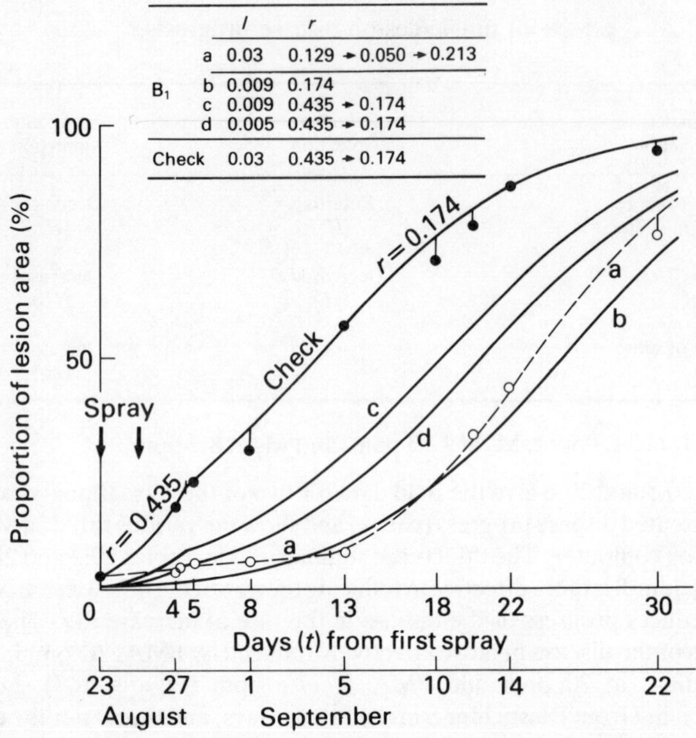

FIGURE 20.5 Analog simulation analysis for the leaf blast progress of the B_1 (Blasticidin-S, 40ppm) plot. The broken line represents field data.

experiment was also repeated for two different rice cultivars, Ginga and Aichi-Asahi (Table 20.5). The results (Figures 20.10 and 20.11) suggest the following relationships:

$$H_r = \frac{G^n}{\phi^n + G^n} \tag{20.6}$$

$$\lg[H_r/(1 - H_r)] = n\lg G - n\lg\phi, \tag{20.7}$$

where ϕ is the dosage giving a 50% reduction in r, i.e., $H_r = 0.5$, and

$$H_{D_{0.5}} = \frac{G^n}{\phi^n + G^n} \tag{20.8}$$

$$\lg[H_{D_{0.5}}/(1 - H_{D_{0.5}})] = n\lg G - n\lg\phi, \tag{20.9}$$

where $H_{D_{0.5}}$, the retardation of $D_{0\cdot5}$

$$H_{D_{0.5}} = 1 \div \frac{D_{0.5-c}}{D_{0.5-f}} \tag{20.10}$$

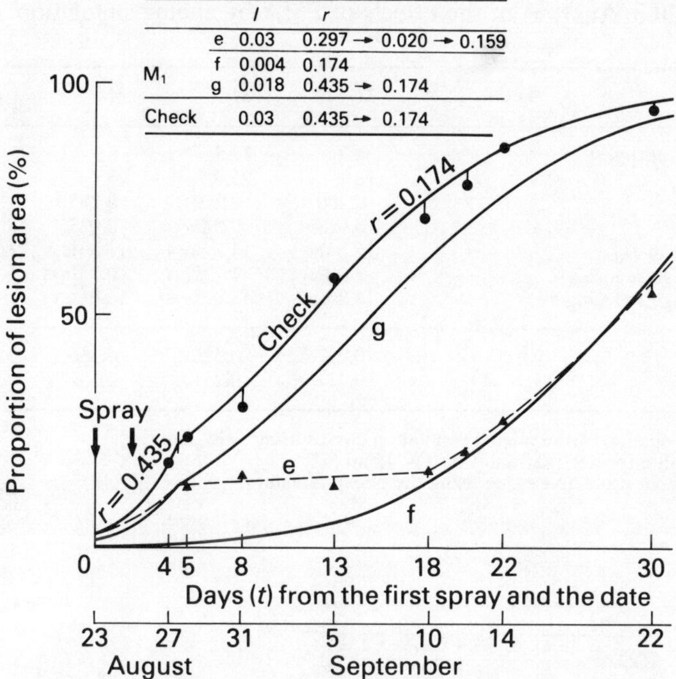

FIGURE 20.6 Analog simulation analysis for the leaf blast progress of the M_1 (PMA, 34.7ppm) plot. The broken line represents field data.

TABLE 20.2 Analysis of the effects of Blasticidin-S by analog simulation.

	Check	B_1(40 ppm)	B_2(20 ppm)
I	0.03	0.005	0.009
H_i		0.833	0.700
$D_{0 \cdot 5}$	11.6 days	23.5 days	22.3 days
Retardation of $D_{0 \cdot 5}$		11.9 days	10.7 days
(i) through reducing I		9.1 days	6.7 days
(ii) through reducing r		2.8 days	4.0 days

TABLE 20.3 Analysis of the effects of PMA by analog simulation.

Plots	M_1	M_2	M_3	Check
Concentration (ppm)	34.7	17.3	8.7	
$D^a_{0\cdot5}$	28.2	25.9	25.0	11.7
H^b	0.400	0.270	0.235	
I^c	0.018	0.022	0.023	0.03
Retardation of $D_{0\cdot5}$	16.5 days	14.2 days	13.3 days	
(i) through reducing I	3.3 days	1.2 days	0.3 days	
(ii) through reducing r	13.2 days	13.0 days	13.0 days	
$r(t = 0–5)$	0.297	0.326	0.356	0.435
$H_r(t = 0–5)$	0.317	0.251	0.182	

[a]$D_{0\cdot5}$(days) obtained from analog simulation curves using field data.
[b]Mean H value (protective value) on t = 4 and 5.
[c]Obtained from protective value using the above H values.

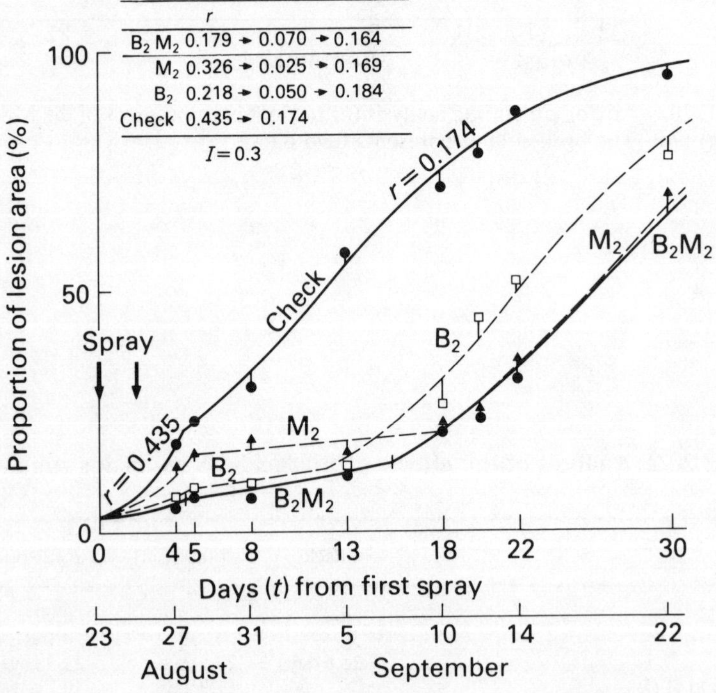

FIGURE 20.7 Analog simulation analysis for the leaf blast progress of the B_2M_2 plots.

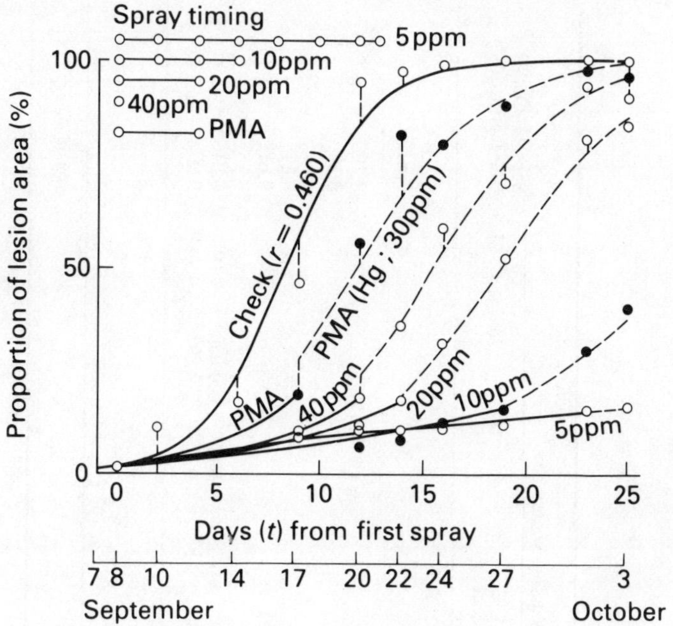

FIGURE 20.8 Analysis of the effects of serial spraying by analog simulation.

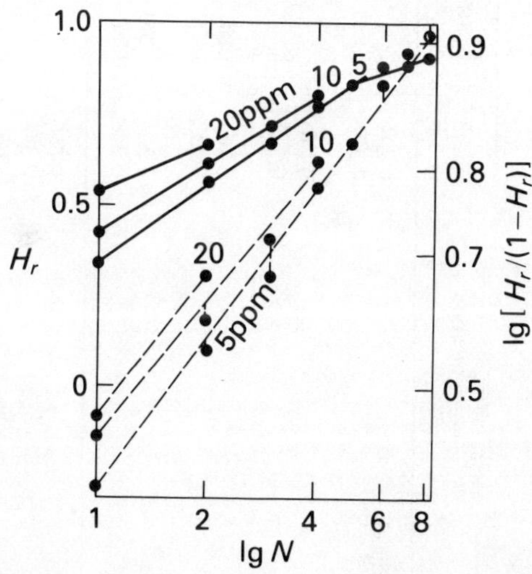

FIGURE 20.9 Relation between H_r (decreasing r) and the spray times N of Blasticidin-S.

TABLE 20.4 Effects of serial spraying on r.

Fungicides	Spray concentration (ppm)		Spraying date—September 8	10	12	14	16	18	20	21	Sept.17–Oct 13 (see Figure 20.2)
Blasticidin-S	40	r	0.204								0.320
		H_r	0.557								0.304
	20	r	0.210		0.150						0.295
		H_r	0.543		0.674						0.359
	10	r	0.260	0.180	0.130	0.090					0.178
		H_r	0.435	0.609	0.718	0.805					0.613
	5	r	0.300	0.200	0.150	0.100	0.080	0.060	0.050	0.045	0.046
		H_r	0.348	0.566	0.674	0.783	0.827	0.870	0.892	0.903	0.900
PMA	30	r	0.300		0.260						0.326
		H_r	0.348		0.435						
Check	—	r	0.460								0.460

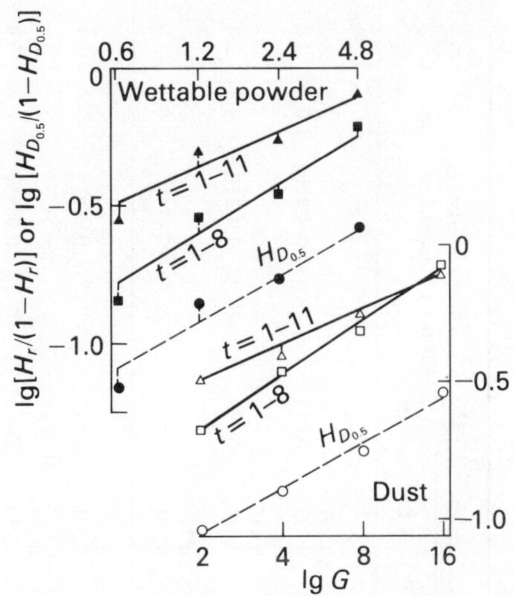

FIGURE 20.10 Relation between lg $[H_r/(1 - H_r)]$, lg $[H_{D_{0.5}}/(1 - H_{D_{0.5}})]$ and lg G according to eqns. (20.7) and (20.9) (KSM; cultivar, Aichi-Asahi). Broken lines, $H_{D_{0.5}}$.

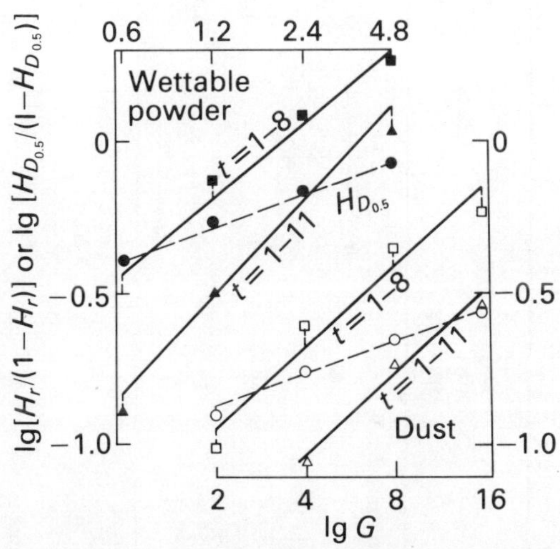

FIGURE 20.11 Relation between lg $[H_r/(1 - H_r)]$, lg $[H_{D_{0.5}}/(1 - H_{D_{0.5}})]$, and lg G according to eqns. (20.7) and (20.9) (KSM; cultivar, Ginga). Broken lines, $H_{D_{0.5}}$.

TABLE 20.5 Effects of antibiotic Kasugamycin on r. Correlation coefficients significant at * = 5% and ** = 1% levels.

	Days from first spraying		Check	Dosage, G(g/1000m^2)				n	ϕ (g/1000m^2)	Correlation coefficient	$\frac{\phi_W}{\phi_D}$	$\frac{\phi_D}{\phi_W}$
				0.6	1.2	2.4	4.8					
Wettable powder(W)												
Aichi-Asahi	8	r	0.474	0.415	0.370	0.351	0.297					
		H_r		0.124	0.219	0.259	0.373	0.653	10.643	0.982*	0.5	2.0
	12	r	0.791	0.617	0.533	0.509	0.445					
		H_r		0.220	0.326	0.357	0.437	0.458	7.969	0.973*	0.3	3.4
		$D_{0\cdot5}$	8.55	9.13	9.73	9.99	10.74					
		$H_{D_{0\cdot5}}$		0.064	0.121	0.144	0.204	0.600	43.758	0.974*	0.3	3.8
Ginga	8	r	0.504	0.459	0.410	0.366	0.317					
		H_r		0.089	0.187	0.274	0.371	0.849	8.067	0.986*	0.3	3.1
	12	r	0.617	0.593	0.562	0.505	0.459					
		H_r		0.039	0.089	0.182	0.256	1.043	11.807	0.989*	0.4	2.4
		$D_{0\cdot5}$	9.01	10.16	10.54	10.98	11.45					
		$H_{D_{0\cdot5}}$		0.113	0.145	0.179	0.213	0.362	168.503	0.998**	0.2	4.2

			Check	2	4	8	16			
Dust(D)										
Aichi-Asahi	8	r	0.474	0.390	0.351	0.316	0.257			
		H_r		0.177	0.259	0.333	0.458	0.643	21.610	0.997**
	12	r	0.791	0.597	0.565	0.508	0.434			
		H_r		0.245	0.286	0.358	0.451	0.449	27.140	0.991**
		$D_{0\cdot5}$	8.55	9.32	9.64	10.09	10.99			
		$H_{D_{0\cdot5}}$		0.083	0.113	0.153	0.222	0.547	168.288	0.997**
Ginga	8	r	0.504	0.459	0.405	0.351	0.317			
		H_r		0.089	0.196	0.304	0.371	0.862	24.769	0.974*
	12	r	0.617	0.611	0.567	0.523	0.477			
		H_r		0.010	0.081	0.152	0.277	1.560	27.922	0.939*
		$D_{0\cdot5}$	8.96	10.11	10.52	10.94	11.31			
		$H_{D_{0\cdot5}}$		0.114	0.148	0.181	0.208	0.343	713.583	0.992**

TABLE 20.6 Arrangement of a factorial experiment.

Factors		L_1	L_2	L_3	L_4
B (block)		1	2		
M (fertilizer)		Low N	High N		
V (cultivar)		Aichi-Asahi (susceptible)	Ginga (resistant)		
F (fungicide)		EDDP	KSM		
C (concentration ratio)		8	4	2	1
(dilution)		(500)	(1000)	(2000)	(4000)
(ppm)	KSM	40	20	10	5
	EDDP	600	300	150	75
(Sticker; Gramin, %)		0.003	0.006	0.012	0.026
D (spray timing)	Sept.	9,11	7,9	4,7	1,4

Treatment levels and time of spraying:

L_1 : spray after disease developed.
L_2 : spray at early stage of disease development.
L_3 : spray at very early stage of disease development.
L_4 : spray before infection.

where $D_{0\cdot5-c}$ and $D_{0\cdot5-f}$ are the days to 50% diseased leaf area for the check and sprayed plots, respectively.

The values of n are little affected by either the formulation or the rice cultivar.

However, ϕ values for the dust were two to three times greater than the wettable powder, indicating a 25–33% saving on chemicals if the wettable powder is used. The ϕ values also changed with the cultivar, the most susceptible cultivar, Ginga, requiring the higher dosage.

20.1.3. Multifactor Effects

A factorial experiment was carried out using an orthogonal array to compare the effects of the timing of two fungicides sprays (KSM and EDDP), each at four concentrations, applied to susceptible (Aichi-Asahi) and resistant rice cultivars (Ginga) at high and low nitrogen levels (Table 20.6). The experiment showed no significant effects of fertilizer level or cultivar on H_i or H_r but, as might be expected, early protective spraying, L_4, had a major effect on H_i, while the later curative sprays, L_2 and L_1, mostly affected H_r (Figure 20.12). The use of the dosage–response relationship for the effect on H_r (Table 20.6) showed that spraying at L_1 and L_3 required 1.24 and 3.6 times the dosage than spraying at L_2. The importance of the timing of curative fungicide spraying is illustrated in Figures 20.13 and 14.

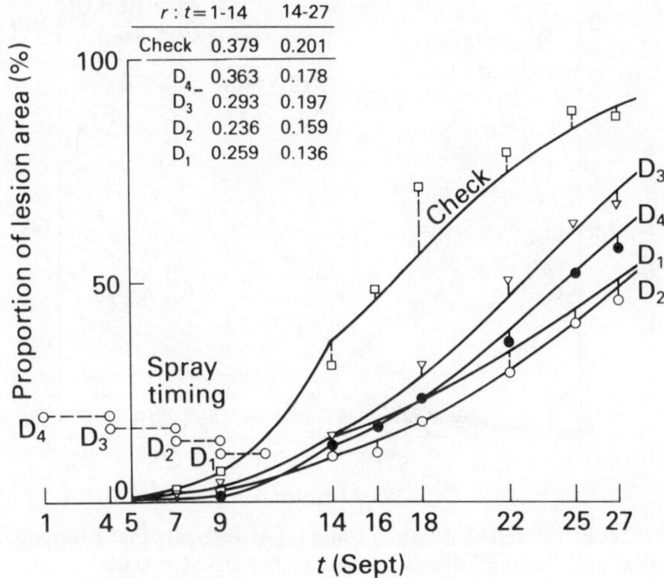

FIGURE 20.12 Disease progress curves for each spray timing plot by analog simulation.

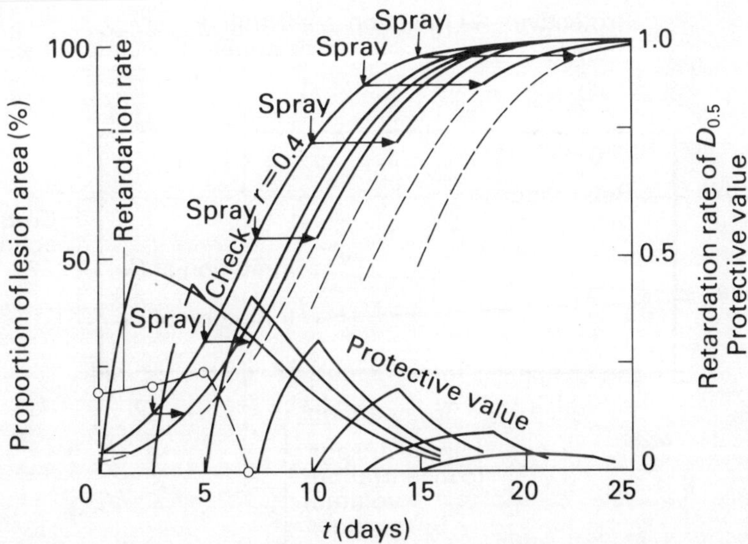

FIGURE 20.13 Effectiveness of spray timing of a curative fungicide that can reduce lesion area to half. (Model, Check; $I = 0.05$, $r = 0.4$). Broken curves, proportion of lesion area (%).

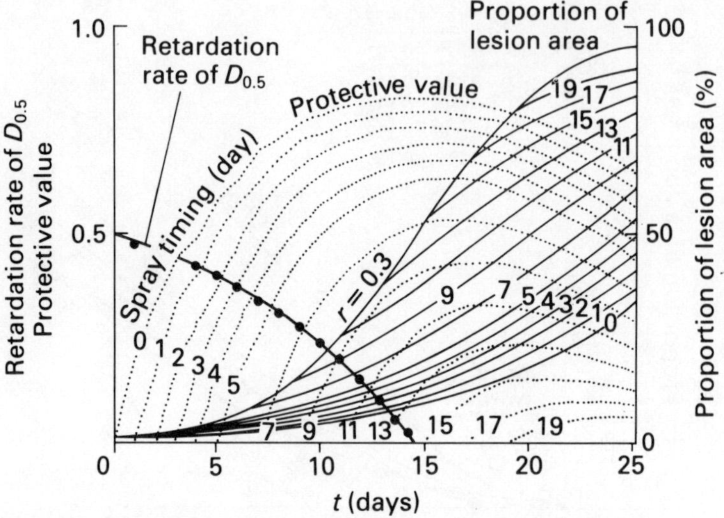

FIGURE 20.14 Effectiveness of spray timing. (A fungicide that can reduce r to 0.15 was sprayed at each stage of disease progress, $r = 0.3$, $I = 0.01$.)

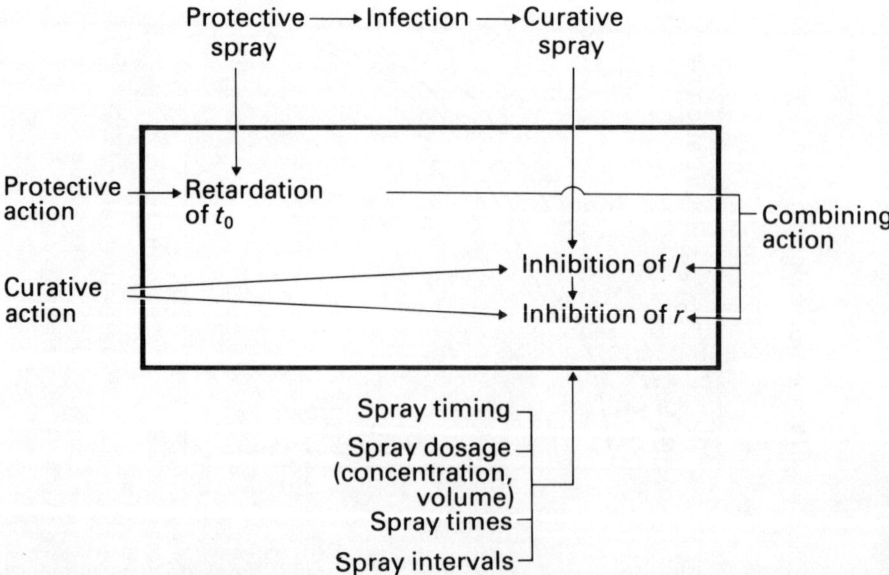

FIGURE 20.15 Schematic illustration of fungicidal action on t_0, I, and r, the characteristics of the disease progress curves.

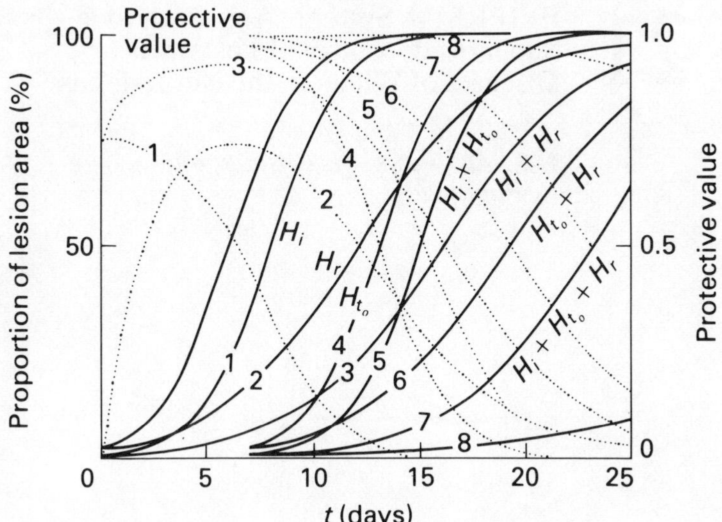

FIGURE 20.16 Comparison of H_{t_0} (retardation of t_0, protective action). H_i, H_r (decreasing rate of I, r, curative action) and combined action.

	t_0	I	r
Check:	$t_0 = 0$	$I = 0.02$	$r = 0.6$
1:	0	0.005	0.6 (H_i)
2:	0	0.02	0.3 (H_r)
3:	0	0.005	0.3 ($H_i + H_r$)
4:	7	0.02	0.6 (H_{t_0})
5:	7	0.005	0.6 ($H_i + H_{t_0}$)
6:	7	0.02	0.3 ($H_{t_0} + H_r$)
7:	7	0.005	0.3 ($H_i + H_{t_0} + H_r$)
8:	7	0.005	0.15 ($H_i + H_{t_0} + H_r$)

20.2. CONCLUSIONS

The interaction between the time of spraying and disease progress is illustrated schematically in Figure 20.15. *Overall, the best fungicide is one that combines reductions in I, r, and t, while the least protective fungicides are those that only reduce I or r* (see Figure 20.16).

REFERENCE

Matsumoto, K. (1979) Analog simulation analysis on the efficiency of chemical control for rice leaf blast epidemics. *Bull. Chugoku Nat. Agric. Exp. Station E*, No. 15, pp1–113.

21 EPIPRE: A Systems Approach to Supervised Control of Pests and Diseases of Wheat in the Netherlands

J. C. Zadoks, F. H. Rijsdijk, and R. Rabbinge

21.1. INTRODUCTION

For 20 years we have been developing models of epidemics of yellow stripe rust (*Puccinia striiformis*), a fungus disease of wheat that causes great damage in the Netherlands and elsewhere, but these studies were of an esoteric nature as long as effective chemical control of the disease was impossible. When such control came within reach, a project was begun in 1977 to utilize the models that had been developed in a disease management system. Implicit in the project is a change from explanatory strategic models to tactical ones.

Scientists are mainly interested in the content matter of a management system—"what", but in this chapter we focus on the introduction and practical execution of a system—"how". Emphasis on "what" at the neglect of "how" ensures the failure of a management system (Kampfrath, personal communication), but shifting emphasis from "what" to "how" is psychologically difficult for the research scientist.

21.2. THE EPIPRE PROJECT

EPIPRE (EPIdemics PREvention) is a cooperative project of some 300 farmers, the Extension Service, the Institute for Plant Protection Research (IPO), the Agricultural University, and various other institutions. EPIPRE is executed by the Laboratory of Phytopathology of the Agricultural University (Rijsdijk, 1982; Zadoks, 1981). The project is largely financed through the Netherlands Grain Centre, a nonprofit foundation funded by the Board for

Grains, Seeds and Pulses. The Board imposes a levy of 1 cent per 100 kg of wheat on farmers to create funds for wheat research.

It was soon realized that an attempt to manage only a single disease was bound to fail for at least two reasons:

(1) Diseases and pests influence each other, directly or indirectly (in our case the powerful chemical triadimefon, Bayleton ®, controlled not only yellow rust but also mildew).
(2) Farmers always have to deal with a variety of pests and diseases, and, accordingly, require a package of advice.

But, as manpower and knowledge were insufficient to work with more than one disease at a time, it was decided to begin in 1978 with yellow rust. In 1979, EPIPRE was extended with a negative forecast of mildew (*Erysiphe graminis*) for the whole country and with warnings against brown leaf rust (*Puccinia recondita*) and the English green aphid (*Sitobion avenae*) in a limited part of the Netherlands.

The objective of EPIPRE is to provide a system of supervised control of diseases and pests in wheat, aimed at minimization of biocide usage and subsequent environmental pollution, and maximization of the value added to the crop by biocide application (within the limits imposed by law). This optimization is not superfluous as EEC wheat prices are such that wheat farmers aim for top yields. Yields up to 10 tonnes per ha have been obtained and the average winter wheat yield of the Netherlands in 1978 was 6.8 tonnes/ha. The average number of biocide treatments per field in the Netherlands was about 1.4 in 1978, seed dressing excepted.

21.3. OUTLINE OF EPIPRE

In advanced agriculture, small causes may lead to large financial effects, and thus even relatively small differences between fields have to be taken into consideration. EPIPRE therefore operates on a field-by-field basis and gives specific recommendations for each of some 400 fields registered. A general outline of EPIPRE is given in Figure 21.1.

Basic data per field are entered into a databank once per year. Initialization is done in late winter and farmers supply basic data as in Table 21.1. Field observations are then solicited from farmers in April/May by means of computer printed postcards. The data comprise: field code number, date of observation, growth stage of wheat, disease and pest assessments, and fertilizer and biocide treatments with types and dosages used. Farmers send their data to the EPIPRE team in Wageningen, and the data are entered immediately into the databank.

The various simulation models and decision systems are stored in the computer and the EPIPRE operator goes through the files daily for updating

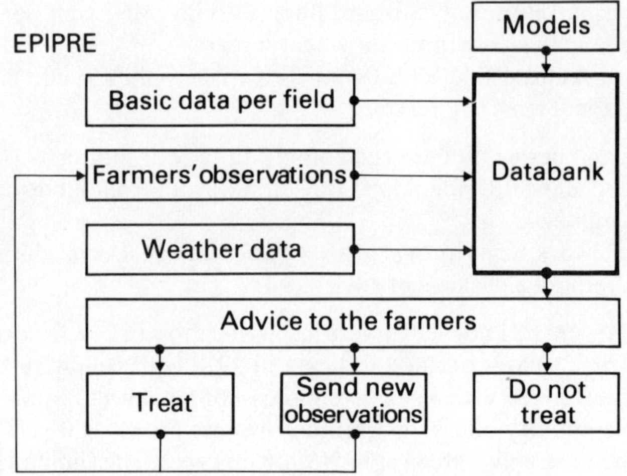

FIGURE 21.1 General outline of EPIPRE.

TABLE 21.1 Initial data asked of the farmers (Anon., 1979).

Cropping plan
Spray equipment (own or hired, beam length)
Size of field
Soil type
Clay fraction (lutum content)
Sowing date
Preceding crop
Yield expectation
Herbicides (dates, chemicals)
Fertiziler (dates, kg N_2/ha)
Cultivar

and advice. The advice falls essentially into one of three classes: "treat", "no treat", or "send new observation", and this information is sent to the farmers. The Extension Service and other interested individuals or institutions receive printouts according to their needs: regional data ranked per cultivar, or cultivar data grouped per region.

The computer used is a DEC 10 with 48 K per job, and the databank is handled through databank management system (DBMS) software. The models used in the operations are not detailed simulation models, but simplified versions in which the growth curves are adjusted to the particular cultivar–race combination present and to the disease fractions x calculated from the observations.

For yellow rust the model is a *mixtum compositum* of epidemic growth functions and damage functions:

$$A = [\exp(BCD) - 1]E, \qquad (21.1)$$

where A is the expected yield loss (kg/ha); B is a function of cultivar resistance; C is a function of N fertilizer applied; D is the field assessment of yellow rust (in number of diseased leaves per 10 m drill length $\simeq$ 25 m^2 of crop); and E is the expected yield (kg/ha) as specified by the farmer. Similar procedures have been developed for aphids and brown leaf rust (Rabbinge and Carter, Chapter 15).

21.3.1. Disease Assessment

The participating farmers carry out their own disease and pest assessment. Since the advice given is based on the farmers' own observations, the procedure places the primary responsibility where it should be. Farmers also find this valuable and instructive. We have endeavored to develop a uniform observation procedure for all diseases and pests. The farmer is required to walk through the field along a diagonal and to look out for the disease(s) or pest(s) to be assessed. At the first finding of yellow rust he is requested to take a sample of rusted leaves and to send it to the EPIPRE team, which then confirms the identification. Yellow rust samples are handed in to IPO for race identification and if a new race is found EPIPRE can be adjusted. The farmer returns along the other diagonal, selects 50 cm of drill length (in different drills) 20 times, and counts the number of leaves with yellow rust. Counts and total are marked on a form. The farmer also takes two stems from each of the 50 cm drill lengths with him. When he has left the field he counts the total number of leaves free from mildew and enters this number on the form. He also enters other relevant information, such as the date of observation, growth stage of wheat, time needed for disease/pest assessment, and biocide and fertilizer treatments.

In essence, disease assessment is a matter of incidence determination. We have found that at low disease severities the log incidence is proportional to severity. Consequently, we can estimate the diseased fraction (van der Plank's x) with sufficient precision for present purposes combining the counts and the growth stage.

21.3.2. Communication with Farmers

Interested farmers all over the country have been invited to participate through the Extension Service. Regional instruction meetings are held in late winter and though the weather in early 1979 was extremely bad and the roads were hardly passable, attendance was over 60%. Early in the season the participants receive instructions for easy symptom recognition in the field. During the season communication is normally by mail, except in the case of aphids where speed is required in communicating data and advice, and information is provided by telephone. Members of the EPIPRE team visit fields to check observations, but are unable to visit all participants and have insufficient time to talk at length with all of them. After harvest, participants receive a printout of their observations and treatments for each field, with the request to check, correct, or complete the data, and to send their yield figures. Amendments and yields are then entered into the databank. The Research Station for Arable Farming and Field Production of Vegetables (PAGV) provides economic analyses, differentiated according to region, soil type, and agricultural activities. These data are used for a standardized calculation of costs and benefits per field due to EPIPRE, or due to deviations from EPIPRE. The cost–benefit analyses are sent to the participants with a request for comments. In general, the farmers agree with our calculations, and believe them to be instructive. Finally, participants and sponsors receive an annual report on the practical aspects of EPIPRE (Rijsdijk and Hoekstra, 1979).

The philosophy of EPIPRE is that the farmer is the master of his own field. EPIPRE gives advice, but advice that is field specific compared to the advice of the Extension Service, which by necessity is more general. The farmer then uses or disregards the advice at his own discretion. EPIPRE only requires that the farmer reports what he has done. As yet, there have been no problems of legal liability.

21.3.3. Cost of Treatment

The cost of treatment consist of four elements: (1) chemical, (2) equipment, (3) labor, and (4) wheel damage. The costs of the chemical are known; use of the farmer's own equipment cost about Dfl 7.50 per ha in 1978; and labor costs depend on the cropping pattern. When cereals are less than 60% of the farmer's acreage and when the farmer grows labor-intensive root crops, he can spend his time more profitably on his root crops. For treatment in cereals he will then hire labor. Tables 21.2 and 21.3 provide data for own and hired labor.

These must be known to determine appropriate damage and action thresholds (Zadoks and Schein, 1979). For yellow rust the damage threshold is: (1) before booting—10% of leaf area covered by disease symptoms; and (2) after booting—5% of flag leaf area covered by disease symptoms (up to three weeks after flowering). For brown rust the damage threshold used is about $x = 0.0005$ at growth stage 10 (Feeke's scale = 45 in decimal code; Zadoks *et al.*, 1974). The actual value depends on the cultivar; information on races is not available. For mildew the damage threshold lies at about two mildew-free leaves per stem, but the actual value depends on cultivar, soil type, and region. This differentiation is essential because mildew is much affected by macrocli-

TABLE 21.2 Approximate costs of yellow rust control in 1978, in kg/ha (wheat price per kg is approximately Dfl 0.48).

Chemical	
Bayleton	95
Bavistin-M	150
Labour	
Hired	65
Own	20
Wheel damage	
1 treatment	150
2 treatments	225

TABLE 21.3 Approximate costs of yellow rust control in 1978, in kg/ha (wheat price per kg is approximately Dfl 0.48).

Chemical used	Number of treatments	Labour Hired	Labour Own
Bayleton	1	310	265
Bavistin-M	1	365	320
Bayleton	2	545	565
Bavistin-M	2	655	565

matic and microclimatic factors. In general the disease damage thresholds quoted are still somewhat tentative, since good experimental evidence is scarce in the Netherlands.

For the aphid *Sitobion avenae*, the 1978 threshold was 15 aphids per ear but this value is subject to change. Preventive schedule treatment against aphids is meaningless, but treatment when needed is highly remunerative, improvements of 1 tonne/ha being obtainable.

If "ear diseases" (mildew, *Septoria* spp., *Fusarium* spp.) and aphids occur together, the damage thresholds of both are lowered as postponement of treatments and mixing of chemicals economize on wheel damage and application costs.

21.4. RESULTS AND PERSPECTIVES

The results for 1978 have been evaluated. Crops were generally healthy but there were localized outbreaks of yellow rust and a late attack of cereal aphids in July surprised farmers and scientists. Yields were unusually high, with an EPIPRE mean of 7.3 tonnes/ha.

Out of a total of 397 fields, 80 fields with yellow rust were treated, of which 36 were treated unnecessarily because the farmers were afraid of yellow rust after bad experiences in 1977, when treatment was not allowed. Of the remaining 44 fields, 18 were treated according to EPIPRE advice. The other 26 fields were treated too early, but they would also have been treated according to EPIPRE. In two cases out of the 317 nontreated fields, the wrong advice was given: in one case due to an incorrect disease assessment by the farmer. In the other, with a late attack on a moderately resistant variety, the loss was still negligible. *Experiments showed that treatment according to the flexible EPIPRE criteria was cheaper than, and equally effective as, a schedule treatment at two predetermined dates.*

Farmers' observations carried out according to instructions were shown to be accurate and adequate. At low disease intensities, farmers had to spend about an hour per field on average, but it was possible to simplify the observation procedure so that in 1979 yellow rust observations took some 30 minutes per field only. In 1978, EPIPRE advised farmers to make three rounds of observations, although in 1979, EPIPRE advised four rounds for yellow rust, brown rust, and mildew together, and one more round for aphids. In the future, more aphid rounds will be needed.

The 1979 data have not yet been evaluated. The winter was long and severe, the summer cool and very long. Yellow rust was relatively unimportant, so that observations and advice appear to have been adequate. Dutch farmers tend to spray early (mid-May) against mildew, but EPIPRE was able to postpone the first treatment, so that a second treatment could be avoided. Warnings against brown rust and *Sitobion avenae* were generally adequate. However, other aphids such as *Metopolophium dirhodum* and *Rhopalosiphum padi* were found. The advisory season was closed around mid-July but this was a mistake in view of late aphid attacks and the possibility of a cool and prolonged summer; it is now clear that EPIPRE should continue until at least the end of July.

In June 1979, the EPIPRE project was reviewed by representatives of the sponsors and the advisory committee. A policy decision was made to extend EPIPRE to 800 participants in 1979 and 3000 in 1980, but these targets could not be achieved. With a long-term perspective we can distinguish the research phase from 1970–77, the present development phase from 1977–80, and an application phase from 1981 onwards. The Laboratory of Phytopathology will take care of the development phase, but will transfer information and equipment to another institution for general application.

REFERENCES

Anonymous (1979) *Instructiemap EPIPRE* Cyclostyled Wageningen.

Rijsdijk, F. H. and S. Hoekstra (1979) *Praktijkverslag EPIPRE*. Laboratory of Phytophatology, Wageningen.

Rijsdijk, F. H. (1982) Decision making in the practice of crop protection. The EPIPRE system. *British Crop Prot. Council Monograph* No. 25: 65–76.

Rijsdijk, F. H. and J. C. Zadoks (1978) *EPIPRE, een poging tot geleide bestrijding van graanziekten met gebruik van een computer. Voordrachten Graanziektendag 1978* (Wageningen: Netherlands Graan-Centrum) pp51–9.

Zadoks, J C. (1981) EPIPRE: a disease and pest management system for winter wheat developed in the Netherlands. *EPPO Bull*. 11: 365–396.

Zadoks, J. C., T. T. Chang, and C. F. Konzak (1974) A decimal code for the growth stages of cereals. *Eucarpia Bull*. No. 7.

Zadoks, J. C. and R. D. Schein (1979) *Epidemiology and Plant Disease Management* (New York: Oxford University Press).

22 Use of On-line Control Systems to Implement Pest Control Models

B. A. Croft and S. M. Welch

22.1. INTRODUCTION

The concept of a closed-loop, on-line control system is well known to electrical engineers and system scientists, but its use in pest control is more recent (Haynes *et al.*, 1973; Giese *et al.*, 1975; Tummala *et al.*, 1976; Croft *et al.*, 1976b). Briefly, on-line control provides feedback and feedforward information on system states and external driving variables, in real time, so as to maintain the desired performance of a system. On-line control for integrated pest management (IPM) is usually based on simulation models of crops, pests, natural enemies, etc., incorporating important biological, environmental, and cultural factors. These models may be expanded into management models incorporating the effect of control tactics used by pest managers and used to identify optimal combinations of control measures.

The components of an on-line pest control system are (Figure 22.1): (1) the pest control model; (2) a biological monitoring system, which measures the changes in the pest's life system; (3) an environmental monitoring component that accesses the real-time weather variables used in driving the pest control model; and (4) a delivery system that gathers the relevant environmental and biological monitoring information, interfaces these with the model and then delivers information on model output to decision makers. Usually, the iterative process of monitoring, model prediction, delivery, action, monitoring etc. (i.e., the closed-loop control) is repeated throughout a growing season, to ensure that the pest population is maintained below the economic damage threshold.

As indicated in Figure 22.1, the use of simulation models in an on-line mode requires additional developments beyond construction of the pest control model itself. While general features of delivery, environmental and, to a lesser extent, biological monitoring may be generally useful for a wide range of pests, specific

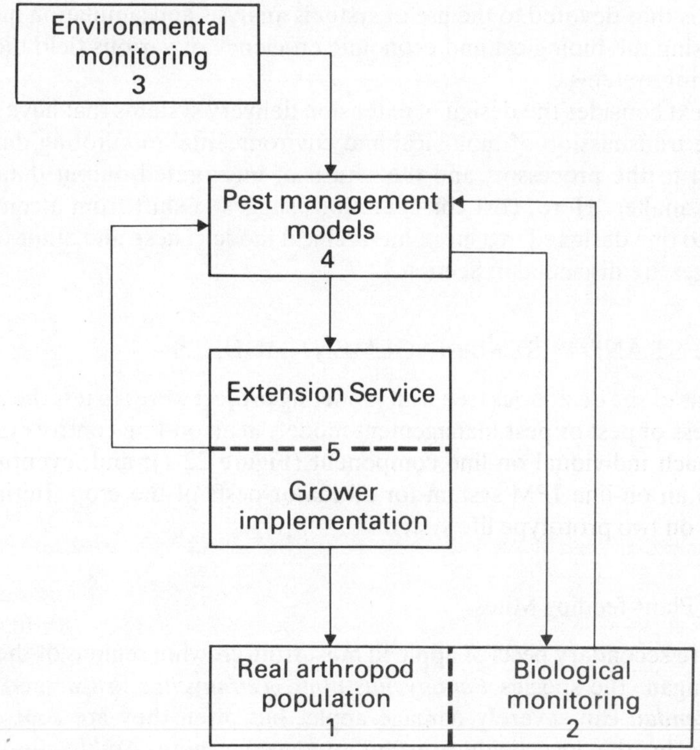

FIGURE 22.1 Components of an on-line pest management system (adapted from Haynes *et al.*, 1973).

unique features have to be developed to make a model for a particular pest or pest complex useful in the field. Each component (Figure 22.1) will also influence the design of the pest control model and other system components and often there is a process of system tuning or adjustment of all components so as to weld them into a useful, efficiently operating system.

In this chapter we describe the evolutionary process of development, implementation and performance evaluation of an on-line, apple pest management program that has been developed in Michigan over the past 8–10 years. We begin with a brief description of the major pests and outline the form of the biological models used. We then turn to the problems of environmental monitoring and, in particular, discuss the need for, and availability of, appropriate weather data, bearing in mind the different levels of resolution required.

At the outset a certain degree of empirical judgment and independent development of system components is inevitable. However, subsequent testing of the system using simulation studies of the performance of alternative components may be used to identify improvements. A major section of this chapter is thus devoted to the use of systems analysis and simulation modeling in assessing the biological and economic efficiency of various field biological monitoring systems.

We next consider the design of extension delivery systems that have to cope with the transmission of biological and environmental monitoring data from the field to the processor, and the return of interpreted output data to the decision maker. Here, cost considerations suggest a shift from a centralized system to one designed in a more hierarchical mode. These and other research challenges are discussed in Section 22.7.

22.2. PEST AND PEST MANAGEMENT MODELS

The goals of the deciduous tree fruit modeling project were: to test the role and usefulness of pest or pest management models in an on-line control system; to refine each individual on-line component (Figure 22.1); and, eventually, to develop an on-line IPM system for all major pests of the crop. Initially, we focused on two prototype life systems.

22.2.1. Plant-feeding Mites

These are secondary pests of apple in most fruit-growing regions of the world. In Michigan, the species *Panonychus ulmi*, *Tetranychus urticae* and *Aculus schlechtendali* can severely damage apple, but often they are controlled in orchards by the insecticide-resistant predaceous mite *Amblyseius fallacis*, when growers use selective sprays for controlling other pests of apple (Croft and McGroarty, 1978). Modeling this predator–prey life system was undertaken first because (1) studies of the population dynamics of both mite groups were well developed (Croft and McGroarty, 1978); (2) the effects on mite species (both prey and predators) of management operations used for controlling other pests, including pesticides, has been extensively studied (Croft and Brown 1975); (3) a preliminary integrated pest control system had been developed for mites (Croft, 1975), but greater predictability in forecasting the effectiveness of control was needed; (4) the ease of sampling mite populations greatly helped experimental studies and model validation (Dover *et al.*, 1979); and (5) the interaction between apple mite predators and prey is limited spatially to a single tree or a portion thereof, thus allowing for considerable simplicity in modeling (Dover *et al.*, 1979).

The specific details of the plant-feeding mite (*P. ulmi*)—predator (*A. fallacis*) research model have been published elsewhere (Croft *et al.*, 1976c; Dover *et al.*, 1979). In essence, the model is a discrete time, state-space model that tracks five cohort life stages of both sexes of each mite at one-day time steps during a single growing season (April–October). Each species model contains features of development, consumption, oviposition, mortality, and dispersal. The model is deterministic relative to most processes, but, a stochastic spatial component mimics predator–prey interactions within and between trees and allows for computation of predator–prey encounter and predator consumption features. Extensive validations of the model are given in Dover *et al.* (1979).

Since the mite predator–prey model was the first attempted by the apple IPM research team, model development and implementation tended to develop sequentially, rather than in parallel (the latter is a more efficient process). In the research phase we developed an intricate and highly descriptive model of the biology of these species, but when the time came to implement the model on-line, it was too detailed and expensive to run on a daily basis for every orchard in the state of Michigan. Several options were open to overcome these problems. One was to simplify the implementation model so that it captured only the major or sensitive features of the research model. A degree of accuracy would then be sacrificed to gain improvement in cost effectiveness. Another was to use the full model to generate simple charts or nomograms of model outputs at relatively frequent intervals (e.g., daily, weekly), so that the benefits of the on-line or real-time interface would not be lost. A third option, was to use a hybrid off-line/on-line approach where the expanded research model was run only on-line when "model-sensitive" environmental or biological conditions were present.

Figure 22.2 shows how the latter compromise solution was accomplished (note that this figure is essentially an expanded version of Figure 22.1; see Croft *et al.*, 1976b for further discussion). Whenever a significant number of spider mites was detected by growers (i.e., densities > three mites per leaf), a monitoring sample of predators and prey mites was taken in the orchard and the predator–prey count then immediately interpreted for the grower, using an off-line index. This index was based on empirical studies of predator–prey interaction under average environmental conditions or derived previously from the model on the basis of expected temperature data. These indices recommended reliance (1) on complete chemical control, if predators were very low in relation to prey; or (2) on complete biological control, if the ratio was in favor of the predators; or (3) on a combination of these in intermediate conditions (Croft, 1975). If model-sensitive environmental conditions (i.e., sustained cool temperatures) were monitored thereafter, or if the predator–prey ratio was in the critical 50–50 region then the simulation model was run to determine the more exact interactions and an updated recommendation was made to the grower. On average, running of the model on-line was necessary only about 10% of the time and this combination of off-line/on-line operation was much less expensive than always using the on-line model.

22.2.2. Codling and Oriental Fruit moth

The codling moth (*Laspeyresis pomonella*) and oriental fruit moth (*Grapholitha molesta*) were studied in the second pest control model. These show life systems

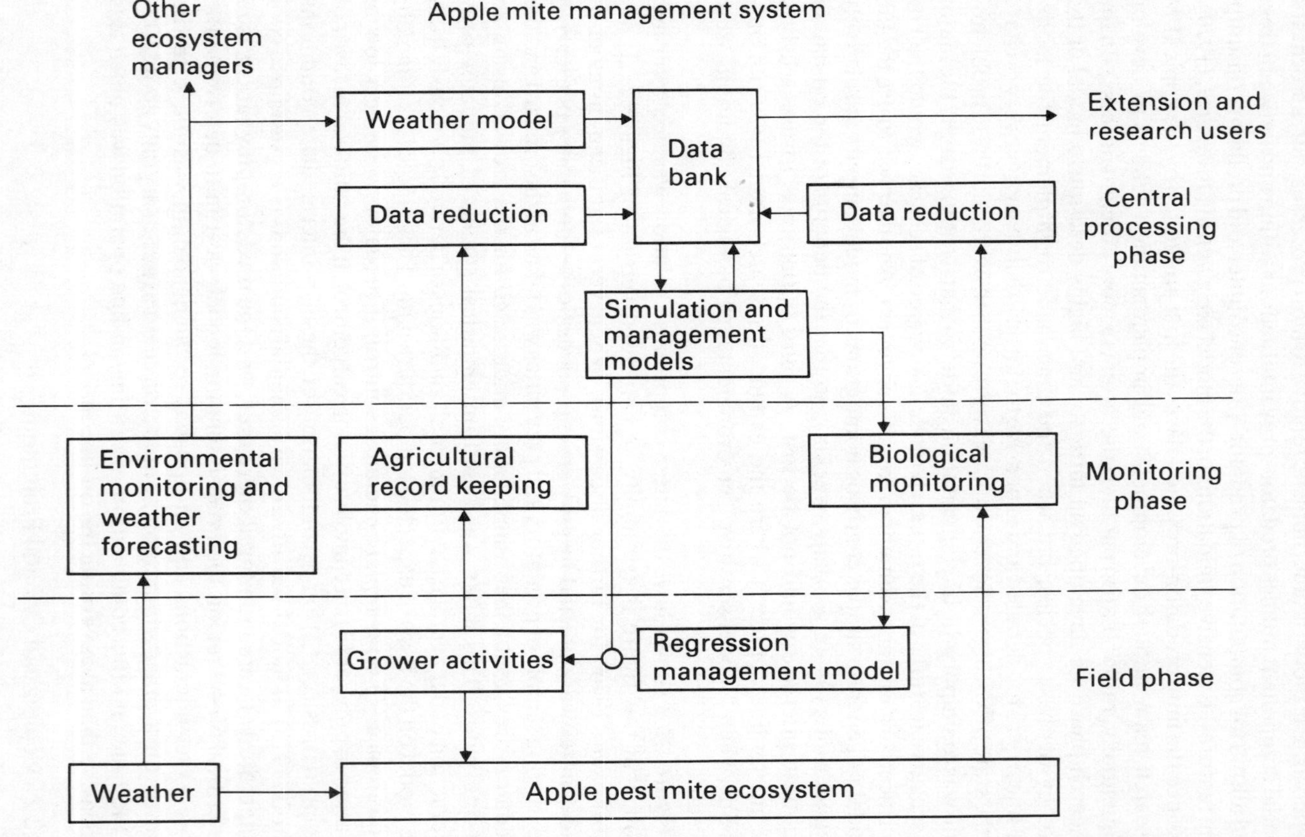

FIGURE 22.2 An implementation system for on-line management of plant-feeding mite pests (adapted from Croft *et al.*, 1976b).

typical of a large group of direct key pests which feed on the apple fruit. These pests occur every year and can only be tolerated at densities causing $\leqslant 1\%$ infestation at harvest. Typically they do not have effective natural enemies and thus chemical pesticides are applied repeatedly during the growing season to suppress them.

The most immediately useful models for pests of this kind are relatively simple phenology models which take accurate biological monitoring estimates (e.g., those obtained from pheromone traps) and provide precise timing of control measures and other management operations (e.g., sampling, pesticide applications, etc.). Riedl *et al.* (1976), Welch *et al.* (1978a) and Croft *et al.* (1980) have described the basic biological components and data requirements for these types of models. It was realized early that a detailed, cohort model run on a daily basis during a growing season and at 50+ weather sites within the Michigan fruit belt would be too elaborate and costly for simulating the development of 20 or more pest species of this kind. Furthermore, the biological monitoring necessary to determine density estimates for species having such low economic thresholds was a practical impossibility.

As an alternative, a generalized, continuous-time phenology or timing model based on two mathematical approaches (i.e., Kolomogorov and *k*th-order distributed delay process; see below) was developed (Welch *et al.*, 1978a). This system, the predictive extension timing estimator (PETE), allows for the construction of developmental models of crops, pests, natural enemies, etc., based on a minimum set of biological parameters (such as development rates, temperature thresholds, initial maturity distributions, oviposition functions, and population variance components). Although specifically designed to simulate the development biology of a wide range of organisms, PETE allows for the incorporation of attritional losses due, for example, to control or natural density-dependent mortality. PETE models can thus be expanded to include pest population dynamics and management factors.

After testing various mathematical approaches to the PETE system and to a range of apple pests, we have come to rely on the distributional delay approach, initially developed by Manetsch and Park (1974). It is an aggregative model of a microprocess where each entity entering the delay, say at time $t = 0$, has a probability $p(\tau)$ of emerging at time $t = \tau$. A $p(\tau)$ of particular interest, the Erlang family of density functions, is characterized by the parameters α and κ (an integer). Specifically,

$$p(\tau) = \frac{(\alpha\kappa)^{\kappa}(\tau)^{\kappa-1}e^{-\kappa\alpha\tau}}{(\kappa - 1)!} . \tag{22.1}$$

The mean of the distribution, $E(\tau)$ is $1/\alpha$ and the variance of τ is given by

$$\text{var}(\tau) = \frac{1}{\kappa\alpha^2} = \frac{\text{MEAN}^2}{\kappa} , \tag{22.2}$$

where κ defines the individual members of the Erlang family. It is apparent from eqn. (22.2) that this function can be used to describe a wide range of processes, with distributions ranging from the exponential ($\kappa = 1$) to a normal with the mean $1/\alpha$ and zero variance ($\kappa \rightarrow \infty$).

The specific details of the construction, parametrization, and validation of the distributed delay developmental model for *L. pomonella* and *G. molesta* have been described elsewhere (Welch *et al.*, 1978a; Croft *et al.*, 1980). To date, timing models using the κth order distributed delay process have also been developed for the tufted apple budmoth (*Platynota idaesalis*), redbanded leaf-roller (*Argyrotaenia velutinana*), white apple leaf hopper (*Typhlocyba pomaria*), apple maggot (*R. pomonella*), San Jose scale (*Quadraspidiotus perniciosus*) and tentiform leafminer (*Lithocoletis blancardella*). Validation of these models and development of additional apple pest timing models are currently in progress. Beyond the Michigan apple project, this same generalized model has been validated in several fruit-producing states of the US (Utah, Washington, California, North Carolina, New York) and has been proposed as the basis for a national apple pest forecasting system (Gillpatrick and Croft, 1979). It has been applied also to rice and corn pests both in the US and other countries with good success.

22.3. ENVIRONMENTAL MONITORING SYSTEMS

On-line IPM models require real-time access to a wide range of environmental parameters. Often, a broader range and increased resolution of weather measurements are required compared to earlier approaches to IPM. More emphasis is also placed on obtaining accurate weather forecasts for estimating feedforward control possibilities. On-line IPM models typically require the standard classes of weather parameter inputs but compared with the other components (i.e., biomonitoring, pest models, implementation), entomologists have made relative little effort to develop unique weather monitoring systems for on-line systems. Rather, we have sought to access existing agricultural sources in real-time and interface this information with existing models.

In the initial phases of the Michigan on-line apple IPM project, weather information was gathered from a variety of sources. Daily maximum and minimum temperatures, precipitation, wind speed and direction, dewpoint, barometric pressure, visibility, runway conditions, etc., for 11 major airport class 1 weather stations in the region were obtained on a daily basis throughout the year. In addition 58 class 2 weather stations located on farms, research facilities, water and sewage plants, and other agriculture-related sites provided similar types of data. Four-day temperature forecasts for four subregions within the state also were obtained, including text advisories of general weather conditions, precipitation probabilities, frost and dew conditions, cloud cover, and storm potentials.

22.3.1. New Systems

These kinds of data were sufficient for the initial demonstration of the benefits of on-line pest control. Today, however, a wide array of additional agricultural weather developments are in progress, in part stimulated by the needs projected by our prototype operation. These include:

(1) AWN (aviation weather network): weather information is delivered over national weatherwire channels on an hourly basis. This is currently being integrated into the on-line weather network at Michigan State University using a microprocessor to eavesdrop on the system and to summarize relevant data. The parameters available include those listed above for the major airport class 1 stations. In total over 60 stations within the north-central region of the US will be available.

(2) *NATS* (national agricultural weather touch tone system): a prototype agricultural weather access and delivery system being tested in selected states in the US (Maryland, Florida, Ohio, Michigan). The system is based on a station from which daily observations of max–min temperatures and other weather variables are collected and transmitted, via touch-tone telephone lines, to a national minicomputer for summarization and output. Currently, data received via this source at the National Weather Center are accessible via the MSU weather acquisition computer using telephone lines and direct hookup.

(3) *Green thumb project*: a national extension delivery system based on a microprocessor box that hooks up to a user's television set and allows for weather data input and weather summary display outputs. Data inputs are linked by telecommunication to a national weather computer via a keyboard console in the microprocessor box. At present this prototype is being tested in Kentucky, with projected use regionally and ultimately throughout the US. For 1979, about 200 experimental units are being tested in two counties in a first-year experiment. Michigan plans to test the system in the second year of operation.

22.3.2. Further Needs

Research on the system has identified needs for greater resolution in several additional aspects of weather monitoring:

(*a*) Evaluation of experimental studies on weather extrapolation to sites between major weather stations (Welch, unpublished; Fulton and Haynes, 1975) reveals that the required level of spatial resolution in macroscale weather monitoring systems is still largely unknown for most IPM or agricultural management operations. Such evaluations are important in explaining patterns of area-wide pest development (see PETE statewide forecasting system) or patterns of movement of highly dispersive/migratory species.

(*b*) Mesoscale environmental monitoring is highly effective for predicting pest dynamics within orchards, using temperature data taken from a nearby station or even from a single instrument located within an orchard in an on-line mode. The effects of slope, topography, surrounding vegetation and elevations are particularly variable in orchard environments, although little is known of their influence and the magnitude of their effects on biological components of the pest ecosystem. Such studies are especially critical in fruit-growing regions of western US, where orchards are often located in watershed drainages or in mountainous regions.

(*c*) At a microscale, both significant and nonsignificant environmental effects on orchard pest popultion dynamics and control have been noted. Differing emergence patterns of moth species corresponding to various pupation habitats on or under apple trees, probably in response to incident solar radiation, have been reported by many authors (e.g., Baker *et al.*, 1980). At the other extreme, Rabbinge (1976), has evaluated the influence of leaf boundary-layer temperatures on predator–pest mite interactions in Dutch apple orchards and showed that the differences between leaves had very little effect on simulation results. A small difference in a factor such as oviposition between the two species was of far greater significance in influencing biological control. Rabbinge concluded that detailed microclimate was not essential to simulate reliably population fluctuations of the fruit tree red spider mite *Panonychus ulmi*, and that mesoscale air temperature was an adequate input variable.

Additional research is clearly needed into weather monitoring requirements at each level of hierarchical resolution in order to improve the predictability and evaluate the sensitivity of on-line IPM models for orchard-inhabiting arthropods.

22.4. BIOLOGICAL MONITORING SYSTEMS

In the initial developmental phases of the system, relatively traditional monitoring techniques for both pest groups (mites and codling moth–oriental fruit moth) were used. For the direct key pests (codling moth, oriental fruit moth) biological fixpoint samples of insect activity or stage distribution density samples were obtained so as to synchronize models with field population development early in the growing season and at periodic intervals thereafter (Riedl and Croft, 1978; Croft *et al.*, 1980). Pheromone traps were especially useful tools, although calibration between trap catches and absolute population measurements was essential (Riedl *et al.*, 1976; Baker *et al.*, 1980).

For plant-feeding mites, the initial approach to biological monitoring was based on a statewide study of mite populations in over 600 orchards over a six-year period (Croft *et al.*, 1976c). Variations in distribution were examined between leaves, trees, and orchards. From these samples, dispersion statistics and optimal sampling procedures for both predatory and prey mites at varying density levels (expressed as error margins relative to the standard error of the true mean density) were obtained and used in the initial phase of implementing the program.

22.5. SIMULATION OF BIOLOGICAL MONITORING SYSTEMS

The biological monitoring system for plant-feeding mites in the apple on-line IPM program has progressed considerably beyond the more traditional level of development and evaluation described above. Evaluation now proceeds by means of simulation studies in the context of overall on-line system operations (Welch, 1977, Welch *et al.*, 1978b, Welch and Croft, 1979). This form of analysis is now reviewed to show the kind of sensitivity testing and evaluation of alternate designs that are needed to determine an optimal configuration.

While the closed-loop system described in Figure 22.1 indicates the feedback between pest population states and control effects via biological monitoring and the bidirectional flow of information, it is a considerable oversimplification of most actual pest management monitoring systems. In Figure 22.3, a more detailed (and realistic) representation of the *A. fallacis*–plant-feeding mite system is presented to emphasize the various aspects of biological monitoring. This system has three hierarchical levels including a decision-making level, a monitoring unit level, and a regional management level. While these levels of function or operation may coincide geographically, more often they relate to progressively larger geographic areas. The decision-making level usually is at some local level of management, such as a farm. The monitoring unit may be a subunit within a farm (e.g., for mite pests of apple trees) or at a multiple farm or area-wide level (as would likely be the case in monitoring moth pest movements). The regional unit level usually includes the umbrella organization for coordinating area-wide management and encompasses a group of farms, a district, a county, or possibly a state.

As can be seen in Figure 22.3, there are a variety of operations going on within each level of the monitoring–management system that must be identified and evaluated in designing an appropriate monitoring system. They include the study of certain biological processes within the ecosystem (e.g., how often should monitoring occur?), stochastic features of the agro-ecosystem and monitoring components (e.g., how precise or of what quality must the monitoring information be?), time delays associated with the various operations of the entire control loop (as indicated in Figure 22.3), and finally

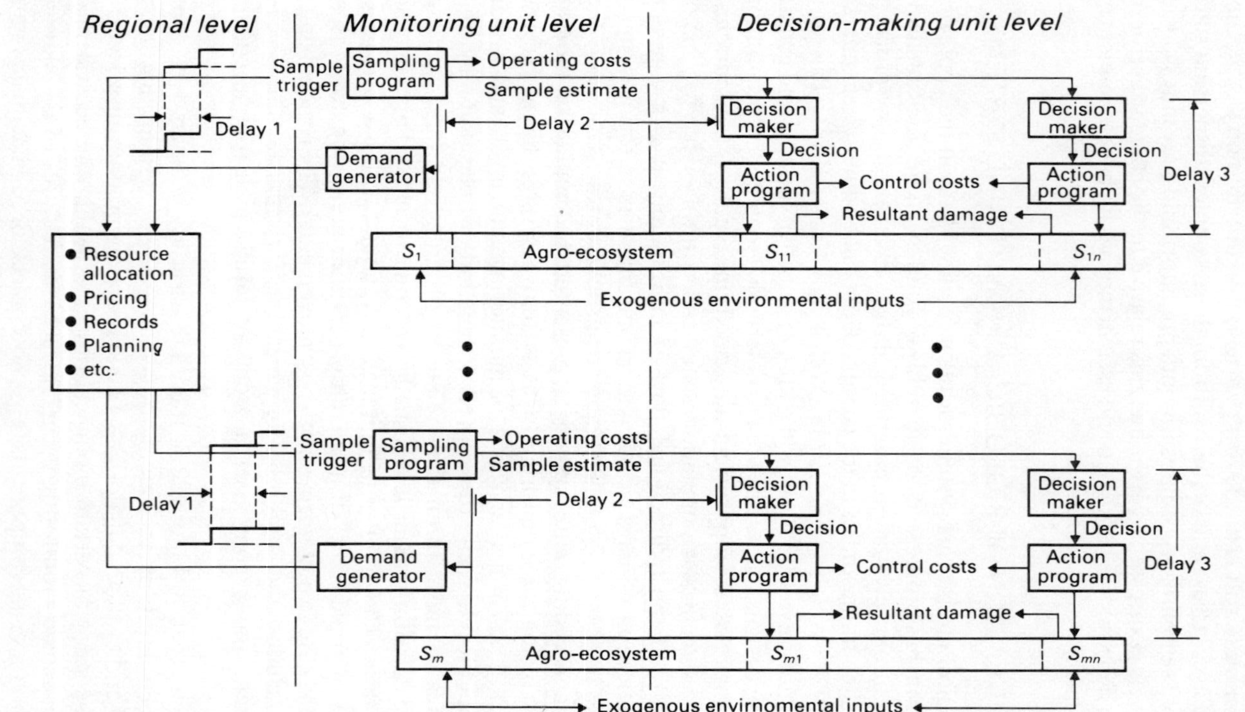

FIGURE 22.3 Components of a detailed biological monitoring system for plant-feeding mites including sampling operations, time delays, decision-making points, and economic factors (adapted from Welch, 1977).

the economics of various alternatives designs. Several of these monitoring system design features as they are applied to three alternative mite monitoring systems are now discussed in greater detail.

22.5.1. Mobile Monitoring Systems

Elements of a traditional IPM scouting system for deciduous tree pests has been described by Croft *et al.* (1976c) and Olsen *et al.* (1974). The traditional system is referred to as reduced system II. To monitor mites in this system a scout takes a leaf sample from the orchard, refrigerates it, and takes it to a centralized laboratory. Samples are then counted, control recommendations determined, and results mailed to the grower. Two alternatives (full system and reduced system I) have also been considered. In the full system a van equipped with a desktop programmable calculator, a computer terminal and laboratory equipment is dispatched to an orchard, when the grower detects mite populations and requests a count. Two scouts take the sample, analyze it, and make a recommendation using the data-processing equipment. Count summaries are automatically transmitted to a central computer via the terminal at the end of the day. These data are then stored and used for assessment in the on-line mode, if they conform to one of the two criteria listed earlier (i.e., subsequent environmental conditions or a critical predator–prey ratio). Reduced system I differs in that data are sent to central computer at the end of the day by telephone and entered by a technician.

Although the mobile monitoring systems (full and reduced system I) are used at the farm level for specific determinations of crop, pest, and natural enemies, their costs are shared on a regional basis by many growers. This permits concentration of equipment for local use that would be too costly for any one producer to acquire. Data obtained via a mobile system may be entered into a hierarchical IPM delivery system (see below) at either the farm or regional level. From there it may be summarized and distributed to other users of the system (researchers, extension personnel in adjacent areas, etc.). Operating management (demand projections, scheduling, accounting, etc.) for the mobile system may reside at the regional or state level.

22.5.2. Analysis of the Mobile Monitoring System

The simulation model used to analyze these various alternative monitoring systems consists of two parts. The first part assesses the operation of the van in the field, while the second analyzes the outputs of the sampling simulation programs. Figure 22.4 illustrates the mobile system and the corresponding activities carried out by each program within the simulation model.

Program QUEGEN generates a series of requests for van service, based on historical distributions of mite outbreaks broken down by date and area of

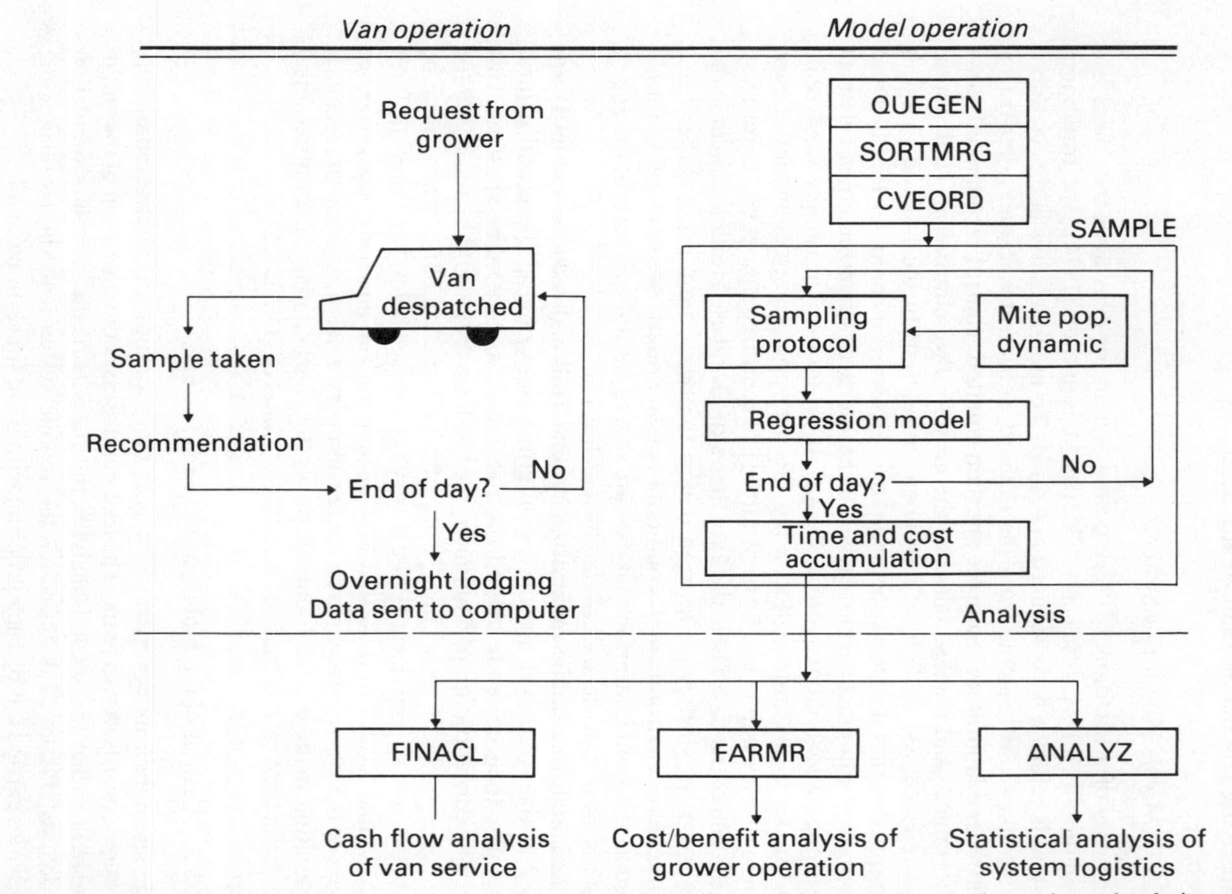

FIGURE 22.4 Operations of a mobile mite monitoring system including activities carried out by various simulation programs (adapted from Croft *et al.*, 1979).

state. SORTMRG then sorts the requests chronologically, and CVEORD reorders each day's requests to minimize travel time using matrices of travel times between areas of the state. SAMPLE simulates the operation of the van itself. Activities requiring fixed amounts of time (e.g., unpacking and repacking equipment, etc.) are simulated by adding fixed constants to an internal clock. Other activities (e.g., sequential sample collection and counting) are functions of mite density and regression equations based on previous scouting experience (Croft, unpublished data), and formulas developed by Welch (1977) are used to determine these times.

For each site, the densities of the predatory mite, *Amblyseius fallacis* (Garman) and the European red mite, *Panonychus ulmi* (Koch), on the date of the sample request are created from cumulative probability functions for both mite types taken from records of mite densities (Croft, unpublished data). These densities represent the "true" state of nature. The waiting time between date of request and data sampled are then calculated. Predator and prey populations are updated daily using a very simplified population model based on that reported by Dover *et al.* (1979). The resulting densities are taken to be those determined in the sample. Control recommendations are made using these densities and a static management model (Croft, 1975). For prey–predator ratios where the static model yields an indeterminate result, the "back-up" model-based procedure discussed earlier is simulated. It is assumed that detailed computer analysis would result in a "no-spray" recommendation and all costs are incremented accordingly. SAMPLE then calculates the date on which the recommendation would be implemented.

The probability of noncontrol (*PNC*) is next calculated for subsequent use in the cost analysis. This is determined from the formula

$$PNC = \frac{1}{0.95}\ CL\ c(a, e, s, t)\, p(a, e \mid A, E)\ \mathrm{d}a/\mathrm{d}e \tag{22.3}$$

where A and E are the estimated densities of the mites *A. fallacis* and *P. ulmi*, respectively, and a and e are the corresponding hypothetical true state of nature. The density function p gives the likelihood of the mite populations being in state (a,e) given that (A,E) has been measured. This distribution is assumed to be an uncorrelated bivariate normal with variances as determined by Croft *et al.* (1976c). CL is a 95% confidence for this distribution, and $c = 1$ if populations in state (a,e) on the date sampled exceed a nominal economic injury level ($e = 15$ mites per leaf) given a control measure s applied at time t; $c = 0$ otherwise. Tabular values for c were determined by extensive computer studies.

Finally, SAMPLE simulates the travel time involved in driving the van to the next sampling site. The total sampling process is repeated until all requests for a given day are completed. The van is then sent to an overnight lodging by entering the travel time and noting the type (i.e., cost) of lodging used. The telephone, computer, and labor charges appropriate to each system's overnight data processing are calculated. For reduced system II the process is similar except that sample counting is simulated as occurring the day after sample collection. All time, cost, and density data generated by the simulation of each sample are combined into information "packets", which provide the input used by all the analysis programs.

FINACL analyzes van activities from the operator's viewpoint and is primarily an accounting system of costs and revenues, and FARMR analyzes the costs to the grower. Tables similar to Table 22.1 are calculated on a per grower basis, per acre basis, and a statewide total basis. Other such tables are calculated under the assumption that the growers apply the recommendations on more than just the 10 acres actually monitored. Questionnaires have revealed that this is common practice.

TABLE 22.1 Cost analysis for growers.

Cost of the strategy	
A. Cost of the control measures	
(1) Cost of material	----
(2) Application cost	----
B. Cost of the pest management system	
(1) Cost of making request	----
(2) Charge for monitoring	----
Expected damage	----
Total cost to grower	------

Finally, ANALYZ tabulates logistical information about the sampling system and calculates a series of distributions based on the simulation parameters and the sample packets. These distributions include number of requests by area and date, number of requests per day, and amount of time spent counting, etc.

22.5.3. Examples of Simulation Results

The analysis programs generate large amounts of data. By examining graphs of this data (Figures 22.5 and 22.6), several important trade-offs between design variables were investigated. Two examples of these are presented here:

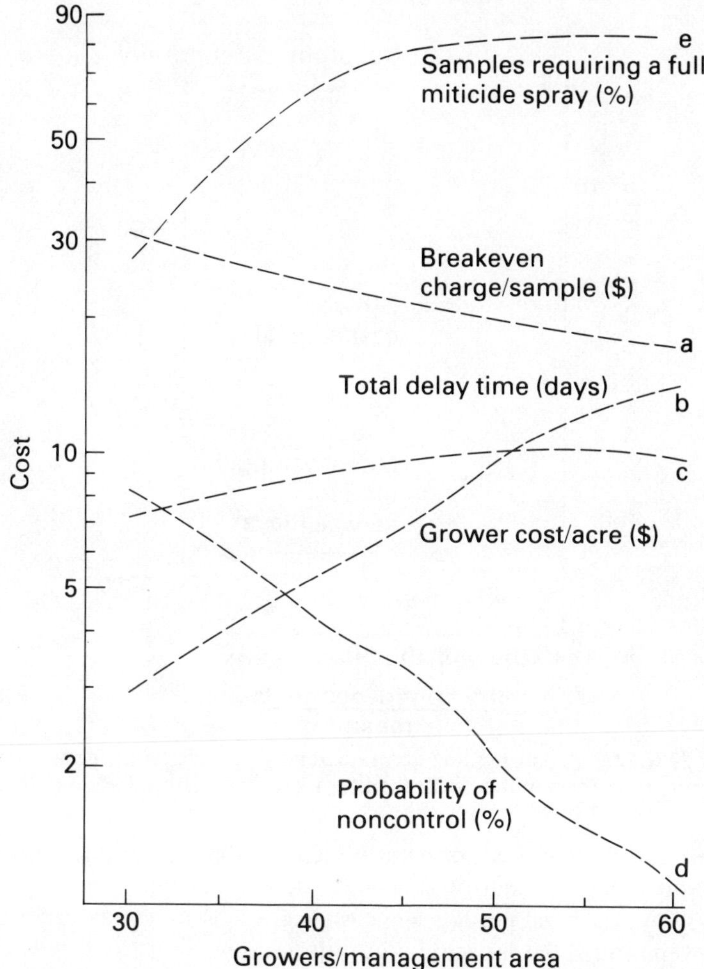

FIGURE 22.5 Output tradeoffs involving the operation of a mobile mite-monitoring system over regional grower population (adapted from Croft *et al.*, 1979).

Numbers of Growers Served

Michigan was divided into seven mite management areas and the number of growers served per area varied. The break-even charge was defined as the payment per visit resulting in zero net profit after taxes to the van operator. Figure 22.5 (curve a) shows that this charge drops from $30.00 to $18.00 as the number of growers served per area increases from 30 to 60. These economies of scale can be passed on to the grower as a lower service charge.

However, this charge is only a component of the cost faced by the grower. As the numbers serviced by a given system increase, the time delay between date

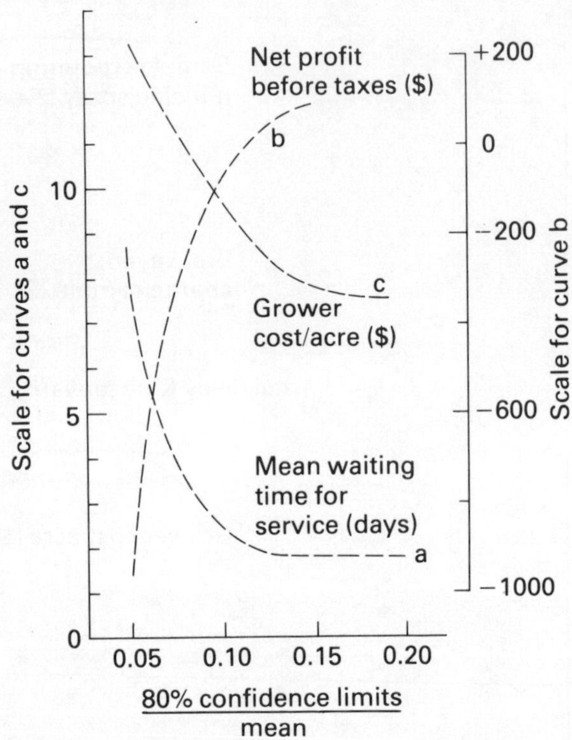

FIGURE 22.6 Effects on sampling accuracy relative to net profit, grower cost, and mean waiting time for a mobile mite monitoring (adapted from Croft *et al.*, 1979).

of request and data of monitoring also increases (curve b); this can result in greater crop damage or control costs due to delay in applying control practices. Curve c shows the total per acreage costs increasing to a peak at 50 growers served, even though the service charge is decreasing.

The reduction on total cost above 50 growers per area seems to result from reduced service charges (curve a) and reduced crop damages (curve d) offsetting increased control costs (curve e). At first curve d may seem paradoxical; one would not intuitively associate better control with longer delays. The answer is that mite increases that might have been controlled with low pesticide dosages earlier, are instead receiving later, more expensive, but more certain, full dosages. This is the real significance of curve e.

Sampling Accuracy

The immediate effects of increasing the accuracy of sampling are shown in Figure 22.6. The abscissa is the width of the 80% confidence limits of the mean population density expressed as a fraction of the mean. As this decreases to 10, mean waiting time is little affected; below this, however, delay time is increased exponentially (curve a).

Curve b shows the effect on the service operator of increasing sample accuracy. Given a fixed service charge, profits decline rapidly as the width of the confidence limits is decreased below 10% of the mean. Curve c indicates that the total costs to growers increases rapidly if accuracy is increased beyond this point due to increased waiting time. *An important conclusion one can draw from these two graphs is that there are levels of accuracy which, far from being helpful to the grower, can actually be detrimental.*

22.5.4. A Simplified Approach

The complex model discussed above required a great investment in time and effort to develop, although it permits system behavior to be analyzed in great detail. To complement this approach, Welch (1977), has developed a simpler procedure that allows a preliminary screening of a variety of biomonitoring design alternatives with reduced effort.

22.6. EXTENSION DELIVERY SYSTEMS

This component of the overall system includes the transmission of biological and environmental monitoring data from the field to the microprocessor-based model and the return of the interpreted model output results to decision makers. As for biological monitoring operations, there are several hierarchical levels of research, extension and field operations personnel involved in each of these phases (see Figure 1 of Croft *et al.*, 1976b). An array of computer telecommunications capabilities is also needed within each level of operation. The delivery component of the system has evolved rapidly in response to model developments and needs, data processing requirements, changing hardware and software technology, and an increasingly larger array of system users. The development of centralized and hierarchically distributed delivery systems are now discussed.

22.6.1. Centralized IPM Delivery Systems

An initial attempt to develop a computer-based IPM delivery system was begun in 1974 using an interactive, centralized mainframe computer (based on a CDC 6500). It was first used by the project in 1975 (Croft *et al.*, 1976b) and has been expanded subsequently to include a similar on-line project on field crop and vegetable IPM (Brunner *et al.* 1979).

PMEX (pest management executive system) is a software program designed to provide users with access to pest management programs and systems level documentation. It makes the information storage and processing capabilities of computers available to users with only a limited understanding of these tools. Detailed computer control sequences and documentation are stored under keyword headings that can be easily used by untrained individuals (see Croft *et al.*, 1976a; Brunner *et al.*, 1979 for a more detailed discussion).

Remote-site send–receive terminals and telephone playback devices, used in combination with the central processing program, provide a communications network linking extension specialists, county agents, pest management personnel, privately employed fieldmen and growers (see Croft *et al.*, 1976a). As a consequence, the lag between research and application is appreciably reduced and there is a marked improvement in the delivery of IPM information to decision makers in the field.

PMEX provides communications features, biological monitoring summaries, environmental monitoring programs, and predictive IPM models (see Table 22.2). Communication is an extremely important component of any delivery system and is especially valuable in the initial development phase, since access to more complex and detailed management models is usually limited. Memoranda and newsletter-type information may be disseminated more rapidly and to a wider spectrum of interested personnel than in a traditional delivery system. Biological information, including data on population states of crops, pests and beneficial populations, can be transmitted quickly and efficiently to decision makers in the field. Environmental monitoring programs serve to collect and summarize weather inputs received from a variety of US weather network sources (see Section 22.3) for use by growers. These programs also are designed to be interfaced, in a real-time mode or with a very short time delay, with predictive IPM models for pest forecasting.

The use of PMEX in linking information from environmental and biological monitoring sources with a pest management model is indicated in Figure 22.7. As can be seen, alert routing is accomplished through the CONTACT program of PMEX.

Different pest management models will use the same environmental and biological monitoring software programs (e.g., INPUTER, DEGREE DAYS, Table 22.2), but output delivery features to users are quite different. For the plant-feeding mite model the essential steps of implementation were outlined in Figure 22.2 and discussed above. Recommendations are provided from this model concerning the probability of biological control without the use of selective acaricides and/or the need to adjust predator–prey ratios in favor of the natural enemy. A variety of output options ranging from detailed density time series outputs, estimates of peak pest density, or regional summaries of general mite distributions are also available. In all cases, these outputs are made in response to an initial set of predator–prey density estimates and appropriate weather data provided or accessed by each user.

TABLE 22.2 Program types and examples of application programs available through the PMEX extension delivery system.

Communications	Biological information	Environmental features	Predictive models
Contact: sends messages of all types and in any format among PMEX users.	*Alfalfa, sugarbeet, asparagus, onion, potato*: provides pest summaries for these crops.	*Forecast*: provides local daily weather forecasts, including four-day temperature predictions.	*Blitecast*: a late blight forecasting model for timing of fungicide applications in potato production.
Alerts: sends pest alerts and control strategy information to extension field staff.	*Blacklight*: summarizes insect information obtained through a blacklight trap program.	*Degree days*: provides daily, weekly, and seasonal precipitation summaries from 58 sites.	*Weevilcost*: economic model of alfalfa production including variable cutting times, insecticide applications, and management of alfalfa weevils.
Automatic scheduling: allows users to receive crop pest information periodically without request.	*Blitesum*: provides summaries of late blight program information generated by BLITECAST.	*Precipsum*: provides daily, weekly and seasonal precipitation summaries from 58 sites.	*Mothmodel*: phenology model describing the development of the codling moth during a growing season.
GETMSG: allows user to obtain alerts and obtain messages through another user's PMEX account.	*Inputer*: A generalized input program for biological data.	*Wthrinput*: inputs and displays raw weather data from 58 sites.	*Potatopest*: simulates potato losses caused by root-lesion nematodes and generalized insect defoliation.
Manual: complete set of PMEX instructions and program descriptions to a user.			*Mitemodel*: prey–predator model for biological control of the European red mite by the predatory mite, *Amblyseius fallacis*.

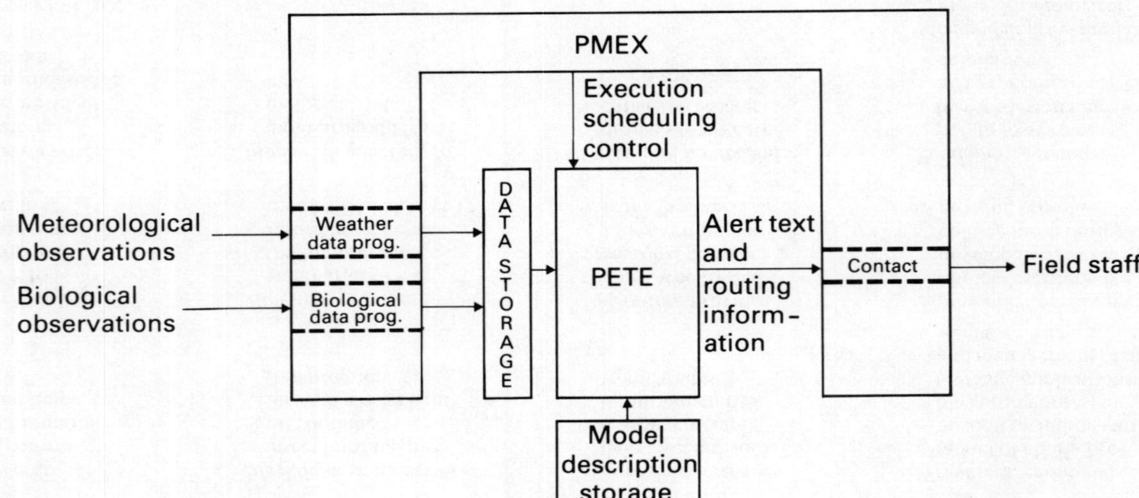

FIGURE 22.7 PMEX–PETE linkages to the agricultural system and pest management personnel (adapted from Welch *et al.*, 1978a).

By contrast, PETE outputs for the codling moth or oriental fruit moth timing model are standardized using PMEX CONTACT programs, and text outputs are generated automatically to meet the needs of a wide range of extension personnel dealing with individual or regional groups of growers (Welch *et al.*, 1979a). Biological monitoring inputs are helpful in improving model outputs, but they are not essential to PETE system operations. Compared with the mite model, PETE has a wider range of communication and data presentation options.

Since the implementation of PMEX in 1975, records of use have been collected and evaluated. Use by extension personnel averages about two times/week/user throughout the April–October growing season. Mean access time is 12–15 minutes per session. The most commonly accessed programs are FORECAST, CONTACT, INPUTER, DEGREE DAYS and WTHRINPUT, accounting for more than 75% of the programs used in 1977. Although use of these PMEX programs probably will remain high in the future, due to their general applicability, it is expected that there will be much greater use of more specific modeling, management, and economic assessment programs as they are more fully developed.

Total costs (including model development and adaptation costs) for PMEX to date have been about \$86 000 and operating costs about \$25 000 per season (Edens and Klonsky, 1977). Since much of the basic system development is completed, it is anticipated that these costs will now decrease relative to operation costs. Even for the 1975 versus the 1976 data, developmental costs declined by some 50%. Most future expenses will be due to new program development and general system improvements and maintenance.

While PMEX has served as an excellent prototype extension IPM delivery system and has been heavily used and tested, it has limitations. In Table 22.3 a listing of the strengths and limitations of this system are identified, based on an economic and use assessment (Edens and Klonsky, 1977, Miller, unpublished data). In the future some PMEX features will be deleted and others added as a less centralized delivery approach is developed (see below). PMEX unit costs can be expected to decline as the number of users increases. Perhaps, more importantly, development and implementation of these systems in the future can be undertaken at a substantially reduced cost because of the work already completed on PMEX.

22.6.2. Hierarchically Distributed Systems

Two major difficulties associated with the types of centralized systems discussed above are high communications costs and low reliability and availability. The high communications costs are due to the high datalink charges (long-distance telephones, dedicated lines, etc.) for even the most trivial uses of the system. In

TABLE 22.3 PMEX strengths and limitations in relation to use, economic and operations assessment.

Strengths	Limitations
Two-way delivery of information with same communication channel useful in both directions	PMEX is a centralized system and as such has potentially reduced reliability and availability
Allows users to be contributors	
Minimum time delay between inquiry and response	The prototype employed a single computer which was excessively large for many applications
Users can interact with other users	Unrealistically high communication costs (exceed computation costs)
PMEX permits interrogation of computer-based data	The system is not capable of dealing with local needs in an economically efficient manner
Permits judicious use of environmental information	The system does not represent an optimum combination of hardware and software
Many biological and environmental events can be easily summarized	
Information about management failures as well as successes is documented	
Permanent machine-readable records of all observations and contributed data are readily available for use by research and extension staff	
Not restricted to pest management—also available for crop and resource management	
Users are not intimidated and make high use of PMEX programs	
Field staff are the strongest supporters of continued development of a similar, more efficient system	

PMEX telephone charges were 50% higher than total computer charges. There are many types of operations which, although otherwise suitable, cannot be economically computerized on such a system, e.g., area or district extension office record keeping or accounting, scheduling of regional events, etc.

The entire system's dependence on the operation of the central processor results in low reliability and/or availability. There are many factors that can interrupt or interfere with this operation; for example, some systems have several hours a day of scheduled downtime; 24-hour staffing may not be economically feasible; time is needed for preventative maintenance, etc. Unscheduled downtime may result from either hardware or software failure. While the former are generally quite rare, the latter are inevitable and often distressingly common for new systems. Another cause of reduced availability is heavy system loads at certain times. All these factors would not be problems if they were uniformly distributed in time. Unfortunately, experience has shown that the hours 8–9 a.m. and 3–5 p.m. are most critical to extension operations. Scheduled downtime or start-up difficulties are likely to interfere with the former; heavy use in mid-afternoon may affect the latter.

An alternative solution to these problems is to replace the centralized system with a hierarchically distributed structure (see Figure 22.8) in which small, cheap data processors are used to distribute processing power to regional, county, and farm levels. These processors (along with certain pencil and paper methods) form a range of data-processing power reaching from low-cost, low-capability units to the high-cost, powerful computers previously discussed.

The key principle of the hierarchical system is this: if a number of people each have to solve some problems of low to moderate complexity, then a set of simple processors, working in parallel, can often do the job more effectively than a single large processor working sequentially. If each individual has his own small processor, he need not worry about the cost of communicating his problem to some distant site. Furthermore, if one of a group of such processors fails, the majority of users are not affected.

Within such a system, institutional computers operate at state, multistate, or national levels, integrating such IPM activities as pesticide registration program, environmental monitoring or national weather summaries. State IPM central processors manage a variety of functions, including environmental network data from peripheral processors and executive systems features similar to those described in Table 22.2. Some subset of the program options at the state level are then available, via telecommunication links, at county or regional offices to meet specific needs. At the lowest level, hardware, software and IPM decision-making tools (computer terminals, microprocessors, calculators, nomograms, charts) assist pest managers in the field. These field units can be updated, serviced, and provide information to

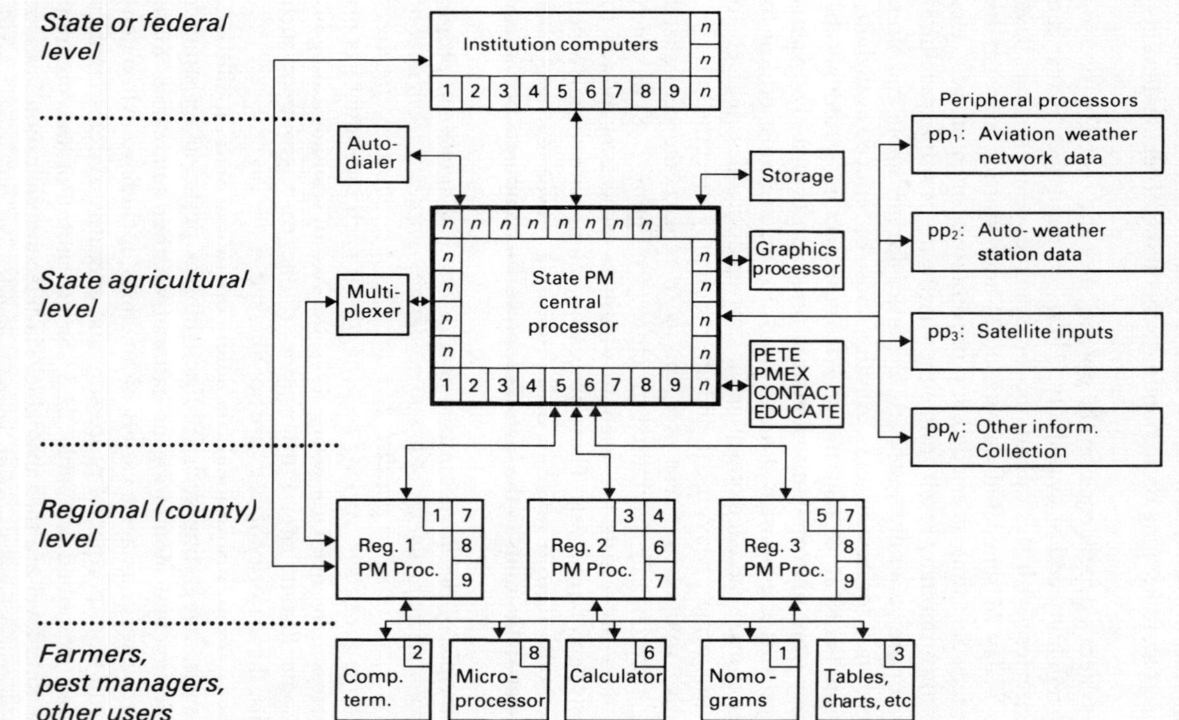

FIGURE 22.8 A hierarchically distributed computer processor based system for communication of IPM information. Numbers within processor illustrate software programs within a given processor (adapted from Croft *et al.*, 1979).

regional processors. They may also serve as input sources back to the state- or higher-level processors. The overall effect is thus to create closed information loops to aid monitoring control in an on-line fashion.

22.6.3. A Prototype Hierarchically Distributed System

While the project has recognized the need for a hierarchically distributed implementation delivery system we have only just begun to evaluate the first stage, using the PETE phenology modeling system for apple pests. The objectives of this research are (1) to develop the necessary software and communications protocols for PETE system operations and interfaces with environmental and biological monitoring at each level; (2) to determine the proper distribution of PETE software in the processor system so as to obtain optimal economic cost/benefits and maximum extension delivery efficiency; and (3) to obtain field validation and user experience with these types of systems.

22.7. ON-LINE IPM SYSTEMS—PROBLEMS AND FUTURE NEEDS

This section addresses our perceived state of the art in the use of IPM models in on-line systems based on experience with the apple IPM project. Some areas of research requiring priority attention in the near future are the following.

(1) We need a better understanding of the types of IPM systems that are most suited to, and those least benefited by, on-line operations. Since many IPM systems are composed of multiple pest complexes, a feasibility study of an on-line system must ultimately include an evaluation of how a total complex of pests can benefit from these technologies. Based on our experience with the apple-project, on-line capabilities are most useful in dealing with IPM of pests that act rapidly (i.e., rapid life cycles) or with pest damage to plants in which the time delay between monitoring (the trigger) and the need to take control action is very short (e.g., 24-hour back-action eradicative properties of fungicides used to control apple scab). If a rapid response or feed-forward predictions are not essential, then more slowly functioning monitoring–response systems are more appropriate and economical. Since on-line systems depend heavily on computer processors and telecommunications technologies, they usually have high initial start-up costs, although in the long term they become much less expensive and may be very cost effective when applied to agricultural management operations beyond IPM (e.g., irrigation, harvesting, thinning). Based on these considerations there is a need for a careful feasibility analysis of the various trade-offs between off-line versus on-line systems development for a particular pest complex or an aggregate of crop–pest complexes.

(2) We need a better understanding of how to develop IPM models for research and extension concurrently so that parallel development of each will be possible and hence resource utilization will be more efficient. For example, what

are the unique demands of each type of model that are compatible or mutually exclusive? How does the coupled feedback from the extension delivery system (including personnel and processor communications sytems) influence the design of extension models and ultimately research models? Is the same level and type of modeling and validation required for each? How can developmental time constraints for these different types of model be reduced? Do generalized modeling approaches apply to either research or application models?

(3) We need to define procedures for on-line system component design and operations testing more explicitly, both at the initial and later stages of development when systems are being evaluated. More specific systems performance criteria for these steps of IPM model implementation would be very useful. These procedures must be based on extensive economic evaluations and should be made in response to the interests of such diverse parties as growers or users, private service operators, research and extension systems, component suppliers (pesticide and natural enemy producers), and ultimately the public.

(4) Technological developments in relation to computer and telecommunications operations are difficult problems for pest-oriented biologists to deal with due to their inexperience with these tools. Thus very early in the development of an on-line control system, agriculture personnel should invest in the help of systems engineers or operations personnel to select and maintain the computer hardware and software components of the system. An ongoing evaluation of new appropriate equipment and operations should be included as part of the long-term maintainence and development of the system. Systems should be designed to provide maximum flexibility in moving from one technological adaptation to another over time. Since on-line systems often include the use of high-speed computer–telecommunications operations, the interface between these tools and the human delivery system element (and ultimately the user) deserves considerable attention.

(5) As noted, each of the component parts of the on-line IPM system are subdivided into hierarchical levels of organization and function. The level of operation or definition in one component often determines the resolution or level of operation needed in another element. In section 22.4, the effect of modeling design and extension delivery features greatly influences the appropriate biological monitoring system and the relevant sampling variability, sampling effort, time delays, and associated economic risk. These interrelationships between hierarchical levels within each component part of an on-line IPM system need further study early in the design process.

22.8. CONCLUSIONS

In summary, in the on-line apple IPM project we have progressed through a first stage iterative cycle of components development (Figure 22.1) as a result of which many new insights into design changes and the trade-offs between components have been realized. We feel that many additional improvements can be realized as we further test other systems elements using simulation analysis, as in the evaluation of the biological monitoring component. We are still far from approaching an optimal design, and this is especially true for such generalizable features as environmental monitoring and implementation delivery systems.

One point has been very clear from these studies: *if model implementation is an eventual goal, then the feedback coupling between the existing on-line components (especially the implementation delivery system) needs to be established early in the research modeling phase so that eventual needs are recognized and incorporated where possible*. Such parallel development of research and implementation models will greatly facilitate efficient utilization of the limited resources available for agricultural research.

ACKNOWLEDGMENTS

Published as an Article of the Michigan Agricultural Experiment Station, East Lansing, MI 48824. This research was supported in part by an EPA grant No. Cr-806277-02-2 to Texas A&M University and Michigan State University.

REFERENCES

Baker, T. C. R. T. Cardé, and B. A. Croft (1980) The relationship between pheromone trap capture and emergence of adult oriental fruit moths *Grapholitha molesta* (Busck). *Can. Entomol.* 112:11–15.

Brunner, J. F., B. A. Croft, S. M. Welch, M. J. Dover, and G. W. Bird. (1980) A computer-based extension delivery system: Progress and outlook. *EPPO Plant Protection Bull.* 10:259–268

Croft, B. A. (1975) Integrated control of apple mites. *Extension Bull. E-825, Michigan State Univ. Extension Service*

Croft, B. A. and A. W. A. Brown (1975) Responses of arthropod natural enemies to insecticides. *Ann. Rev. Entomol.* 20:285–335.

Croft, B. A., J. L. Howes, and S. M. Welch (1976a) A computer-based, extension pest management delivery system. *Environ. Entomol.* 5:20–34.

Croft, B. A. and D. L. McGroarty (1978) The role of *Amblyseius fallacis* in Michigan apple orchards. *Michigan Agric. Expt. Station Rep.* 333.

Croft, B. A., M. F. Michels, and R. E. Rice (1980) Validation of a PETE timing model for the Oriental fruit moth in Michigan and central California. *Great Lakes Entomol.* 13:211–217

Croft, B. A., R. L. Tummala, H. W. Riedl, and S. M. Welch (1976b) Modeling and management of two prototype apple pest subsystems *Proc. 2nd USSR/USA Symp. on Integrated Pest Management, October 1974* (East Lansing: Michigan State University Press) pp97–119.

Croft, B. A., S. M. Welch, and M. J. Dover (1976c) Dispersion statistics and sample size estimates for populations of the mite species *Panonychus ulmi* (Koch) and *Amblyseius fallacis* (Garman) on apple. *Environ. Entomol.* 5:227–34.

Croft, B. A., D. J. Miller, S. M. Welch and M. Marino (1979) Developments in computer-based IPM extension delivery and biological monitoring systems design, in D. J. Boethel and R. D. Eikenbury (eds) *Pest Management Programs for Deciduous Tree Fruits and Nuts*. (New York: Plenum) pp223–50.

Dover, M. J., B. A. Croft, S. M. Welch, and R. L. Tummala (1979) Biological control of *Panonychus ulmi* (Acarina: Tetranychidae) by *Amblyseius fallacis* (Acarina: Phytoseiidae): A prey–predator model. *Environ. Entomol.* 8:282–92.

Edens, T. C. and K. Klonsky (1977) The pest management executive system (PMEX): Description and economic summary. *Michigan State Univ. Agric. Exp. Station Rep.*

Fulton, W. C. and D. L. Haynes (1975) Computer mapping in pest management. *Environ. Entomol.* 4:357–60.

Giese, R., R. M. Peart, and R. M. Huber (1975) Pest management *Science* 187:1045–52.

Gillpatrick, J. D. and B. A. Croft (1979) *Development of a comprehensive unified, economical and environmentally sound system of integrated pest management for major crops: Apples* (EPA/USDA Grant) (Texas: A&M University).

Haynes, D. L., R. K. Brandenburg and P. D. Fisher (1973) Environmental monitoring network for pest management systems. *Environ. Entomol.* 2:889–99.

Manetsch, T. J. and G. L. Park (1974) *System analysis and simulation with applications to economic and social systems*. Dept Electrical Engineering and Systems Science. (East Lansing, MI: Michigan State University Press).

Olsen, L. G., C. F. Stephens, J. E. Nugent, and T. B. Sutton (1974) Michigan apple pest management annual report: 1973. *Michigan Extension Service Rep.*

Rabbinge, R. (1976) Biological control of fruit-tree red spider mite. *Center Agric. Publ. and Doc.* (Wageningen)

Riedl, H. and B. A. Croft (1978) Management of the codling moth in Michigan. *Michigan State Univ. Tech. Bull.* 337.

Riedl, H., B. A. Croft, and A. J. Howitt (1976) Forecasting codling moth phenology based on pheromone catches and physiological time models. *Can. Entomol.* 108:449–60.

Tummala, R. L., D. L. Haynes, and B. A. Croft (eds) (1976) *Modeling for Pest Management: Concepts, Applications and Techniques* (East Lansing: Michigan State University Press).

Welch, S. M. (1977) *The Design of Biological Monitoring Systems for Pest Management* PhD Thesis, Michigan State University.

Welch, S. M. and B. A. Croft (1979) *The Design of Biological Monitoring Systems for Pest Management*. Simulation Monogr. Ser. PUDOC (Wageningen).

Welch, S. M., B. A. Croft, J. F. Brunner, and M. F. Michels (1978a) PETE: An extension phenology modeling system for management of a mite-species–pest complex. *Environ. Entomol.* 7:487–94.

Welch, S. M., B. A. Croft, J. F. Brunner, M. J. Dover and A. L. Jones (1978b) The design of biological monitoring systems for pest management. *EPPO Plant Protection Bull.*

23 Implementation Models: The Case of the Australian Cattle Tick

G. A. Norton, R. W. Sutherst, and G. F. Maywald

23.1. INTRODUCTION

The value of systems analysis for improving our general understanding of pest problems and for showing how to tackle them is well illustrated in Part I of this volume. Undoubtedly, an awareness of a system's behavior, resulting from the often complex dynamics of the ecological processes involved, does provide a basis for deriving general management principles. However, it has been argued elsewhere that this is only valid "...provided the purpose is educational in communicating this 'feel' [for a system's behavior] from one group to another" (Charlton and Street, 1975). Considering agricultural systems generally, these authors continue, "To anyone familiar at the practical level... with all the factors which can distort the real behaviour of such systems, such modelling efforts are extremely simplistic and naive."

While this may be overstating the case, it is nevertheless desirable that attempts should be made to respond to Way's (1973) observation that "...mathematical modelling and systems analysis... have not yet proved their practical value in controlling pests." What we now require are models that are capable of giving "...guidance on specific issues, or of producing information directly useable by decision makers" (Charlton and Street, 1975).

In the context of pest management, systems analysis techniques have two major roles to play: (1) to synthesize, analyze, and disseminate "real-time" information concerning the status and development of pest populations; and (2) to assess and evaluate the performance of available pest control measures. For those situations where monitored information is available and where the major questions concern the tactics of when and where curative control measures are to be applied, real-time modeling can provide a variety of services (Benedek *et al.*, 1974; Teng *et al.*, 1979; Zakoks *et al.*, Chapter 21; Croft and Welch, Chapter 22). In this case, since only short-term predictability is required, albeit of high accuracy, simple pest models are likely to suffice (Tummala and Haynes, 1977; Zadoks *et al.*, Chapter 21).

In other situations, where monitoring is not generally feasible or where questions concerning integrated pest management strategies are being considered, real-time modeling is clearly inappropriate. Here, the emphasis will swing more to the use of systems analysis techniques for assessing and evaluating control strategies. The remainder of this chapter describes the approach that has been developed for the Australian cattle tick (*Boophilus microplus*), a pest problem that falls into this second category on both counts.

23.2. THE CATTLE TICK PROBLEM

The life cycle of the cattle tick is shown in Figure 23.1. In southeast Queensland, where there are four generations a year, the direct reduction in liveweight gain (LWG) caused by engorging females is the major concern. In tackling this problem, the approach described in Figure 23.2 has been adopted. Following many years of empirical research, a simulation model of tick population processes has been developed, allowing the evaluation of various integrated control strategies (Sutherst and Dallwitz, 1979; Sutherst *et al.*, 1979). Through liaison with beef producers and extension agents, strategies

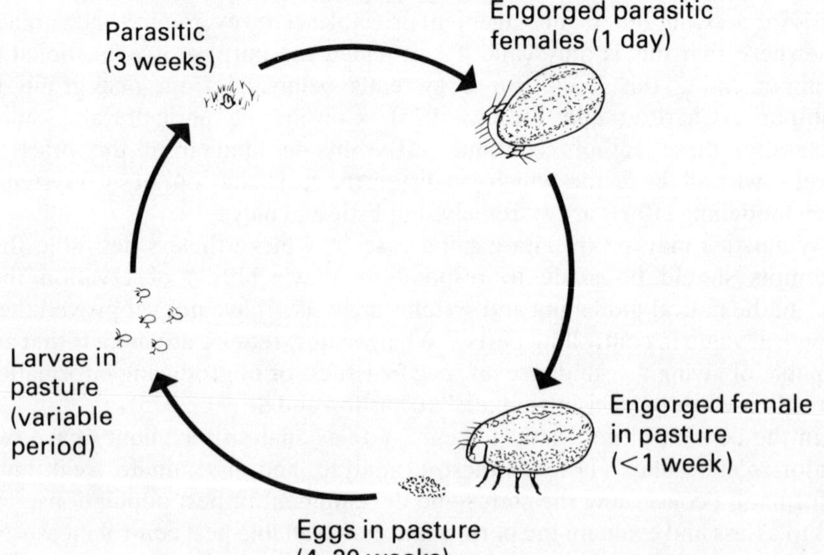

FIGURE 23.1 Life cycle of the cattle tick (*Boophilus microplus*).

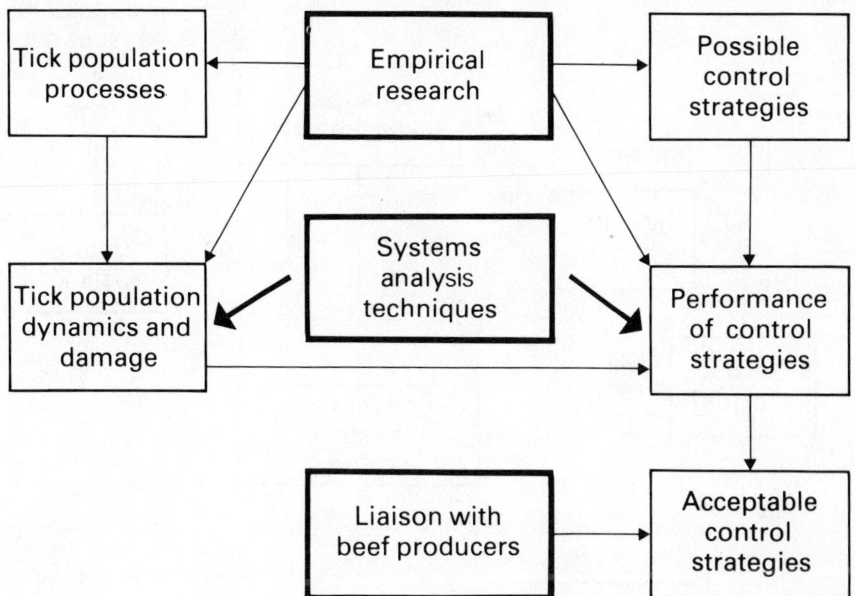

FIGURE 23.2 The systems analysis approach to cattle tick control in Australia.

that meet particular objectives and constraints can then be identified, serving as a basis on which to make control recommendations.

The problem that extension agents face in making such recommendations is that cattle-raising properties (farms) differ from research sites, and from each other, in three respects:

(1) *Intrinsic characteristics*—including the climatic, topographical, and biological features of the farm.
(2) *Design and management characteristics*—including cattle breed and herd structure, stocking density, the grazing system, and the tick control strategies employed.
(3) *Objectives and constraints*—which affect the ability and desire to modify existing design and management characteristics.

Thus, to make a recommendation, the extension agent has to identify the feasible strategies of control from those available (Figure 23.3) and account for other biological and management features of the farm concerned.

With a population model available, at least for conditions in southeast Queensland (Sutherst *et al.*, 1979, 1980), implementation can be tackled in two ways. The first approach is to concentrate on particular farms, using appropriate parameter values to determine the complete, multidimensional, "protection possibility surface" for a particular situation. On this basis the most suitable control strategy for that particular case can be identified. *An*

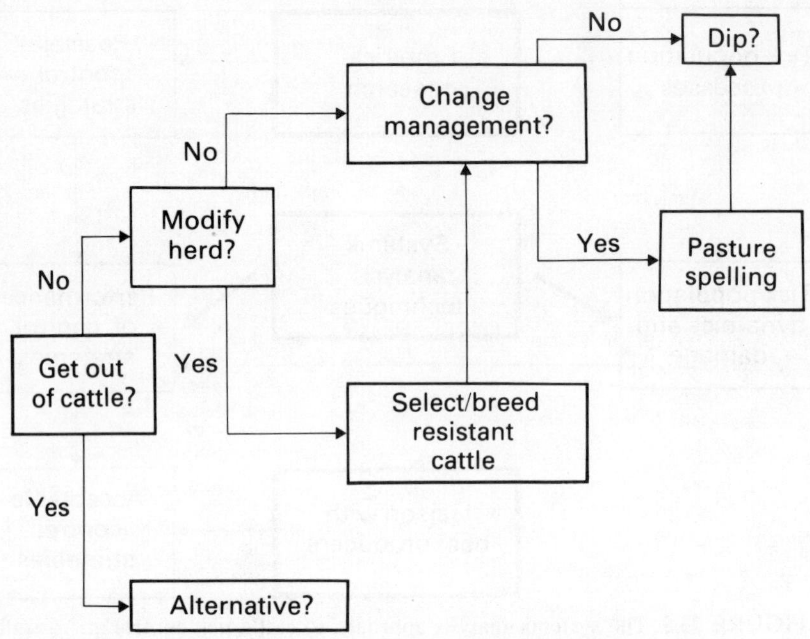

FIGURE 23.3 Choices available for integrated tick control.

alternative approach is to search for "robust" strategies that perform adequately over a wide range of parameter values. Instead of finding the optimal strategy for a specific situation or occasion, emphasis is placed on finding the boundaries within which "robust" strategies perform well. Any individual farms can then be identified with respect to these boundaries, and recommendations made accordingly. This second approach is the one we have adopted (see Norton *et al.*, 1983); here we outline the approach.

23.3. THE MODEL

Conceptually, the cattle tick problem can be simplified by focusing on the three main processes in its life cycle (Figure 23.1)—the development of free-living stages, host finding, and parasitic feeding (Figure 23.4). On this basis, a tick population model has been constructed, allowing assessment of the combined impact of producer's decisions and climatic conditions.

The simulation model used here has 12 age classes, representing eggs (four classes), free-living larvae (four), parasitic ticks (three), and engorged females in the pasture (one). Starting in the spring, in week 0, with an initial input of overwintering eggs, the model updates the 12 age classes each week, according to parameter values for egg and larval survival, hatching rate, host-finding rate, density-dependent mortality of parasitic ticks (caused by

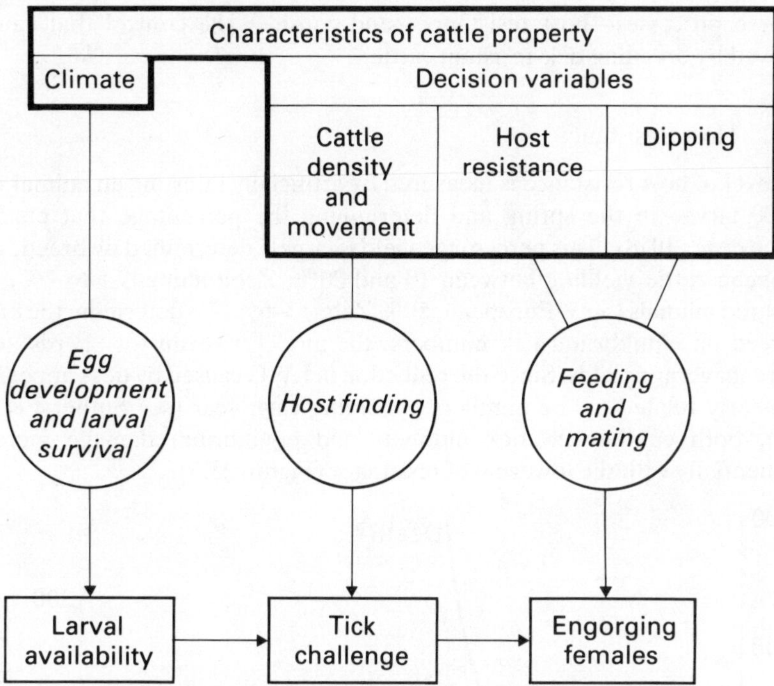

FIGURE 23.4 Key processes in the cattle tick life system.

host resistance), and finally, fecundity. All parameter values have been obtained from field experiments and, where appropriate, these values change to reflect differences that occur during the year. In this way, the model simulates a sequence of four generations that progressively overlap and finally produce overwintering eggs, the success of which determines the size of the initial population in the subsequent year. Further details of the initial model and later modifications can be found in Sutherst *et al.* (1979) and (1980), respectively.

Since the cattle tick is virtually sessile and can show a rapid increase in numbers within a farm, it was decided to assess control strategies by running the model for four years, taking the "equilibrium" outcome in the fourth year as the main criterion of performance. To cover the range of microclimatic conditions likely to be met in southeast Queensland, in different years, locations, and paddocks (e.g., river flats or hill ranges), parameter values for egg development and larval survival, associated with three climatic scenarios—a sequence of wet, average, or dry years—were used in the model. Other differences between farms, such as herd structure, breeding and culling policy, and supplementary feeding, as well as stocking and grazing policy, can be accounted for in the parameters describing host resistance and host-finding, respectively. To show how control strategies are investigated, we start with one of these processes—host resistance—and consider the control that can be achieved by breeding tick-resistant cattle.

23.3.1. Host Resistance

The level of host resistance is measured by artificially infesting an animal with 20 000 larvae in the spring and determining the percentage that engorge (Utech *et al.*, 1978). This percentage yield is largely determined by breed, with European cattle yielding between 10 and 20%, Zebu animals 1 to 2%, and crossbred animals (50% European, 50% Zebu) 3–6%. To determine the effect of breed on equilibrium tick numbers, the model was run for "herds" with different average yields. Since the reduction in LWG caused by ticks appears to be linearly related to the numbers engorging each year (see Sutherst *et al.*, 1979), both equilibrium tick numbers and equilibrium damage increase exponentially with the lowering of resistance (Figure 23.5).

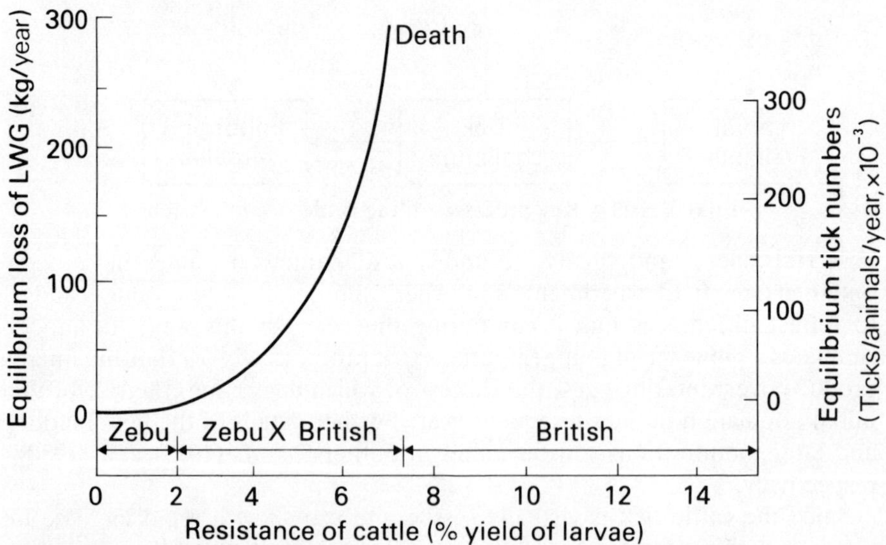

FIGURE 23.5 Effect of host resistance (expressed as % yield of engorging ticks) on "equilibrium" tick numbers and damage simulation results for average seasons in southeast Queensland.

Under average climatic conditions in southeast Queensland, the risk for cattle that yield 7% or more is very high (Figure 23.5), thus explaining why European cattle have been subjected to frequent acaricide dipping. It would also appear from Figure 23.5 that the best policy would be to run pure Zebu animals. Unfortunately, pure Zebus have relatively slow growth rates and low fertility, and their benefits for beef production are best obtained by crossbreeding with European animals, a practice the majority of those cattlemen introducing Zebu blood into their herds have chosen (Elder, 1979). Since the proportion of Zebu genes, through its influence on the level of resistance (percentage yield), has a large effect on the tick damage crossbred cattle are likely to incur (Figure 23.5), breeding policy will clearly influence the control strategies adopted.

23.3.2. Acaricide Dipping

Although the use of crossbred cattle with an average yield of (say) 4% yield reduces the tick hazard considerably, sufficient damage can still be caused for dipping to be required, particularly in years favorable to ticks. Since crossbred cattle are still a fairly recent innovation in southeast Queensland, there is little practical experience to indicate what constitutes an efficient and reliable dipping strategy for these animals. Many who have switched to crossbreds still use dipping strategies which are more appropriate for European breeds: indeed, Elder (1979) found that almost half the crossbred herds in Queensland were being dipped more than six times per year. In this particular instance, the model provides "instant experience", enabling tentative recommendations to be made without undue risk.

While the adoption of a particular dipping strategy, or indeed any tick control strategy, is likely to be based on a number of considerations, three aspects are likely to predominate—profitability, acaricide resistance, and management constraints. It is on these three criteria that simulated dipping strategies have been assessed.

(1) *Profitability*. To assess the short-term performance of dipping strategies, the tick model was used to assess their effect on the combined costs of "equilibrium" damage and control. Damage is priced at A$0.70 per kg liveweight lost and control costs A$0.20 per beast per dipping for acaricide plus A$0.60 per beast per dipping for mustering (Powell, 1977)). Assuming each dipping achieves a 97% kill of parasitic ticks on the herd, a level of control that can be expected when a near complete muster is obtained and a plunge dip is used (Powell, 1977), the performance of three dipping strategies—single, two and three dippings per year—were assessed. With more than one dipping per year, a dipping interval of three weeks was adopted, as recommended by Norris (1957), and confirmed by the model to be the most effective interval. The outcome of these three strategies is shown in the left-hand column of Figure 23.6 for two starting times. The open histograms represent strategies starting at a "bad" time (as determined by previous runs), the shaded histograms representing "well timed" strategies.

Although the expected level of profit associated with a particular strategy is an important consideration, equally important is the reliability with which this profit is maintained. While the difference between good and bad starting times to some extent indicates how robust each strategy is, further evidence is

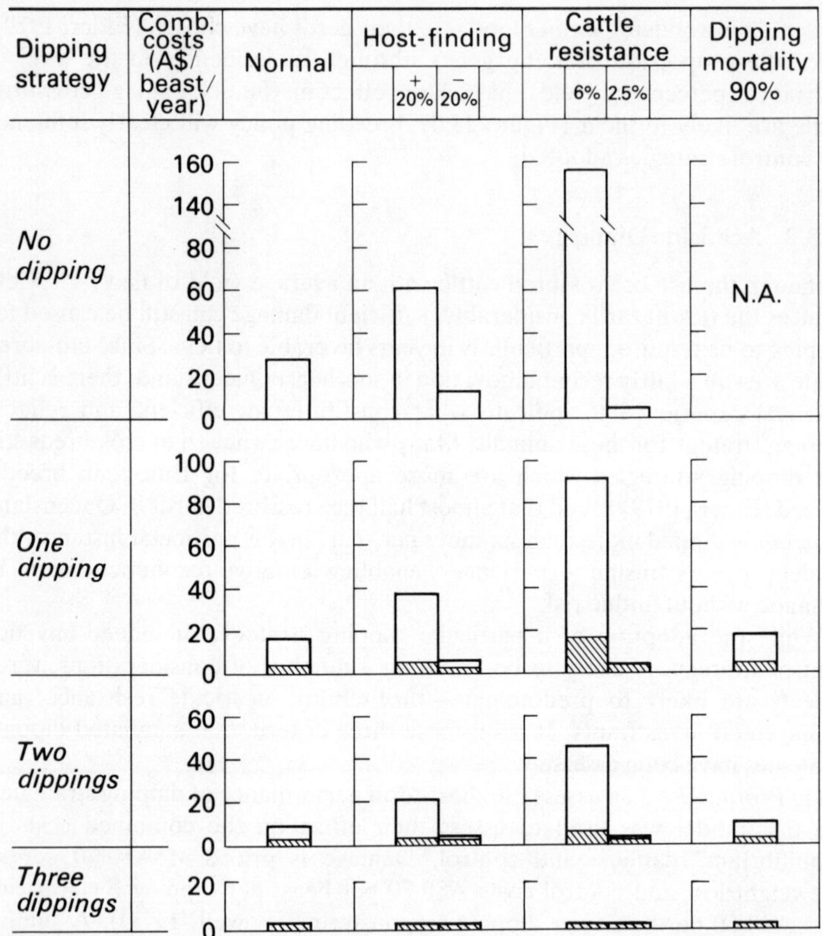

FIGURE 23.6 Effect of parameter values on "equilibrium", combined (dipping and damage) costs (A$/beast/year) associated with three dipping strategies (one, two, and three dippings per year) starting at a good time (week 14; shaded); and a bad time (week 17; unshaded).

obtained by changing parameter values associated with host finding, host resistance, and dipping mortality. *Clearly, the results shown in Figure 23.6 demonstrate that single and two-dipping strategies can be as profitable as the three-dipping strategy but they are far less reliable, three-dippings per year being a highly robust strategy that performs well over a wide range of conditions.*

(2) *Acaricide resistance*. Sutherst and Dallwitz (1979) pointed out the dangers of using population models to derive control strategies that are attractive in the short term but ignore long-term effects. In line with this, the model was also used to assess the consequences for acaricide resistance of

using various dipping strategies—such as the three-dipping strategy—on a regular basis.

Employing mathematical models to explore the factors which delay resistance, Sutherst and Comins (1979) conclude that apart from the advantage that tick-resistant breeds of cattle have in reducing dipping frequencies and so reducing selection pressure, they also allow cattle to be dipped in late autumn and early winter. In southeast Queensland this can have two advantages:

(i) Low temperatures at this time prevent those ticks which survive dipping from producing viable progeny. This enables the benefits of dipping with acaricides to be obtained whilst minimizing the selection of resistance genes that can occur when ticks are exposed to acaricides at other times of the year.

(ii) It allows the use of a simple monitoring scheme, which, in combination with an economic threshold decision rule, could be expected to further reduce the number of occasions on which resistant cattle are dipped.

(3) *Management constraints*. Once a set of dipping strategies that meet particular objectives has been identified, a further subset has to be considered, based on the extent to which these strategies meet a variety of management constraints. For instance, the economic threshold strategy mentioned above is only practicable where regular inspection for ticks can be carried out. Similarly, the adoption of the efficient, three-dipping strategy, in spring or summer, may be constrained by calving or the sowing of crops (Elder, 1979), or by summer rain upsetting the dipping interval. A further constraint can occur where producers wish to maintain host resistance by culling animals with low resistance in early autumn, a period when ticks on undipped animals are at their most numerous, and animals with low resistance can be easily identified (Sutherst and Utech, 1981).

In terms of practical significance, the results of the model, interpreted in the light of detailed ecological knowledge and an awareness of the requirements of the producers themselves, have provided a thorough understanding of the relative advantages and disadvantages of different dipping strategies. In providing this background, and with good communication between research and advisory personnel, the modeling approach now plays an integral part in producing new dipping recommendations for crossbred cattle.

23.3.3. Pasture Spelling

One of the major spurs for changing to resistant cattle has been the fear of acaricide resistance, resulting from the long history of acaracide-resistant strains in Australia (Wharton, 1976). Since there is a real risk that strains resistant to all available acaricides could develop in the future, the development of control strategies that do not involve acaricides would appear, at the very least, a desirable option.

At various times in the past, attempts have been made to design a control strategy that relies on the movement of cattle from one paddock to another, to break the life cycle of the tick (Wilkinson, 1957; Harley and Wilkinson, 1971; Waters, 1972). These pasture spelling strategies, designed for use with European cattle and relying on paddocks being destocked for lengthy periods, adversely affected the pastures and found little acceptance. With the introduction of crossbred cattle, a manageable strategy appears far more likely. In this case, the model was used to identify and explore the potential of a novel spelling strategy, involving a single, short spelling period per year, for herds of 4% yield.

To investigate spelling strategies, two tick populations had to be modeled in parallel, one representing the population in the main paddock and the other representing that in the holding paddock, where the cattle are kept for the spelling period. Assuming that the holding paddock is tick-free, a population starts to develop in this paddock only when engorged females are introduced on the cattle. By this means, spelling strategies were simulated, starting each week throughout spring and summer, with a duration ranging from one to five weeks. The effect of these strategies on equilibrium tick damage is shown in Figure 23.7.

As we have already seen, one of the reasons why pasture spelling has not been widely adopted in the past is the deleterious effect it can have on pasture productivity and digestibility, resulting from over- and undergrazing respec-

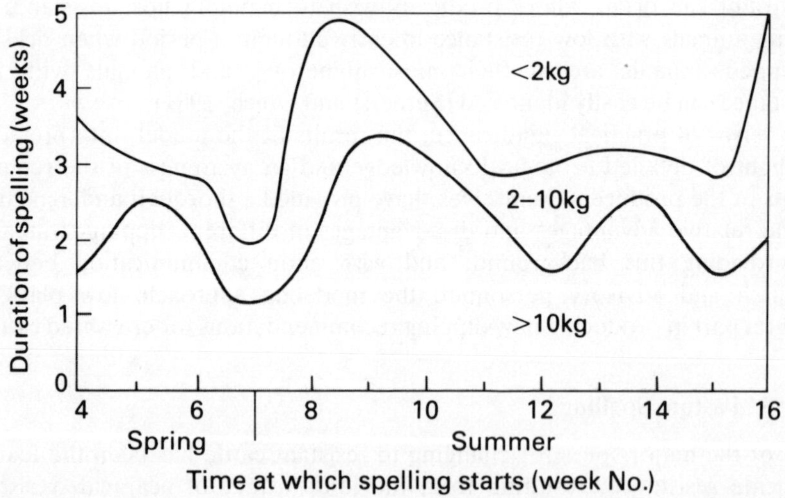

FIGURE 23.7 Effect of timing and duration of pasture spelling on "equilibrium" tick damage (kg/beast/year).

tively. *Referring to Figure 23.7, the model shows that effective tick control can be achieved with spelling periods of only four weeks' duration when carried out in spring or summer*. Damage to pastures is unlikely to occur in this time, particularly in summer, so that the growth rate of cattle should not be deleteriously affected.

This unexpectedly high level of control can be explained, with the help of simulation output (Figure 23.8), by three factors:

(*a*) Cattle removed from the main paddock do not contaminate it for the duration of the spelling period.
(*b*) When cattle are returned to the main paddock four weeks later, they do not carry ticks because the progeny of the ticks introduced into the holding paddock would have had insufficient time to develop to the host-seeking larval stage. Consequently, no contamination of the main paddock with engorged females will occur for another three weeks.
(*c*) Since free-living larvae do not encounter hosts when cattle are removed from the main paddock, many will die before the cattle return.

As shown in Figure 23.8, the net result in the first year is a population less than one-third of that on unspelled cattle. At this stage it would be premature to recommend single spelling as a complete alternative to dipping. The next step is to continue the modeling investigations, in combination with field trials (on experimental stations and with cooperating cattlemen), with the purpose of identifying the practical problems that may arise and assessing how they might be overcome.

23.4. CONCLUSIONS

It has been argued elsewhere that a succession of systems analysis techniques has an important and evolving role to play in the development of applied ecological research (Walker *et al*., 1978; Norton, 1979). From our experience with the modeling approach adopted for the Australian cattle tick, the potential for practical management appears equally important. In this particular case, "implementation" takes two forms:

(1) Answering Specific Management Questions

As new extension problems arise, associated with changes in technology or management, modeling can provide the "experience" on which to base initial recommendations; "experience" that would otherwise be too expensive or impractical to achieve by conventional field investigation. The recent widespread adoption of crossbred cattle in southeast Queensland is a case in point; it was only by means of the model that the full implications of various dipping strategies could be assessed.

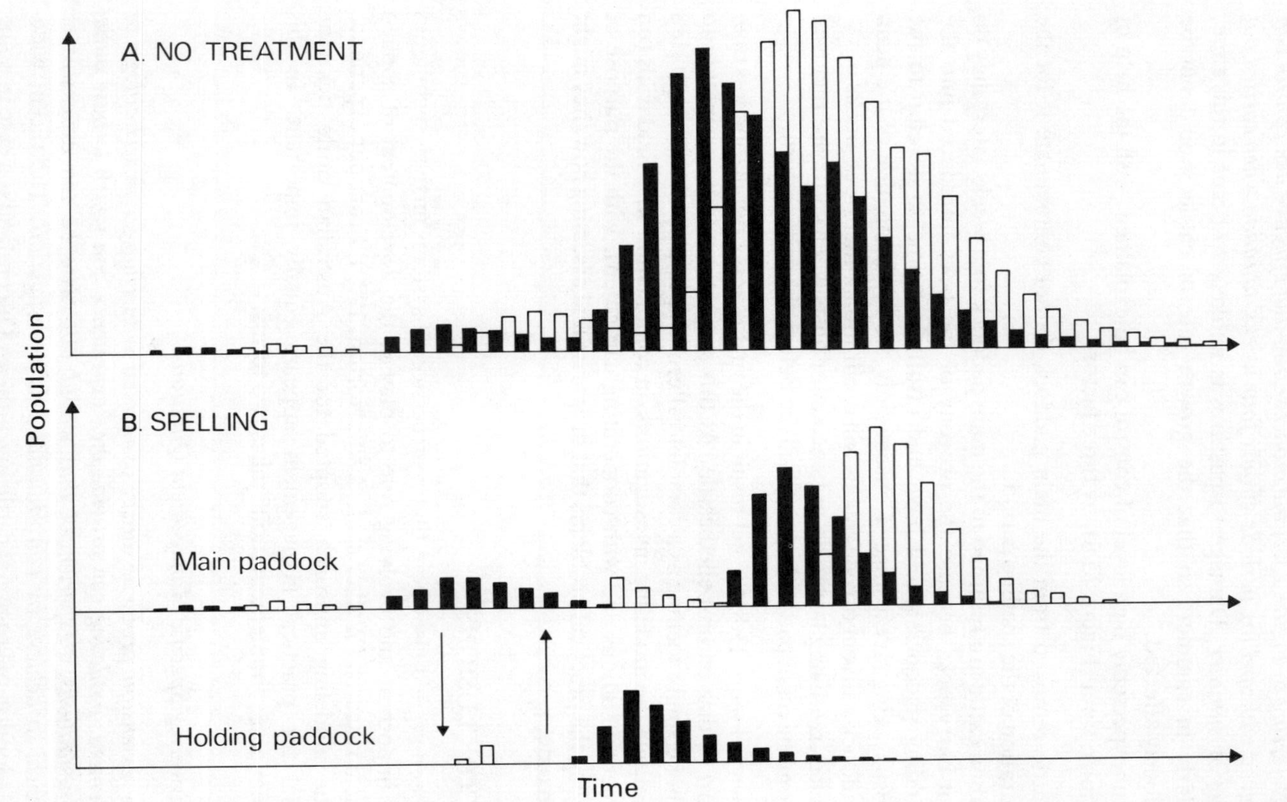

FIGURE 23.8 Numbers of free living larvae/beast in the paddock (shaded) and engorged female ticks/beast (unshaded) in the first year, associated with A, no treatment; and B, a spelling period of four weeks, starting on week 14. Arrows indicate the movement of cattle between the main paddock and the holding paddock.

The model has also been used to show that a single spelling period of four weeks in summer can achieve good control of ticks without resort to acaricides, a strategy that could have importance in the future in the event of serious acaricide resistance. In conjunction with field investigations, further modeling could be of value in developing and adapting this strategy for practical use.

(2) Providing a Framework For Tick Control Recommendations

Since the model cannot feasibly be "validated" on all individual farms, a comprehensive analysis has been undertaken to identify the parameter space (both biological and management) within which control strategies can meet specific objectives and constraints. Using empirical information, such as the range of percentage yield that can be expected for particular breeds (Utech *et al.*, 1978), each farm can be identified in this space, providing some guidance for specific recommendations. *By placing the emphasis on robust strategies, that give a satisfactory performance over a wide range of values, inevitable, real-world variations are accounted for and allowance is made for errors that may arise in estimating parameter values for particular farms*. As experience is gained, we would expect this explicit, extension "handbook" to be up-dated and improved.

ACKNOWLEDGMENTS

The authors wish to acknowledge the Australian Meat Research Committee for providing some of the funds for the cattle tick ecology group. Support for Dr Norton has come from the Office of Resources and Environment of the Ford Foundation and from the Wellcome Trust; the CSIRO Division of Entomology provided the position of Visiting Scientist.

REFERENCES

Benedek, P., J. Surján, and I. Fésűs (1974) *Növényvédelmi Előrejelzés* (Budapest: Mezőgazdasági Kiadó).

Charlton, P. J. and P. R. Street (1975) The practical application of bio-economic models, in G. E. Dalton (ed) *Study of Agricultural Systems* (London: Applied Science Publishers). pp. 235–65.

Elder, J. K. (1979) *Cattle Tick Control Survey Results* 1977–78. Veterinary Services and Pathology Branches Tech. Bull. No. 1. (Queensland: Dept of Primary Industries).

Harley, K. L. S. and P. R. Wilkinson (1971) A modification of pasture spelling to reduce acaricide treatments for cattle tick control. *Austral. Vet. J.* 47:108–11.

Norris, K. R. (1957) Strategic dipping for control of the cattle tick, *Boophilus microplus* (Canestrini) in south Queensland. *Austral. J. Agric. Res.* 8:768–87.

Norton, G. A. (1979) Systems analysis and pest management: a pragmatic approach. *Proc. 3rd Australian Applied Entomology Conf.* (Gatton) pp17–37.

Norton, G. A., R. W. Sutherst, and G. F. Maywald (1983) A framework for integrating control methods against the cattle tick *Boophilus microplus* in Australia. *J. Appl. Ecol.* 20:489–505.

Powell, R. T. (1977) *Project Tick Control.* Advisory leaflet No. 856. (Queensland: Division of Animal Industry, Dept of Primary Industries).

Sutherst, R. W. and H. N. Comins (1979) The management of acaricide resistance in the cattle tick, *Boophilus microplus* (Canestrini). (Acari: Ixodidae) in Australia. *Bull. Entomol. Res.* 69:519–40.

Sutherst, R. W. and M. J. Dallwitz (1979) Progress in the development of a population model for the cattle tick *Boophilus microplus. Proc. 4th Int. Congr. Acarology, Prague 1974*, pp559–63.

Sutherst, R. W., G. A. Norton, N. D. Barlow, G. R. Conway, M. Birley, and H. N. Comins (1979) An analysis of management strategies for cattle tick (*Boophilus microplus*) control in Australia. *J. Appl. Ecol.* 16:359–82.

Sutherst, R. W., G. A. Norton and G. F. Maywald (1980) Analysis of control strategies for cattle tick on Zebu X British cattle, in L. A. Y. Johnston and M. G. Cooper (eds) *Ticks and Tick-borne Diseases*, Proc. Symp. at 56th Ann. Conf. Austral. Vet. Assoc., Townsville, 1979. (Sydney: Australian Veterinary Association) pp. 46–51

Sutherst, R. W. and K. B. W. Utech (1981) Controlling livestock parasites with host resistance, in D. Pimentel (ed) *CRC Handbook of Pest Management in Agriculture, Vol II* (Florida, CRC Press) pp. 385–407

Teng, P. S., M. J. Blackie and R. C. Close (1979) Simulation modelling of plant diseases to rationalise fungicide use. *Outlook on Agriculture* 9:273–77.

Tummala, B. W., G. W. Seifert and R. H. Wharton (1978) Breeding Australian Illawara shorthorn cattle for resistance to *B. W., G. W. Seifert and R. H. Wharton (1978) Breeding Australian Illawara shorthorn cattle for resistance to Boophilus microplus*, I: Factors affecting resistance. *Austral. J. Agric. Res.* 29:411–22.

Utech, K. B. W., R. H. Wharton, and J. D. Kerr (1978) Resistance to *Boophilus microplus* (Canestrini) in different breeds of cattle. *Austral. J. Agric. Res.* 29:885–95.

Walker, B. H., G. A. Norton, G. R. Conway, H. N. Comins, and M. Birley (1978) A procedure for multidisciplinary ecosystem research: with reference to the South African Savanna ecosystem project. *J. Appl. Ecol.* 15:481–502.

Waters, K. S. (1972) Grazier's finding on pasture spelling for tick control. *Queensland Agric. J.* 98:170–75.

Way, M. J. (1973) Objectives, methods and scope of integrated control, in P. W. Geier, L. R. Clark, D. J. Anderson and H. A. Nix (eds) *Insects: Studies in Pest Population Management,* (Canberra: Ecol. Soc. Australia, Memoirs 1) pp 137–52.

Wharton, R. H. (1976) Tick-borne livestock diseases and their vectors, 5: Acaricide resistance and alternative methods of tick control. *World Animal Rev.* 20:8–15.

Wilkinson, P. R. (1957) The spelling of pasture in cattle tick control. *Austral. J. Agric. Res.*, 8:414–23.

PART THREE

POLICY

24 Policy Models

Gordon R. Conway

24.1. INTRODUCTION

Whereas strategic and tactical models are intended to help the farmer or medical practitioner, policy models address the needs of decision makers at regional, national or even international level. They thus deal explicitly with large-scale ecological and economic issues and with medium- and long-term questions. Explicitly or implicitly, they also address issues often sensitive in nature, of social value and political priority. It is becoming increasingly accepted that government intervention is essential for successful long-term control of many pests and pathogens, and models have an important role in illustrating the need for this intervention and the form it should take.

24.2. SOCIOECONOMIC MODELS

Policy issues arise most dramatically in the case of human diseases. Many diseases are of national significance and some indeed are globally important because of the effect of increasing international travel. Governments may thus, as a matter of policy, set up checks and barriers to the invasion of disease pathogens. But the most important policy questions arise in setting national priorities for control among a range of indigenous diseases and in establishing strategies of control that, since the habits and beliefs of people are so intimately involved, are socially and politically acceptable, and hence are likely to succeed.

All too often in disease modeling, the human variables are neglected relative to other biological factors in the disease cycle. The many studies of schistosomiasis (bilharzia) offer a good case in point (see Chapter 25). Transmission of the disease in the cycle depends critically on human excreta (urine or feces) entering the habitat of the intermediate snail vector and, in turn, on the probability of the free-living, water-dwelling infectious stage of the parasite encountering new, uninfected human hosts. In each case the probabilities depend on the habits and ways of life of the human population that reflect the underlying prevalent education, beliefs, and socioeconomic status. Rosenfield shows how variables, such as population movement, the availability of sanitary facilities and domestic water supply and the location of dwellings and schools in relation to snail habitats, can be appropriately quantified and included in the transmission equations.

In terms of control, however, such variables are difficult to influence. Change may be costly and meet with social resistance. Thus it is being recognized increasingly that the social and economic aspects of transmission in diseases such as schistosomiasis are the key to successful control and policy models are required that take a realistic view of the social and economic costs, and benefits of different strategies. The costs are usually straightforward to calculate, but for purposes of comparison with other proposed programs of national investment that are competing for funds, it is necessary to calculate the net present value (*NPV*) of these costs:

$$NPV = K_0 + \frac{c_1}{1+i} + \frac{c_2}{(1+i)^2} + \frac{c_3}{(1+i)^3} + \dots, \tag{24.1}$$

where K_0 is the current capital or start-up costs, and c_1, c_2, c_3 are the running costs of the control program in subsequent years. The choice of the discount rate i then becomes critical in determining the attractiveness of a particular disease-control program. It is also usually relatively straightforward to set control objectives in terms of reduction in disease incidence. For schistosomiasis the number of case years presented is commonly used as the measure of effectiveness and a cost-effectiveness analysis can then be performed in terms of case years prevented per unit cost of control (Figure 24.1). Rosenfield (Chapter 25) shows that for a number of countries chemotherapy is the most effective strategy, at least in the short term, although a longer horizon that copes with reinfection requires a strategy combining a number of techniques.

A full cost–benefit analysis (Figure 24.1) is more difficult, however:

$$\frac{B}{C} = \frac{[b_1/(1+i)] + [b_2/(1+i)^2] + [b_3/(1+i)^3] + \dots}{K_0 + [c_1/(1+i)] + [c_2/(1+i)^2] + [c_3/(1+i)^3] + \dots}, \tag{24.2}$$

where b_1, b_2, b_3 are the benefits in each year, and B/C is the ratio of net present values of benefits and costs. This is primarily because of the problem of quantifying the benefits in monetary terms. Rosenfield discusses attempts to relate symptoms of the disease to work capacity and hence productivity (usually agricultural productivity in the case of schistosomiasis) but argues that for a full assessment of benefits an attempt should also be made to estimate the political savings in domestic, educational, and social/civic impairment. Ideally,

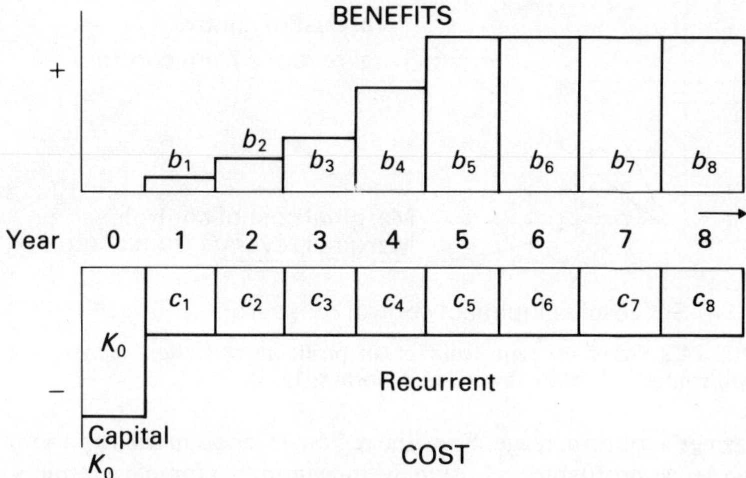

FIGURE 24.1 A schematic comparison of costs and benefits for a human disease control program. Benefits include reduced disease incidence (i.e., case years prevented), or economic return (i.e., from increased labor productivity).

perhaps, control of such diseases should be justified solely on humanitarian grounds but, in practice, harder economic justification is frequently required and thus attempts will have to be made to quantify the less immediately tangible benefits.

In the case of crop pest (and pathogen) control, economic assessments are easier, primarily because the benefits in terms of increased crop yields can be fairly readily quantified in economic returns. In Chapter 11 we saw how economic thresholds can be established to provide an indication of the level of control that is profitable. Such analysis can be carried further by examining each successive increment in control to determine the level at which profit is maximized. In mathematical or graphical terms this occurs at the level of input where marginal cost just equals the marginal return (Figure 24.2). Essentially this is the result achieved by Shoemaker's use of dynamic programming to optimize pesticide applications (Chapter 18). The problem, however, becomes more complicated if the farmer is not a profit maximizer (see, for example, Norton 1976; Norton and Conway, 1977), or if significant externalities are involved (see Section 24.3), or if regulatory or fiscal distortions are imposed on the farm from outside. Here, as in human disease control, human variables become explicitly or implicitly significant in determining appropriate and successful control programs.

In Chapter 26 Carlson and Rodriguez examine the situation of externally imposed distortion as it arises in regional and national mandatory pest control programs. They examine a program aimed at eradicating the cotton boll weevil in North Carolina, the costs of which are borne by a mandatory levy based on

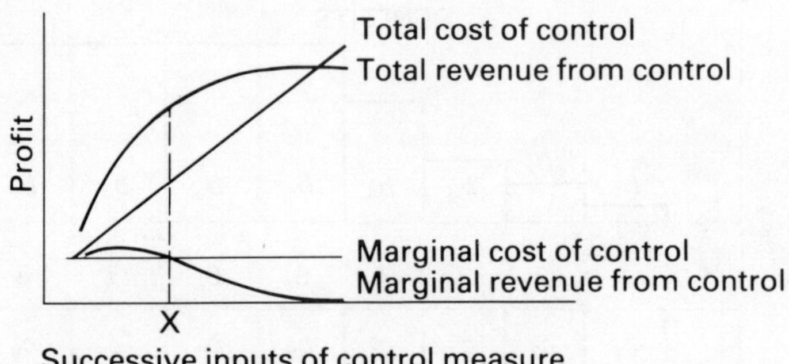

FIGURE 24.2 Graphical representation of profit maximization analysis of pest or pathogen control (after Southwood and Norton, 1973).

each farmer's cotton acreage. From the regional viewpoint such a program may be regarded as profitable, at least in the long term, but for many farmers with a lower than average pest incidence the levy may represent a level of expenditure above that they would normally expend to maximize profit. Their response may then be to readjust inputs that are supplementary or complementary to pest control or, as happened in North Carolina, to reduce the acreage of crop subject to the levy and invest in alternative crops, thus reestablishing their maximum-profit position. This problem is likely to be particularly acute in the case of eradication programs, since costs tend to rise dramatically as extinction of the pest population approaches. Nevertheless, significant adjustments by farmers will occur to most mandatory control programs, and Carlson and Rodriguez provide a model for anticipating these and so improving the likelihood of success.

24.3. USE AND MISUSE OF PESTICIDES

Undoubtedly the most important contemporary policy issue in pest and disease management is the misuse of pesticides and its consequences (Coaker, 1977; Corbett, 1978; Perring and Mellanby, 1977; Pimentel *et al.*, 1977; Southwood, 1979). If (1) pesticides only affected the target pest or pathogen, (2) their efficiency was unimpaired however frequent their use, and (3) pest and pathogen damage could be accurately predicted, then profit-maximization analysis, as outline earlier, would produce not only short-term but also long-term maximum profits for farmers and other users; furthermore, it would lead to an overall maximum profit for the community as a whole. Adam Smith's notion of the invisible hand whereby, in a situation of perfect knowledge, each individual maximizing his own satisfaction thereby contributes to the optimum satisfaction of the community, would then apply.

Unfortunately, as Regev demonstrates (Chapter 27), pesticides do not behave in this way. First, pesticides not only attack target organisms but also kill a variety of natural enemy species that would otherwise regulate pest and pathogen populations. Hence, short-term control may be at the expense of a worsening problem in the long-term (see Chapter 3). Secondly, although beneficial to the users, they may have detrimental effects on valued components of wildlife and on the health of human populations. Thirdly, with frequent and intensive pesticide use, there is almost inevitably a long-term evolution of pesticide resistance in the target pest and pathogen populations. Finally, because the incidence and magnitude of imminent pest or pathogen damage tends to be so uncertain, pesticide users frequently resort to heavy pesticide applications by way of insurance. The net effect of all this is, in Regev's view, a vicious cycle of pesticide use generating problems which in turn require further resort to pesticides, so leading to an "addiction" to pesticides which, in the absence of outside regulation, harms both the user and the community as a whole.

24.3.1. Pesticide Resistance

The particular problem of pesticide resistance, which also applies generically to antibiotics and other medicinal drugs, is dealt with in more detail by Comins in Chapter 28. Resistance raises important policy issues because it is a phenomenon largely outside the control of the rational pesticide user (Conway, 1982; Georghiou, 1980). The individual user has little chance of detecting the growth of resistance before it is too late; moreover, resistance usually develops over a regional scale as a result of the sum total of the actions of individual users. It is thus a classical "common property problem" where the individual has little incentive or ability to influence the course of events. As Hueth and Regev (1974) put it, pesticide tolerance is essentially a free resource of regional and national, if not global, significance and its maintenance can only be accomplished by action at the appropriate level.

The policy issues of pesticide resistance are also ideally suited to mathematical modeling because of the very considerable difficulties involved in field experimentation. Comins, in a classical demonstration of the power of mathematical modeling in this field, shows how a few realistic simplifying assumptions permit policy guidelines to be developed, of considerable robustness, that may serve at least to delay the development of resistance.

Perhaps the most intuitively obvious guidelines are those concerned with reducing pesticide selection pressure. The widespread overuse of pesticides is clearly particularly undesirable if much of the pesticide mortality is substituting for mortality that would have naturally been generated by intraspecific competition or by the action of predators or parasites. Comins also illustrates the considerable moderating influence of pest population refuges and population migration in the development of resistance and hence the necessity for restricting, as far as possible, the spatial extent of pesticide application.

However, the greatest value of such models is in helping to assess current suggestions for policies advocating the use of pesticide dosages giving high kills or of "cocktails" of pesticides to delay resistance. These may seem, at first glance, to be sensible propositions, but Comins shows that in most situations such policies will be counterproductive. High kills or cocktails may work in special conditions, for example, where the dosage of the pesticide to the individual pest can be precisely controlled, and providing that "super-resistant" or cross-resistant genotypes are very rare. But such situations are uncommon, and this approach is in practice very sensitive to any lack of precision in pesticide application. It thus represents a considerable gamble and lacks the robustness of a good resistance policy.

24.4. RESILIENCE AND ROBUST POLICIES

The most significant work in recent years on policy modeling in pest control has been that of Holling and his co-workers (see References in Chapter 29), which is characterized by a set of powerful concepts or hypotheses ("myths" in their lexicon) that have evolved from consideration of a small number of key case studies. The earliest of these studies was of the spruce budworm (*Choristoneura fumiferana*) described at the previous IIASA conference (Jones, 1979; Holling *et al.*, 1979) and clearly summarized by Peterman *et al.* (1979). The spruce budworm is a defoliator of several North American forest trees and while usually rare throughout its range is subject to periodic, epidemic and serious outbreaks affecting vast tracts of forest. The core process in its dynamic appears to be bird predation, characterized by a type III functional response (see Chapter 2), resulting in two stable equilibria for the budworm population—a lower, endemic equilibrium where the birds regulate budworm numbers, and an upper, epidemic equilibrium set by the availability of budworm food resources in the forest. However, the boundary between these two stable states is not fixed but varies with the degree of forest maturity (Figures 24.3 and 24.4). In practice, as the forest matures the budworm population increases and above a certain threshold value of maturity the budworm numbers are large enough that escape from the bird predators becomes possible. The population flip across the breakpoint may then be purely a stochastic event or may be triggered by a number of factors, including weather and the sudden immigration of budworm from elsewhere. The traditional management response has been to spray budworm outbreaks. This has certainly had the effect of saving the trees (and indeed in some regions has saved the pulp and paper industry), but, as indicated in Figure 24.4, spraying essentially holds the budworm system in a state of incipient outbreak, both in terms of budworm numbers and forest maturity, which, should it break down, would create an epidemic with consequences far more severe than before.

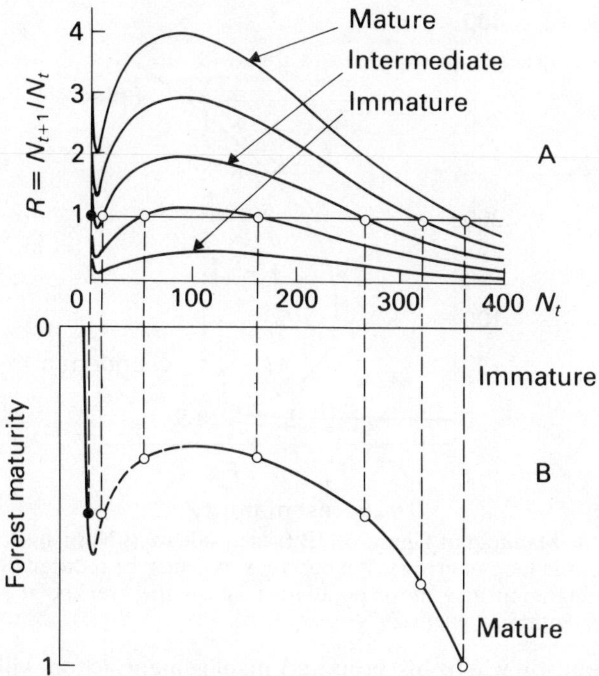

FIGURE 24.3 A, Recruitment curves for spruce budworm at different levels of forest maturity (measured by branch density). B, Projected manifold of equilibrium points: • stable and ○ unstable equilibria (after Peterman *et al.*, 1979).

The budworm model can be used to generate management schemes based on a pattern of logging strategies that reduces the effective forest maturity and hence the risk of calamitous outbreak—but the implications for policy from this model are considerably broader.

In 1973 Holling applied the term resilience to the property of ecological systems to absorb disturbance (Holling, 1973). In particular, his work drew renewed attention to systems that are characterized by more than one stable state and stressed the need to focus both analysis and management on the boundaries between states rather than on the details of behavior in the neighborhood of the equilibria. Traditionally, ecological management (and this particularly includes pest management) has aimed at close targeting on desired equilibrium conditions. Often this is highly effective, at least in the short-term, but, as the pesticide-resistance phenomenon illustrates, may produce partial or total collapse in the medium to long-term.

The fundamental problem is lack of knowledge. Taking Figure 24.4 as our case, the manager's ignorance may take many forms. He may not know that multiple stable states exist, what process creates their existence, where the thresholds lie in the control variables, where his system lies in the complex

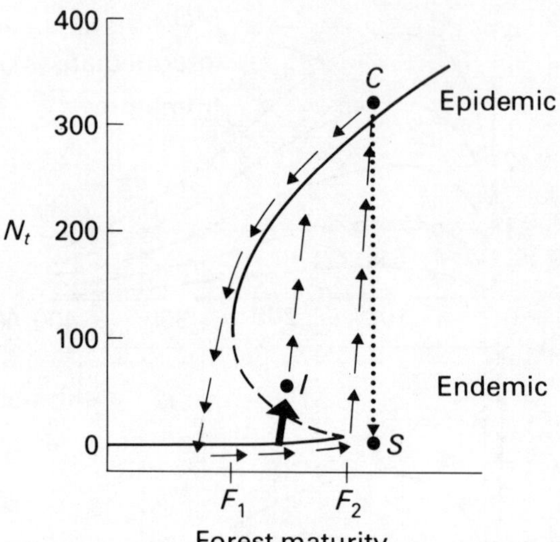

FIGURE 24.4 Manifold of Figure 24.3B turned sideways. Normal oscillatory pattern shown by outside loop of arrows. An outbreak at *C* may be reduced by insecticides to point *S*. Immigration may move population across the breakpoint (dotted line) to point *I* (after Peterman *et al.*, 1979).

configuration, or where his proposed management action will transport it. Most important, however, he is likely to be unaware that his control action may in turn change the basic structure of the system, shifting thresholds and boundaries so that the likelihood of collapse is more imminent rather than less.

More research, more analysis and measurement may help to provide relevant knowledge and so improve the decision making. But, as Walters and Holling argue in Chapter 29, the uncertainty in the real world is formidable in practice. For there are not only the surprises contained in the target managed system, but also great uncertainty in the behavior of the managing institutions and in the broader all-encompassing environment. Coupled together, as is the reality, these constitute a "supra" system with its own stability and resilience properties, and our knowledge of these is very slender.

So stated the problem is fairly clear, as is the need for robust policies of pest management that acknowledge, anticipate, and cope with such uncertainty. But as yet, as Walters and Holling point out, there are few indications of the way ahead. One starting point, however, is the growing realization that resilience is primarily an evolutionary phenomenon of ecological systems. Indeed, Holling now defines resilience as the capacity of systems to absorb and benefit from change and the unexpected. A corollary is that, in order to maintain and improve their resilience, systems must experience continually challenging inputs and hence become, in some sense, preadapted to withstand future serious perturbation. Walters and Holling argue that this experimental probing and evolutionary strategy needs to be a characteristic, not only of the target system, but also of the controlling policy and management institution.

Finally Walters and Holling also argue for a style of analysis that squarely faces the problem of uncertainty. To acknowledge and understand the nature of uncertainty in pest management is in itself a major challenge for both the analyst and the manager. They argue from practical experience that interdisciplinary workshops, brainstorming, and free modeling exercises can provide settings within which better awareness, and clearer understanding, can be attained, and hence better overall policies evolved.

REFERENCES

Coaker, T. H. (1977) Crop pest problems resulting from chemical control, in J. M. Cherrett and G. R. Sagar (eds) *Origins of Pest, Parasite, Disease and Weed Problems*, Proc. 18th Symp. Br. Ecol. Soc. (Oxford: Blackwell) pp313–28.

Conway, G. R. (ed) (1982) *Pesticide Resistance and World Food Production* (London: Imperial College Centre for Environmental Technology).

Corbett, J. R. (1978) The future of pesticides and other methods of pest control, in T. H. Coaker (ed) *Applied Biology* Vol. 3 (New York: Academic Press) pp229–330.

Georghiou, G. P. (1980) Insecticide resistance and prospects for its management. *Residue Rev.* 76:131–45.

Holling, C. S. (1973) Resilience and stability of ecological systems. *Ann. Rev. Ecol. Systems* 4:1–23.

Holling, C. S., D. D. Jones, and W. C. Clark (1979) Ecological policy design: A case study of forest and pest management, in G. A. Norton and C. S. Holling (eds) *Pest Management*, IIASA Proceedings Series Vol. 4 (Oxford: Pergamon) pp91–156.

Heuth and U. Regev (1974) Optimal agricultural pest management with incresing pest resistance. *Am. J. Agric. Econ.* 56:543–52.

Jones, D. D. (1979) The budworm site model, in G. A. Norton and C. S. Holling (eds) *Pest Management*, IIASA Proceedings Series Vol. 4 (Oxford: Pergamon) pp13–90.

Norton, G. A. (1976) Analysis of decision making in crop protection. *Agro-Ecosystems* 3:27–44.

Norton, G. A. and G. R. Conway (1977) The economic and social context of pest, disease and weed problems, in J. M. Cherrett and G. R. Sagar (eds) *Origins of Pest, Parasite, Disease and Weed Problems*, Proc. 18th Symp. Br. Ecol. Soc. (Oxford: Blackwell) pp205–26.

Peterman, R. M., W. C. Clark, and C. S. Holling (1979) The dynamics of resilience: Shifting stability domains in fish and insect systems, in R. M. Anderson, B. D. Turner, and L. R. Taylor (eds) *Population Dynamics*, Proc. 20th Symp. Br. Ecol. Soc. (Oxford: Blackwell) pp321–41.

Perring, F. H. and K. Mellanby (eds) (1977) *Ecological Effects of Pesticides*, Linnean Society Symp. Ser. No. 5 (London:Academic Press).

Pimentel, D., E. C. Terhume, W. Dritshilo, D. Gallahan, N. Kinner, D. Nalus, R. Peterson, N. Zareh, J. Misiti, and O. Harber-Shaim (1977) Pesticides, insects in foods and cosmetic standards. *Bioscience* 27:178–85.

Southwood, T. R. E. (1979) Pesticide usage, prodigal or precise? *Bawden Lecture*, Proc. Br. Crop Protection Conf. on Pests and Diseases, 1979. Vol. 3 pp603–19.

Southwood, T. R. E. and G. A. Norton (1973) Economic aspects of pest management strategies and decisions, in P. W. Geier, L. R. Cark. D. J. Anderson, and H. A. Nik (eds) *Insects: Studies in Pest Population Management, Ecol. Soc. Austral., Canberra, Memoirs 1*, pp168–84.

25 Schistosomiasis Transmission and Control: The Human Context

Patricia L. Rosenfield

25.1. INTRODUCTION

The human element in schistosomiasis transmission and control is commonly considered only implicitly in mathematical modeling efforts (Cohen, 1976). Far greater attention has usually been given to the other relevant factors: the schistosome and the snail. The human life cycle—birth, maturation, and death—has been accounted for by epidemiological measurements of intensity, prevalence, or incidence. The human life events that result in the contamination of snail habitats with schistosome eggs, or in humans exposing themselves to cercariae by water contact, have rarely been linked to these epidemiological indicators. Other characteristics of human life, such as housing, sanitary, and employment conditions, nutritional and educational levels, or distribution of human settlements and water on the landscape, have thus been silent variables in the transmission models.

In addition to the lack of explicit inclusion of the human element in mathematical models is the lack of use of these variables in planning and operating schistosomiasis control programs. In only rare instances have such programs taken into account the social, economic, and epidemiological characteristics of the population at risk (Farooq, 1973). The incomplete nature of transmission models may explain the lack of acceptance of models by the public health profession; and the incomplete design of control programs may at least partially explain why schistosomiasis continues to be a public health problem. In this chapter, I first show how the human element can be explicitly taken into account in modeling by an example of a particular transmission model. Secondly, I describe the social and economic aspects of control program design.

25.2. HUMAN VARIABLES IN TRANSMISSION

In the several excellent reviews of the mathematical models of schistosomiasis, the ecological context of disease transmission has been carefully described (Cohen, 1977; Fine and Lehman, 1977). Most of the models that have been developed focus either on the ecological system as a whole (Nasell and Hirsch, 1973; Hairston, 1965; Macdonald, 1965; Hairston, 1962), or the particular link in the transmission cycle of concern to the model builder (Sturrock *et al.*, 1975) and the human element, i.e., the social and economic context, is usually cited as an implicit concern (Fine and Lehman, 1977). A detailed comparison has been made (May, 1977) to show the conceptual links between the theoretical models of schistosomiasis transmission and the empirical approach described in this chapter.

For schistosomiasis transmission to take place, the schistosome parasite must find, at different stages in its cycle, a human host and an appropriate snail host (Figure 25.1). (Animal vectors are not discussed here, although infected ones can contribute to the transmission cycle by passing eggs into snail habitats.) For three human-infecting species of schistosomes—*Schistosoma haematobium, S. mansoni*, and *S. japonicum*—this requirement is satisfied when a number of social and economic factors affecting the behavior of the human host are present, and when the ecological factors are satisfactory for the survival of the snail host. The first set of factors is discussed in detail here.

The social factors influencing human transmission reflect the economic status of an individual and of the community. The economic status affects living conditions, the amount of food that an individual can buy (besides subsistence crops grown), and possibly, though not always, educational status. In addition, it may be directly through economic behavior that humans have contact with the snail host. Other, noneconomic influencing factors are age, sex, religious practices, and population density. Although these factors are often interrelated, and certainly some are difficult to separate in any model, it is possible to consider how they affect transmission.

My own modeling efforts have attempted to relate some of these factors directly to predicted changes in prevalence over time in the community. These efforts are reported elsewhere (Rosenfield *et al.*, 1977; Rosenfield and Jordan, 1977; Rosenfield and Gestrin, 1978; Rosenfield, 1979), but are briefly reviewed here to show how such characteristics of human life can be included in transmission modeling.

25.3. THE BASIC MODEL

The basic differential equation used in these studies to describe transmission is

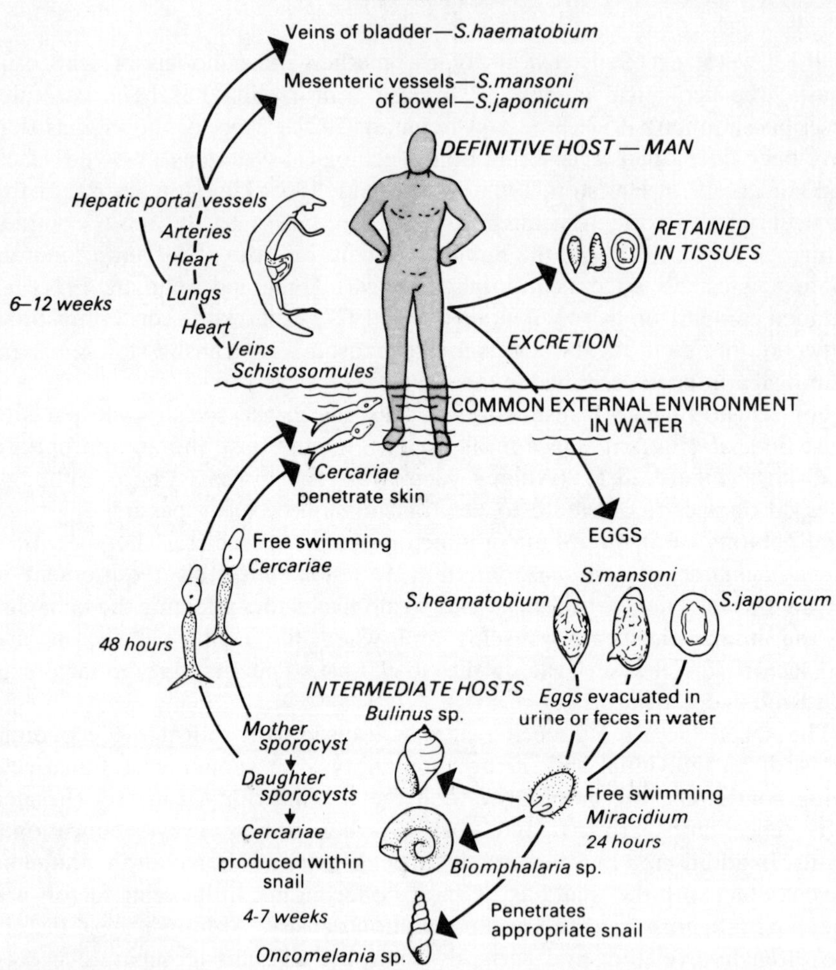

FIGURE 25.1 Life cycle of schistosomiasis (after Jordan and Webbe, 1969 and from Homans, J. (1945), *A Textbook of Surgery*. 6th ed. Courtesy of Charles C. Thomas, Publisher, Springfield, Illinois).

$$dY/dt = A(1 - Y) - BY, \tag{25.1}$$

where Y is the fraction of the population infected (prevalence), A is the infection rate (incidence), B is the loss rate (reversion), and t is time (Muench, 1959). This has been solved for use in the modeling attempts as follows:

$$Y_{t+\Delta t} = \left(Y - \frac{A}{A + B} \right) \exp[-(A + B)\Delta t] + \frac{A}{A + B}, \tag{25.2}$$

where $Y_{t+\Delta t}$ is the fraction of the population infected in the following time period (Rosenfield *et al.*, 1977).* This deterministic form has been used because it can be verified with field data and may be considered an approximation to stochastic models that have yet to be confirmed with data (Fine and Lehman, 1977).

25.3.1. The Relevant Social Variables

Social aspects of transmission can be incorporated into the infection and loss rate estimates, and can include such variables as:

(1) Presence of safe domestic water supplies, or its absence, leading to human exposure to snail habitats and possibly cercariae, through bathing, washing, collecting water, and playing.
(2) Presence and use of sanitary facilities for excreta and other waste disposal, or their absence, leading to infected individuals passing schistosome eggs into snail habitats.
(3) Lack of human awareness as to how water contact behavior influences disease transmission—a function of education.
(4) Nutritional status, possibly affecting egg passing, worm death rate, and severity of infection.
(5) Proximity of human settlements to snail habitats.
(6) Religious rituals using water in or from snail habitats.
(7) Workplace activities that lead to contact with snail habitats e.g., siphoning water from irrigation canals, fording, rice growing.
(8) Attendance at school.
(9) Proximity of schools to snail habitats.
(10) Population movements—short, medium, and long term.

Recent studies have demonstrated that human water contact (Dalton, 1976; Dalton and Pole, 1978), geographical distribution of human settlements (McCullough *et al.*, 1972), and migration (Ruyssenaars *et al.*, 1973) can be expressed quantitatively in relation to schistosomiasis transmission indicators. Work is under way to estimate the role of religious rituals in equally quantitative terms (Kloos *et al*, 1982). Awareness of disease transmission, educational status, and nutritional status has yet to be quantitatively related to schistosomiasis epidemiological indicators.

For eqns. (25.1) and (25.2) it was postulated that the infection rate A was a function of some measure of snail populations, parasite status of the human, and human behavior. Although snail populations can be modeled independently, and linked to human population data, it is often not possible to find

*Revisions to this basic equation used to generate prevalence over time have been personally suggested by I. Nasell (1976, 1979), R. May (1977), and K. Dietz (1979). These suggestions are being tested with data sets described below.

detailed snail population data combined with similarly detailed parasite or human behavioral data.

25.3.2. Data Sources, Equations, and Results

Working with schistosomiasis control project data from the Bilharziasis Control Project, Dezful, Iran (Arfaa *et al.*, 1970),* and the WHO/Tanzania Schistosomiasis Pilot Control and Training Project, Mwanza District, Tanzania (McCullough and Eyakuze, 1973)**—in both instances with *S. haematobium*—a surrogate variable was used to account for snail populations. The variable was "accessible snail habitats" (measured in feet or meters) within easy walking distance of a household (one mile or less in Iran, half a mile or less in Tanzania), which were used for domestic or economic purposes. The human parasite load variable, in this case, was simply the number of infected persons, the surrogate variable for the mean worm load. With data on *S. mansoni* from the Research and Control Department in St Lucia/Rockefeller Foundation, in a third study the behavioral element was directly introduced into the infection rate estimation equation (Jordan, 1977).***

The general infection rate equation for use within the first two studies was described as

$$A = b_0(H^{b1}P^{b2}), \tag{25.3}$$

where H is the distance of accessible snail habitats (feet or meters), P is the number of infected persons, and b is the regression parameter. The non-linear, multiplicative nature of the infection process was hypothesized to be the most appropriate relationship (Rosenfield, 1975). This equation was calibrated with data from Iran and Tanzania, resulting in the following estimations:

$$\textit{Iran}: \quad A = 5.67 \times 10^{-6} \quad (\underset{\substack{t=4.42\\ P=<0.01}}{H^{1.11}} \quad \bullet \quad \underset{\substack{t=5.12\\ P=<0.01}}{P^{0.45}}) \tag{25.4}$$

$$R^2 = 0.69$$

$$F(2,19) = 21.54 \qquad (P = < 0.01);$$

$$\textit{Tanzania}: \quad A = 1.6 \times 10^{-6} \quad (\underset{\substack{t=1.85\\ P=<0.001}}{H^{0.91}} \quad \bullet \quad \underset{\substack{t=2.67\\ P=<0.05}}{P^{0.36}}) \tag{25.5}$$

$$R^2 = 0.86$$

$$F(2,5) = 15.13 \qquad (P = < 0.01),$$

*Data were generously provided by the staff of the Bilharziasis Control Project, Dezful, Iran, and from their numerous articles and 1963–74 Quarterly Reports.

**The 1967–73 Quarterly and Annual Reports of WHO/Tanzania Schistosomiasis Pilot Control and Training Projects, Mwanza District, Tanzania. Data were generously provided by WHO and project staff.

***The 1965–76 Annual Reports of the Research and Control Department, Ministry of Education and Health, Castries, St Lucia, and from numerous publications. Data were generously provided by project staff.

where H in eqn. (25.4) was linear meters of accessible snail habitat, and in eqn. (25.5) was the volume of accessible snail habitat in cubic feet; the other variables are in the same units as described above. Equation (25.4) was developed on a village-specific basis for ages 0–10; eqn. (25.5) on a household-and age-specific basis, for ages 0–9. Comparisons between predicted and observed infection rates are given in Figure 25.2.

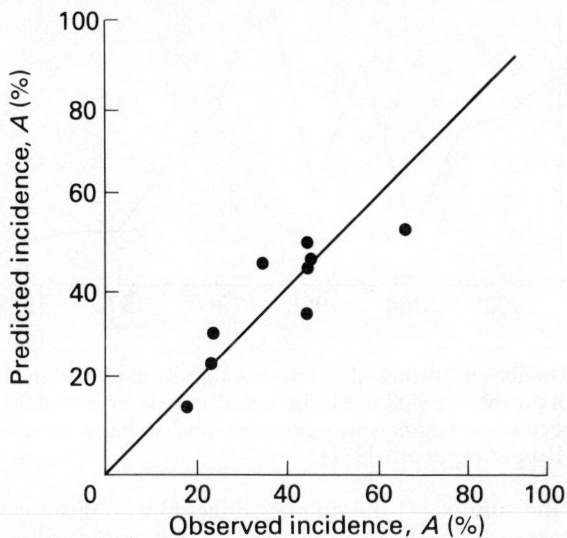

FIGURE 25.2 Predicted incidence (infection rate) of schistosomiasis obtained with eqn. (25.5), compared with observed incidence, under precontrol conditions in WHO/Tanzania Pilot Schistosomiasis Control and Training Project, Mwanza District, Tanzania.

When eqn. (25.4) was used in the estimation of prevalence combined with a constant loss rate term, a good fit between predicted and observed prevalence was obtained (Figure 25.3; Rosenfield *et al.*, 1977). The predictions were also modified to account for control measures and the natural rate of human population increase.

Equation (25.5) was also used to predict prevalence over time with data from the project in Tanzania. In this study, data were available for the first time on migration into and out of a schistosomiasis project area in Misungwi (Rosenfield and Gestrin, 1978).*

Prevalence predictions were modified to incorporate values for migration (Figure 25.4).** The population used as the basis for the next year's

*This work has been developed with P. Gestrin and F. McCullough.

**Baseline (pre-control program) data from the years prior to the start of the analysis were used to set the parameters of the model. The initial input data of pre-control prevalence, snail habitat values, total population, and loss rate are then used to start the simulation, as indicated in Figure 25.4 and as described in more detail elsewhere (Rosenfield, 1975; Rosenfield and Jordan, 1977; Rosenfield and Gestrin, 1978).

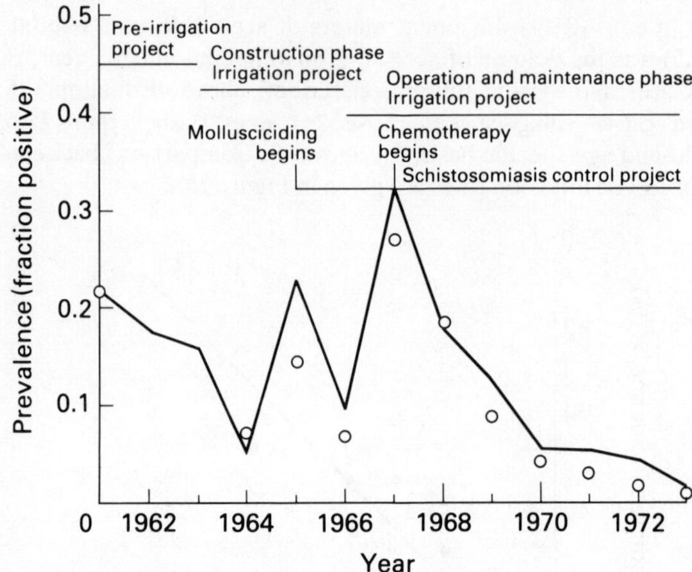

FIGURE 25.3 Results of testing the schistosomiasis model with Iranian data. Predicted values of prevalence (full line), compared with observed values (○) for years of irrigation project construction and operation, and Bilharziasis Control Project operations (after Rosenfield *et al.*, 1977).

predictions was the sum of (1) the number infected who did not migrate, (2) the number of infected immigrants, (3) the number uninfected who did not migrate, and (4) the number of uninfected immigrants. P was the sum of those infected who did not migrate, and infected immigrants. In order to calculate the appropriate correction factors for migration, the following were assumed to be constant:

(*a*) the fraction of the sector that emigrated;
(*b*) the prevalence of infection in the immigrant group;
(*c*) the fraction of the pool that emigrated; this is produced by assuming that the size of the population pool from which the immigrants come will increase at the same rate as the population in the sector; and
(*d*) the prevalence of infection in the emigrant group.

Due to the fact that migration data for only one year were available, it was necessary to assume that the immigration and emigration rates remained constant over time. This certainly affected the reliability of the prediction of prevalence. The comparison of predicted with observed values of prevalence is given in Figure 25.5. The model also incorporated control measure use, seasonal variation, and the natural rate of population increase, as shown in Figure 25.4. The sources of error in these predictions are being explored through sensitivity analysis emphasizing the structure of the infection rate

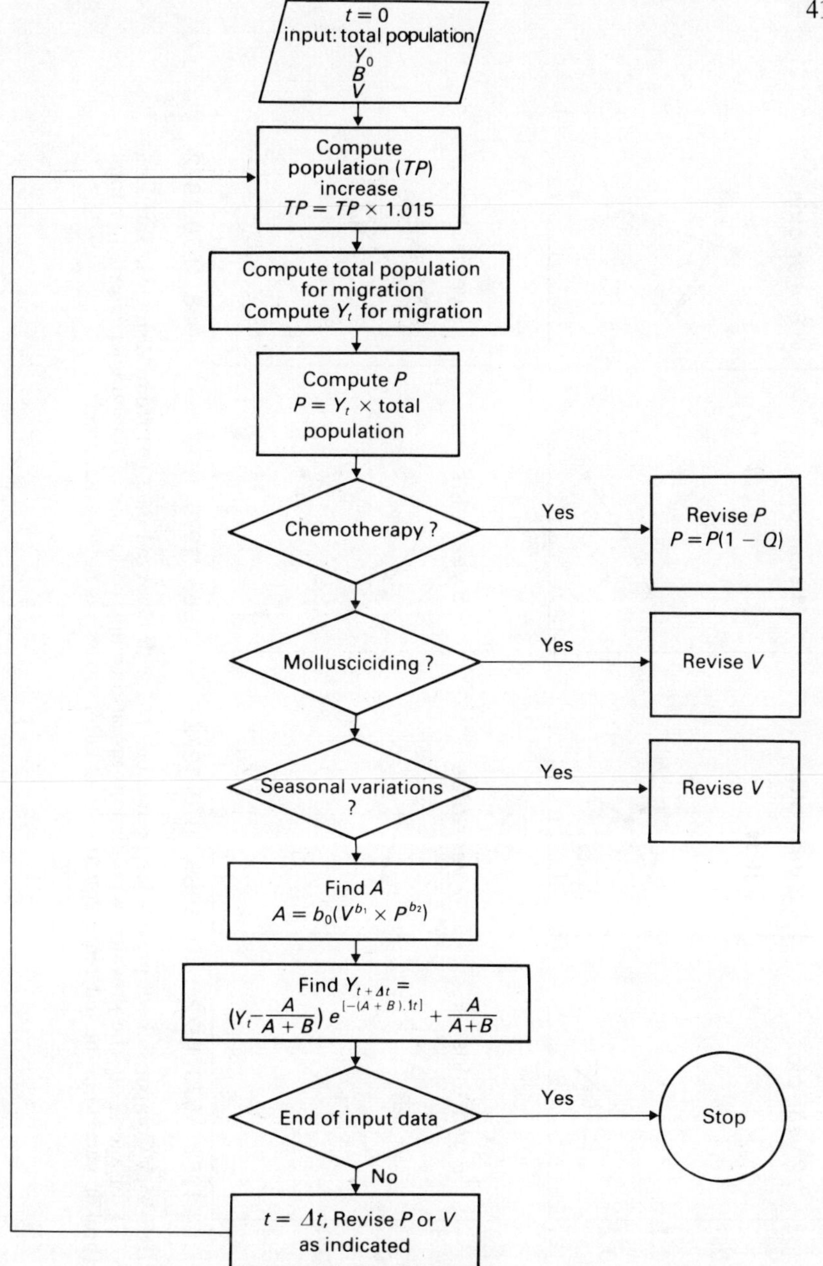

FIGURE 25.4 The logic of the model, as applied to the Tanzania study. Y = baseline prevalence or fraction positive; A = incidence; B = loss rate; V = length (in feet) of snail habitats within half a mile of households/age group; P = number of infected persons/age group; Q = chemotherapy correction factor; t = 2 years; and b_1, b_1, and b_2 are the regression-estimated parameters.

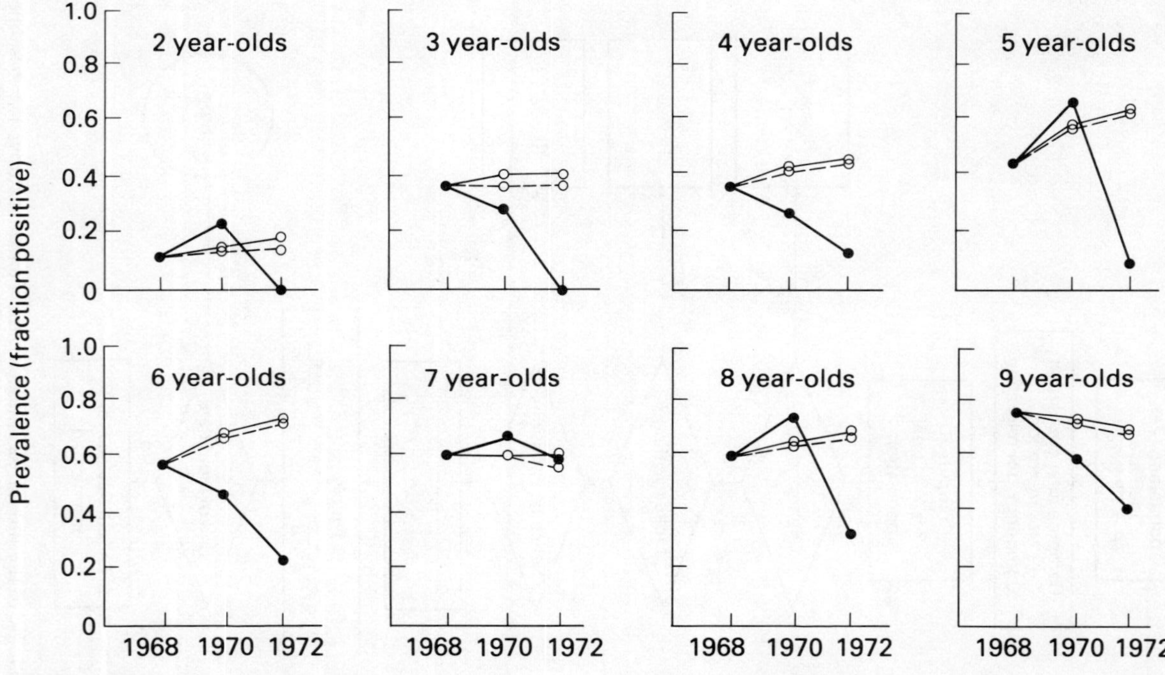

FIGURE 25.5 Comparisons of prevalence predictions (○) with observed data (●) from Misungwi, Tanzania (1968–72) for ages 2–9. The prevalence predictions are also compared for use of the number infected or total eggs (broken curves) in the indicence equation (this latter analysis is not discussed here).

equation and the method of accounting for migration data. Complete field data for two years only are available for verification.

A third infection rate equation that explicitly incorporated human behavior was developed with data from St Lucia (Rosenfield and Jordan, 1977). In this case, quantitative water contact and epidemiological data were available (Jordan *et al.*, 1975), and it was thus possible to estimate the infection rate as a function of water contact W and the number of infected persons P:

$$A = 2.7 \times 10^{-2} (\underset{t\,=\,1.47}{W^{0.36}} \cdot \underset{t\,=\,1.97}{P^{0.31}}) \tag{25.6}$$

$$R^2 = 0.63$$

$$F(2,4) = 3.47.$$

The water contact term was calculated on an age group basis as the relative distribution of water contact minutes per year per person, summed for the age group and for all water contact activities, such as washing clothes, bathing, swimming, fording, and carrying water. They were observed for all ages over a 15-month period, corrected in the analysis for one year (Dalton, 1976). A comparison of predicted and observed infection rates is given in Figure 25.6.

The model predictions of prevalence were made over time using eqn. (25.6) and an age-specific loss rate term. The model also accounted for the installation of water supplies and natural rate of population increase. The results are given in Figure 25.7.

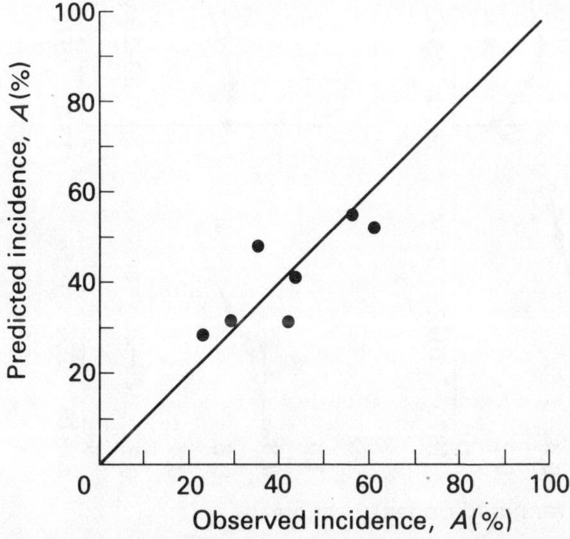

FIGURE 25.6 Predicted incidence (infection rate) obtained with eqn. ((25.6), compared with observed incidence under pre-control conditions (Riche Fond, St Lucia data; after Rosenfield and Jordan, 1977).

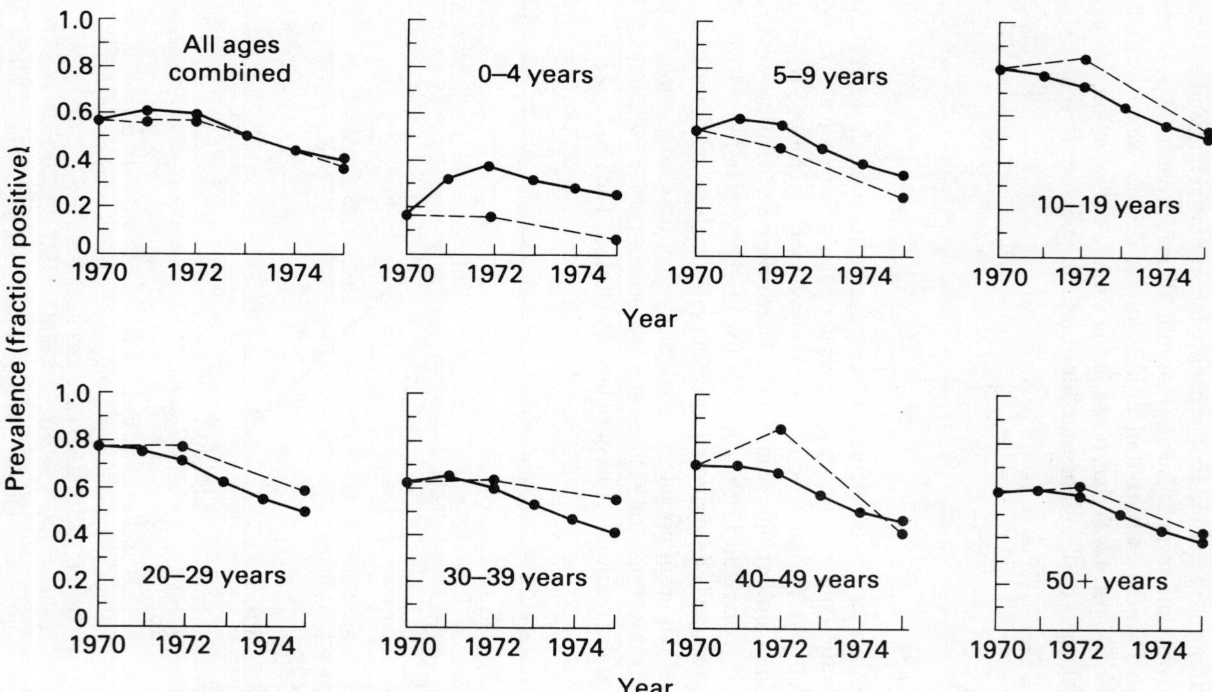

FIGURE 25.7 Results of transmission model testing with Riche Fond, St Lucia data. Predicted prevalence values (full curves) are compared with observed values (broken curves) for various age groups. Water supply installation was begun in 1970 and completed by 1973 (after Rosenfield and Jordan, 1977).

These equations, all based on site-specific considerations, demonstrate the potential for including human variables in transmission modeling. It is interesting to note that the explicit inclusion of the human water contact term dramatically changed the value of the exponents of the nonparasite term, bringing the regression-estimated exponents closer together in weight. Field studies that are now under way to collect more detailed age-specific quantitative water contact data will enable further testing of such results, and perhaps the verification of a more completely defined transmission model—one in which snail populations are included, together with expressions for water contact and human parasite load.

Finally, it is important to stress that such models should be developed by cooperative efforts between epidemiologists, modelers, clinicians, and social scientists, and should be verifiable with field data. If a model is to be used to design a control program, then it must reflect both the life cycle of the disease and the related human elements.

25.4. SOCIAL AND ECONOMIC ASPECTS OF CONTROL

Designers and operators of control programs are becoming increasingly aware that the social and economic aspects of transmission may be a key to successful control, and the reason why such programs have often not achieved their goals in the past. All of the social characteristics discussed above affect the implementation of control programs, and a transmission model incorporating these social concerns is one way of ensuring the choice of socially appropriate control measures. However, when resources are scarce—the prevailing situation in most areas in which schistosomiasis is found—economic realities will also have to be taken into consideration. The time-frame for a control program must be determined, as well as the cost of various control strategies in terms of supplies, equipment, manpower, transport, buildings, etc., in relation to other community needs. Two methods of economic analysis are available to help make these choices:

(1) *cost-effectiveness analysis*, permitting comparison of different strategies when objectives are the same; and
(2) *cost–benefit analysis*, assisting the decision maker to decide whether or not, and how much to invest in schistosomiasis control.

25.4.1. Cost-effectiveness Analysis

Detailed discussions of these techniques as applied to schistosomiasis control are given in Prescott (1979), Jobin (1979), Bekele and Golladay (1979), and Paulini (1977). Briefly, cost effectiveness analysis can be used with the transmission model developed above by measuring the success of different control strategies in preventing case years of infection (Rosenfield *et al.*, 1977; Rosenfield and Gestrin, 1978; Rosenfield, 1979). Case years of infection prevented are calculated as the difference between the incidence of cases with the use of a particular control strategy, and the incidence without controls, summed over the years of analysis to estimate the total number of case years of infection prevented by the use of a given control strategy. The case years of infection prevented are obtained from the predictions of prevalence.

The effectiveness of different control measures is reflected in the infection rate equation. The snail habitat term is affected by molluscidíng, which reduces the snail population, or by engineering measures that eliminate a given snail habitat. The number of infected persons is reduced by chemotherapy. Human water contact is reduced by provision of safe domestic water supplies, laundry facilities, showers, excreta/waste disposal facilities, and location of human settlements. Health education programs are likely to increase the acceptance of chemotherapy, water supplies, and excreta/waste disposal facilities. These changes will modify the input data for the infection rate equation; for example, a reduction in number of infected persons from chemotherapy will reduce the infection rate so that there will be fewer new cases predicted for the next year. Results from the different studies where case years of infection prevented were calculated are given in Table 25.1.

Estimating the cost of a particular control measure involves the relatively straightforward calculation of supply needs, salaries, transport needs, and capital requirements. Unit costs per control measure (e.g., cost of one chemotherapy session per person) can be calculated. For the above three *ex-post* analyses, the total costs of control expenditures for use of one or combined control measures were divided by case years of infection prevented, to yield cost per case year of infection prevented, as shown in Table 25.2.

Such comparisons can also provide the basis for comparing choices of schistosomiasis control strategies over time, but the values calculated in Table 25.2 are not sufficient for this. When making comparisons into the future, it is also necessary to estimate the cost over time and to calculate the present value of future costs in order to account for changes in the cost of money. This requires estimating the appropriate interest or discount rate. The replacement costs must also be estimated. In addition, for sound budgeting purposes, the foreign exchange costs and accounting prices (shadow prices) should be calculated for all resources needed. The techniques of financial analysis should thus be utilized in calculating the total costs of various schistosomiasis control programs before comparing resulting costs per case year of infection prevented (Valentine and Mennis, 1971).

In Tables 25.1 and 25.2 it may be noted that *chemotherapy appears to be the most effective cost-effective strategy of all the analyses. This may be a short-term phenomenon; in considering control measure use over time, it is necessary to be concerned with preventing reinfection, which none of the drugs*

TABLE 25.1 Comparison of effectiveness estimations from analyses of case years of infection prevented in Iran, Tanzania, and St Lucia. These results are only estimates since they are based on model predicted case years of infection prevented (Rosenfield *et al.*, 1977; Rosenfield and Gestrin, 1978; Rosenfield 1979).

Control measure used	Iran[a]	Tanzania[b]	St Lucia[c]
Chemotherapy	1637	162	3517
Molluscibiding	433	86	1257
Engineering measures	484	—	—
Water supplies	—	—	2732
Chemotherapy, mollusciciding and engineering	1596	—	—
Molluscicides and chemotherapy	—	232	—
Water supplies and chemotherapy	—	—	3652
Case years of infection (no controls)	2637	552	9430

[a]Estimated over seven years of control measure use.
[b]Estimated over seven years of control measure use.
[c]Estimated over eight years of control measure use.

available for use today can accomplish. Such a concern suggests that the combined control measure approach, including prevention of human contact with snail habitats, must be considered. In each of the project analyses the second most cost-effective strategy was this particular combination.

Cost-effectiveness analyses are appropriate when the decision has been made to invest in schistosomiasis control. They cannot directly help make that decision, since this requires a commitment that may have many bases. The use of cost–benefit analyses may assist in that decision, although the experience with such analyses for schistosomiasis is not yet encouraging (Prescott, 1979). Some suggestions as to how cost–benefit studies could be more appropriately carried out are given below.

TABLE 25.2 Comparison of cost-effectiveness estimate (US$) from analyses in Iran, Tanzania, and St Lucia by comparing estimated costs per case year of infection prevented. Here only estimates are given since unit costs were calculated and case years of infection prevented were based on model projections (Rosenfield *et al.*, 1977; Rosenfield and Gestrin 1978; Rosenfield 1979).

Control measure used	Iran[a]	Tanzania[b]	St Lucia[c]
Chemotherapy	1.26	9.05	4.38
Mollusciciding	4.76	60.73	41.21
Engineering measures	4.26	—	—
Water supplies	—	—	20.50
Chemotherapy, mollusciciding and engineering measures	1.29	—	—
Mollusciciding and chemotherapy	—	28.83	—
Water supplies and chemotherapy	—	—	16.54

[a]Estimated over seven years of control measure use.
[b]Estimated over seven years of control measure use.
[c]Estimated over eight years of control measure use.

25.4.2. Cost–Benefit Analysis

Cost–benefit analyses of schistosomiasis have rarely been based on the level of the infection or disease in a community. Costs have been calculated as above, but in most studies the economic benefits have been related to directly measurable economic impacts, such as reduced work output, increased work absenteeism, or increased medical care costs (Prescott, 1979). Little attention has been paid to the severity of the disease in the community, or to the disease symptoms that may result in both social and economic impacts. Such studies of schistosomiasis that have been done have led to confusing and sometimes contradictory results. A complete cost–benefit study should be able to relate disease levels to social and economic impairment to investments needed to reduce such impairments.

One of the few studies where economic impacts have been estimated as a function of the severity of schistosomiasis was carried out by Farooq (1963) in the Philippines for cases of *S. japonicum*. He separated the cases into four classes, and described the disability resulting from each class.

I—Mild: no absence from work;
II—Moderate: reduced capacity to work;
III—Severe: frequent absence from work;
IV—Very severe: total absence from work.

Actual disability in the population was estimated on the basis of individual questionnaires and medical examinations. These values were then costed per class on the basis of drug, laboratory, doctor, and health center costs, as well as costs from income losses. Although the estimation of costs, especially output losses, could have been greatly refined, Farooq demonstrated the importance of establishing a clinical gradient or some classification of disease conditions before attempting an economic study. As Farooq concluded:

> The quantitative assessment of disease and disability is always difficult, whatever the infection, but a beginning must be made. However imperfect the present study may be, we feel that at least it provides a rational basis on which an estimate may be made.

Indicators of schistosomiasis infection range from fever, weakness, and pain on urination, to swollen spleen or liver to central nervous system involvement or bladder cancer (Jordan and Webbe, 1969; Warren and Mahmoud, 1978). In Table 25.3, various physical impacts from three species of human schistosomiasis at different stages of the infection can be postulated as a function of frequency of reinfection, length of infection, intensity of infection, infection with other diseases, and nutritional status. Each of these variables can be measured quantitatively. Reinfection is a function of behavior and relates closely to the transmission model variables. Estimating the length of infection requires detailed knowledge of the age prevalence pattern in a community, and this has been discussed in the context of transmission modeling (Hairston, 1965; Kloetzel and Da Silva, 1967). The intensity of infection is measured by "egg counts" (see below), which have been shown to be positively related with disease development in different body organs (Cheever, 1968). Infection with other diseases and nutritional status, each of which affects the death rate of the schistosome worm as well as the development and impact of schistosomiasis symptoms, could be estimated by levels of disease prevalence and some anthropormophic measurements recommended by nutritionists. These last two topics have not been examined in detail in this context (De Witt *et al.*, 1964).

The likelihood of disease development could be estimated as a function of these five characteristics, through regression analysis. The relative importance of each of the body burden-activity variables could be estimated on an age- and sex-specific basis. For example, one might be able to develop a quantitative measure of disease based on the extent of swollen spleen, swollen liver, fibrosis

TABLE 25.3 Symptoms of schistosomiasis in humans (see Macdonald, 1973).

Symptoms in different stages of parasite in man	*Schistosoma haematobium*	*Schistosoma mansoni*	*Schistosoma japonicum*
Invasive	"Swimmer's itch"	"Swimmer's itch"	"Swimmer's itch"
Acute	Mild allergic reaction	Mild allergic reaction	Katayama fever: possibly serious allergic reaction (headache, cough, dysenteric symptoms); mild duodenal and spleen enlargement
Chronic	Bladder lesions hematuria (bloody urine)	Granulomas of the colon (from trapped eggs)	
	Increased micturation, cystitis, reduced bladder capacity		Granulomas of the colon (from trapped eggs)
	Obstruction of the ureter, stones, fibrosis of uretal wall.	Fibrosis of liver liver and spleen enlargement and portal hypertension, anemia	Calcified eggs in colon, intestinal damage
	Cancer of the bladder		Polyps leading to obstruction of lumen of colon
	Cor-pulmonale (heart disease due to pulmonary hypertension)	Fibrosis of the liver	Fibrosis of the liver
	Lesions in spinal cord from eggs	Lesions in spinal cord (transverse myelitis) from eggs	Liver and spleen enlargement and portal hypertension, necrosis of liver cells, brain lesions
General chronic symptoms	Painful urination	Fatigue, abdominal pain intermittent diarrhea	Fatigue, abdominal pain intermittent diarrhea

of the liver, or central nervous system disorders. It is not immediately apparent what units should be used to measure the extent of disease. The number of eggs passed per unit volume of urine or feces is assumed to be the quantitative measure of intensity of infection; i.e., the number of worm pairs in the individual. Autopsy studies have shown that the number of eggs found in organ tissues correspond with (i) disease status of the organs, and (ii) worm pairs in the body (Smith *et al.*, 1974; Ongom and Bradley, 1972). In some instances clinical gradients have been developed to describe the relationship between disease symptoms and eggs passed per unit excreta (Kloetzel, 1967), while some economic studies have used egg counts as the basis for their impact comparisons (Weisbrod *et al.*, 1973).

Nonetheless, several problems remain with the interpretation and use of egg counts in modeling and economic studies. It is not yet possible to measure worm burdens in living individuals, and thus the relationship between worm pairs and eggs passed out of the body requires further laboratory study. In addition, many factors affect worm longevity and egg production over time that require further field studies, such as acquired immunity, passage of eggs from the body in cases of fibrosis of bladder and intestinal tissues, impact of infection, and worm crowding in the body (Warren, 1973; May and Anderson, 1979; Bradley and May, 1978).

If egg counts are used as a measure of disease severity, then the body burden/activity variables equation would require respecification. Whatever measure of disease extent is used, there is likely to be a nonlinear relationship between that measure and the body burden/activity variables; i.e., there may be some threshold above which the intensity of the variables becomes so great that disease symptoms develop rapidly, leading directly to death. In addition, there may be a serious problem of multicollinearity since it is conceivable that each of the body burden variables affects the other, and it may be useful to develop an appropriate weighting procedure for the regression equation so that combinations of variables may be used.

It should be noted that by basing the index of disease development on measurable clinical symptoms, other physical, but not easily measurable impacts that could significantly impair the individual will be unaccounted for—the analysis will underestimate the impact of the disease (B. Cvejetanovic, 1979, personal communication).

The above concerns underline the necessity for economists to work closely with epidemiologists and clinicians to establish the range of disease indicators, and to estimate the body burden-activity variables for each situation. The results from this kind of cooperation would provide the basis for estimating social and economic impairment.

25.4.3. Definitions of Impairment

Social and economic definitions of impairment are needed prior to costing the "impact" of schistosomiasis in a community or a country. The highest level of physical, social, and economic impairment resulting from schistosomiasis is death. More common, however, are less dramatic classes of impairment that seriously affect the lives of millions in the following ways:

- *Economic impairment*: reduced output at work; absenteeism.
- *Domestic impairment*: reduced contribution to home chores; complete inability to contribute to the home.
- *Educational impairment*: reduced output at school; absenteeism.
- *Social/civic impairment*: reduced ability to participate in community activities.

The level of impairment in each of these classes could be measured quantitatively, e.g., reduction in the amount of sugarcane cut, fewer trips to the water source, increased number of classes missed, and/or absence from community events. Previous economic studies have mainly emphasized the purely economic losses without considering social losses (Prescott, 1979). The establishment of a relationship between level of disease and social and economic impairment can provide the basis for estimating the direct and indirect costs of the disease, and ultimately the benefits of its control.

However, at this stage of the analysis, other factors must be considered that further complicate the study. Conditions in society that also affect the economic and social evaluation of the disease include: employment/unemployment/underemployment (e.g., the availability of substitute labor for "impaired" persons); seasonal aspects of employment; demographic situation; migration (see above); urban–rural differences in living conditions; economic development projects planned (which could change economic, social, and epidemiological conditions); and the debt situation of the country (especially if foreign exchange costs are involved). The list could extend for several paragraphs.

It may seem overwhelming, but these factors all play a role in the development of one of the economist's basic tools for planning, the "production function", which is essentially a statement of factors affecting output at the national or community level. For analyzing the impacts of schistosomiasis the production function must be defined appropriately to reflect the economic, social, and epidemiological situation. Some research is already under way on the definition of production functions appropriate for schistosomiasis studies (Prescott, 1979); while others are concentrating on the "domestic" and "educational" aspects of production (Cooper and Rice, 1976; Becker, 1965; Caldwell, 1979).

The results of the studies described in this chapter will vary from site to site as much as results from the application of transmission models. Yet a general model or guidelines—modified to account for country- and site-specific conditions—could be used by schistosomiasis control program planners and managers. The results of such efforts would:

(1) Enable the specific impacts of schistosomiasis to be evaluated.
(2) Suggest the appropriate level of investment in schistosomiasis prevention and control.
(3) Assist in the design and implementation of control programs.

25.5. CONCLUSIONS

With regard to specific applications of models to developing schistosomiasis prevention and control programs, this chapter has stressed a particular gap in existing transmission models: the human element. By not explicitly accounting for human behavior and other relevant social variables, major influences on transmission are missing, thus detracting from any potential practical use of a model in preventing or controlling schistosomiasis. Control program design must also include the human–social elements since, without an understanding of the individual's awareness of the disease or the role human behavior plays in disease transmission, appropriate control measures cannot be applied, and they will not be accepted and used by the population at risk.

The choice of whether or not to invest in schistosomiasis control is rarely made on purely humanitarian grounds. If it were, there would perhaps be no need for debate over whether to aid in the control of schistosomiasis. Unfortunately, the burden to prove a need for investment in control has fallen upon economists and economically inclined public health research workers. To date, their success has been limited, due in part to the fact that their economic models have been inappropriately defined. *There is no existing model, to my knowledge, that relates disease development to physical, social, and economic impairment and then incorporates such inputs into an epidemiologically, economically, and socially appropriate production function*. Until such studies have been carried out, the claim that there is no measurable economic affect from schistosomiasis will continue to be made. Thus, millions may continue to suffer due to lack of proof that their suffering contributes to economic loss.

ACKNOWLEDGMENTS

The inputs provided by Dr Norman Bailey, WHO are acknowledged with great gratitude. Other persons generously contributed useful ideas for this paper, especially Dr Andrew Davis, WHO and Mr N. Prescott, Oxford University. The Schistosomiasis Program Staff from Iran, Tanzania and St Lucia have been exceptionally generous with their data and advice, without which these studies could not have been done. The advice and comments of Mr A. Thomas, WHO, Mrs Ann Gregory, WHO, Mr Scott Shuster, and the anonymous reviewer are also appreciated. The paper was expertly and patiently typed by Mrs Joy Calo, Mme Lise Ducret, and Mrs Sue Pruzensky. The ideas and opinions expressed in this paper are those of the author and do not necessarily reflect official policy of the UNDP/World Bank/WHO Special Programme for Research and Training in Tropical Diseases.

REFERENCES

Arfaa, F., G. H. Sahba, I. Farahmardian, and H. Bijan (1970) Progress towards the control of bilharziaris in Iran. *Trans. R. Soc. Trop. Med. Hyg.* 64:912–17.

Becker, G. S. (1965) A theory of allocation of time. *Econ. J.* 75(3):493–517.

Bekele, A. and F. Golladay (1978) The Economics of Schistosomiasis Control, paper presented to Session on Rural Health in Developing Nations at Operations Research Society of America Annual Meeting May 1978, New York.

Bradley, D. J. and R. M. May (1978) Consequences of helminth aggregation for the dynamics of schistosomiasis. *Trans. Roy. Soc. Trop. Med. Hyg.* 72:262–73.

Caldwell, J. C. (1979) Education as a factor in mortality decline: An examination of Nigerian data, Paper No. DSI/SE/WP/79.3 Rev. 1, presented at *WHO Conf. on Socioeconomic Determinants of Mortality*, Mexico City, 19–25 June.

Cheever, A. W. (1968) A quantitative post-mortem study of *Schistosomiasis mansoni* in Man. *Am. J. Trop. Med. Hyg.* 17:38–64.

Cohen, J. (1976) Schistosomiasis: A human host–parasite system, in R. M. May (ed) *Theoretical Ecology: Principles and Application* (Philadelphia, PA: W. B. Saunders) pp237–56.

Cohen, J. E. (1977) Mathematical models of schistosomiasis. *Ann. Rev. Ecol. System.* 8:209–33.

Cooper, B. S. and D. P. Rice (1976) The economic cost of illness revisited. *US Social Security Bull.* (Washington: US Dept of Health, Education and Welfare) pp21–36.

Dalton, P. (1976) Sociological approach to the control of *Schistosoma mansoni* in St Lucia. *WHO Bull.* 54:587–95.

Dalton, P. R. and D. Pole (1978) Water-contact patterns in relation to *Schistosoma haematobium* infection. *WHO Bull.* 56(3):417–26.

De Witt, W. B., J. Oliver-Gonzalez, and E. Medina (1964) Effects of improving the nutrition of malnourished people infected with *Schistosoma mansoni. Amer. J. Trop. Med. Hyg.* 13:25–35.

Farooq, M. (1963) A possible approach to the evaluation of the economic burden imposed on a community by schistosomiasis, *Ann. Trop. Med. Parasitol.* 57:323–31.

Farooq, M. (1973) Review of national control programmes, in N. Ansari (ed) *Epidemiology and Control of Schistosomiasis* (*bilharziasis*) (Baltimore, MD: University Park Press) pp388–421.

Fine, P. E. and J. S. Lehman, Jr. (1977) Mathematical models of schistosomiasis: Report of a workshop, *Am. J. Trop. Med. Hyg.* 26(3):500–4.

Hairston, N. (1962) Population ecology and epidemiological problems, in G. E. W. Wolstenholme and M. O'Connor (eds) *CIBA Foundation Symp. on Bilharziasis* (London: Churchill) pp36–80.

Hairston, N. (1965) An analysis of age and prevalence data by catalytic models. *Bull. WHO*, 33:163–75.

Hairston, N. G. (1965) On the mathematical analyses of schistosome populations. *WHO Bull.* 33:45–62.

Jobin, W. (1979) The cost of snail control. *Am. J. Trop. Med. Hyg.* 28:142–54.

Jordan, P. (1977) Schistosomiasis—research to control. *Am. J. Trop. Med. Hyg.* 26(5):877–86.

Jordan, P. and G. Webbe (1969) *Human Schistosomiasis* (Springfield, IL: Charles C. Thomas).

Jordan, P., L. Woodstock, G. O. Unrau, and J. A. Cook (1975) Control of *Schistosoma mansoni* transmission by provision of domestic water supplies. *WHO Bull.* 52:9–21.

Kloetzel, K. (1967) Mortality in chronic splenomegaly due to *Schistosomiasis mansoni*: Follow-up study in a Brazilian population, *Trans. R. Soc. Trop. Med. Hyg.* 61:803–5.

Kloetzel, K. and J. R. Da Silva (1967) *Schistosoma mansoni* acquired in adulthood: Behaviour of egg counts and the intradernal test. *Am. J. Trop. Med. Hyg.* 16:167–9.

Kloos, H., W. Sidrak, A. A. M. Michael, E. W. Mohareb, and G. I. Higashi (1982) Disease concepts and practices relating to *Schistosomiasis haematobium* in Upper Egypt. *J. Trop. Med. Hyg.* 85:99–107.

McCullough, F., and V. M. Eyakuze (1973) *Final Report, AFR/SCHIST 129*, (Geneva: WHO).

McCullough, F. S., G. Webbe, S. S. Baalawy, and S. Maselle (1972) An analysis of factors influencing the epidemiology and control of human schistosome infections, Mwanza, Tanzania. *East Afr. Med. J.* 49:568–82.

Macdonald, G. (1965) The dynamics of helminth infections with special reference to schistosomes. *Trans. R. Soc. Trop. Med. Hyg.* 59(5):489–506.

Macdonald, G. (1973) Measurement of the clinical manifestations of schistosomiasis, in N. Ansari (ed) *Epidemiology and Control of Schistosomiasis* (Baltimore, MD: University Park Press) pp354–87.

May, R. M. (1977) Togetherness among schistosomes: its effects on the dynamics of transmission. *Math. Bios.* 35:301–343.

May, R. M. and R. M. Anderson (1979) Population biology of infectious diseases: Part II. *Nature* 80:455–61.

Muench, H. (1959) *Catalytic Models in Epidemiology* (Cambridge, MA: Harvard University Press).

Nasell, I. and W. Hirsch (1973) The transmission dynamics of schistosomiasis. *Comm. Pure Appl. Math.* 26:395–453.

Ongom, V. L. and D. J. Bradley (1972) The epidemiology and consequences of *Schistosoma mansoni* infection in West Nile, Uganda, 1: Field studies in a community at Panyogor. *Trans. R. Soc. Trop. Med. Hyg.* 66(6):835–57.

Paulini, E. (1977) *The Use of Mathematical Models in Formulation Strategies for Schistosomiasis Control*, Paper presented at Symp. on schistosomiasis modelling, Int. Epidemiol. Ass. Ann. Mtg, Puerto Rico, September 1977.

Prescott, N. (1979) Schistosomiasis and development. *World Devel.* 7(1):1–14.

Rosenfield, P. L. (1975) *Schistosomiasis Transmission Model* (Washington, DC: Agency for International Development).

Rosenfield, P. L. (1979) *The Management of Schistosomiasis* (Baltimore, MD: Johns Hopkins Press for Resources for the Future).

Rosenfield, P. L. and P. J. Gestrin (1978) *Socio-Economic Analysis of Impact of Water Projects on Schistosomiasis*, Report to Agency for International Development, Washington, DC Contract No. 931–113 (unpublished).

Rosenfield, P. L. and P. Jordan (1977) Testing of a schistosomiasis transmission model with field data. *Bull. Inter. Stat. Inst.* 47(2):31–60.

Rosenfield, P. L., R. A. Smith, and M. G. Wolman (1977) Development and verificationa of a schistosomiasis transmission model. *Am. J. Trop. Med. Hyg.* 26(3):505–16.

Ruyssenaars, J., G. Van Etten, and F. McCullough (1973) Population movement in relation to the spread and control of schistosomiasis in Sukumaland, Tanzania. *Trop. Geog. Med.* 25:179–86.

Smith, D. H., K. S. Warren, and A. A. F. Mahmoud (1979) Morbidity in *Schistosomiasis mansoni* in relation to intensity of infection: A study of a community in Kisumu, Kenya. *Am. J. Trop. Med. Hyg.* 28(2):220–9.

Sturrock, R. F., J. E. Cohen, and G. Webbe (1975) Catalytic curve analyses of schistosomiasis in snails. *Ann. Trop. Med. Parasitol.* 69:133–4.

Valentine, J. L. and E. A. Mennis (1971) *Quantitative Techniques for Financial Analyses*, (Homewood, IL: Richard D. Irwin).

Warren, K. S. (1973) Regulation of the prevalence and intensity of schistosomiasis in man: Immunology or ecology. *J. Infectious Diseases* 127(5):595–609.

Warren, K. S., and A. A. F. Mahmoud (1978) *Geographic Medicine for the Practitioner* (Chicago, IL: University of Chicago Press) pp107–8; and (London: Heinemann) pp12–15.
Weisbrod, B., R. L. Andreano, R. E. Baldwin, E. H. Epstein, A. A. Kelley, and T. W. Helminiak (1973) *Disease and Economic Development* (Madison, WI: University of Wisconsin Press).

26 Farmer Adjustments to Mandatory Pest Control

Gerald Carlson and Ricardo Rodriquez

26.1. INTRODUCTION

There is increasing interest in mandatory, area-wide pest control. Among factors contributing to this are: more information on the mobility of pests, increasing costs (both to the farmer and others) of individual farmer suppression methods, and improved monitoring and pest suppression technology for area-wide programs. Mobility of pests decreases individual farmer incentives to prevent pesticide resistance development, preserve beneficial insects, and engage in long-term pest suppression efforts. Examples of important forms of area-wide, mandatory pest control are abatement districts, eradication or quarantine programs, legally required production practices, and community or cooperative pest management.*

There are distinct economic gains and costs to mandatory versus individual farmer pest control. Scale economies in pesticide application and decision making, and prevention of spatial spread are often cited as gains (Tullock, 1969). As pest management becomes more complex and requires more skills and capital, there may be gains to collectively performing this management phase of agricultural production, but there may also be additional costs of centralizing control related to unequal pest problems, unequal pest management skills, and other resource adjustment costs.

This chapter concentrates on short-term farmer adjustments to mandatory pest control. Several individuals have proposed methods to measure incremental benefits of pest eradication (Carlson, 1975; Power and Harris 1973;

*Community pest control efforts may be voluntary, but many of the problems of uniform service and financing are similar (see Pridgen, 1980). Taxing schemes to internalize costs of mobile pests have been analyzed by Regev *et al.* (1976).

Johnston, 1975). However, there does not seem to be a general economic model of producer response to mandatory pest controls. If producer response is ignored, planning and financing area-wide programs can have problems because group costs are not proportional to acreage. It is important to understand what adjustment costs might be involved with spatially uniform pest control.

The economic model presented here is specific to a uniform manner of financing and providing service to farmers. This appears to be a common organizational form for many pest eradication and other centralized pest control programs in the US and elsewhere.* The effect of variable pest populations and crop yields is thus of special interest. After the economic model is presented, some results from an analysis of a trial eradication of the cotton boll weevil (*Anthonomus grandis*) are given.

26.2. THE GROWER MODEL

A profit-maximizing firm is assumed to represent the decision framework of the grower desiring to control pests. In the simple case, the grower has a production process in which only pest control resources X_1 and all other resources X_2 are used to produce yield Y of a single target crop C. Given competition such that product price for the crop P_C and input prices r_1 and r_2 (for X_1 and X_2) are given, the farmer's problem of how many pest control resources to use is solved by finding where marginal produce values saved from using pest control inputs just equal their marginal costs:

$$P_C(\partial Y/\partial X_1) = r_1. \qquad (26.1)$$

More complete statements of this problem are needed when multiple treatments are involved, when pest populations grow, when crop susceptibility changes, or when more complex cost structures are accounted for (Headley, 1972; Hall and Norgaard, 1973).

Pest species often damage one crop type much more than others grown on the same farms and a farmer can thus avoid damage by switching the crops he grows. The grower will maximize profits if he adjusts acreage until the marginal returns to land X_3 in each crop (C or D) are equalized (Carlson, 1979). This can be seen in terms of the above model as

$$P_C(\partial Y_C/\partial X_{3C}) = P_D(\partial Y_D/\partial X_{3D}). \qquad (26.2)$$

This follows from the assumption that there are various levels of pest damage to

*See Regev *et al.* (1976) for a proposed tax to internalize and finance an area pest suppression program. Carlson and DeBord (1976) investigate the effect of pest populations and various socioeconomic variables on demand and expenditures for mosquito abatement districts. Johnston (1978) examines some of the public finance issues when pests can develop pesticide resistance and are migrating.

crop $C(X_{3C})$, and that this can be reduced by shifting some or all of the acreage to crop $D(X_{3D})$. Other inputs are held constant.

26.2.1. Profit Maximation with Uniform Pest Control

Suppose it now becomes possible to control pests by a group of growers at a reduced cost per unit area relative to what has been expended by individual growers. Assume that control provided by a group is uniform to all producers of crop C in a region and no group or area-wide control of pests is provided on crop D.

The group pest control program is mandatory for all producers of crop C. The producers of crop C must pay a fixed fee per acre of crop C grown equivalent to an equal acreage share of the total cost of the program for that year. Assume r_g is the price per unit paid by the group for pest control resources X_1, that Z is the level of pest control per acre chosen by the group, and that O is overhead expenditures. Then $\sum_i^m r_g Z X_{3i} + O$ is the total pest control expenditure for a group of m farmers on crop C. The fee for farmer j producing X_{3j} acres of crop C is:

$$r_g Z X_{3j} / \left(\sum_i^m r_g Z X_{3i} + O \right). \tag{26.3}$$

A farmer is usually allowed to add to the group level of pest control Z by spending his own pest control resources $(X_1' r_1)$ if he chooses. Total pest control on crop $C(X_1)$ for a given farmer is:

$$X_1 = X_1' + Z, \tag{26.4}$$

where X_1' is self-applied pest control. However, because pest control of level Z is applied by the group, there is no way to reduce total pest control below Z. This becomes a lower bound on pest control effort for each acre of crop C.

Optimization of the use of pest control resources X_1, other inputs X_2, and land in crop $C(X_{3C})$ and crop $D(X_{3D})$ is now more complex. A fixed pest control charge per acre of crop C grown $(r_g Z X_{3C})$ and a minimum level of pest control Z on each acre of crop C may lead to misallocation of pest control and other resources for certain farmers. These misallocations can result in income losses to producers.

The group level of pest control Z can either be too high or too low relative to individual farmer pest control. If a particular farmer was using a high level of pest control resources prior to the group program, he can easily retain his level by adding to the group level. There will be no misallocation of resources in this case. Because of lower unit cost of pest control $(r_g < r_1)$ this may appear attractive to these farmers. For the farmer who has low pest problems (optimal $X_1 < Z$), it is possible that paying the fixed fee for a high level of group pest control is inefficient. (This will be strictly true for the case when $r_g \geqslant r_1$.) This producer can only reach equilibrium by reducing acreage of crop C grown, by increasing the use of substitute inputs X_{2A}, or by decreasing the use of complementary inputs (see Rodriquez, 1979, for details of the pest control optimization problem with group constraints).

26.2.2. Area-wide Pest Control Effects on Individual Farmers

To illustrate some of the features of farmer adjustments to group pest control, consider the situation shown in Figure 26.1. The horizontal axis shows pest control effort per acre, which would include expenditures for insecticides, monitoring, and decision making. Incremental benefits and costs of the control effort are measured on the vertical axis. The farmers' share of group costs per acre is shown as r_g, and individual farmer costs for the same number of pest control units are shown as r_1. The marginal benefit curves from pest control are downward sloping, reflecting the diminished productivity of increased units of pest control for crop production. It is assumed that crop production benefits of a given level of pest control on a unit area rise as pest infestation or potential yield rises. (Pesticides are the major input whose productivity rises as infestation rises. Monitoring inputs are most productive at intermediate infestation levels, while decision-making inputs probably are most productive at intermediate and high infestation levels. A combination of these input productivities is implied by the marginal benefit curves.) This is shown for two typical marginal benefit curves for farms commonly experiencing low (LL')

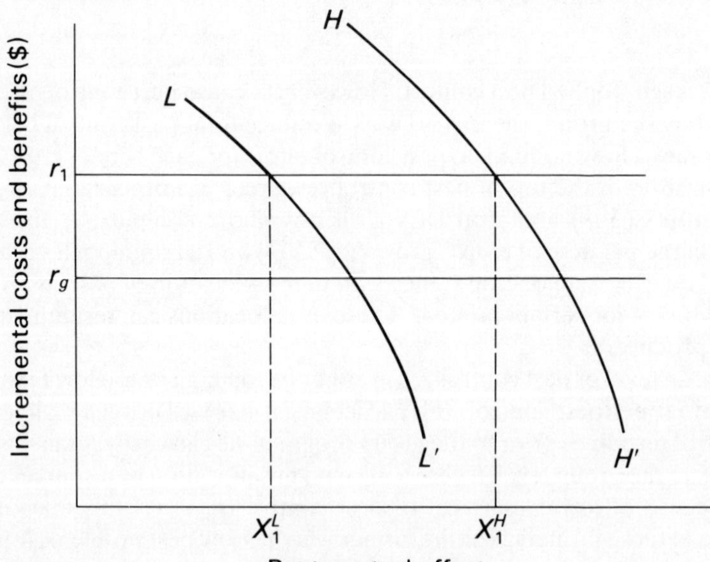

FIGURE 26.1 Optimal pest control effort for various pest populations.

and high (HH') pest populations. These same curves can also be considered to be for high- and low-yield farmers, respectively.

At this point group use of a set of resources is assumed to have equal productivity to that when the same resources are used by individual producers. This assumption can be relaxed. Optimal individual choices are at X_1^L and X_1^H for low and high pest populations (yields), respectively.

Consider the case in which the farmer's per acre cost for group control r_g is equal to his individual control cost per acre r_1 (Figure 26.2). Optimal group or government pest control is at X_g^L or X_g^H for low and high pest populations, respectively. However, the centralized pest control effort is announced to be at the uniform level Z, regardless of potential yields or pest populations. The low pest population farmer has benefits above costs equal to area A for effort levels up to X_g^L, but if he produces crop C, he must pay r_g times Z for each acre grown. He receives total benefits equal to the area under his marginal benefit curve L and will have a welfare loss equal to the difference in the shaded areas $B - A$ if he produces crop C. The high yield or high pest population farmer (shown as curve H) is more likely to produce crop C. He can pay the government fee r_g for Z units of pest control and add $(X_g^H - Z)$ extra units on his own to reach his equilibrium level X_g^H.

At very high levels of group pest control effort beyond X_g^H, even farmers exposed to very high pest populations will prefer growing another crop to participating in the group program. This may be typical of pest eradication programs as opposed to pest control for immediate crop production. Eradica-

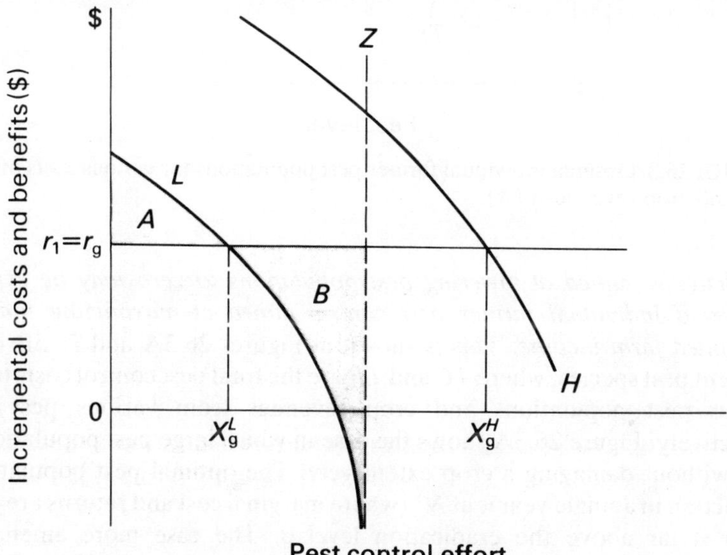

FIGURE 26.2 Pest control effort with eradication.

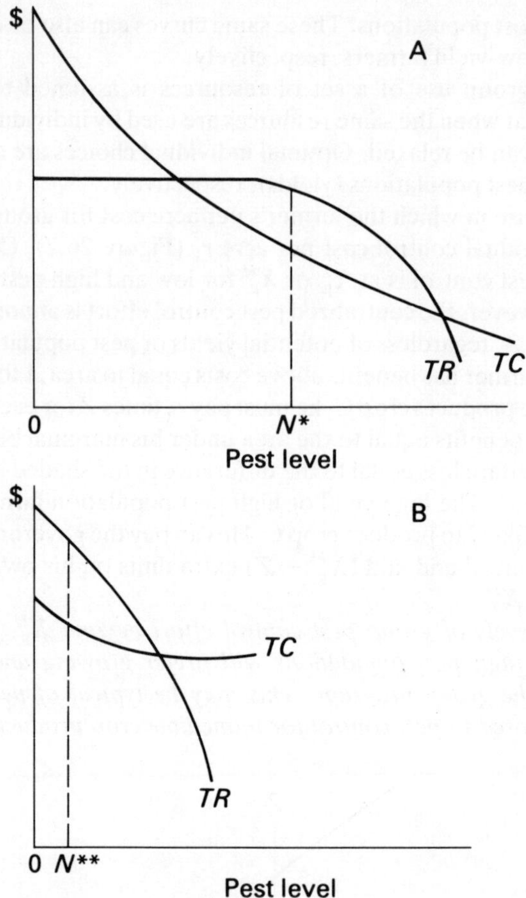

FIGURE 26.3 Optimal individual farmer pest populations for various costs of control (*TC*) and crop revenues (*TR*).

tion activities aimed at lowering pest populations to zero may be expensive relative to individual farmer pest control aimed at maximizing individual discounted farm income. This is shown in Figures 26.3A and 26.3B for two different pest species, where *TC* and *TR* are the total pest control costs to reach various pest populations and crop revenues from various pest levels, respectively. Figure 26.3A shows the case in which large pest populations can exist without damaging a crop extensively. The optimal pest population for production in a single year is at N^* (where marginal cost and returns are equal), which is far above the eradication level 0. The case more amenable to eradication is that shown in Figure 26.3B, where the individual farmer optimal crop production population is near zero at N^{**}.

Of course, the other major feature which affects the pest population choice is how rapidly the costs TC of driving the population to extinction rise as extinction is approached. Some species in some locations and climates will resist very high suppression efforts in refuges or by mutation, whereas other species, especially recently introduced ones, can more easily be eradicated when their reproduction rates are low or alternative hosts are limited. The difference between individual farmer optima and eradication is critical in acceptance of eradication programs.

To summarize, the area-wide option becomes more attractive, relative to pest control by the individual farmer, under various economic and biological conditions. These include:

(1) More uniform pest populations.
(2) More uniform yield per unit area.
(3) When area-wide inputs are similar to crop protection inputs for the current year.
(4) Where economies of scale in pest control arise from area-wide effort.
(5) When individual farmers can supplement the area-wide effort.
(6) When area-wide payments can be closely aligned with the individual farmer's present value of crop production benefits.

26.3. ANALYSIS OF ACREAGE RESPONSES TO ERADICATION

To see how individual farmers adjust to mandatory pest control fees, responses of cotton farmers in North Carolina were investigated. A trial boll weevil eradication program in North Carolina in 1978 required payment of fees on each acre of cotton grown, as shown in eqn. (26.3). Farmers were expected to adjust cotton acreage or complementary inputs to maximize their profits. A survey of grower practices in 1978 showed no adjustment in fertilizer as a complementary input. Major attention was directed to crop acreage adjustments to reduce area-wide pest control fees.

The actual decrease in cotton acreage planted in 1977 to 1978 in the North Carolina eradication area was larger than any previous adjustment, and also larger than that observed in other cotton areas in North Carolina for the same period. However, the acreage response to prices, weather, and other variables needed to be held constant to determine if there were acreage adjustments to the eradication program and its costs.

To estimate the adjustment in county cotton acreage to the eradication program in North Carolina, an acreage response model was estimated. The variables for this equation came from the demand for cotton land function, which was found by maximizing profits of cotton and a substitute crop subject to their production functions and the eradication program constraints (see Rodriquez, 1979) for a discussion of the derivation of the model).

Most agricultural supply analysis has been based on Nerlove's (1958) approach of partial adjustment or adaptive expectations, which leads to the use of the lagged dependent variable as one of the regressors. Those models usually yield large R^2 and statistically significant coefficients. Just (1974) suggested that the Nerlovian response models have possibly performed well for estimation purposes because in reduced form they can implicitly explain risk response, and he recommends explicit use of risk measures rather than lagged dependent variables. Such risk variables in the spirit of Just are added in the model to follow.

To account for price risk it was assumed that farmers learn from their experience in using futures quotations for expected price. If the farmer selects his planted acreage based on future prices, he would be able to compare that quotation with the actual price he gets at the harvest season. The deviation of the expected from the actual price received should enter his planting decision for the following period. To measure the response to risk on cotton yield, a moving standard deviation of cotton yield was included.

On the grounds of rational expectations, Gardner (1976) hypothesized that, at least for some crops, available forward information with respect to prices, such as that provided by futures market quotations, may be an important source of information to producers. In estimating cotton and soybean acreage response models, he obtained reasonably good results using futures prices for expected price. In this study futures prices, deflated by an index of prices paid by farmers, are used for own and substitute prices (soybeans). These were three-day averages of closing futures prices of the last week in April (planting season) for December (harvest season).

Some of the substitute crops other than soybeans, such as tobacco, peanuts, are limited by acreage allotments and so were not considered.

Extraordinarily unfavorable weather conditions were observed during the 1978 cotton planting season. Specifically, the excessive rainfall during that period was considered to have prevented some of the cotton from being planted in that year. To account for this possible effect on cotton acreage planted, a weather variable was specified as the positive departure from normal rainfall, weighted by the number of days of rain accumulated for the last half of April.

Eight North Carolina counties were pooled in a single estimate assuming the same responsiveness to prices and other variables by farmers within a given county. Differences between counties were accounted for with dummy variables. To measure the eradication effect in two dimensions, space and time, four counties in the eradication trial area and four outside that area were included, which accounted for more than 80% of the cotton in North Carolina in recent years. A 0–1 dummy variable was used to identify those counties and years during which eradication was in effect. A richer set of responses might have been estimated if there were different fees charged in different areas over time. This will be possible when more years of this experiment are completed.

A semi-logarithmic form or exponential model was chosen since large differences in acreage among counties would imply a serious problem of heteroskedasticity in a linear model. The statistical model is then specified as:

$$\begin{aligned} A_{it} = {} & \beta_0 + X_i D_i + \beta_1 P_{ct} + \beta_2 P_{st} + \beta_3 YV_{it} + \beta_4 RPC_{t-1} \\ & + \beta_5 WD_{it} + \beta_6 EP + e_{it}, \end{aligned} \tag{26.5}$$

where

A_{it} = natural logarithm of cotton acreage ($i = 1, 2, \ldots, 8$) in year t (year 1 = 1967; year 12 = 1978);
D_i = 1 for county i and 0 for other counties;
P_{ct} = expected cotton price; futures quotations at planting season for harvest season deflated with index of prices paid by farmers;
P_{st} = expected soybean price (time period the same as for cotton prices);
YV_{it} = four-year moving standard deviation of cotton yield;
RPC_t = $FC_t - ACP_t$ = measure of price risk, where
FC_t = cotton futures price quotations at planting season;
ACP_t = average cotton price plus support payments received by farmers;
WD_{it} = $(W_{it} - \bar{W}_i)/\bar{W}_i$ for $W_{it} > \bar{W}_i$ (mean annual rainfall at one location), = 0 for $W_{it} \leqslant \bar{W}_i$;
W_{it} = $(RA_{it})(NA_{it})$, where
RA_{it} = accumulated rainfall in last half of April in county i in year t,
NA_{it} = number of days of rain in last half of April;
EP = 1 for counties in eradication area in 1978,
= 0 for others, and
e_{it} = random disturbance with zero mean and constant variance.

The error term assumes no autocorrelation and cross-section independence when the ordinary least squares (OLS) estimation procedure is used. An initial OLS estimate suggested the existence of positive, serial correlation in some of the counties. Also, the cross-section independence was thought to be a weak assumption. Thus, a generalized least squares (GLS) procedure was used to correct for these problems. The GLS method followed here was that developed by Fuller and Battese (1974), which is applicable for cross-section, time-series models. The error components approach is followed in that method to estimate the covariance matrix. For comparison purposes both the OLS and GLS estimates are presented in Table 26.1. The magnitudes of most of the parameters remain at similar levels with both estimation methods. The only two parameter values that show significant differences between the OLS and GLS procedures are the weather variable *WD* and the eradication variable *EP*.

The acreage adjustment model gave most of the results expected. The only unexpected result was the positive sign of the cotton yield moving standard

TABLE 26.1 Estimates of the pooled time-series county model.[a]

		OLS[b]	GLS[b]
Intercept		9.605 (28.20)	9.50 (11.66)
Cotton price, P_{ct}		2.033 (2.59)	2.224 (1.13)
Soybean price, P_{st}		−0.448 (−4.09)	−0.399 (−1.50)
Yield standard deviation, YV_{it}		0.003 (3.94)	0.0024 (4.00)
Price risk, RPC_{t-1}		−0.0064 (−2.53)	−0.007 (−1.18)
Rainfall, WD_{it}		−0.0785 (−1.88)	0.0093 (0.25)
Eradication dummy, EP		−1.014 (−3.51)	−0.4158 (−1.90)
County dummies	$D1$	−0.217 (−1.74)	−0.245 (−0.89)
	$D2$	0.576 (4.67)	0.537 (1.96)
	$D3$	−0.435 (−3.56)	−0.471 (−1.72)
	$D4$	0.541 (4.36)	0.511 (1.86)
	$D6$	1.074 (8.72)	1.048 (3.82)
	$D7$	0.391 (3.19)	0.357 (1.30)
	R^2	0.627	

[a] *t*-statistics are in parentheses.
[b] OLS designates ordinary least-squares estimates; GLS designates generalized least-squares estimates.

deviation YV. The positive relation between planted acres and yield variability suggests a risk-taker attitude for cotton farmers. The more likely explanation is that omitted variables such as management levels and soil quality, which decrease yield variability, will increase mean yields. However, the use of mean cotton yields in the model distorted the eradication variable. The model can be rewritten as

$$\hat{A}_{it} = \exp(\hat{\alpha} + \Sigma\hat{B}_i X_i + \hat{E}P); \tag{26.6}$$

then

$$\hat{A}_{it} - \hat{A}_{it-1} = \exp(\hat{\alpha} + \hat{B}X_i + E\hat{P}) - \exp(\Sigma X\hat{\alpha} + \hat{B}X_i) \qquad t = 1978 \tag{26.7}$$

$$(\hat{A}_{it} - \hat{A}_{it-1})/\hat{A}_{it-1} = [\exp(E\hat{P}) - 1]. \tag{26.8}$$

Thus the acreage response to the eradication program in percentage terms with respect to the previous year (1977), can be computed as:

$$\%EP = [\exp(E\hat{P}) - 1]\cdot 100 = 34\%. \tag{26.9}$$

The estimated impact of the program is therefore a 34% decline in cotton acreage.

Individual farmer characteristics such as pest management skills, pest infestation levels, and usual yield levels will also affect participation in group pest control. An analysis of individual growers was made both inside and outside the eradication area of North Carolina for 1978. They confirm the notions that those farmers who gain most by area-wide pest control (high yield, high pest pressure, low risk) are the most likely to grow the target crop. More analysis of data from the second year of the eradication trial is in process.

26.4. SUMMARY AND CONCLUSIONS

Evaluating returns to area-wide, mandatory pest controls has typically ignored the possibility of farmers adjusting crop acreages or other inputs. This is critical because the financing base for group pest control has usually been a fixed fee per unit area. If many producers do not choose to produce the host crop, then the remaining participating producers will have a higher fee. An exception to this might be the case where eradication of a pest is achieved by not growing a single host crop for a sufficiently long period. Estimates of likely acreage adjustments will be helpful in planning area-wide pest control resources.

Variability in demand for pest control across farms in a region can lead to high incentives for non-participation in area-wide control programs. The models investigated here showed that high-yield, high pest control expenditure farms tend to benefit relative to those with low yields, low pest populations, and certain other characteristics. Further, there can be quite different levels of pest control needed for current profit maximization as opposed to pest eradication. Areas with close substitute crops are likely to experience considerable crop switching to avoid or take advantage of area-wide programs for a target crop.

The model developed and estimated here indicated that there was likely to be a decrease in crop acreage during the trial boll weevil eradication program. This model accounted for other sources for acreage adjustments. Also, it was possible for complementary inputs such as fertilizer to be used by participating farmers at increased levels to capture higher net returns to the eradication resources. The change in other inputs was not detected for the producers in the program area.

Many factors influencing relative attractiveness of area-wide pest management programs were not considered here. Among those needing further economic analysis are: factors affecting area-wide control costs, availability of group pest management information to nongroup members, perception of growers of the difference in marginal productivities of group and individually applied pest controls, and others.

Group pest control imposes some costs of conformity to achieve economies of scale. Eradication planners and evaluators should consider acreage response to uniform pest control. Eradication and profit maximization of producers will often be at odds and will result in adjustments by growers.

Scaling eradication fees closer to individual farmer benefits should be considered. Finally the analysis of group pest control needs the attention of scientists, managers, and policy developers.

REFERENCES

Carlson, G. A. (1975) Control of a mobile pest: The imported fire ant. *Southern J. Agric. Econ.* 7:35–41.

Carlson, G. A. (1979) Pest control risks in agriculture, in J. A. Roumasset, *et al.* (eds) *Risk, Uncertainty and Agricultural Development* (New York: Agricultural Development Council) pp200–9.

Carlson, G. A. and D. V. DeBord (1976) Public mosquito abatement. *J. Environ. Econ. Management* 3:142–53.

Fuller, W. A. and G. E. Battese (1974) Estimation of linear models with crossed-error structure. *J. Econometrics* 2(1):67–78.

Gardner, B. (1976) Futures prices in supply analysis. *Am. J. Agric. Econ.* 58(1):81–4.

Hall, D. C. and R. B. Norgaard (1973) On the timing and application of pesticides. *Am. J. Agric. Econ.* 55(2):198–201.

Headley, J. C. (1972) Defining the economic threshold, in R. L. Metcalf (ed) *Pest Control Strategies for the Future* (Washington, DC: National Academy of Sciences) pp100–8.

Johnston, J. H. (1975) Public Policy on cattle tick control in New South Wales. *Rev. Marketing Agric. Econ.* 43(1)3:39.

Johnston, J. H. (1978) *Optimal Public Versus Private Control Strategies for Migrating Pests with Pesticide Resistance and Pigovian Taxes and Subsidies,* Paper presented at 7th Conf. of Economists, Macquarie University, Sidney (unpublished).

Just, R. E. (1974) *Econometric Analysis of Production Decisions with Government Intervention: The Case of the California Field Crops*. Giannini Foundation Monograph No. 33 (berkeley, CA: University of California Press).

Nerlove, M. (1978) *The Dynamics of Supply: Estimation of Farmers' Responses to Price* (Baltimore, MD: Johns Hopkins University Press).

Power, A. P. and S. A. Harris (1973) A cost–benefit evaluation of alternative control policies for foot and mouth disease in Great Britain. *J. Agric. Econ.* 24:573–600.

Pridgen, S. G. (1980) *Farmer Selection of Cotton Insect Control Services: Spray Groups and Scouts*. PhD Dissertation Dept of Economics and Business, NC State University, Raleigh, North Carolina (unpublished).

Regev, U., A. P. Gutierrez, and G. Feder (1976) Pests as a common property resource: A case study of alfalfa weevil control. *Am. J. Agric. Econ.* 58(2):186–97.

Rodriquez, R. (1979) *Acreage Response to Government Pest Control Programs: The Case of Boll Weevil Eradication*, PhD Dissertation, Dept of Economics and Business, NC State University (unpublished).

Tullock, G. (1969) Social cost and government action. *Am. Econ. Rev. Proc.* 59(2):189–97.

27 An Economic Analysis of Man's Addiction to Pesticides

Uri Regev

27.1. INTRODUCTION

The history of the struggle between man and his many agricultural and other pests is longer than civilization itself. This struggle for life and for high and stable food supply has taken many forms. While in the past cultural techniques were predominant, since the early 1940s massive quantities of chemical pesticides have been the major means of controlling agricultural and other pests. At present more than 1500 varieties of pesticides are in common use in the world, and their volume has grown more than tenfold in the last 25 years (van den Bosch, 1978). However, the increasing spiral of both use and cost of pesticides has also been accompanied by increasing pest-related losses (Luck *et al.*, 1977). Furthermore, man is now faced with many negative effects of chemical pesticides most notable of which are: increasing pest resistance to pesticides; suppression of natural enemies and consequently pest and secondary pest resurgence; and various types of environmental pollution. Pesticides are used to prevent losses, yet we still incur them — this is obviously a contradiction. The accumulated evidence for increasing pesticide-caused damage clearly implies that man has become addicted to pesticide use and we find ourselves in a vicious circle of pests–pesticides–more pests–more pesticides, and so on.

The main thesis of this chapter is that economic forces coupled with the institutional framework of the free economy necessarily lead to pesticide addiction in pest management. This addiction is "pushed" and enhanced by a misguided selection of control means, fueled by the notion that if a little is good, more is better (overuse of chemical pesticides), and by our ignorance of when and how to apply them. This may or may not be surprising to many who do pest management research, but as will be shown, the amazing result is, that on all the above three decision aspects of pest control, the *individual* decision maker is acting in *his own self-interest*. This implies that no matter how well research is able to show that pesticide overuse increases costs and damage and leads to a "doomsday" in crop production and other pest control problems, the individual user—housewife, farmer, forester, or a user of antibiotics—will continue to overuse chemicals, and will do so in his "best interest" as he or she perceives it. In principle this is a special case of the general problem of the overexploitation of a common-property resource, that has so eloquently been called the "tragedy of the commons" (Hardin, 1968; Hardin and Baden, 1977). Here the common-property resource is the susceptibility of pests to control. For a recent treatment of the analogous situation in fish harvesting, see Clarke (1981).

The line of the argument is as follows: *A benchmark of optimal pest management is defined as a strategy with respect to the combination of controls used, their quantities and timing, which yields maximum present value of the net benefits, with due consideration to resistance, natural enemies, and environmental pollution*. It will be shown that a disregard of these factors leads to wrong decisions, the most notable being pesticide overuse, that are arrived at in the best, possibly short-sighted, interests of each decision maker. Then because of the problems of resistance, natural enemies, etc., this in turn leads to overuse of chemicals, and hence "addiction". This sequence is reinforced by the uncertainties facing the decision maker, who is presumably risk averse, and uses pesticides as a form of insurance against pest damage.

Section 27.2 unfolds the economic basics necessary for the analysis. Section 27.3 constitutes an economic analysis of four of the major problems in pest management: pest resistance to pesticides, prey–predator relationships, environmental pollution, and behavior towards risk. Finally, Section 27.4 suggests some possible methods of avoiding addiction to chemical pesticides.

27.2. ECONOMIC BASIS

The economic framework is defined in terms of crop production, where a decision maker is faced with the problem of the choice of an input vector $X = (X_1, X_2, \ldots, X_n)$, which maximizes a net benefit function $\pi = \pi(X)$. In terms of the farmer, his decisions in general involve the choice of crops, and the choice of inputs, including labor, irrigation, etc., together with specific pest control inputs. The interrelationships of cultural, biological, and chemical means of controlling pests are taken into consideration while he makes the choice on both the crops and inputs. To simplify the analysis we consider a problem of only one crop, though some remarks on the choice of crops and its contribution to compounding the addiction problem will be given later.

In the following, $Y = Y(X)$ is a single crop production function; $X = (X_1, X_2, \ldots, X_n)$ are the production inputs, including pest control actions; $C(X)$ is a cost function; and $\pi(X) = P_y Y(X) - C(X)$ is the net benefit (profit) function. If the profit function is differentiable, then the necessary conditions for its maximization are:

$$P_y(\partial Y/\partial X_j) = \partial C/\partial X_j \qquad j = 1,2,\ldots,n. \tag{27.1}$$

These conditions imply that the choice for each quantity of each of the inputs is based on the following marginal rule: Increase X_j so long as its next unit costs less than the benefits resulting from such an addition. (It should be noted that this necessary condition leads to a local extreme point, but if the profit function is concave it also guarantees a global maximum.)

This rule determines for the user the optimal input mix, namely, the quantities of each of the inputs in X that will yield maximum profits. A general result from the maximization procedure is well known in economics and given here without proof.

Proposition: An increase in a given input related cost implies reduction of its proportion in the inputs mix. This straightforward and sensible proposition is the basis for comparing the "optimal" benchmark strategy with the behavior of the individual user who seeks profit or utility maximization. It should be noted that the model definition of the production function is very general, but it could, in principle, incorporate the dynamic properties of the problem, given that the input vector is interpreted as including each input with time index repeatedly over as many (though finite) time periods as required. The dynamic dimension of the problem is not well presented in the model, however, and an optimal control formulation is a better analytical tool for that purpose. Considered in a discrete-time framework the problem is:

$$\max_{\{X_t\}} \sum_{t=1}^{\infty} \pi(X_t)/(1+r)^t, \tag{27.2}$$

where $X_t = (X_{1t}, X_{2t}, \ldots, X_{nt})$, is as defined above but related to time t, and r is a fixed interest rate used to discount future benefits and costs. Posed in this way the necessary conditions for maximization expressed by eqn. (27.1) remain intact. Thus, apparently, short- and long-term considerations make no difference in the decision strategy; the difference lies in the fact that model (27.2) is incomplete since it does not incorporate the dynamic behavior of the system. This should be done via difference or differential (for continuous-time models) equations describing the system dynamics. These equations serve as constraints on the objective (profit) function and they alter the conditions for maximum as well as the decisions. The general problem is thus:

$$\begin{aligned} &\max_{\{X_t\}} \sum_{t=1}^{\infty} \left(\frac{1}{1+r}\right)^t [P_Y Y(X_t, Z_t) - C(X_t)] \\ &\text{s.t. } Z_{t+1} = g(X_t, Z_t, t) \end{aligned} \tag{27.3}$$

where $Z_t = (Z_{1t}, Z_{2t},...,Z_{mt})$ *is a state vector*, incorporating the important components of the system, and $g(X_t,Z_t,t)$ is a vector function, each element of which describes the time change of a single component of Z_t. Examples for components of the state vector could be tree size, pest and predator populations, pesticide resistance, residues of pesticides and so forth. Details of the solution are beyond the scope of this chapter (see Burt and Cummings, 1970; Regev *et al.*, 1976), but the main result needed for our discussion is that necessary conditions for a solution of eqn. (27.3) included:

$$P_Y \frac{\partial Y}{\partial X_{jt}} = \frac{\partial C}{\partial X_{jt}} + \sum_{i=1}^{m} \lambda_{it+1} \frac{\partial Z_{i,t+1}}{\partial X_{jt}}, \tag{27.4}$$

where λ_{it} are Lagrangean multipliers, sometimes called shadow prices, which serve to evaluate the contribution (or loss) to the objective (profits) due to a marginal change in the ith component of the state vector at time t. Comparing eqns. (27.4) and (27.1) the difference is an additional term that is added to the cost. This final term in eqn. (27.4) is known as the *user cost*, and is interpreted as an addition to the (marginal) costs due to long-term considerations. Recalling the proposition at the beginning of this section it is concluded that for every input for which user costs are positive, long-term considerations tend to reduce its usage as compared with short-term actions.

With these analytical tools we now turn to analyze the effects of several pest-related problems on the behavior and strategies in pest management. The problems of resistance, biological interactions, and environmental pollution are analyzed separately for expository reasons, but some remarks regarding their accumulative effects are reserved for further discussion.

27.3. ANALYSIS OF THE ADDICTION PROBLEMS

27.3.1. Pest Resistance to Pesticides

Decreasing efficiency of chemical pesticides following their continuous use has been long recognized as a result of pests developing resistance to pesticides. There is much evidence for increasing pesticide resistance, but since resistance develops over a relatively long period of time and on a regional scale, the exact relations are not clearly understood. However, it is sufficient for our qualitative analysis to adopt the tenable assumption that resistance to a given chemical is an increasing function of the use of the chemical. The measure of resistance poses another problem since resistance mechanisms may differ according to the chemicals used. One way to estimate it empirically would be by an efficiency (or disefficiency) parameter of the dosage response function (e.g., Hueth and Regev, 1974). Another way is to measure it by the frequency of the resistance genes in the pest population (Gutierrez *et al.*, 1979; Regev *et al.*, 1981). This formulation works well for simple Mendelian inheritance, and has the advantage that the time motion equation could be simplified to the Hardy–Weinberg law. In any case the production function could be written as $Y(X,R)$, where R is the resistance "stock", and

$$R_{t+1} = g(X_t,R_t); \qquad \partial R_{t+1}/\partial X_t \geqslant 0, \qquad \partial R_{t+1}/\partial R_t \geqslant 0 \tag{27.5}$$

is the motion equation of the resistance level.

Simplifying the problem (27.3) for expository ease, it is assumed for the moment that resistance is the only state variable to be considered, so that profit maximizers should (would they?) have the following problem:

$$\max \sum_{t=1}^{\infty} \left(\frac{1}{1+r}\right)^t [P_Y Y(X_t,R_t) - C(X_t)] \tag{27.6}$$

$$\text{s.t. } R_{t+1} = g(X_t,R_t), \qquad X_t \geqslant 0,$$

which results in the following *decision rule*: for $X_{jt} > 0$, decrease the quantity of the input X_{jt} if:

$$P_Y(\partial Y/\partial X_{jt}) < (\partial C/\partial X_{jt}) + \lambda_{t+1}(\partial R_{t+1}/\partial X_{jt}), \tag{27.7}$$

and increase X_{jt} if the inequality in eqn. (27.7) is reversed. The second right-hand term in eqn. (27.7)—the user costs—is interpreted as the losses (cost) due to future decreased efficiency of the control input X_j resulting from increase in resistance caused by one unit of X_j applied at time t. Further insight into this term is gained through the following equation, obtained by differentiating the Lagrangean of eqn. (27.6) and using the transversality condition

$$\lim \lambda_s/(1+r)^s = 0:$$

$$\lambda_t = -\sum_{s=t}^{\infty} \left(\frac{1}{1+r}\right)^s P_Y \frac{\partial Y(X_s,R_s)}{\partial R_t} \geqslant 0, \tag{27.8}$$

i.e., λ_t is the present value of all future (negative) changes in benefits resulting from a marginal increase in resistance at time t. Since the term $\partial R_{t+1}/\partial R_t$ in eqn. (27.7) could be either positive (when X_j is a chemical pesticide) or zero (if X_j is any other input), the second right-hand term in eqn. (27.7) is positive or zero. Comparison of eqn. (27.7) with eqn. (27.1) yields the following result: if resistance effects are taken into account, this leads to an increase in the marginal costs of inputs that cause resistance. In view of the *proposition,* resistance-raising inputs should be reduced by profit-maximizing decision makers when knowledge of resistance development reaches them (see Gutierrez *et al.*, 1979). Although this result seems logical and sound, the end result in many pest control problems is that eqn. (27.1) rather than eqn. (27.7) is the appropriate rule for individual decision makers, resulting in increased pesticide use.

In general, individual users are not accountable for resistance development in their domain. Resistance is an example for a whole class of problems known in the literature as *common property resources*. The mobility of most pests carries them outside the field of the user, and this implies that most pests have no tenure to a single user. In other words, the pests that appear in a field at the beginning of the season are not (except by rare chance or special circumstance of pest biology) those that left the field last season. Thus, the resistance level in each field is not a result of the individual user's application of pesticides, but a result of the cumulative actions taken by all users in the relevant area. When the individual user is relatively small, he would not (and should not) relate the resistance level to his actions of last season, and therefore would not consider resistance effects in his pest control strategy. Thus, while a central government should adopt a decision strategy following eqn. (27.7), which implies reduced resistance-causing inputs, an individual user who seeks to maximize his (even long-term) profits would disregard these considerations and follow his old rules of (27.1) implying: (i) choose the most deadly pesticide, and (ii) apply large quantities so as to maximize the instant profits disregarding user costs. An individual grower's long-term profits are not affected by *his* current decisions, and he therefore disregards these effects.

The main conclusion is that in a free economy, each individual user of chemical pesticides has no economic motive to consider their long-term effects on resistance. This outcome when compared to the long-term optimal strategy results in the use of excessive amounts of wrongly chosen pesticides. However, since the individual user cannot benefit in the future from a change in his strategy today, it is in his best interests to ignore the possibility of change. This is only one reason, even though it may be crucial, for the overuse of pesticides, resulting in addiction. The solutions necessary to solve this problem are beyond the scope of this paper, but clearly, a collective action by all the users in the relevant region is called for, either by enforcement of standards or through a penalty system.

27.3.2. Environmental Pollution

Environmental pollution by pesticides takes a wide range of forms, such as health damage to workers and neighbors, damage to wildlife, residues of chemicals passed to man and animals through the food chain, and many more. From an economic point of view, the common denominator to all forms of environmental pollution is that the user of pesticides is not the only one who carries the burden of the damage, and in most cases he suffers from it the least. The pollution damages are inflicted by him on others and he is not forced to account for these losses. Hence pesticide applications can be categorized as having *external effects*. A well known result in economic theory is that external effects, or externalities, imply inefficiencies and misallocation of resources.

Furthermore, when the externalities are causing damage, the misallocation would in general imply overuse of those resources that cause the highest external damage.

My argument in this case runs along the lines following eqn. (27.7), namely, that *the marginal costs of environmental damage are not considered by pesticide users; hence they chose inappropriate controls and overuse them, and in this respect the direction of misallocation is similar to that caused by the response of growers to the development of resistance*. However, whereas in resistance the user himself pays the costs in the long-term (when the efficiency of the pesticide decreases), here the society pays the price in terms of lower environmental quality. In contrast to the former case, group action by growers is not enough here because their actions and interests still conflict with the rest of society. What is needed is some form of government regulation or other form of intervention such as penalties or subsidies. However, without being forced, the free and competitive pesticide users tend to overuse pesticides as compared with optimal strategies. As we move to the next problem of biological interactions the story becomes more complicated, but the above results are strengthened.

27.3.3. Biological Interactions: Prey–Predator Relationships

In a natural, untapped ecosystem, each biological species has, in addition to food limitations, natural enemies (parasites, predators, etc.) that may prevent its population exploding. In trying to enhance and stabilize health and food sources, in the past 30–40 years man has adopted pest control measures that in many instances do not distinguish between beneficial organisms and those that cause damage. Biocides such as DDT, other organochlorines, and the organophosphates have deadly effects not only on pests but also on their natural enemies.

The concept of a locally stable, natural ecosystem equilibrium of one prey and one predator is a simplification, but is assumed here to sharpen the focus of the discussion. The idea of a stable equilibrium does not imply that in reality, the two populations are at any time at that state, but simply indicate the direction (i.e., point) toward which they move following perturbations of a stochastic or other nature. The simplest model we can devise containing the basic assumptions for the time motion of the prey–predator relationships is:

$$\frac{1}{N(t)}\frac{\mathrm{d}N(t)}{\mathrm{d}t} = g[N(t), P(t); X(t)] \tag{27.9}$$

$$\frac{1}{P(t)}\frac{\mathrm{d}P(t)}{\mathrm{d}t} = f[N(t), P(t), X(t)] \tag{27.10}$$

where $N(t)$ is the pest (prey) population at time t; $P(t)$ is the predator population at time t; and $X(t)$ is the rate of pesticide application. It is further assumed that in

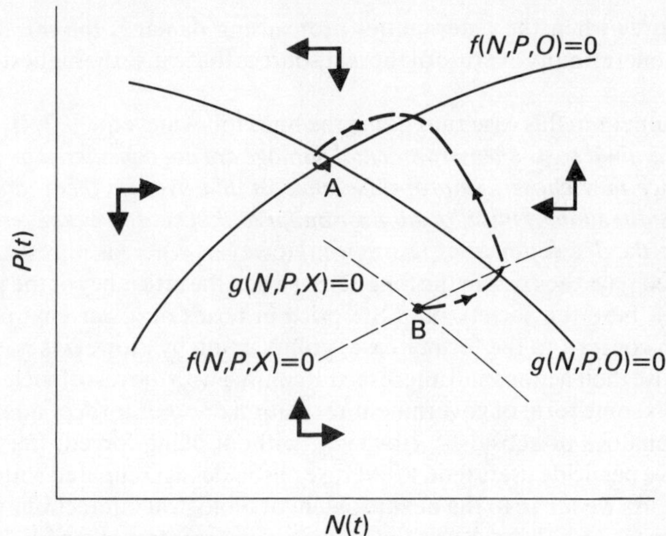

FIGURE 27.1 Model of predator–prey interactions based on eqns. (27.9) and (27.10), showing the natural steady state A, and a new steady state B, following a fixed rate of pesticide application.

the relevant region all partial derivatives are nonpositive, except for $\partial f/\partial N \geqslant 0$, and that conditions for a stable, steady-state equilibrium are satisfied (for details, see Feder and Regev, 1975).

In this study, it is of interest to examine the effect of a fixed pesticide application rate on the equilibrium point. To see this, we compare the undisturbed natural equilibrium ($X = 0$) with another point for which $X > 0$ (Figure 27.1). Starting from the natural steady state A, if we employ some fixed rate of pesticides for sufficiently long period we reach another steady state B. Under the model assumptions the predator steady-state population is necessarily lower the higher the rate of pesticides applied. However, this may be reversed for the pest population. It turns out that the new steady-state pest population is higher the higher is X, if an *impairment index* I is greater than unity, where this index is defined as (Feder and Regev, 1975):

$$I = \frac{f_X g_P}{g_X f_P} \substack{< \\ >} 1 \rightarrow \begin{cases} N(\mathrm{B})\ N(\mathrm{A}) \\ N(\mathrm{B})\ N(\mathrm{A}) \end{cases} \qquad (27.11)$$

where the subscripts denote partial derivatives. To see the derivation of eqn. (27.11), take the total differentials of eqns. (27.8) and (27.10) and equate them to 0. This results in

$$\begin{pmatrix} \mathrm{d}N/\mathrm{d}X \\ \mathrm{d}P/\mathrm{d}X \end{pmatrix} = \frac{1}{g_N f_P - f_N g_P} \begin{pmatrix} f_X g_P - g_X f_P \\ g_X f_N - f_X g_N \end{pmatrix} . \qquad (27.12)$$

Note that the denominator must be positive to satisfy local stability conditions.

The index I is referred to as an impairment index, since it measures the extent to which the predator, as a natural pest regulator, is impaired by the application of pesticides. This can be seen as follows: f_X/g_X is the ratio of pesticide effectiveness on predators versus pests, where a higher ratio implies a greater pesticide-induced reduction of predators compared to pests. The term g_P could be interpreted as the efficiency of the predator in suppressing the pest, and f_P measures the recovery capacity of predator from pesticide applications. The higher f_P is, the lower the damage caused by the predator's suppression by the control measures. Thus if $I > 1$, the steady-state pest population is increased rather than reduced by pesticide application. This possibility, of increased pest population following pesticide use, is borne out by many studies (see Ehler *et al.*, 1974; Gutierrez *et al.*, 1975, and Chapter 5).

Although the model is very simple, it captures three essential aspects of the pest management problems: (1) The model describes the specific way in which wide range chemical pesticides (biocides) can cause essential increases in pest populations. (2) Equilibrium points such as A and B in Figure 27.1 are long-term steady-state points and the immediate effect of pesticide application is obviously to cause pest reduction. Thus, as in the case of resistance and pollution, the long-term effects are likely to be beyond the individual user's control, and the same common-property problem is faced here, which implies an overuse of pesticides. (3) If, however, after a steady-state point B is reached, pesticide applications are abruptly stopped, pest population will first increase as a result of the low predator population, and only later will it decrease to reach point A. This path is shown by the heavy broken line in Figure 27.1. Consequently, a government policy of banning the pesticide under consideration, could possibly cause heavy losses at the beginning and cannot be recommended, as might be concluded from this model.

Pesticide user's disregard of natural enemies, may possibly be also due to lack of information but whatever the reason, long-term considerations are similar in principle to the arguments for resistance, and lead in the same direction of addiction because the ecosystem is a common-property resource. It should be noted that in the absence of predators, the common-property argument leads to an opposite conclusion, where the pesticide user could adopt a policy of a "free rider" and underuse pesticides (see Regev *et al.*, 1976). The disastrous results of such strategies, which ignore natural enemies, are the well known phenomena of pest and secondary pest resurgence. Ignorance, disbelief, and uncertainty may be additional reasons for lack of concern for the ecosystem by pesticide users, and this is discussed next.

27.3.4. Risk and Uncertainty

Uncertainty affects many facets of pest management decision making: the size of pest population, its time motion, immigration factors, resistance development, and the efficiency of pesticides and the damages or losses caused by pests. Uncertainty involves both subjective elements (lack of information due to inability to measure various key parameters), and objective factors (such as climatic conditions). While the first part could be reduced in principle by research and information services, the second is inherent in any natural ecosystem.

The reaction of farmers to the sight of a few bugs (some may not be pests) in his field is to immediately apply pesticides. Often he views this action as a kind of insurance paid to avoid total destruction of his crop, which he believes could occur if he did not spray. Even if this possibility is, objectively, rare or even negligible, it is irrelevant since he assigns this possibility a subjectively high value, and acts accordingly. Psychological factors such as panic may further explain his behavior.

This situation can be analyzed within the general framework of decisions under uncertainty and attitudes towards risk, where the individual is assumed to maximize expected utility (or possibly other criteria). A person who is risk-averse will assign greater importance to a marginal increase in income when it is low, than when it is high, which can be expressed by a concave utility function. The decision problem is illustrated in Figure 27.2, where the problem is simplified to assume only two possibilities for income—low (B_0) and high (B_1). Suppose further that the decision maker assigns a (subjective) probability P to the event of low income. With the possibility of insuring

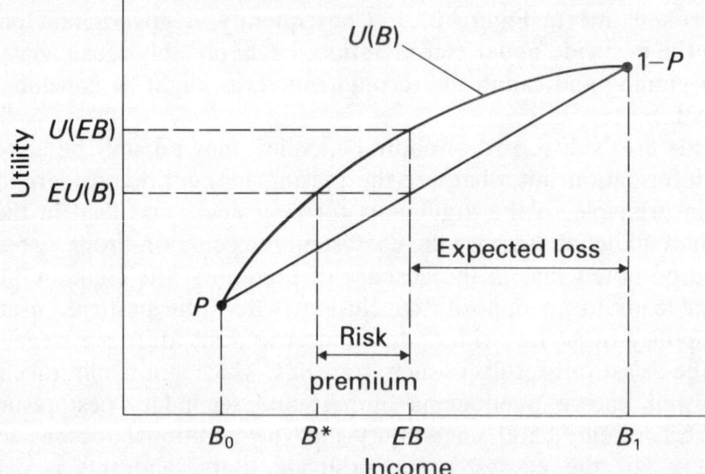

FIGURE 27.2 Risk-averse utility function, assuming low (B_0) and high (B_1) possibilities for income with probabilities P and $1 - P$, respectively.

against low income, the question is how much he is willing to pay in order to avoid this uncertainty.

From Figure 27.2 it is seen that a concave utility function implies $U(EB) > EU(B)$, namely, that a risk-averse person will be indifferent between a certain income B^* and a stochastic income with expected income EB, since the certain income B^* ($B^* < EB$) gives him the same utility as $EU(B)$. Thus any certain income that is higher than B^* will leave him better off than the stochastic income. The difference between EB and B^*, called the *risk premium*, is the maximum (insurance) cost he is willing to pay in order to ascertain his income at level EB. Before this tool is applied to our problem it should be noted that: (1) the higher the risk averseness, the higher is the risk premium; (2) the higher is P—the disaster probability—the lower are EB and B^* and (3) the difference $B_1 - EB$ is the expected loss due to the undesirable conditions that yield B_0.

Certain pesticide applications could be regarded as insurance spray when given early in the season either "just to make sure", or at the sight of the first insects. Two factors affect the magnitude of this action: (*a*) the degree of risk averseness of the decision maker; and (*b*) the subjective probability (chance) he assigns to the "disaster possibility". While the first is a personality matter, the probability of disaster is affected by knowledge and information that decrease the errors made in insurance applications. Pest and secondary pest resurgence often imply that applications early in the season will be followed by continual applications throughout the season. *All in all, uncertainty tends to lower the economic thresholds for pesticide applications and consequently increases their usage (i.e., keeps growers addicted to excessive pesticide usage).*

27.4. SUMMARY AND CONCLUSIONS

Pest management involves decisions with many effects beyond the domain of the decision maker. The externality effects that are inherent in the problem lead to a disparity between the optimal strategy and that employed by users of pesticides. While the individual user considers only the direct monetary costs of pesticides, the "optimal strategist" adds also the costs of future damages from resistance and suppression of natural enemies, as well as costs of environmental pollution. This difference in costs leads to overuse of chemical pesticides in most cases and such overuse is also enhanced by the user's uncertainty and risk averseness. Mainly resistance, but also the dwindling of beneficial populations, lead to diminishing effectiveness of chemical pesticides, which in turn further increases their usage. This vicious circle is certain to lead to addiction and ecological catastrophe unless ...

Solutions to the problem should be the subject of another paper, but a few remarks cannot be avoided here. First, research has a most important role in understanding the ecosystem and learning to cope with the problems of human intervention in the natural system. Dissemination of knowledge about beneficial species, and the mechanism of resistance development could itself alter the behavior of pesticide user. In addition it would affect the user in reducing excessive "insurance" pesticide applications. Second, development of effective biological alternatives such as enhanced use of biological controls and plant breeding are promising ways of fighting addiction to pesticides. These methods, coupled with sound pest management advice, could help to avoid most of the harmful effects mentioned above, but the methods are difficult to implement. The integrated pest management road is not easy: many obstacles lie ahead, and much research is needed to make it a practical and efficient control, but we have few alternatives. Third, as shown above, even with all the knowledge that could be possibly gathered, farmers, gardeners, housewives, and other users will continue to overuse pesticides because there exists a basic gap between their viewpoint and that of society. This difference implies that users of pesticides see the cost of chemicals as much lower than it should be in terms of long-term damage and pollution. What is called for in such a case is government intervention. This could be done in one of two basic ways: (i) taxes on or subsidies to pesticide users to inhibit its use; (ii) regulations and standards. But government intervention has its own disadvantages and costs, such as enforcement and bureaucratic costs, the adverse influences of political pressures, and the fact that bureaucrats seldom wish to expose their flanks. Finally, the role of the pesticide industry has not been considered. The producers of chemical pesticides definitely have a vested interest, and their pesticide salesmen are known to play an important role in the decision-making process of pesticide usage. The research of some companies is more and more oriented to finding selective pesticides, thus hoping to avoid hindrance of beneficial species. But we might ask, would they also advertise in a way similar to electric companies encouraging their customers to save energy? This pinpoints a basic conflict of interests.

A basic economic paradigm of *laissez-faire* states, that the profit motive, if not inhibited by government or other institutional impediments, is the driving force to efficiency in economic activities. But it has been made clear here that this paradigm does not apply to pest management, which is riddled with numerous external effects. Since pesticide users and the agrochemical industry can not regulate themselves in their own interests for the interests of the society, government, or some other form of centralized regulatory mechanisms are called for in order to block the road leading to total addiction.

ACKNOWLEDGMENT

I would like to acknowledge the help, encouragement, and unlimited time for discussions given by A. P. Gutierrez who is also largely responsible for my addiction to these problems.

REFERENCES

van den Bosch, R. (1978) *The Pesticide Conspiracy* (New York: Doubleday).

Burt, O. R. and R. G. Cummings (1970) Production and investment in natural resource industries. *Am. Econ. Rev.* 60:576–90.

Clarke, C. W. (1981) Bioeconomics, in R. M. May (ed) *Theoretical Ecology: Principles and Applications* 2nd edn (Oxford: Blackwell) pp387–418.

Ehler, L. E., K. G. Eveleens, and R. van den Bosch (1974) An evaluation of some natural enemies of cabbage looper in cotton in California. *Environ. Entomol.* 2:1009–15.

Feder, G. and U. Regev (1975) Biological interactions and environmental effects in the economics of pest control. *J. Environ. Econ. Management* 2:75–91.

Gutierrez, A. P., L. A. Falcon, W. Loew, P. A. Leipzig, and R. van den Bosch (1975) An analysis of cotton production in California: a model for Acala cotton and the effects of defoliators on its yields. *Environ. Entomol.* 4(1):125–36.

Gutierrez, A. P., Regev, U., and H. Shalit (1979) An economic optimization model of pesticide resistance: alfalfa and egyptian alfalfa weevil—an example. *Environ. Entomol.* 8:101–7.

Hardin, G. (1968) The tragedy of the commons. *Science* 162:1243–8.

Hardin, G. and J. Baden (1977) *Managing the Commons* (San Francisco: Freeman).

Hueth, D. and U. Regev (1974) Optimal agricultural pest management with increasing pest resistance. *Am. J. Agric. Econ.* 56:543–51.

Luck, R. F., R. van den Bosch, and R. Garcia (1977) Chemical insect control: A troubled pest management strategy. *Bioscience* 27:606–11.

Regev, J., Gutierrez, A. P., and G. Feder (1976) Pests as a common property resource: A case study of alfalfa weevil control. *Am. J. Agric. Econ.* 58:186–7.

Regev, U., Shalit, H., and A. P. Gutierrez (1983) On the optimal allocation of pesticides: The case of pesticide resistance. *J. Environ. Econ. Management* 10:86–100.

28 The Mathematical Evaluation of Options for Managing Pesticide Resistance

Hugh N. Comins

28.1. INTRODUCTION

Historically, pesticide resistance has been an annoying but not particularly expensive trait of certain pest species. This was due to the great variety of pesticides, available at modest cost, which allowed the easy replacement of a chemical to which resistance had developed. Although most pests can still be controlled, we now have the situation where the species that have developed resistance fastest are outrunning the supply of new pesticides (some examples are houseflies, tetranychid mites, malarial mosquitoes (Sawicki, 1979), and the Australian cattle tick, *Boophilus microplus* (Stone, 1972)). If the pests continue to gain ground in this "technological race" we may expect serious effects on world food supplies and on the control of disease, for example, if we were no longer able to control an insect pest of a major cereal crop (Conway, 1982).

The central problem of resistance management is thus to try to eliminate pesticide resistance or, failing that, to extend the life of existing pesticides in the hope that new chemicals can be introduced before all the current ones become unusable. Completely avoiding resistance, however, is difficult since the phenomenon is a special case of evolution, akin to natural selection. An evolutionary adaptation requires only three conditions: that there be a genetic variation in the fitness of different members of the population; that the species does not become extinct; and (in a sense which depends on the mating structure of the population) that the fitness of the adapted type is higher on average than that of the original population.

The first condition in general requires the organism to have some preadaptation; for example, in the evolution of birds, the development of feathers from existing scales. Many arthropod pests have also clearly demonstrated that they have several adaptive methods for surviving in the presence of very high pesticide levels, possibly they are preadapted by their continued exposure to host defensive substances (Anon, 1975). Although it may happen that future advances in biochemistry will provide an unassailable metabolic poison, it must be assumed for the present that many species will harbor genotypes capable of surviving even the largest pesticide dosage that can reliably be delivered without causing environmental damage.

The condition that the resistant type must have a greater average fitness is the basis for some suggestions for avoiding resistance. Possible methods are to ensure the existence of a large reservoir of untreated pests (assuming the resistance gene is deleterious in the absence of pesticide, Taylor and Georghiou, 1979), or to introduce specially nurtured susceptible pests (Curtis *et al.*, 1978). Unfortunately the complete avoidance of resistance using these techniques is usually incompatible with adequate control. The latter method also may be expensive and technically difficult (note that the quoted proposal of Curtis *et al.* involves the release of male mosquitoes, which are not pests *per se*).

Thus, failing complete eradication of the pest, it is almost certain that continued exposure to a pesticide will eventually produce resistance, although a number of options are available for *delaying* pesticide resistance while retaining adequate control. Adopting some of these methods would improve our chances of remaining ahead of pest species in the deployment of effective control agents.

Since field trials of resistance development present insuperable difficulties, the logical method for evaluating control options is to construct mathematical models of the selection process using known genetic principles (Comins, 1977a,b, 1979a,b, 1980; Conway and Comins, 1979; Georghiou and Taylor, 1977a,b; Hueth and Regev, 1974; Taylor and Georghiou, 1979). Such models face the difficulty that it is not known initially what resistance genes are present in a population, nor their frequencies or dominance characteristics. However, this situation can be alleviated to some extent by adopting the "heterozygote selection approximation" (Comins, 1977b); this is discussed in Section 28.2. Even so, resistance models can only provide comparisons of different control options, and are intrinsically incapable of predicting absolute time scales for the development of resistance.

Later sections discuss models of the pesticide strategies that seem to have the greatest potential for delaying resistance. Detailed economic considerations are beyond the scope of this chapter and the strategic aim may be regarded as delaying resistance as long as possible, consistent with reducing the pest population to a desired level.

28.2. HETEROZYGOTE SELECTION

In order to introduce the heterozygote selection approximation we note that the selection of a resistance gene can be divided into four phases (Figure 28.1).In

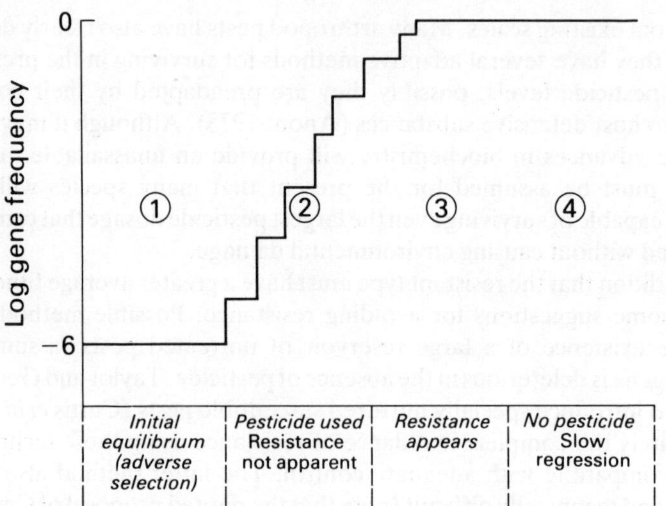

FIGURE 28.1 Phases in the development of pesticide resistance.

the first phase, when the pesticide has not yet been applied, the gene is at very low frequency and is maintained in the population by a balance between mutation and adverse selection. In the second phase, once pesticide selection has begun, the resistance gene spreads rapidly through the population, although it is still too rare to have any noticeable effect on control efficiency. Eventually, however, the crisis occurs: control measures fail, resistance is detected and, after perhaps a period of increasing dosages, the pesticide is abandoned (the third phase). The fourth and final phase is the regression of resistance once the use of the pesticide is discontinued. This seems to occur very slowly, due to the selection of modifying genes that improve the fitness of the resistance type once it becomes connon. Regression has never been sufficient for an old pesticide to be "recycled" for any useful period.

The period of interest for resistance models is largely coincident with the second phase of selection, since the third phase is typically of very short duration and corrective or delaying action is limited. Thus it can be assumed for modeling purposes that resistance genes are only present at undetectably low frequencies. This implies that resistant homozygotes (individuals having two resistance-conferring alleles) are very rare compared with heterozygotes, and consequently almost all the selection of the resistance gene occurs by the preferential survival of heterozygotes, even though homozygotes may be much less susceptible to pesticide. Figure 28.2 illustrates the generality of this result, showing a case in which the resistance gene is almost recessive (relative survival in each generation of RR : RS : SS = 1 : 0.1 : 0.05). If the gene were present only in heterozygotes the gene frequency would be increased by a factor of 2 in each generation. As can be seen from Figure 28.2 the actual

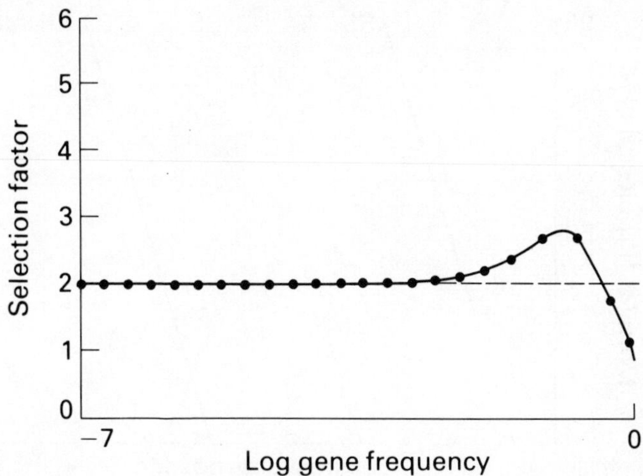

FIGURE 28.2 Ratio of gene frequencies in successive generations (selection factor) versus initial frequency, for a discrete generation pest when the ratios of survivals in each generation are 1.0 : 0.1 : 0.05 for RR : RS : SS. The dots show successive gene frequencies for a simulation with initial frequency 10^{-7}.

factor differs very little from this value, and is never remotely near 20, which is the ratio of RR to SS survival.

It is thus possible, to a good approximation, to consider the pest population as being made up of a large population of susceptible individuals, with a small resistant (RS) subpopulation having a somewhat slower rate of population growth, and the same rates of dispersal and survival (except for pesticide survival). Since unions of a susceptible and a resistant parent produce the same proportion of resistant offspring (half of parents RS gives half of offspring RS), the interaction of the populations at mating can generally be ignored, although the result is more complicated if pesticide exposure depends on sex (in this approximation double RS matings are ignored).

We can thus imagine that in the second phase of resistance the pest population harbors a number of noninteracting subpopulations of heterozygotes for various different resistance genes. These populations vary in initial size, but will all start to increase relative to the susceptible population when pesticide is applied. The rate of increase is given in each case by the ratio of heterozygote to susceptible survival, which depends on the distribution of pesticide dosages and on the dosage–mortality relationship (Figure 28.3). Eventually one of the genes will become common enough to produce a noticeable loss of control; shortly afterward the other resistance genes become superfluous and stop increasing in frequency even if the pesticide continues to be used (Comins, 1980: Figure 6). In designing resistance management strategies it is important to remember that the resistance gene that finally becomes apparent is only the winner in a race between a number of

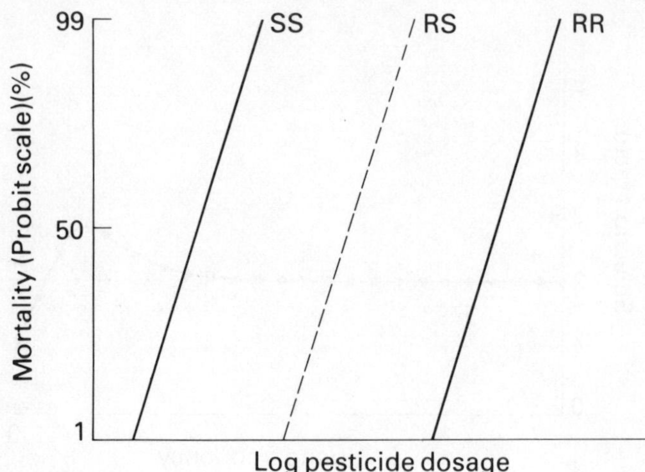

FIGURE 28.3 Typical dosage–mortality relationship for a laboratory experiment with homozygous resistant (RR), heterozygous resistant (RS) and susceptible (SS) strains of an insect pest.

alternatives. Thus strategies aimed at a subclass of resistant genes will only delay resistance if the alternative genes are initially relatively rare, or confer smaller resistance factors or greater reproductive handicaps.

An additional consequence of low resistance gene frequencies is that resistance mechanisms that absolutely require two R-alleles (i.e., recessive), or polyfactorial combinations of genes, are realized in an exceedingly small proportion of animals. Thus unless pesticide kill rates are extraordinarily high, selection for these genotypes is slow. Given alternative resistance genes that are effective as heterozygotes, it is most unlikely that the eventual apparent resistance gene would be of this type. This conclusion is in accord with the fact that most observed resistance mechanisms are monofactorial and involve partially dominant genes (Brown, 1971).

28.3. RESISTANCE MANAGEMENT

28.3.1. Manipulation of Population Dynamics

Consider a large, freely interbreeding pest population in which each individual inflicts equal damage on a crop. In this case the heterozygote selection approximation says that the rate of increase in the frequency of a resistance gene is equal to the ratio of the rate of growth of the small resistant heterozygote subpopulation to the rate of growth of the susceptible population. We know, however, that in the long term the growth of the susceptible population must average out to unity (assuming the pest is not eradicated). There are three main density-dependent mechanisms that can act to maintain this constancy: intraspecific competition, predation (including parasitism), and pesticide mortality.

If we presume that there is an intrinsic rate of increase (depending on environmental conditions) that would be exhibited by a small pest population in the absence of predators or pesticide, then the pesticide mortality, plus the reproductive inhibition and mortalities of the density-dependent mechanisms, must be just sufficient to cancel this increase. Thus the necessary rate of pesticide mortality and consequently the rate of resistance selection are determined by the extent to which intraspecific competition and predation fail to keep the susceptible population constant.

As a simple example we consider the case of a continuously breeding pest population with no predators and with intraspecific competition of the "scramble" form (maintaining an exact upper limit on the population, but having no effect on survival if the population is below the limit). Figure 28.4 illustrates the changes in the susceptible and resistant populations following a nonresidual pesticide application. The logarithmic increase in resistance gene frequency is given by $A - B$. This value can be interpreted in two ways: as the negative logarithm of the ratio of heterozygote to susceptible survival of the pesticide treatment, or as the reduction in intraspecific competition mortality multiplied by the length of time for which it is reduced.

The latter interpretation becomes useful when we consider control strategies aimed at keeping the pest population below the crop injury threshold for some period of time. In the simple model of Figure 28.4, it is irrelevant how this is done because intraspecific competition is always decreased by the same amount whenever the population is below its maximum, but in general

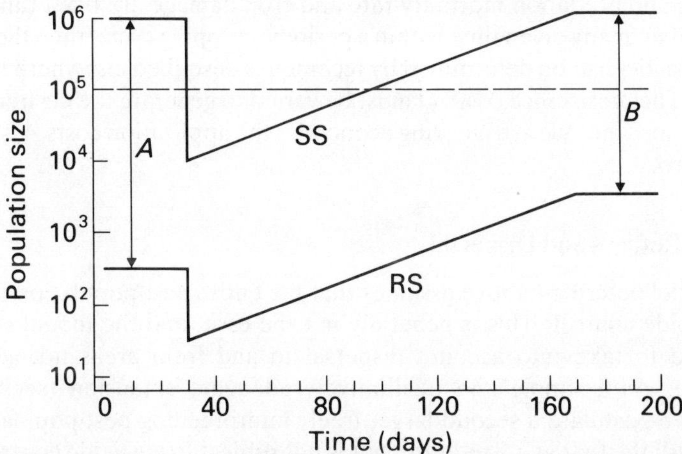

FIGURE 28.4 Growth of susceptible population (SS) and resistant (RS) subpopulation following a nonresidual pesticide application. The broken line is the crop injury threshold. For clarity the constant back-selection effect of resistance gene reproductive disadvantage is not shown. This can always be scaled out when making relative comparisons of single pesticide strategies acting on a single resistance gene.

intraspecific competition may be partially effective for populations below the crop injury threshold. In this case we would be selecting resistance needlessly by maintaining the population at any lower level. Such considerations apply even more to predation, since the saturation points of a predator's functional and numeric responses will often be very much below the equilibrium level produced by intraspecific competition (Southwood, 1975). If these effects mean that more pesticide is required to achieve a lower pest population, there would be an additional requirement due to direct pesticide mortality in arthropod predators, since their destruction would reduce the pest's predation mortality rate.

In summary, we observe that an estimate of resistance selection rate can be obtained from the extent to which the pest population is not maintained constant by processes other than pesticide mortality (in particular, resistance selection tends to be fast in species with a high intrinsic growth rate i.e., r*-pests; see Chapter 2). It is possible to adversely affect these processes without a corresponding gain in yield if the pest population is unnecessarily reduced.*

Another way in which a greater than necessary pesticide mortality can be wasteful is illustrated in Figure 28.5a, where we assume that the crop is seasonal but the pest is still able to multiply in its absence. Note that there is a considerable period following harvesting in which the pest population continues to grow. A much less severe pesticide treatment (Figure 28.5b) would give adequate control with less selection of resistance. If it were practicable it might even be worthwhile introducing susceptible insects after the cropping season (Figure 28.5c).

If the pest population at any time can be adequately described by a single number, and predation mortality rate and crop damage are fixed functions of this number at any given time within a periodic cropping cycle, then the optimal control tactics can be determined by techniques described elsewhere (Comins, 1979b). The "resistance cost" Q must be varied to generate the desired level of control, and since we are ignoring economics the application costs A_i should be set to zero.

28.3.2. Refuges and Dispersal

The model described above assumes that the entire pest population is subject to pesticide control. This is generally not the case, and the model should be extended to take into account dispersal to and from areas not subject to pesticide. As a simple and qualitatively adequate extension (see Comins, 1977a) we postulate a second large, freely interbreeding pest population that mixes with the first at a fixed rate, but is not subject to pesticide control.

The effect of such mixing is always to reduce resistance selection. The application of pesticide in the treated area causes a "vacuum" in the susceptible population, which results in a net inflow of susceptible individuals. On the

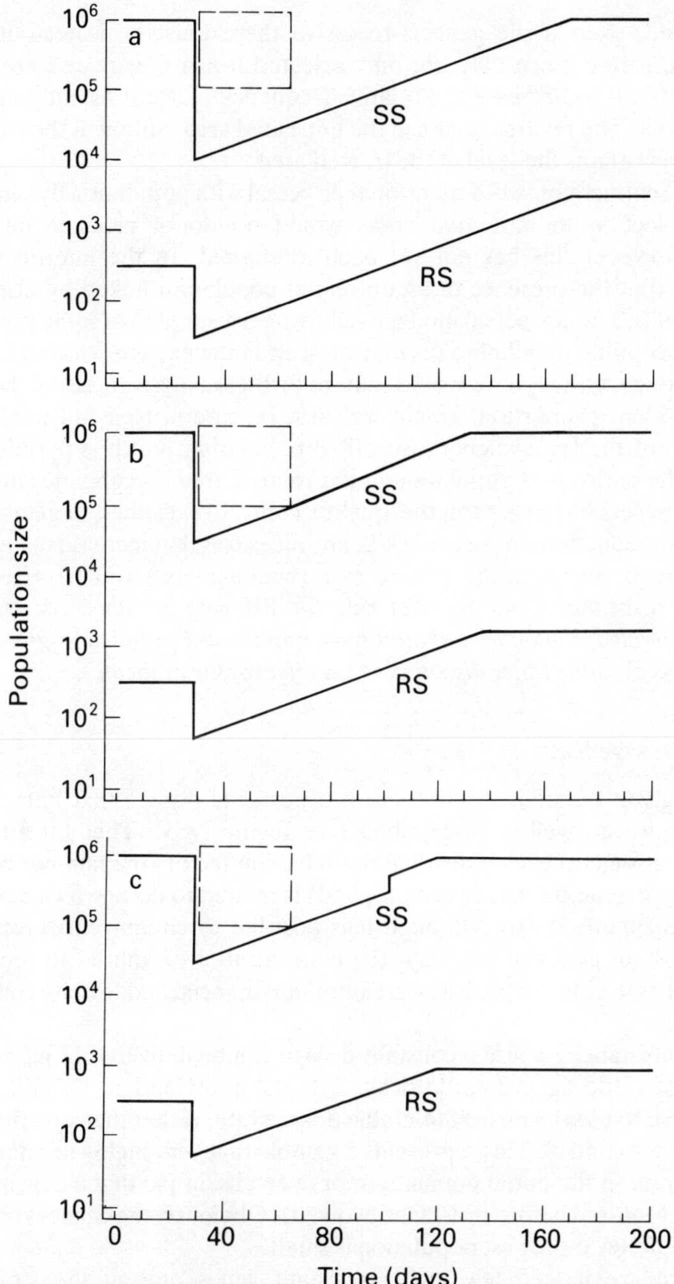

FIGURE 28.5 As Figure 28.4, but with seasonal crop (denoted by infinite crop–injury threshold during noncropping periods: (a) treatment as in Figure 28.4; (b) more efficient treatment; (c) post-cropping addition of susceptible pests.

other hand, even if the gene is recessive there must be a net outflow of resistance genes, since they are only selected for in the treated area. Thus dispersal tends to decrease the relative frequency of the resistant gene in the treated area. The reverse occurs in the untreated area, although the frequency there never attains the level of the treated area.

A full treatment of two-dimensional dispersal with population dynamics and counterselection in untreated areas would no doubt produce interesting results; however this has not yet been attempted. In the interim we may conclude that the presence of an untreated population linked by dispersal is equivalent in a nondispersal model to allowing the survival of some proportion of the susceptible population (as may be seen in the extreme case of infinitely rapid mixing), although we would require further analysis to assess the size of the equivalent proportion. Again we must be careful to avoid unnecessary reduction of the (equivalent) susceptible population, but it is permissible to destroy the entire pest population in the treated area if economic constraints make it necessary (e.g., for the quality control of canned vegetables). In addition to reductions in pesticide kill, any measures that increase the reservoir population or increase the mixing rate (within limits) will be effective in delaying resistance. *The essential rule for delaying resistance is again that pesticide-treated populations should have unbalanced population growth rates as rarely as possible (after discounting the effect of the pesticide).*

28.3.3. Dosage Rates

For any given resistance gene there is a pesticide dosage that kills resistant heterozygotes as well as susceptibles (see Figure 28.3). Thus for a constant pesticide dosage to each individual the selection factor (the number by which the resistant gene frequency is multiplied) is related to dosage by a curve such as that in Figure 28.6a. Although it is possible to circumvent selection for some resistant genes in this way, the concentrations required to prevent all resistance selection are probably greater than financial and toxicity constraints would allow.

In circumstances where a constant dosage can be delivered (e.g., acaricide dipping of cattle against cattle ticks, Sutherst and Comins, 1979) it may be worthwhile to use the highest tolerable dosage rate, rather than one that is just adequate for control. This represents a gamble that very highly resistant genes are very rare in the initial population, or even absent (so that a new mutation would be required before resistance evolved). Obviously the chances of success are much better if the pest population is small.

If there are indeed few highly resistant genes present the strategy of consistently using high dosage rates is superior to that of using a low dose and switching to a high dose when weak resistance becomes apparent. This is because a gene originally selected on its low heterozygote resistance can have a

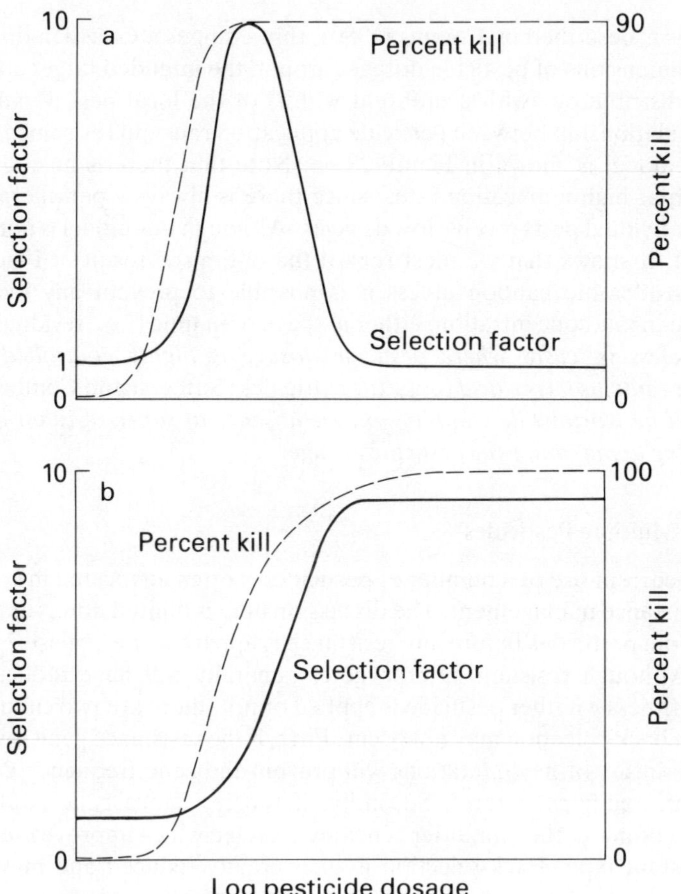

FIGURE 28.6 Resistance selection factor and percentage of population killed as a function of the dosage received by each pest individual in a single pesticide application: (a) 90% of individuals receive the standard dose, while 10% escape altogether; (b) diffusion model of relationship between application rate and dosage (Comins, 1980). Scale shows dosage at target point, dosages elsewhere are smaller.

much higher homozygote resistance, and because of the possibilities of high resistance factors arising by the combined effect of weak resistance genes with independent physiological mechanisms (Comins, 1980). Such combinations of genes can only be significantly selected once the components have been brought to high frequencies by selection with small doses of pesticide.

It must be remembered that the above discussion applies only to the case where pesticide dosage can be precisely controlled. In the general agricultural situation, it is possible that increasing application rates would increase the proportion of pests exposed to a low dosage, rather than subjecting a particular proportion to a higher one. A conceptual model that lies between these extremes is described by Comins (1980); this assumes a Gaussian distribution in two dimensions of pesticide dosage (around the intended target area) and a similar distribution (with a different width) of the local pest population. A typical relationship between pesticide application rate and resistance selection for this model is shown in Figure 28.6b. Note that there is no reduction of selection at high application rates, since there is always a peripheral zone in which individual pests receive low dosages. Although this model is not accurate in detail, it shows that we must regard the optimistic result of Figure 28.6a with considerable caution unless it is possible to prevent any leakages of pesticide in low concentration, either in space or in time (i.e., residual effects). *Nevertheless in cases where pesticide dosage is highly controlled and the resistance situation is critical* (e.g., the cattle tick; Sutherst and Comins, 1979) *it may well be worthwhile gambling on the absence of super-resistant genes and attempting to maximize the pesticide dosage.*

28.3.4. Multiple Pesticides

The concurrent use of a number of pesticides is often advocated in connection with resistance management. The discussion here is limited however to the use of multiple pesticides before any resistance is apparent (i.e., phase 2 of Figure 28.1). Although resistant heterozygotes generally will have reduced reproductive fitness whether pesticide is applied or not, there are two circumstances in which back-selection may not occur. First, if the resistance gene is very rare, then the influx of new mutations will prevent the gene frequency decreasing below an equilibrium level. Secondly, if the resistance gene is allowed to become common, then modifier genes are selected which improve the fitness of the resistant type. Back-selection is then greatly reduced and may even be totally eliminated.

Thus in order to take maximum advantage of back-selection it is necessary to avoid both extremes of gene frequency. The low-frequency extreme can be avoided by using relatively expensive pesticides to make up part of the required pest mortality even though no resistance to a cheaper chemical has been observed. Avoiding the selection of improved fitness of resistant types would require monitoring of very low levels of resistance, and immediate withdrawal of the pesticide if resistance was observed. That is, one would have to forego the residual degree of control attainable in the final stages of resistance development in the hope that the pesticide could be reused some time in the future.

The use of two or more pesticides adds an additional variable to resistance management strategies, since the total selection (defined to be the sum of the changes in log gene frequency for two hypothetical independent resistance genes) is not the same for different treatments which cause the same overall pest mortality. The simplest case is that in which there is no mortality of either homozygote and no synergism. *It may then be shown that a two-pesticide control strategy selects less for resistance than a one-pesticide strategy with the same total kill if, and only if, the two proportional mortality rates in various sections of the population are positively correlated* (see Appendix).

This result is easy to understand since it derives from the fact that with positive correlation a resistant individual that has survived treatment with pesticide A is more likely to be exposed to pesticide B (to which it is susceptible) than an A-susceptible individual which happens to have survived a lower dosage (and vice versa for negatively correlated applications). A similar trend is expected to hold when homozygote mortality occurs, although it is not then possible to compare control strategies solely on the basis of total mortality.

An extreme example of a positively correlated strategy is to treat (say) 50% of the population with a dose of pesticide A sufficient to kill 99.5% of susceptibles, and to treat the *same* 50% with a dose of pesticide B having a similar effect. The remaining 50% of the population receives no treatment whatever. It can be seen that this treatment reduces the survival of resistant individuals in the treated area by a factor of 200, and the selection of resistance is slowed by a similar factor. This tactic is related to the "overkill" tactic described in Section 28.3.3, and has the same sensitivity to leakage of small amounts of pesticide. The equivalent to the super-resistant genotype in this case is a cross-resistant type. Thus in cases suitable for this tactic (where dosages could be closely controlled) it would represent a gamble that cross-resistant genes were only present in very low frequency. In such cases the choice between high dosages of one pesticide and a pesticide "cocktail" would depend on the mammalian or plant toxicities of the required high dosages, and on the apparent likelihood of cross-resistance.

28.4. CONCLUSION

Because of the wide variety of pest genetic responses, it appears that there is no resistance management strategy that can provide a certainty that resistance will not develop. However, in appropriate circumstances there are a number of methods which either guarantee a relative delay in resistance or provide a reasonable chance of a delay.

Due to the efficacy and cheapness of synthetic pesticides, pest populations are probably often reduced to levels considerably below those which would give adequate control of crop damage. This can lead to more rapid development of resistance if intraspecific competition and predation are density-dependent for low pest densities, or if the pest population is able to recover during intercropping periods. In order to delay resistance it is imperative that we recognize the expensiveness of such practices and attempt to eliminate them. Also any contribution to control using nonpesticide methods allows the use of pesticide to be cut back, resulting in less selection of resistance.

Population refuges and dispersal have a moderating effect on resistance development when the treated population is subjected to large pesticide mortalities. Thus pesticide should never be applied over a wider area than is necessary for adequate control. In situations where pesticide dosages can be precisely controlled, the use of high dosages of a single pesticide, or of cocktails of two or more pesticides, can result in resistant heterozygotes being killed. This produces a substantial delay in resistance development, provided super-resistant or cross-resistant genotypes are very rare. This strategy is very sensitive to leakage of pesticide (including residual effects) and cannot be recommended for general use.

The extent of back-selection against resistant genes due to lower reproductive fitness cannot easily be quantified, although it is probably underestimated by post-resistance genetic studies, due to the selection of modifier genes. *If back-selection is significant, it can best be exploited by interspersing the use of relatively expensive new pesticides with the currently preferred chemical. Also, the reappearance of susceptibility can probably be hastened by withdrawal of pesticides immediately resistance is detected at very low levels* (thus preventing the selection of fitness modifiers). Other pesticides with cross-resistance effects must also be avoided.

ACKNOWLEDGMENTS

The author is supported by a Queen Elizabeth II Fellowship in the Environmental Biology Department, Research School of Biological Sciences, Australian National University, Canberra.

APPENDIX

Consider a pest population, the control of which requires a pesticide application every τ days to kill a proportion m of the population. We suppose that the pesticide dosage varies spatially, but that the population has time to redistribute during τ days, so that there is negligible correlation between exposure to a high dosage in one application and receiving a high dosage in the next. Assuming that dosages are never high enough to kill homozygotes, the frequency of a rare resistance gene is increased by $(1 - m)^{-1}$ by each control event. Thus N such events increase the logarithm of the gene frequency by:

$$-N \ln (1 - m). \tag{28.1}$$

Let us consider two pesticides, for one of which N_1 control events are sufficient to bring the resistance gene R_1 to a damagingly high frequency, whereas for the other N_2 control events will bring the resistance gene R_2 to dangerously high frequency (we assume R_1 and R_2 are different, i.e., no cross-resistance). Thus:

$$\Delta_1 = -N_1 \ln(1-m) \qquad \Delta_2 = -N_2 \ln(1-m), \tag{28.2}$$

where Δ_1 and Δ_2 are the differences between the initial and final log frequencies of the resistance genes R_1, R_2. This interpretation can be modified if necessary to include back-selection—there is no effect on the analysis provided back-selection is constant, as it will be if use of the two pesticides is interspersed to some extent.

Assuming constant back-selection the sequence of use of pesticides 1 and 2 can be varied arbitrarily without affecting the total time for which control is obtained, namely $(N_1 + N_2)\tau$ days. This is because selection of the two resistance genes is completely independent.

Let us now consider an alternative strategy in which K applications are of both pesticides at once, while there are N_1' of pesticide 1 and N_2' of pesticide 2. Suppose that this set of applications is just sufficient to cause resistance to both pesticides. Then we wish to know under what circumstances $K + N_1' + N_2' > N_1 + N_2$ (i.e., when does using mixed applications extend the duration of control).

Suppose the mixed application increases the logarithm of the frequency of R_1 by L_1, and that of R_2 by L_2 and let $L = -\ln(1-m)$. Then, by analogy with eqn. (28.3), we have:

$$\Delta_1 = N_1'L + KL_1 \qquad \Delta_2 = N_2'L + KL_2. \tag{28.3}$$

Combining eqns. (28.2) and (28.3) we get:

$$N_1'L + KL_1 = N_1 L \qquad N_2'L + KL_2 = N_2 L. \tag{28.4}$$

Adding the two parts of eqn. (28.4) and rearranging gives:

$$(K + N_1' + N_2') - (N_1 + N_2) = (K/L)[L - (L_1 + L_2)]. \tag{28.5}$$

Therefore mixing pesticides gives an extension of control if and only if $L_1 + L_2 < L$. Assuming no synergism, the combined mortality for the mixed application is:

$$m = 1 - \iint \Psi(x,y)[1 - m_1(x)][1 - m_2(y)]\mathrm{d}x\mathrm{d}y, \tag{28.6}$$

where m is as in eqn. (28.1), $\Psi(x,y)$ dxdy is the proportion of the pest population receiving a dose x of pesticide 1 and a dose y of pesticide 2, $m_1(x)$ is the mortality rate following exposure to a dose x of pesticide 1, and $m_2(y)$ is the mortality rate following exposure to dose y of pesticide 2. Equation (28.6) may be written more succinctly as:

$$1 - m = \langle\langle(1 - m_1)(1 - m_2)\rangle\rangle, \tag{28.7}$$

where the probability distribution $\Psi(x,y)$ is understood.

Continuing with the assumption that resistant heterozygotes are not killed by the corresponding pesticide, the survival rate for the R_1 heterozygote is $\langle 1 - m_2\rangle$. Therefore the logarithm of the increase in R_1 frequency is

$$L_1 = \ln[\langle 1 - m_2\rangle / \langle (1 - m_1)(1 - m_2)\rangle]. \tag{28.8}$$

Using $L = -\ln[\langle (1 - m_1)(1 - m_2)\rangle]$, from eqn. (28.7), we have

$$L_1 + L_2 - L = \ln[\langle 1 - m_1\rangle\langle 1 - m_2\rangle / \langle (1 - m_1)(1 - m_2)\rangle]. \tag{28.9}$$

Therefore mixed application is beneficial if and only if

$$\langle (1 - m_1)(1 - m_2)\rangle > \langle 1 - m_1\rangle\langle 1 - m_2\rangle;$$

i.e.,

$$1 - \langle m_1\rangle - \langle m_2\rangle + \langle m_1 m_2\rangle > 1 - \langle m_1\rangle - \langle m_2\rangle + \langle m_1\rangle\langle m_2\rangle$$

i.e.,

$$\langle m_1 m_2\rangle > \langle m_1\rangle\langle m_2\rangle. \tag{28.10}$$

That is, mixed application is beneficial if and only if the mortalities from the two pesticides are positively correlated.

REFERENCES

Anonymous (1975) *Pest Control: An Assessment of Present and Alternative Technologies* Vol. 1 (Washington, DC: US National Academy of Sciences).

Brown, A. W. A. (1971) Pest resistance to pesticides, in R. White-Stevens (ed) *Pesticides in the Environment* (New York: Dekker) vol 1 (2):457–552

Comins, H. N. (1977a) The development of insecticide resistance in the presence of migration. *J. Theor. Biol.* 64:177–97.

Comins, H. N. (1977b) The management of pesticide resistance. *J. Theor. Biol.* 65:399–420.

Comins, H. N. (1979a) Control of adaptable pests, in G. A. Norton and C. S. Holling (eds) *Pest Management*, Proc. Int. Conf., IIASA Proceedings Series No. 4 (Oxford: Pergamon) pp217–25.

Comins, H. N. (1979b) Analytic methods for the management of pesticide resistance. *J. Theor. Biol.* 77:171–88.

Comins, H. N. (1980) *The Management of Pesticide Resistance: Models* Working Paper (New York: Rockefeller Foundation).

Conway, G. R. (ed) (1982) *Pesticide Resistance and World Food Production* (London: Imperial College Centre for Environmental Technology).

Conway, G. R. and H. N. Comins (1979) Resistance to pesticides: Lessons in strategy from mathematical models. *SPAN* 22:53–55.

Curtis, C. F., L. M. Cook, and R. J. Wood (1978) Selection for and against insecticide resistance and possible methods of inhibiting the evolution of resistance in mosquitoes. *J. Ecol. Entomol.* 3:273–87.

Georghiou, G. P. and C. E. Taylor (1977a) Genetic and biological influences in the evolution of insecticide resistance. *J. Econ. Entomol.* 70:319–23.

Georghiou, G. P. and C. E. Taylor (1977b) Operational influences in the evolution of insecticide resistance. *J. Econ. Entomol.* 70:653–8.

Hueth, D. and U. Regev (1974) Optimal agricultural pest management with increasing pest resistance. *Am. J. Agric. Econ.* 56:543–52.

Sawicki, R. M. (1979) Resistance of insects to insecticides. *SPAN* 22:50–2.

Southwood, T. R. E. (1975) The dynamics of insect populations, in D. Pimentel (ed) *Insects, Science and Society* (New York: Academic Press) pp151–99.

Stone, B. F. (1972) The genetics of resistance by ticks to acaricides. *Aust. Vet. J.* 48:345–50.

Sutherst, R. W. and H. N. Comins (1979) The management of acaricide resistance in the cattle tick *Boophilus microplus* in Australia. *Bull. Entomol. Soc.* 69:519–40.

Taylor, C. E. and G. P. Georghiou (1979) Suppression of insecticide resistance by alteration of gene dominance and migration. *J. Econ. Entomol.* 72:105–9.

29 Resilience and Adaptability in Ecological Management Systems: Why Do Policy Models Fail?

Carl J. Walters and C. S. Holling

29.1. INTRODUCTION

It is generally supposed that good management will result from the development of more accurate ecological monitoring schemes, better models of response to control actions, and more careful attention to the objectives (including economic dimensions) of management. We expect nature to produce surprises, but we hope that these will become more unusual as we learn from mistakes to consider more details, or multiple equilibria and catastrophes, or whatever. We expect that better models and policy analyses will be used by appropriate institutions, once we overcome some obvious problems of communication and, often, bureaucratic inertia.

This final chapter argues that such views contains great dangers for the ecological policy analyst. First, we contend that nature's store of surprises is not limited, because we deal with open, arbitrarily defined systems. Second, we point out that a managed system and the institutions that deal with it form part of a larger, coupled system that has its own stability properties (and therefore whole new sources of surprise). Finally, we try to identify approaches to management and policy design that explicitly account for these difficulties.

29.2. CONCERNS FOR THE POLICY ANALYST

Figure 29.1 suggests we should worry about nine basic issues: the internal features of natural systems, their controlling institutions, and their environments, and the six "flows of information" between these components.

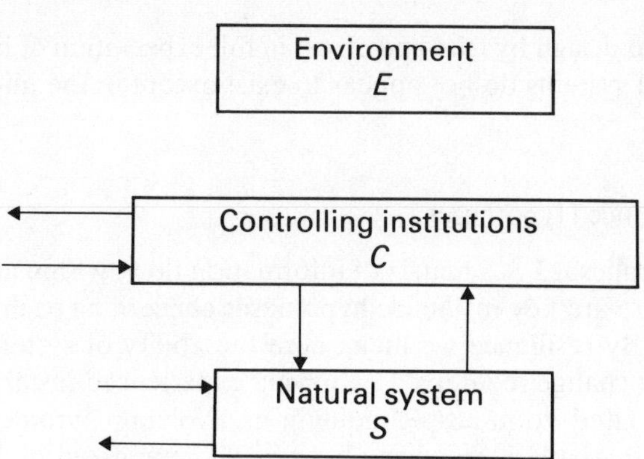

FIGURE 29.1 The ecological policy analyst should be concerned with flows of information besides those occurring within managed systems.

However, the components first need to be defined more clearly, and for what follows it will be important to adopt *relativistic* definitions:

Natural system, S: A collection of variables and relationships perceived by controlling institutions as having "internal dynamics" and couplings to the environment through "forcing functions". This is an open system definition, and there is no pretense that the controlling institutions have chosen the arbitrary boundaries between system and environment in a perfectly knowledgeable or even intelligent fashion.

Controlling institutions, C: A system that (1) acquires information about *S*; (2) develops models (or has some model) for *S* and filters incoming information so as to identify the model(s); and (3) exerts controlling actions on *S* that are chosen with regard to the model(s) and to objectives that are at least partly established by the environment *E* (as inputs to *C* from *E*).

Environment, E: A collection of relationships that affect and are affected by both *S* and *C*, yet are not clearly perceived as "part of the problem" by *C*. In a very real sense, *E* can be viewed as "everything that goes on in the world". *E* has economic, biological, and physical dimensions.

These definitions are not entirely precise, but perhaps the only crucial point is that "natural systems" are arbitrary objects of analysis. One obvious inference from the definitions is that most natural systems will exhibit plenty of nasty surprises to the controlling institutions. Unless *S* is expanded to include all of *E* (a practically impossible task), there will be feedbacks between various components of *S* and *E* such that *S* will appear to behave unpredictably.

Another way to define natural systems is by contrast with "designed systems" whose characteristics are supposedly established entirely by the controlling institutions. Examples are planned economies, agricultural ecologies, and engineered industrial plants. However, even the simplest mechanical system must communicate in some way with a complex environment; engineers have learned to design by trial and error, in full expectation of large surprises. Fully designed systems do not appear to exist except in the minds of systems theorists.

29.2.1. Resilience Hypotheses

Figure 29.1 implies a 3 × 3 matrix of information flows within and between *E*, *S*, and *C*. There are key resilience hypotheses concerning each of these flows (Table 29.1). By resilience we mean here the ability of systems to adapt to disturbance or change, conferred by having experienced instabilities. System hypotheses related to multiple equilibria, evolving parameters, and the importance of variable inputs have been well summarized by Holling (1973, 1978a) and Peterman *et al.* (1979). The control institution hypotheses are existing topics in organization theory (Cyert and March, 1963; Wildavsky, 1974, 1979). The coupling between system performance, expectations established in *E*, and constraints on action in *C* has not been fully explored, although Walters (1975) has sketched out part of the problem.

The original resilience idea came from "internal" studies of natural systems, and the first testable hypothesis was that ecological (and probably other)

TABLE 29.1 Statements underlying resilience hypotheses about managed systems.

FROM	TO Environment	Controlling institution	System
Environment	Complex, unpredictable, strongly coupled to controller and system	Control resources and expectations that change with system performance	Cause apparent changes in parameters and states of *S*, on several time scales
Controlling institution	Often unreasonable promises of success, incomplete information on performance of *S*	Goals and constraints from environment, incomplete information from *S*, simplified model of *S*, tendency to self-simplify	Constrained actions on small subset of state variables, parameters
System	Valued outputs influence expectations for future performance, many feedback effects	Unpredicted changes in monitored states, sharp changes in control costs	Essentially nonlinear (multiple)equilibria, evolving parameters due to internal heterogeneity, persistence requires variable inputs from *E*

systems will exhibit domains of stability rather than a single, globally stable equilibrium. It was further hypothesized that boundaries between domains (in state space) are a function of (1) parameters that are not really constant but rather are influenced by the environment (there is obvious contact here with the "control variable" definitions of catastrophe theory), and (2) evolutionary changes within the system. An evolutionary hypothesis that has never been carefully developed is that systems lose their ability to respond to variable inputs (desirable stability domains become "smaller") when these inputs are held fixed—in other words, a system must "explore" its state space in order to remain capable of similar responses (same state-space configuration) in the future (Holling, 1978a).

Then it was noted that institutions have some analogous properties. Their responses evolve (adapt) in the face of changing inputs from E and S, and they may lose their capability to respond to surprises (become less adaptable) when their inputs are held steady for some time: a "successful" stabilizing policy contains the seeds of its own disaster. The Kalman filter of system identification theory (Kalman, 1960) is a simple model of this process: successive observations lead to a reduction in the Kalman gain matrix, until even large prediction errors (innovations) have only a small effect on state estimates; only if states are assumed *a priori* to change rapidly can the filter adapt fast enough. Pretending that natural systems have fixed parameters (no evolution, small environmental effects) is analogous to assuming *a priori* in Kalman filtering that the state disturbance covariance matrix has small elements.

29.3. A PROTOTYPE EXAMPLE

Perhaps a simple example (adapted from Walters, 1978) will help to clarify the system–institution analogy.* Suppose it is desired to manage an ecological population with dynamics believed to follow:

$$\mathrm{d}x/\mathrm{d}t = p(x) - h \qquad \text{where } x(0) > x_c \tag{29.1}$$

$$\mathrm{d}h/\mathrm{d}t = u \qquad \text{where } \delta^- < u < \delta^+, \delta^- < 0, \text{ and } h \geqslant 0, \tag{29.2}$$

where x is the population, h is the harvest (or removal from the population), and the control u is a rate of change in harvest. By considering harvest a state variable (representing cumulative effects of past decisions) rather than a control, we presume that the controlling institution is constrained to incremental changes in policy. Suppose the population has been observed to follow the natural production function $p(x)$ shown in Figure 29.2. Suppose

*Although the example here is drawn from fishery theory the arguments being propounded are equally applicable, with simple modification, to pest and pathogen control.

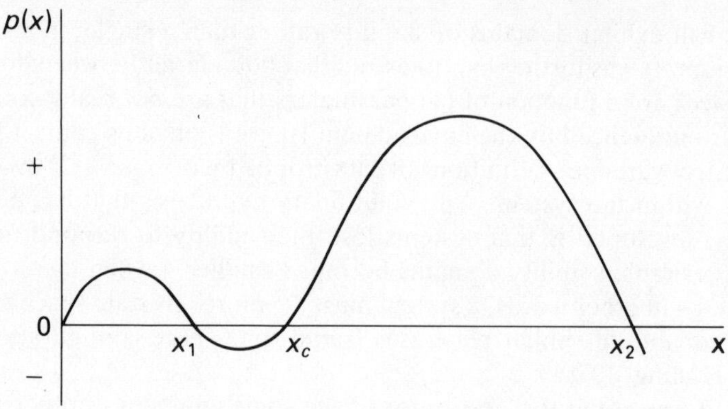

FIGURE 29.2 A simple, multiple equilibrium population model. $p(x)$ is net rate of population growth in the absence of harvest.

the objective is simply to maximize the discounted harvest $\int_0^\infty h \exp(-\lambda t)dt$, where $\lambda > 0$ is a discount rate. Here the natural system has stable equilibria at x_1 and x_2, with the point x_c dividing the state space ($x \geqslant 0$) into two domains.

The optimal policy in this example will be bang-bang (Clark, 1976), with u always equal to δ^-, δ^+, or 0. The solution will look something like Figure 29.3, where the broken curves separate regions where it is best to use δ^- and δ^+; $\delta = 0$ should be optimum only at the (x,h) points (x_a,h_1) and (x_b, h_2) where $0 < x_a < x_1, x_c < x_b < x_2$. The optimal policy is not clear for the region marked $\delta^\pm$. In this case there is no hope of recovering the system to h_2, x_b, but it may still be best to try to stay in the upper stability domain ($x > x_c$) as long as possible.

The separatrices A and B are found by solving (dx/dt, dh/dt) backward in time from the "stable management points" (x_a, h_1) and (x_b, h_2), where $p(x) = h$, while holding $u = \delta^-$ for $x >$ equilibrium, and $u = \delta^+$ for $x <$ equilibrium. The separatix C represents points where it is not longer possible to drive h to zero before x drops below x_c. This separatrix is critically dependent on δ^-; for $\delta^- << 0$, C is very steep and intersects A almost directly above x_c. For δ^- small negative, and δ^+ small, B and C may intersect at some nonzero h, and the upper equilibrium (h_2, x_b) becomes unstable.

Obviously (x_b, b_2) is a desired equilibrium for the system. But the *managed* system has a small domain of stability around (x_b, h_2) if δ^- is small, in the sense that disturbances in x may push the state across the separatrix C such that recovery is impossible. The best action may then be to use δ^+ and δ^- so as to move as rapidly as possible to (x_a, h_1).

A key "institutional resilience" hypothesis is that success at holding the system near (x_b, h_2) will lead to δ^- *becoming smaller*. The beneficiaries of h, as part of the unmodeled "environment" of this problem, will come to expect

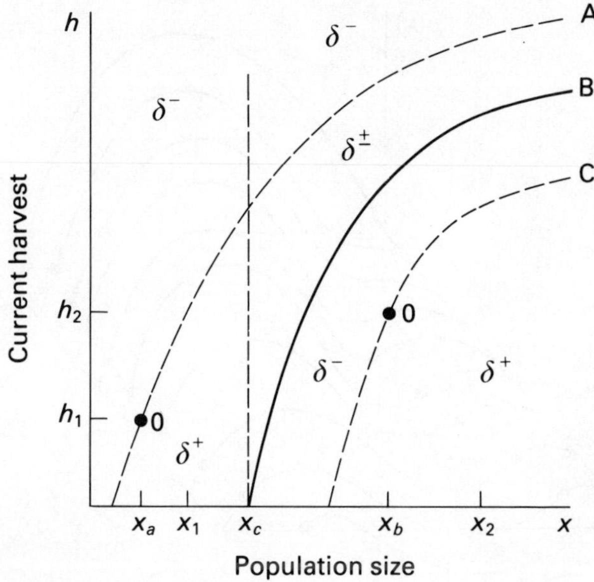

FIGURE 29.3 Regions in the population–harvest state space where different incremental actions are optimal.

stable h and will influence the controlling institution to adopt a smaller value of δ^-.

Constraints on controlling actions introduce structures (separatrices) into problem state spaces that are striking analogous to the stability domains of early resilience theory. Indeed, it may turn out that constraints are the 'essential nonlinearities' of institutions viewed as dynamical systems. In a broader setting, examples like this are a serious condemnation of "incrementalist" management policies, at least if we view incrementalism as a process taking small steps always *toward* a desired goal such as (x_b, h_2). The myopic, nonmodeling incrementalist will assume a simple, globally stable state space around (x_b, h_2), and will behave in exactly the wrong way whenever separatrices like C are crossed.

Now suppose that the controller has failed to recognize some key "slow variable" y that affects $p(x)$ in Figure 29.2. Suppose the effect of y is as shown in Figure 29.4 (such an effect could arise, for example, if y is a predator on x). x can then be modeled as a cusp catastrophe, with "slow variables" y and h (Jones and Walters, 1976).

Uncontrolled and unmonitored variations in y will obviously have various surprising effects. Even if the controller comes to realize that y exists and revises his model (moves y from the "environment" into his "natural system" in the general terms presented above), there will probably be another environmental variable Z that he has failed to recognize. Z may in turn generate sharply nonlinear system responses.

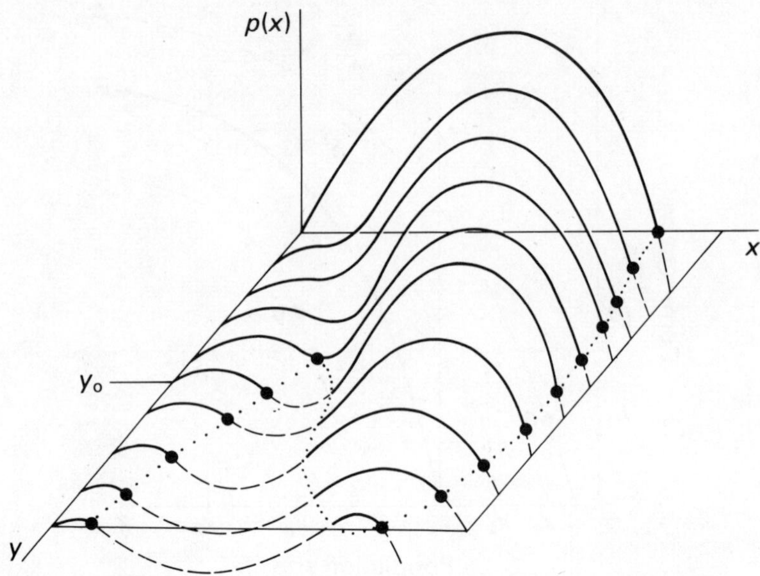

FIGURE 29.4 Variation in the production function of Figure 29.2 in relation to another variable *y*.

One might assert at this point that variables like *y* and *z* are mathematical curiosities that rarely occur in "real" natural systems, and the fact that we can demonstrate the *possibility* of perverse effects should not be a determinant of practical policy design. All we can say now about this assertion is that we have found many case examples to the contrary: natural systems do display multiple equilibria and sharp changes in dynamic behavior (see Chapter 2).

A pervasive model in systems analysis is that of fixed dynamical relationships that are *perturbed* by *small* "process noises". Resilience hypotheses, if correct, imply that this model is fundamentally incorrect and not even a "reasonable approximation" to nature. Resilience argues for changing dynamical relationships *and large, often irreversible perturbations*.

If big structural changes and large process errors are the rule rather than the exception, we need some radically new approaches to control system design. Traditional systems analysis may be much more deceptive than most practitioners would be willing to admit.

29.4. STYLES OF CONTROL

Faced with nonlinear and apparently unpredictable system behavior, at least three types of response are possible on the part of controlling institutions. First, they may try to design away the variability by deliberately simplifying the system and/or its environment. Thus some economists are fond of laws that enforce economies to act more like perfect markets; some wildlife managers are fond of predator control programs; agriculturists and foresters would like to do away entirely with many insects and weeds.

Secondly, they may try to extend the boundaries of definition of the natural system, so as to include "all relevant factors" in their analyses. Environmental agencies are spending more and more money on economists and economic modeling; water resource agencies are hiring ecologists; large interdisciplinary teams are formed to analyze all sorts of problems.

Thirdly, they may simply try to find ways to live with (and in Holling's words, "profit from") high variability. There are at least three design possibilities for living with surprise. (i) The controlling institutions may attempt to influence the "clients" component of the environment to adopt different objectives regarding stability/predictability, by asking questions like "why are stability and/or predictability such basic goals in the first place?" This design approach is fundamentally stupid. (ii) the institutions may attempt to design some means to stabilize system *outputs* without stabilizing system states, by finding some way to store up outputs and release them in a more or less steady stream. Individuals hedge against uncertainty by storing money in savings accounts; water is stored in dams for release in dry periods; fisheries harvest can be taxed in good years to provide subsidies in poor years. This design approach is the most promising, in terms of social acceptability, that we have uncovered so far. (iii) The institutions may attempt to spread risks by disaggregating the system into "operational units", each with a relatively low cost of failure. Operational units might consist of spatial areas, industries producing similar products, or simply general types of investments. The trick in this design alternative is to discover ways to minimize interdependencies (or behavioral correlations) among the operational units. For example, the salmon manager who tries several enhancement (culture) techniques on different stocks must be able to prevent the fish being harvested together in some ocean area where the stocks are mixed. The energy planner must be able to design parallel development "coal options", "nuclear options", etc., such that failure of one does not also drag the others down.

29.4.1. An Adaptive Approach: Brainstorm, Probe, and Monitor

There must exist some sensible compromise among the approaches outlined above. We should be able to do better than just living with surprise, even if we cannot anticipate or design away all uncertainties. First, we know that modeling can be used to "brainstorm" problems in interdisciplinary settings, so as to deliberately push out or broaden the perceived boundaries of natural systems. Our modeling workshops serve this function (Walters, 1974; Holling, 1978b). Broadening perceptions is a creative process that apparently has to involve some temporary suspension of analytical criticism. Modeling as brainstorming is in sharp contrast to the usual view of systems analysis as a fully rational and systematic process of dissecting problems for more detailed study.

Secondly, we can design policies that probe the behavior of natural systems, by deliberately applying variable control actions and conducting "diagnostic experiments". The idea here is to explore the local topology of the system state space, and to reveal key relationships that may influence its global structure. Unfortunately, such activity is directly counter to objectives related to stabilizing system outputs; therefore stabilizing storage systems for outputs become a key design ingredient for the probing manager.

Thirdly, we should be able to use modeling and analysis to identify more informative variables to include in monitoring programs. This raises an interesting control problem: given limited monitoring resources and a highly nonlinear system dynamic, do there exist "key variables" such that feedback policies based on monitoring these variables will give essentially the same results as monitoring the whole system state?

It is not enough to design a brainstrom/probe/monitor (BPM) process that meets theoretical criteria of improved performance given "rational" controlling institutions. We should further address the question of how to implement such a process in real institutions, with their complex and often self-seeking objectives and constraints. We have had some experience with the development of rational control schemes using BPM concepts, but almost none of our results have been implemented (Hilborn, 1979). We apparently have some fundamental misunderstanding about the dynamics of controlling institutions, so studies on "institutional resilience" should be a top research priority.

REFERENCES

Clark, C. W. (1976) *Mathematical Bioeconomics: The Optimal Management of Renewable Resources* (New York: Wiley/Interscience).

Cyert, R. M. and J. G. March (1963) *A Behavioral Theory of the Firm*. (Englewood Cliffs, NJ: Prentice-Hall).

Hilborn, R. (1979) Some failures and successes in applying systems analysis to ecological systems. *J. Appl. Syst. Analysis* 6:25–31.

Holling, C. S. (1973) Resilience and stability of ecological systems. *Ann. Rev. Ecol. Syst.* 4:1–23.

Holling, C. S. (1978a) Myths of ecological stability: resilience and the problem of failure. *J. Business Admin.* 9:97–109.

Holling, C. S. (ed) (1978b) *Adaptive Environmental Assessment and Management* (New York: Wiley).

Jones, D. D. and C. J. Walters (1976) Catastrophe theory and fisheries regulation. *J. Fish. Res. Board. Can.* 33:2829–33.

Kalman, R. E. (1960) A new approach to linear filtering and prediction problems. *J. Basic Engng*, 82D:35–45.

Peterman, R. M., W. C. Clark, and C. S. Holling (1979) The dynamics of resilience: shifting stability domains in fish and insect systems, in R. M. Anderson, B. D. Turner, and L. R. Taylor (eds) *Population Dynamics*. Proc. 20th Symp. Br. Ecol. Soc. (Oxford: Blackwell) 321–41.

Walters, C. J. (1974) An interdisciplinary approach to development of watershed simulation models. *Technological Forecasting and Social Change* 6:299–323.

Walters, C. J. (1975) *Foreclosure of Options in Sequential Resource Development Decisions*. Research Report RR-75-12 (Laxeburg, Austria: International Institute for Applied Systems Analysis).

Walters, C. J. (1978) Some dynamic programming applications in fisheries management, in M. Puterman (ed) *Dynamic Programming and Its Applications* (New York: Academic Press) 233–46.

Wildavsky, A. (1974) *The Politics of the Budgetary Process* 2nd edn (Boston, MA: Little, Brown & Co.).

Wildavsky, A. (1979) No risk is the highest risk of all. *Am. Sci.* 67:32–7.

Index